Systems of partial differential equations and Lie pseudogroups

Mathematics and Its Applications

A Series of Monographs and Texts Edited by
Jacob T. Schwartz, Courant Institute of Mathematical Sciences,
New York University

Volume 1
Jacob T. Schwartz LECTURES ON THE MATHEMATICAL METHOD IN ANALYTICAL ECONOMICS

Volume 2
O. A. Ladyzhenskaya THE MATHEMATICAL THEORY OF VISCOUS INCOMPRESSIBLE FLOW (*Second English Edition. Revised and Enlarged*)

Volume 3
R. L. Stratonovich TOPICS IN THE THEORY OF RANDOM NOISE (*in 2 volumes*)

Volume 4
S. Chowla THE RIEMANN HYPOTHESIS AND HILBERT'S TENTH PROBLEM

Volume 5
Jacob T. Schwartz THEORY OF MONEY

Volume 6
F. Trèves LINEAR PARTIAL DIFFERENTIAL EQUATIONS WITH CONSTANT COEFFICIENTS

Volume 7
Georgi E. Shilov GENERALIZED FUNCTIONS AND PARTIAL DIFFERENTIAL EQUATIONS

Volume 8
I. I. Pyatetskii-Shapiro AUTOMORPHIC FUNCTIONS AND THE GEOMETRY OF CLASSICAL DOMAINS

Volume 9
I. Colojoră and C. Fioaş THEORY OF GENERALIZED SPECTRAL OPERATORS

Volume 10
M. N. Barber and B. W. Ninham RANDOM AND RESTRICTED WALKS

Volume 11
N. Nakanishi GRAPH THEORY AND FEYNMAN INTEGRALS

Volume 12
Y. Wallach STUDY AND COMPILATION OF COMPUTER LANGUAGES

Volume 13
T. S. Chihara AN INTRODUCTION TO ORTHOGONAL POLYNOMIALS

Volume 14
J. F. Pommaret SYSTEMS OF PARTIAL DIFFERENTIAL EQUATIONS AND LIE PSEUDOGROUPS
Additional volumes in preparation

Systems of partial differential equations and Lie pseudogroups

J. F. POMMARET

Ancien élève de l'Ecole Polytechnique
Collège de France, Paris

with a preface by

André Lichnerowicz

Collège de France, Paris

GORDON AND BREACH SCIENCE PUBLISHERS

NEW YORK LONDON PARIS

Gordon and Breach Science Publishers, Inc.
One Park Avenue
New York, NY10016

Gordon and Breach Science Publishers Ltd.
42 William IV Street
London WC2N 4DF

Gordon & Breach
7–9 rue Emile Dubois
Paris 75014

Library of Congress Cataloging in Publication Data

Pommaret, J F 1945–
Systems of partial differential equations and Lie pseudogroups.

(Mathematics and its application)
Bibliography: p.
Includes index.
1. Differential equations, Partial. 2. Lie groups.
3. Pseudogroups. I. Title.
QA374.P64 515′.353 77-21457
ISBN 0-677-00270-X

To Professor M. JANET
For His 88th Birthday

CONTENTS

CHAPTER 4

CHAPTER 5

SECOND PART

CHAPTER 6

CHAPTER 7

PREFACE

For a century, partial differential equations have provided mathematicians and theoretical physicists with one of their most important tools and fields of work. Originally, p.d.e. formed a fundamental part of what was called "*analysis*." However, with the contribution of Sophus LIE, our points of view on p.d.e. have developed along new lines. On the one hand, our main point of interest may be in a "*well stated*" problem in the sense of HADAMARD, and in the existence theorems concerning such problems within a given differential field: we are still dealing here with "*analysis*." On the other hand, our interest may be in the study of the structure of p.d.e. systems themselves and all that can be derived from them by operations such as derivations or prolongations. In this case we have the "*formal theory*" of p.d.e. systems, and we leave the field of "*analysis*" proper in order to enter that of "*algebra*" and "*differential geometry*" with their associated cohomologies. This separation is already perfectly discernible in the work of Sophus LIE as also in that of Elie CARTAN, and the close relationship between the algebraic theory of p.d.e. systems and that of Lie pseudogroups bears witness to this.

During recent years, Donald SPENCER has pioneered the formal theory, stating the problems and developing many of the essential tools; at the same time, he has been the leader of a constellation of mathematicians, well known to everyone, whose names are continuously recalled throughout this book. The collected works of these different authors constitute a distinctive part of the contemporary mathematical scene. There are however certain difficulties in the modern setting: the considerable number of recent important papers makes it difficult today for any non-specialist to enter the subject, despite the existence of the book by KUMPERA and SPENCER. Moreover, the point of view adopted—one may say that it is the formal inheritance of that of Elie CARTAN—is more theoretical than truly operational and, whenever geometry and physics offer a p.d.e. system which is natural in some sense, then the general theory does not often provide the help one would expect in analysing the fine structure of "*this particular system*."

The aim of the author of this book is to overcome these difficulties. The book explicitly seeks to provide a progressive initiation into the formal theories of p.d.e. systems and Lie pseudogroups. After a careful reading of it, any scientist should be able, easily and profitably, to understand the main original papers. But it is far from just being that: it adopts a fundamentally different view point from that in general use, a view point which might be described as operational, reintegrating the somewhat old-fashioned approach of people like VESSIOT or JANET into present day science and inviting the reader to think about particular systems in order to confront each step of their development with SPENCER's theoretical approach. It is probably to KUMPERA that POMMARET must feel closest. Certainly, the richness of the examples treated here will help the reader to master the techniques of many standard proofs. But, at the same time, the operational view point adopted suggests new results, new foci, new problems and new interpretations. It seems to me that many specialists will be led to think about this and take it into account.

Among the original and important features, apart from the adopted view point, one may quote the involutiveness criterion of Chapter 2, the detailed study of SPENCER's cohomology in Chapter 3, the analysis of the first and second SPENCER's sequences in Chapter 5, and virtually the whole of Chapter 7, with its introduction of the bundle of geometric objects associated with a pseudogroup, the "*integrability conditions*" the so-called "JACOBI's *conditions*" and the analogue for transitive LIE pseudogroups of the third fundamental LIE theorem. This chapter ends with the application to formal deformations of infinite dimensional LIE algebras, touching on some of my own work but in a completely different framework.

Because of the view point adopted, this book is far from being written along the standard lines and some will perhaps be put off by this. Of necessity it has imperfections and is not an exhaustive work, nor could it be, since it deals with a theory which is far from being completed, as it is one which is still actively evolving. But I am sure that a careful reader of this deliberately different book will gain considerable benefit from his efforts and achieve a more comprehensive grasp of this wide and important field, with a greater potential for effectively using the knowledge acquired.

André LICHNEROWICZ

PRÉFACE

Depuis un siècle, les équations aux dérivées partielles fournissent aux mathématiciens et aux théoriciens de la physique un de leurs champs de travail et un de leurs outils les plus importants. A l'origine, les équations aux dérivées partielles constituaient une partie fundamentale de ce qu'on appelait *l'Analyse*. Mais, dés l'œuvre de Lie, nos points de vue sur les équations aux dérivées partielles se sont différentiés: l'objet principal d'intérêt a pu étre un probléme « *bien posé* » au sens d'Hadamard et les théorèmes d'existence concernant de tels problèmes dans un champ fonctionnel déterminé — nous demeurions dans *l'Analyse*; mais il a pu étre aussi l'étude de la structure du système aux dérivées partielles lui-même et de tous ceux qu'on en déduit par certaines opérations, dérivations, prolongements etc., et l'on a *la théorie formelle* des systèmes aux dérivées partielles qui sort, à proprement parler, du champ de *l'Analyse* pour entrer dans celui de *l'Algèbre* et de la *géométrie différentielle* et des cohomologies qui s'y rattachent. Cette scission est déjà parfaitement perceptible dans l'œuvre de Sophus Lie comme dans celle d'Elie Cartan et la liaison étroite entre la théorie algébrique des systémes aux dérivées partielles et celle des psuedogroupes de Lie en est le signe.

Dans les années récentes, Donald Spencer a été le pionnier de la théorie formelle, posant les problèmes et créant beaucoup des instruments nécessaires, et il a été en même temps le leader d'une pleiade de mathématiciens de premier ordre, connus de tous, dont les noms sont rappelés au long de cet ouvrage. L'ensemble des travaux de ces différents auteurs constitue un véritable volet du paysage mathématique contemporan. La situation contemporaine présente cependant certaines difficultés: le nombre notable d'articles récents importants rend difficile à quiconque n'est pas un spécialiste d'entrer vraiment dans le sujet, malgré l'existence du livre de Kumpera et Spencer. D'autre part le point de vue trés généralement adopté (dont on peut dire qu'il est l'héritier formel du point de vue d'Elie Cartan) se révèle plus théorétisant que vraiment opératoire et, lorsque la géométrie ou la physique nous fournissent un système aux dérivées partielles en quelque sorte naturel, la théorie générale ne nous offre pas souvent l'aide souhaitée pour analyser la structure fine de *ce système particulier*.

Le but de l'Auteur de ce livre est d'amener chaque lecteur à surmonter ces difficultés. Cet ouvrage se veut explicitement d'abord une initiation pédagogique aux théories formelles des systèmes aux dérivées partielles et des pseudogroupes de Lie. Tout scientifique qui en aura approfondi la lecture abodera aisément et fructueusement la lecture des grands articles originaux. Mais il est loin de n'être que cela: il adopte un point de vue profondément différent du point de vue le plus usuellement répandu, un point de vue, qu'on peut qualifier d'opératoire, réintégrant l'apport un peu oublié d'un Vessiot ou d'un Janet dans le présent de la Science, invitant à réflechir sur des systèmes particuliers et à confronter chaque étape de leur maniement avec la théorie à la Spencer. C'est probablement de Kumpera que Pommaret doit se sentir le plus proche. Certes la richesse des exemples traités ici aidera le lecteur à mieux surmonter la technicité de bien des démonstrations usuelles. Mais aussi le point de vue opératoire adopté suggère de nouveaux résultats, de nouvelles précisions, de nouvelles questions ou de nouvelles interprétations est il m'apparait que bien des spécialistes seront amenés à y réfléchir et à en tenir compte.

Parmi les approches originales et importantes quel que soit le point de vue adopté on doit citer le critère d'involution du Chapitre 2, les précisions concernant la cohomologie de Spencer du Chapitre 3, l'analyse des première et seconde suites de Spencer du Chapitre 5 et la quasi-totailité du Chapitre 7, avec l'introduction du fibré des objets géométriques associé à un pseudogroupe, l'apparition des « *conditions d'intégrabilité* » et des « *conditions de Jacobi* », l'analogue pour les pseudogroupes de Lie du troisiéme théorème fondamental de Lie. Le chapitre se termine par l'application à la théorie des déformations formelles qui recoupe certaines de mes approches, dans un cadre complétement différent, de l'étude des déformations des algébres de Lie infinies.

A cause du point de vue adopté, cet ouvrage est loin de se présenter selon les standards usuels et certains en seron peut-étre rebutés. Avec ses nécessaires imperfections, il n'est pas un achèvement, et ne saurait l'étre pour une théorie qui ne l'est pas, à beaucoup prés, parce qu'elle est une théorie bien vivante. Mais je suis sûr que le lecteur attentif de ce livre volontairement différent retirera de son travail beaucoup de fruit: une intelligence plus large d'un champ vaste et important, une possibilité plus grande d'utiliser effectivement ses connaissances.

André LICHNEROWICZ

INTRODUCTION

Roughly speaking, a "*Lie pseudogroup*" of transformations of a differentiable manifold is a differentiable group of transformations, solutions of a "*system of partial differential equations*". Lie pseudogroups were formerly known as "*infinite groups*" when the "*Lie groups*" were known as "*finite groups*".

Of course a Lie group can be considered as a particular kind of pseudogroup of transformations of itself, depending on a finite number of parameters.

The reader will see from the following picture that the development of the theory of Lie pseudogroups has paralleled that of systems of p.d.e.:

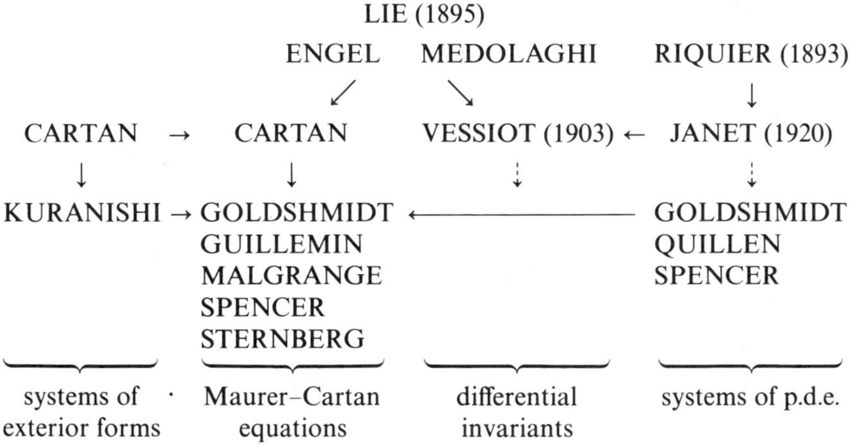

The purpose of this book is to fill in the two dotted arrows in the picture and to give a self-contained treatment of both theories and their relations. We have chosen a didactic and operational approach and we hope that the methods described in the two parts of this book will become accepted as useful tools in many branches of mathematics and theoretical physics.

Numerous examples and problems will help the reader in applying his new knowledge to concrete cases. He is only assumed to be familiar with the implicit function theorem and certain elementary definitions on differentiable manifolds that can be found in any relevant graduate textbook. Other basis material that is needed for our purpose is introduced throughout the book as it is required, since some differential and algebraic results are not easy to find in the literature.

We will now sketch briefly the history of the two theories.

A lot of physical theories deal with systems of p.d.e. comprising a great number of equations and many variables.

The first problem, before solving these systems, is to know "*a priori*" if any solution exists. One can differentiate the given equations thousands of times and make use of any "*elimination process*" one desires in order to get new equations of a given order (which is, by definition, the greatest order of the derivatives involved in the equations). This elimination process may result in "*incompatible equations*" that clearly admit no solution, and we can then say that the original system is not "*compatible*".

If a system is compatible, the Frenchmen C. Riquier (1893) and E. Cartan (1901) have shown, using quite different methods, how it is possible to determine what they called intuitively the "*degree of generality*" of the solution of a system of p.d.e.

The method of Cartan is based on the use of the "*exterior calculus*" which he created at the same time in order to study infinite groups by means of their defining equations. He was thus able to give an intrinsic meaning to the compatibility conditions of the systems of defining equations and these differential identities are known as "*Maurer–Cartan equations*". They involve some constants called "*structure constants*". Moreover, the systems made up of these equations are themselves compatible, in the above sense, if and only if the structure constants satisfy quadratic polynomial identities known as "*Jacobi conditions*".

The reader will find a detailed treatment of the method with examples in the books of Cartan himself (5, 6) or in reference (26).

Cartan was the first to point out some formal algebraic properties of the systems of exterior forms and, in order to describe them, he introduced integers called "*characters*" that are ordered by certain inequalities.

It has to be noticed that these numbers are defined intrinsically, that is to say independently of the system of coordinates.

The system is said to be "*involutive*" if and only if certain equalities hold between these characters.

However, even if the method, culminating in the famous Cartan–Kähler theorem for analytic systems, is useful, it is very difficult to work with, even in the case of linear equations. In particular it is not easy to distinguish between the "*independent and dependent variables*" or to study the "*polongations*" of a given system, obtained by successive differentiations.

The methods of Riquier (41, 42), used again by M. Janet (19–23), a student of D. Hilbert, in 1920 and modernised by Thomas (47, 48) and Ritt (43) are quite different and give an operational process that can be accomplished in a finite number of steps in order to study any linear or non-linear system of p.d.e.

The key idea is that of a "*cut*". In fact, supposing that it is always possible to use the implicit function theorem, they consider the infinite set of equations obtained by differentiating as many times as necessary a given finite number of p.d.e. This gives a way of computing certain derivatives called "*principal*" as functions of the variables and of the other derivatives called "*parametric*". The method can be used both in the differentiable and analytic cases but we have to suppose that certain "*regularity conditions*" are fulfilled.

The problem is thus to know what are the principal and parametric terms in the Taylor formal expansion of any solution in a neighbourhood of a given point.

Janet used a "*total ordering*" of the derivatives by means of sets of integers, called "*cotes*". Then, to any principal derivative, he associated "*in a non-intrinsic way*", depending on the coordinate system, certain of the variables called "*multiplicative variables*" (the other ones being called "*non-multiplicative*"). This being done in such a way as to obtain all the principal derivatives, each one separately, by differentiating repeatedly (with respect to the multiplicative variables), the equations solved with respect to the principal derivative concerned. If this was done with respect to the non-multiplicative variables, the elimination of the same principal derivative, obtained twice, was used in order to start the elimination process and look for the compatibility of the system. If two such computations done by any method gave the same expression for any given principal derivative, the system was said to be "*passive*".

Moreover, in the case of inhomogeneous linear systems with second members, Janet employed a very important construction (19, 20). He proved in fact that, if the system was passive with zero second member, then it was also passive with non-zero second member, if and only if the second member itself satisfied certain differential conditions called "*integrability conditions*" that could be considered as a new passive linear system of p.d.e.; and so on. His fundamental result was to show the possibility of constructing a chain of *no more than n* other systems, each of them expressing in some sense the integrability conditions of the inhomogeneous preceding one, n being the number of independent variables.

The work of this author was forgotten.

During the last ten years, a formal theory, starting with an abstract algebraic basis, has been developed, mainly in America, by H. Goldschmidt (12–15), D. Quillen (39) and D. C. Spencer (44).

In fact the situation was similar to that found by Cartan at the beginning of the century and Spencer had to bring out this new tool in order to apply it to the study of the systems of p.d.e. defining the transformations of Lie pseudogroups (25, 45).

The main method used is that of "*diagram chasing*". It is a new kind of process to become used to and our aim has been to develop it in a systematic way. Broadly speaking we can use the following analogy. If the reader looks at some classical textbooks in continuum mechanics written at the beginning of this century, he will get lost among pages and pages filled up with formulas. The use of tensor calculus and vector geometry made it possible to condense these formulas to a few lines and even to bring out identities that would have been difficult to see otherwise. However, this only works for certain particular kinds of systems. Our aim has been to show that it is possible to look at systems of p.d.e., whether they are linear or non-linear, in quite a new spirit.

We have concentrated mainly on non-linear systems. The key result is a process that allows one to study the formal solutions of any sufficiently regular system of p.d.e. by means of a finite number only of differentiations and eliminations.

We may now remark that there is no book available today on such systems and that there are only two important papers in the literature on these topics (14, 26).

One of them, by M. Kuranishi, uses classical methods and mixes the work of Janet on integrability conditions with that of Cartan on involutiveness. However it does not explicitly use any diagram chasing.

The other one, by H. Goldschmidt, is very difficult for the non-specialist reader to follow. There are two reasons for this. First of all the idea of the proof cannot be used in practice as it is completely different from the "*natural procedure*" that anybody would tend to use intuitively. Second, the expositions of the proofs are very difficult to follow because of their generality, particularly when they deal with affine bundles.

For this reason we have given new proofs that are well adapted to any concrete application and we have used local coordinates when this was the best and shortest way to clarify a proof. We ask the reader always to have in mind an easy example while reading the proofs. By this way he will quickly become acquainted with this new symbolic way of thinking.

Finally, for linear systems, we have first summarized the main results to be used. Their proofs can be found in the study of non-linear systems with some slight changes as we have been using similar notations for this purpose.

The reader who is interested mainly in that topic, can trust them but they will be of constant use in the second part of this book.

To summarise, the idea of Janet is translated into modern language and gives the key for the construction of a "*finite length sequence*" of involutive differential operators, each of them being related to the integrability conditions of the linear system associated with the preceding one.

All the formal properties of any given system of p.d.e. may be studied now by means of this sequence and the related ones. For example, all the intrinsic integers that can be associated with a given system of p.d.e. may be computed from the dimensions of the vector bundles appearing in these sequences and vice versa.

We also relate this sequence to similar sequences used by Spencer that will not however be found in the sequel.

We call this sequence a "*P-sequence*" because its existence has been suggested to us by the study of integrability conditions analogous to the ones that can be found in the "*Poincaré-sequence*" for the exterior derivative.

Having now in mind the first part of this book, the reader will understand that it may be interesting to look at the properties of the *P*-sequence when its first operator has some invariant properties similar to that of a Lie derivative; in particular when there is a group structure behind, as is the case in most of the applications.

This remark will serve as a transition to the subject matter of the second part of this book. But first we need to recall some historical facts that are not all well known.

S. Lie was the pioneer who, after his study of the finite groups, gave the basis of the theory of infinite groups in his Leipziger Berichte (1895) (27). He just studied the construction of "*differential invariants*", that is to say, functions of the derivatives "*invariant in form*" by a transformation of the pseudogroup acting on the unknown variables.

He was followed by Engel (10) and Medolaghi (30) who are almost unknown now.

In France, only J. Drach (8, 9) and E. Vessiot (1903) (49–51), who had been a student of Lie, took the same way, studying the defining equations of a pseudogroup directly without transforming them into a system of exterior forms as Cartan did. The main tool was the use of the differential invariants in order to study the integrability conditions for a family of systems of equations associated with a given pseudogroup. Vessiot was of course using the non-rigorous theory of systems of p.d.e. that existed at the time, and his work also was forgotten after a time.

Today Goldschmidt (15), Guillemin (16), Malgrange (28, 29), Spencer (45) and Sternberg (16) have since 1957 been retracing the ideas of Cartan and have tried to exhibit some invariant forms in order to describe the "*first and second Spencer sequences*". However, the methods, involving numerous brackets, Lie derivatives and quotients, are very sophisticated and cannot be applied easily.

The Brazilian Kumpera (24) followed the ideas of Vessiot and Lie concerning the differential invariants, but in the general case, that is to say when the number of variables transformed by the Lie pseudogroup is different from the number of independent variables with respect to which the differentiations are performed. His way is to construct a "*fundamental set*" of such invariants, that is to say a minimum set of functionally independent ones by means of which one can get all the others by differentiations and functional relations. He does not study at all any integrability condition, and this is the heart of our book.

Finally, the deformation theory of Lie algebras and other algebraic structures was initiated by Inonu and Wigner in 1953 (18). One tries to modify the structure constants of an algebra by introducing a parameter. For example one can pass from the inhomogeneous Lorentz group in two variables to the Galilean group in two variables. The work of Gerstenhaber (1964) (11) gives the best account of the theory. The idea in the case of finite dimensional Lie algebras is to deform the structure constants. According to the Lie fundamental theorems, this gives a deformation of the Maurer–Cartan forms and thus of the group.

Many authors have tried in vain to generalise these methods to infinite Lie algebras. S. Piper (31) created a "*deformation cohomology*". D. S. Rim (40) tried to use the same framework and unhappily, in this case, conjectured that it was possible to deal only with finite dimensional spaces. Some of his results are quoted in the last pages of this book but with a very different background.

After this brief historical preliminary, we give notes on the different chapters.

The first chapter introduces the basic material on fibered manifolds and jet bundles that will be used in the sequel. The reader aware of these topics can begin directly with the second chapter.

The second chapter is the heart of the study of non-linear systems of p.d.e. It is essential to all that follows. The main result is the "*criterion of involutiveness*" which has been proved in such a way that it can be used directly in any concrete case. There are some algebraic results on the Spencer cohomology that are needed throughout this chapter but are independent of it. For this reason we have collected them separately in the third chapter which must therefore be referred to at the same time. The reader will see clearly that the purely differential or purely algebraic aspects of the formal theory of systems of p.d.e., though they are interdependent, have separate features.

The third chapter deals with the Spencer cohomology. This is a powerful piece of machinery and we have tried to give practical help to the reader by doing all the computations completely in local coordinates and introducing the P-table trick which is very convenient when one has just to compute the dimensions of the vector bundles one is dealing with. We apologise for the fact that the proofs are very technical in general and we hope that the reader will feel the necessity of using diagrams and chases.

In the fourth chapter we go on creating some families of vector spaces that will be useful for the study of linear systems. Our aim is to show that some constructions that were initiated in the preceding chapters, can be continued in order to get finite length sequences.

The fifth chapter deals exclusively with linear systems of p.d.e. We construct essentially the P-sequence of inductively related integrability condition by two different ways. Then we construct the first and second Spencer sequences and exhibit the links between those three sequences. Finally we study the algebraic properties of the principal top order part of a linear system. Though we give interesting results, they are not essential in the sequel and the reading can be delayed. The elements of algebra that are necessary, are recalled as they are used.

The sixth chapter is concerned with the differential tools that must be known in order to study the local theory of Lie pseudogroups. Some of them are well known but it is difficult to find them properly in the literature. Moreover we have treated them, using the simplest notations and proofs. In particular we recall the fundamental theorems of Lie for Lie transformation groups, that is to say the special kind of pseudogroups of transformations of a manifold resulting from the action of a Lie group on that manifold.

We then explain how to look for the prolongations of both finite and infinitesimal transformations, giving explicit formulas.

We also give a proof of the Frobenius theorem that will be of constant use thereafter. The study of invariant submanifolds under a Lie group of transformations is also a key method in the sequel and is difficult to find in the literature. We have given a detailed treatment involving many examples.

The seventh and last chapter is really the culminating point of the application of the theory of systems of p.d.e. and differential invariants. We give an original and comprehensive treatment of the theory of Lie pseudogroups. It is written progressively in such a way that the reader can start at its beginning with any one of the problems stated at the end of the chapter and go ahead, reading the chapter and computing the example at the same time.

We define a Lie pseudogroup and give the properties of its defining finite and infinitesimal equations. We compute the differential invariants and use them in order to associate with any Lie pseudogroup a "*bundle of geometric objects*". Conversely, to any section of such a bundle, we can associate a Lie pseudogroup. But, in order to get some constancy of rank in the prolongation of the equations, the section must satisfy "*integrability conditions*", looking like the Maurer–Cartan equations as they involve some constants.

The reader may look at any known example in the literature in order to understand why the proofs must be and are sometimes technical. However we hope that, using our methods, the reader will be able to apply the formal machinery to any concrete problem.

The constants, called "*structure constants*", must satisfy some polynomial relation, called "*Jacobi conditions*", that are the integrability conditions of the integrability conditions(!).

The analogy existing between these results and the similar ones existing in the theory of Lie transformation groups is then developped in an exhaustive way.

The results on the deformation theory and in particular the computation and interpretation of the finite dimensional cohomology groups of an infinite Lie algebra are quite new. The methods we use are far from the usual approaches that are to be found today in the literature. The reader should for example, compare our third fundamental theorem with the one proposed by Goldschmidt (15).

The main point is that we generalise the deformation theory and the Hochschild cohomology (17) of finite dimensional Lie algebras, proving that the other generalisations and links that have been proposed by many authors up to now are not the proper ones.

Finally we give here an outline of the structure of this last chapter:

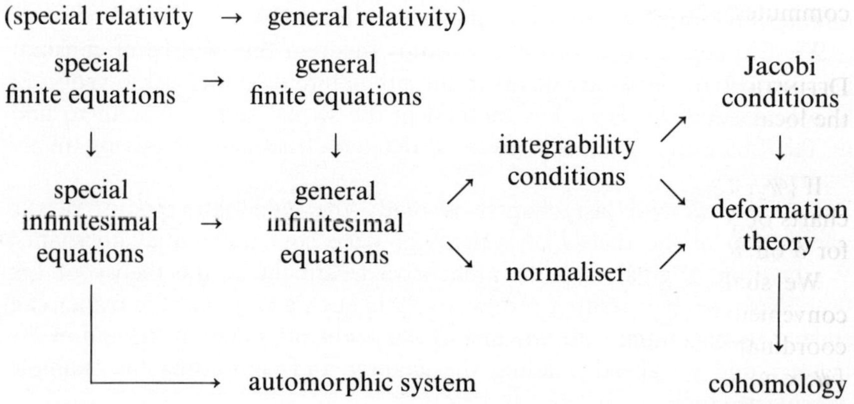

It is a pleasure to express my deep sense of gratitude to Professor A. Lichnerowicz for all the personal interest he devoted to this work and to Professor D. C. Spencer for many helpful conversations.

CHAPTER 1

1 Fibered manifolds

Let X and $\mathscr{E}$ be manifolds with $\dim X = n$ and $\dim \mathscr{E} = m + n$.

Let $\pi : \mathscr{E} \to X$ be a surjective submersion.

DEFINITION 1.1 We say that $\mathscr{E}$ is a "*fibered manifold*" over X, with projection π, if, for any point of $\mathscr{E}$ there exist a coordinate neighbourhood $\mathscr{U}$ of this point in $\mathscr{E}$, a local chart $\{\mathscr{U}, \Phi\}$ of $\mathscr{E}$ and a local chart $\{U, \varphi\}$ of X, with $U = \pi(\mathscr{U})$, such that the diagram:

$$\begin{array}{ccc} \mathscr{U} & \xrightarrow{\ \Phi\ } & \mathbb{R}^n \times \mathbb{R}^m \\ {\scriptstyle \pi}\big\downarrow & & \big\downarrow \\ U & \xrightarrow{\ \varphi\ } & \mathbb{R}^n \end{array}$$

commutes, where $\mathbb{R}^n \times \mathbb{R}^m \to \mathbb{R}^n$ is the projection onto the first factor.

DEFINITION 1.2 Such a local chart $\{\mathscr{U}, \Phi\}$ of $\mathscr{E}$ is said to be "*fibered*" over the local chart $\{U, \varphi\}$ of X.

If $\{\mathscr{U}_\alpha, \Phi_\alpha\}$ is an atlas of local charts of $\mathscr{E}$, then $\{U_\alpha, \varphi_\alpha\}$ is an atlas of local charts of X. We shall adopt local coordinates (x^i_α) for X on U_α and (x^i_α, y^k_α) for $\mathscr{E}$ on $\mathscr{U}_\alpha$.

We shall denote a point of $\mathscr{E}$ by its local coordinates (x, y) on any convenient neighbourhood of this point. Its projection, which has local coordinates (x) will be simply denoted by x. Similarly we often denote by $\{\mathscr{U}, x, y\}$ and $\{U, x\}$ the above local charts which are also called "*coordinate systems*".

It follows that the "*transition functions*", or "*coordinate transformations*" of $\mathscr{E}$ on $\mathscr{U}_\alpha \cap \mathscr{U}_\beta$ have the following form:

$$\begin{cases} y^k_\beta = \psi^k_{\beta\alpha}(x_\alpha, y_\alpha) \\ x^i_\beta = \varphi^i_{\beta\alpha}(x_\alpha) \end{cases}$$

where the transition functions of X on $U_\alpha \cap U_\beta$ are $x^i_\beta = \varphi^i_{\beta\alpha}(x_\alpha)$.

In general we shall not write the index α of the open coverings and consider coordinate transformation of the following kind:

$$\begin{cases} \bar{y} = \psi(x, y) \\ \bar{x} = \varphi(x) \end{cases}$$

where all the indices are omitted.

DEFINITION 1.3 X is called the "*base manifold*", $\mathscr{E}$ is called the "*total manifold*". For any $x \in X$, the set $\mathscr{E}_x = \pi^{-1}(x)$ is a closed submanifold of $\mathscr{E}$ called the "*fiber*" over x.

It is clear that $\mathscr{E} = \bigcup_{x \in X} \mathscr{E}_x$ and that two distinct fibers $\mathscr{E}_{x_1}$ and $\mathscr{E}_{x_2}$ have no common point if $x_1 \neq x_2$.

DEFINITION 1.4 $\mathscr{E}$ is called a differentiable (analytic) fibered manifold if $\mathscr{E}$ is a differentiable (analytic) manifold, X is a differentiable (analytic) manifold and π is a differentiable (analytic) map.

In the sequel we shall assume that X is a C^∞ (differentiable), connected, paracompact manifold and we shall deal only with C^∞ fibered manifolds, unless stated otherwise.

In the same way, all the manifolds and maps between them, to be considered, will be supposed to be C^∞.

If now X' is a submanifold of X, then we note $\mathscr{E}|_{X'}$ the restriction of $\mathscr{E}$ to X', that is to say the fibered manifold $\pi : \pi^{-1}(X') \to X'$. We also say that $\mathscr{E}|_{X'}$ is the fibered manifold "*induced*" by $\mathscr{E}$ on $X' \subset X$.

EXAMPLE 1.5 For any open set $U \subset X$, we have $\pi^{-1}(U) = \mathscr{E}|_U$.

DEFINITION 1.6 A "*local section*" of $\mathscr{E}$ over an open set $U \subset X$, is a map $f : U \to \mathscr{E}$, such that $\pi \circ f(x) = x$, $\forall x \in U$. We write $\pi \circ f = id_U$, and we call U the "*domain*" of f, noted $U = \operatorname{dom} f$.

In particular, if $\operatorname{dom} f = X$, then f is called a "*global section*" of $\mathscr{E}$ over X.

EXAMPLE 1.7 A local section f of $\mathscr{E}$ is a global section f of $\mathscr{E}|_{\operatorname{dom} f}$.

REMARK 1.8 If f is any section of $\mathscr{E}$ over X, we shall denote it by $\pi : \mathscr{E} \overset{f}{\leftrightarrows} X$.

Using local coordinates we write:

$$f : (x^i) \to (x^i, f^k(x)) \quad \text{or simply} \quad f : (x) \to (x, f(x)).$$

Moreover, if $\mathscr{E}$ is a fibered manifold, we shall also denote by $\mathscr{E}$ the set of germs of sections of $\mathscr{E}$. In fact our study will be mainly local and the context will always make clear when differentiations arise.

Now let Y be another manifold with $\dim Y = m$.

PRELIMINARY DEFINITION 1.9 A fibered manifold $\pi : \mathscr{E} \to X$ is called a "*bundle*" over X with fiber Y if, for any open covering $\{U_\alpha\}$ of X, there exist "*trivialising*" homeomorphisms $\Phi_\alpha : \pi^{-1}(U_\alpha) \to U_\alpha \times Y$, such that the following diagram is commutative:

$$\begin{array}{ccc} \pi^{-1}(U_\alpha) & \longrightarrow & U_\alpha \times Y \\ {\scriptstyle \pi}\downarrow & & \downarrow \\ U_\alpha & = & U_\alpha \end{array}$$

REMARK 1.10 As above, the bundle is called a "*differentiable* (*analytic*) *bundle*" if the manifolds and maps involved in the definition are differentiable (analytic).

EXAMPLE 1.11 Let X and Y be two manifolds. Then $\pi : X \times Y \to X$ with projection π onto the first factor is a fibered manifold over X.

DEFINITION 1.12 $\pi : X \times Y \to X$ is called a "*trivial fibered manifold*". As it is also a bundle over X, it will be valled a "*trivial bundle*" over X.

In particular, if $Y = \varnothing$, then we can consider $X \times \varnothing$ as a fibered manifold over X with projection $id_X : X \to X$.

The "*graph*" of any map $f : X \to Y$, is the corresponding section also noted $f : X \to X \times Y$.

If Y is a copy of X, we can thus speak equivalently of map of X into itself, or of map of X into Y, or of section of $X \times Y$ over X.

EXAMPLE 1.13 Moebius band.

This is a bundle over S^1 with fiber an open segment of $\mathbb{R}$, which is obtained by twisting $S^1 \times Y$ according to the following picture:

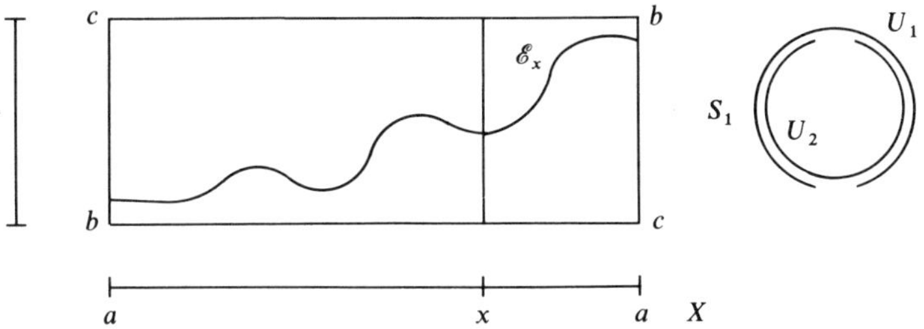

We can cover S^1 with coordinate neighbourhoods U_1 and U_2 as in the picture. The homeomorphism $\Phi_1 : \pi^{-1}(U_1) \to U_1 \times Y$ is the identity and the homeomorphism $\Phi_2 : \pi^{-1}(U_2) \to U_2 \times Y$ is obtained by means of a symmetry with respect to the middle point of Y.

EXAMPLE 1.14 Hopf bundle:

Let $\mathscr{E} = S^3$ be identified with the unit sphere of $\mathbb{C}^2$:

$$\mathscr{E} = S^3 = \{(z_1, z_2) \in \mathbb{C}^2 | z_1\bar{z}_1 + z_2\bar{z}_2 = 1\}.$$

Let $X = S^2$ be identified with the projective line $P_1(\mathbb{C})$.

$$X = S^2 = \{(w_1, w_2) \in \mathbb{C}^2 - \{0, 0\} | (w_1, w_2) \sim (\lambda w_1, \lambda w_2), \forall \lambda \neq 0\}.$$

Let $Y = S^1$ be identified with the unit circle in $\mathbb{C}$

$$Y = S^1 = \{\lambda \in \mathbb{C} | |\lambda| = 1\}$$

Let us define $\pi : S^3 \to S^2$ by the local formulas:

$$\pi(z^1, z^2) = (w_1, w_2) \quad \text{with} \quad w_1 = z_1, w_2 = z_2.$$

Then π is a surjective submersion because:

$$(w_1, w_2) \sim (w_1(w_1\bar{w}_1 + w_2\bar{w}_2)^{-1/2}, \qquad w_2(w_1\bar{w}_1 + w_2\bar{w}_2)^{-1/2})$$

In order to show the local triviality of S_3 we take an open covering $\{U_1, U_2\}$ of S^2 with $U_1 = S^2 - \{1, 0\}$ and $U_2 = S^2 - \{0, 1\}$.

Local coordinates $(w, 1)$ can be taken on U_1 and $(1, w')$ on U_2. Define

$$\Phi_1 : \pi^{-1}(U_1) \to U_1 \times S^1 : (z_1, z_2) \to \left(\left(\frac{z_1}{z_2}, 1\right), \left(\frac{z_2}{\bar{z}_2}\right)^{1/2}\right)$$

and

$$\Phi_2 : \pi^{-1}(U_2) \to U_2 \times S^1 : (z_1, z_2) \to \left(\left(1, \frac{z_2}{z_1}\right), \left(\frac{z_1}{\bar{z}_1}\right)^{1/2}\right)$$

It is then easy to check that Φ_1 and Φ_2 trivialise S^3. Finally, on $\pi^{-1}(U_1 \cap U_2)$ the transition functions are:

$$w' = \frac{1}{w}, \qquad \lambda' = \left(\frac{w}{\bar{w}}\right)^{1/2} \cdot \lambda$$

It follows that S^3 is a bundle over S^2 with fiber S^1.

We notice that in each of the three preceding examples, there exists a group of homeomorphisms of the fiber, which is called the "*structure group of the bundle*".

- It is the identity in the example 1.11.
- It is a cyclic group of order 2 in the example 1.13.
- It is a Lie group in the example 1.14 as appears from the fact that $|w/\bar{w}| = 1$ in the transition function.

More generally, we can state:

DEFINITION 1.15 A fibered manifold $\pi : \mathscr{E} \to X$ is called a bundle over X, with fiber Y and structure group G if there exist a manifold Y, a Lie group G acting effectively on Y and maps $\Phi_{\beta\alpha} : U_\alpha \cap U_\beta \to G$ such that, with a slight change of notations, $\forall x \in U_\alpha \cap U_\beta$ and $\forall y \in Y$, then

$$\Phi_{\beta\alpha}(x)(y) = \Phi_{\beta\alpha}(x, y) = \Phi_\beta \circ \Phi_\alpha^{-1}(x, y).$$

EXAMPLE 1.16 A bundle is called a principal bundle if its fiber coincides with its structure group acting on itself by left translation.

REMARK 1.17 The reader who is not familiar with Lie groups can refer to the beginning of Part II where he will find precise definitions. He will then easily be able to use local coordinates in order to interpret the above definitions locally.

As G acts effectively on Y, we have:

1) $\forall x \in U_\alpha \cap U_\beta \cap U_\gamma : \Phi_{\gamma\beta}(x) \circ \Phi_{\beta\alpha}(x) = \Phi_{\gamma\alpha}(x)$
2) $\forall x \in U_\alpha \qquad : \Phi_{\alpha\alpha}(x) = id_Y$
3) $\forall x \in U_\alpha \cap U_\beta \qquad : \Phi_{\beta\alpha}(x) = \Phi_{\alpha\beta}^{-1}(x)$

Conversely, we have the following theorem:

THEOREM 1.18 Let X and Y be manifolds, G be a Lie group acting effectively on Y and $\{U_\alpha\}$ be an open covering of X. If $\forall_{\alpha,\beta}$, $\exists$ maps $\Phi_{\beta\alpha} : U_\alpha \cap U_\beta \to G$ such that $\Phi_{\gamma\beta}(x) \circ \Phi_{\beta\alpha}(x) = \Phi_{\gamma\alpha}(x)$, $\forall x \in U_\alpha \cap U_\beta \cap U_\gamma$, then there exists a bundle $\mathscr{E}$ for which the $\Phi_{\beta\alpha}$ are transition functions.

Proof Taking $\alpha = \beta = \gamma$ it follows that $\Phi_{\alpha\alpha}(x) = id_Y$, $\forall x \in U_\alpha$. Then taking $\alpha = \gamma$, it follows that $\Phi_{\alpha\beta}(x) = \Phi_{\beta\alpha}^{-1}(x)$, $\forall x \in U_\alpha \cap U_\beta$.

Now let $\tilde{\mathscr{E}}$ be the disjoint union of the products $U_\alpha \times Y$. We shall identify two points $(x, y) \in U_\alpha \times Y$ and $(\bar{x}, \bar{y}) \in U_\beta \times Y$ if and only if $x = \bar{x} \in U_\alpha \cap U_\beta$ and $\bar{y} = \Phi_{\beta\alpha}(x)(y)$.

It is easy to check that this is an equivalence relation defined on $\tilde{\mathscr{E}}$ and we call $\mathscr{E}$ the quotient of $\tilde{\mathscr{E}}$ by this equivalence relation.

We take on $\mathscr{E}$ the topology induced by the topoloty of $\tilde{\mathscr{E}}$ and we note $[x, y]$ the equivalence class of a point $(x, y) \in \tilde{\mathscr{E}}$.

We define a unique projection $\pi : \mathscr{E} \to X$ setting $\pi([x, y]) = x$.

Now we define $\Phi_\alpha : \pi^{-1}(U_\alpha) \to U_\alpha \times Y$ by means of $\Phi_\alpha([x, y]) = (x, y)$ where (x, y) is the unique representative of $[x, y]$ in $U_\alpha \times Y$. In fact if $(\bar{x}, \bar{y})$ is another representative of $[x, y]$, we must have $x = \bar{x} \in U_\alpha$ and $\bar{y} = \Phi_{\alpha\alpha}(x)(y) = y$. It follows that Φ_α is injective.

Now if $(x, y) \in U_\alpha \times Y$ then its equivalence class $[x, y]$ is such that $\Phi_\alpha([x, y]) = (x, y)$ by definition and Φ_α is also surjective. It is then easy to show that Φ_α is a homeomorphism which defines a trivialisation of $\mathscr{E}$ on U_α, and that $\Phi_\alpha^{-1}(x, y) = [x, y]$.

Finally, for $x \in U_\alpha \cap U_\beta$, x fixed, $\Phi_\beta \circ \Phi_\alpha^{-1}$ is a homeomorphism of Y onto itself.

Let $(x, \bar{y}) = \Phi_\beta \circ \Phi_\alpha^{-1}(x, y)$, then $\Phi_\beta^{-1}(x, \bar{y}) = \Phi_\alpha^{-1}(x, y)$ that is to say $[x, \bar{y}] = [x, y]$ and thus $\bar{y} = \Phi_{\beta\alpha}(x)y$. This concludes the proof. C.Q.F.D.

EXAMPLE 1.19 If the structure group of a bundle is reduced to the identity, then this bundle is a trivial bundle.

EXAMPLE 1.20 Tangent and cotangent bundles of X.

Let $\{U_\alpha\}$ be an open covering of X and $x \in U_\alpha \cap U_\beta$. Take $Y = \mathbb{R}^n$ with local coordinates u^i. Take $G = GL(n, \mathbb{R})$ acting on $\mathbb{R}^n$ in a natural way. Define $\Phi_{\beta\alpha} : U_\alpha \cap U_\beta \to GL(n, \mathbb{R})$ by means of the formula:

$$\Phi_{\beta\alpha}(x) = \left[\frac{\partial \varphi^i_{\beta\alpha}(x)}{\partial x^j_\alpha}\right], \qquad \forall x \in U_\alpha \cap U_\beta$$

where (x^i_α) are local coordinates for X on U_α, (x^i_β) local coordinates for X on U_β and $x^i_\beta = \varphi^i_{\beta\alpha}(x_\alpha)$ are the transition functions of X on $U_\alpha \cap U_\beta$.

Then on $U_\alpha \cap U_\beta \cap U_\gamma$ we have by definition

$$\varphi_{\gamma\beta}(\varphi_{\beta\alpha}(x_\alpha)) \equiv \varphi_{\gamma\alpha}(x_\alpha)$$

It follows that

$$\frac{\partial \varphi^k_{\gamma\beta}(x)}{\partial x^j_\beta} \cdot \frac{\partial \varphi^j_{\beta\alpha}(x)}{\partial x^i_\alpha} = \frac{\partial \varphi^k_{\gamma\alpha}(x)}{\partial x^i_\alpha}$$

and we can apply the preceding theorem in order to construct a bundle denoted by $T(X)$ or simply T.

DEFINITION 1.21 $T = T(X)$ is called the "*tangent bundle*" of X.

Similarly we can take $\Phi'_{\beta\alpha}(x) = \Phi_{\alpha\beta}(x)$. The bundle thus obtained is denoted by $T^*(X)$ or simply T^*.

DEFINITION 1.22 $T^* = T^*(X)$ is called the "*cotangent bundle*" of X.

DEFINITION 1.23 A section ξ of T is called a "*vector field*" on X. Similarly, a section χ of T^* is called a "*covector field*" or "*1-form*" on X.

EXAMPLE 1.24 As in the case of example 1.20, it is now possible to introduce bundles on X, the sections of which are mixed tensor fields on X.

We denote $\Lambda^r T^*$, $S_r T^*$ and $\overset{r}{\otimes} T^*$ respectively the r-exterior product, the r-symmetric product, and the r-tensor product of copies of T^*.

These are bundles over X, the sections of which are respectively antisymmetric, symmetric and ordinary contravariant tensor fields on X. The reader will use the method of example 1.20 in order to construct them by patching together local trivialisations using the same open covering $\{U_\alpha\}$ of X.

We shall write simply $\xi = \xi^i(x)\partial/\partial x^i$ and $\chi = \chi_i(x)\, dx^i$ using classical notations.

An r-form ω is a section of $\Lambda^r T^*$ and we shall also write simply $\omega = \sum_{i_1 < \cdots < i_r} \omega_{i_1 \cdots i_r}(x)\, dx^{i_1} \wedge \cdots \wedge dx^{i_r}$ or in condensed notation $\omega = \omega_I(x)\, dx^I$.

EXAMPLE 1.25 Vector bundles.

DEFINITION 1.26 A bundle is called a "*vector bundle*" if the fiber is a finite m-dimensional vector space, identified with $\mathbb{R}^m$ and if the structure group is a subgroup of $GL(m, \mathbb{R})$ acting in a natural effective way on $\mathbb{R}^m$.

We shall write vector bundles using capital letters.

If E is a vector bundle over X, we define $\dim E = m$ as the dimension of E.

Using matrix notation, and local coordinates (x^i_α, v^k_α) for $E|_{U_\alpha}$ where $\{U_\alpha\}$ is an open covering of X, then the transition functions of E on $U_\alpha \cap U_\beta$ are:

$$v_\beta = A_{\beta\alpha}(x)v_\alpha$$

$$x_\beta = \varphi_{\beta\alpha}(x_\alpha)$$

Let $\{U_\alpha\}$ be an open covering of X and consider the two vector bundles E and F with local coordinates (x_α, v_α) for E and (x_α, w_α) for F on U_α.

Let the transition functions of E and F be given on $U_\alpha \cap U_\beta$ by the matrices $[A_{\beta\alpha}(x)]$ and $[B_{\beta\alpha}(x)]$.

DEFINITION 1.27 The tensor product $E \otimes F$ of the two vector bundles E and F is a vector bundle on X, the transition functions of which are given on $U_\alpha \cap U_\beta$ by the tensor product of the matrices $[A_{\beta\alpha}(x)]$ and $[B_{\beta\alpha}(x)]$.

We have the relation $\dim E \otimes F = \dim E \times \dim F$.

EXAMPLE 1.28 Affine bundles.

DEFINITION 1.29 A bundle is called an "*affine bundle*" if the fiber is an affine space and if the structure group is a subgroup of the corresponding affine group.

If $\{U_\alpha\}$ is an open covering of X trivialising the affine bundle $\mathscr{E}$, and if we take local coordinates (x_α, y_α) for $\mathscr{E}$ on U_α, then the transition functions are given on $U_\alpha \cap U_\beta$ by the formulas:

$$\begin{aligned} y_\beta &= A_{\beta\alpha}(x)y_\alpha + B_{\beta\alpha}(x) \\ x_\beta &= \varphi_{\beta\alpha}(x_\alpha) \end{aligned}$$

where $[A_{\beta\alpha}(x)]$ is a square matrix.

By means of the preceding existence theorem it is easy to see that we can associate with $\mathscr{E}$, a vector bundle E with local coordinates (x_α, u_α) on U_α and transition functions given on $U_\alpha \cap U_\beta$ by the matrix $[A_{\beta\alpha}(x)]$.

DEFINITION 1.30 We say that the affine bundle $\mathscr{E}$ is "*modelled*" on the vector bundle E and we denote it by a dotted arrow:

$$E \dashrightarrow \mathscr{E} \longrightarrow X.$$

2 Fibered morphisms

DEFINITION 2.1 By a "*morphism*" between a fibered manifold $\pi : \mathscr{E} \to X$ and a fibered manifold $\pi' : \mathscr{E}' \to X'$, we mean a couple (φ, Φ) of maps $\varphi : X \to X'$ and $\Phi : \mathscr{E} \to \mathscr{E}'$ such that the following diagram is commutative:

$$\begin{array}{ccc} \mathscr{E} & \xrightarrow{\Phi} & \mathscr{E}' \\ \Big\downarrow{\scriptstyle \pi} & & \Big\downarrow{\scriptstyle \pi'} \\ X & \xrightarrow{\varphi} & X' \end{array}$$

We shall simply say that the morphism $\Phi : \mathscr{E} \to \mathscr{E}'$ is "*fibered*" over the map $\varphi : X \to X'$, or simply that Φ is over φ.

EXAMPLE 2.2 Let X, Y be manifolds and $f : X \to Y$ be a map. If $x \in X$, we can choose a coordinate neighbourhood U of x in X and a coordinate neighbourhood V of $y = f(x)$ in Y with $f(U) \subset V$ and such that $y^k = f^k(x^i)$ is a local expression of f.

We define a map $T(f): T(U) \to T(V)$, using local coordinates (x, u) for $T(U)$ and (y, v) for $T(V)$ by the formulas $y^k = f^k(x^i)$, $v^k = (\partial f^k(x)/\partial x^i)u^i$. The reader will check easily that this map is compatible with all changes of coordinates and that we can define in this way a map $T(f): T(X) \to T(Y)$.

In fact, if $\bar{U}$ is another coordinate neighbourhood of x in X and $\bar{V}$ another coordinate neighbourhood of $y = f(x)$ in Y with $f(\bar{U}) \subset \bar{V}$, we adopt new coordinates $\bar{x}^j = \varphi^j(x^i)$ on $\bar{U}$ and $\bar{y}^l = \psi^l(y^k)$ on $\bar{V}$.

Then, by definition, we have:

$$\bar{v}^l = \frac{\partial \psi^l}{\partial y^k}(y)v^k = \frac{\partial \psi^l}{\partial y^k}(y)\,\frac{\partial f^k}{\partial x^i}(x)u^i \qquad \text{and} \qquad \bar{u}^j = \frac{\partial \varphi^j(x)}{\partial x^i}\,u^i$$

but $\bar{y}^l = \bar{f}(\bar{x}^j)$ is such that $\psi^l(f^k(x^i)) \equiv \bar{f}^l(\varphi^j(x^i))$ and it follows that $\bar{v}^l = (\partial \bar{f}^l/\partial \bar{x}^j)(x)\bar{u}^j$.

Finally we have the commutative diagram:

$$\begin{array}{ccc} T(X) & \xrightarrow{T(f)} & T(Y) \\ {\scriptstyle \pi_X}\downarrow & & \downarrow{\scriptstyle \pi_Y} \\ X & \xrightarrow{f} & Y \end{array}$$

DEFINTION 2.3 The map f is said to be "*regular*" at x if rank $[\partial f^k(x)/\partial x^i] = n$. It is said to be regular if it is regular at x, $\forall x \in X$.

REMARK 2.4 When $X = X'$ and $\Phi: \mathscr{E} \to \mathscr{E}'$ is a morphism of fibered manifolds over X, fibered over id_X, we say that Φ is an X-morphism, or simply a morphism when there is no confusion.

Let Φ be a morphism fibered on φ, then $\Phi(\mathscr{E}_x) \subset \mathscr{E}'_{\varphi(x)}$ and the map $\Phi|_{\mathscr{E}_x}: \mathscr{E}_x \to \mathscr{E}'_{\varphi(x)}$ is denoted by Φ_x or simply Φ when there is no confusion. When Φ is an X-morphism, then, to any section f of $\mathscr{E}$ over X, there corresponds by Φ the section f' of $\mathscr{E}'$ over X' such that $f' = \Phi \circ f$, according to the diagram:

$$\begin{array}{ccc} \mathscr{E} & \xrightarrow{\Phi} & \mathscr{E}' \\ {\scriptstyle f}\nwarrow\searrow{\scriptstyle \pi} & & {\scriptstyle \pi'}\swarrow\nearrow{\scriptstyle f'} \\ & X & \end{array}$$

3 Fibered submanifolds

DEFINITION 3.1 We say that a fibered manifold $\mathscr{R} \to X$ is a "*fibered submanifold*" of the fibered manifold $\pi: \mathscr{E} \to X$, if $\mathscr{R}$ is a (open or closed) submanifold of $\mathscr{E}$ and if the inclusion $\iota: \mathscr{R} \to \mathscr{E}$ is a X-morphism.

REMARK 3.2 The fibered manifold $\mathscr{R}$ has projection $\pi \circ \iota$. However we shall, in general, identify $\mathscr{R}$ with its image by ι in $\mathscr{E}$ and get a fiber manifold $\pi: \mathscr{R} \to X$ where π is written for $\pi \circ \iota = \pi|_{\mathscr{R}}$.

If $\mathscr{R}$ is a closed submanifold of $\mathscr{E}$, then for any point $(x_0, y_0) \in \mathscr{R} \subset \mathscr{E}$, there exists a coordinate neighbourhood U of (x_0, y_0) in $\mathscr{E}$ such that $\mathscr{R} \cap U$ is described by the local equation $\Phi^\tau(x^i, y^k) = 0$ with

$$\operatorname{rank}\left[\frac{\partial \Phi^\tau(x, y)}{\partial y^k}\right] = \operatorname{rank}\left[\frac{\partial \Phi^\tau(x_0, y_0)}{\partial y^k}\right] = \operatorname{codim} \mathscr{R}$$

where $\operatorname{codim} \mathscr{R} = \dim \mathscr{E} - \dim \mathscr{R}$.

In fact, as $\pi: \mathscr{R} \to X$ is a fibered manifold, then π is surjective. It follows that we cannot obtain, using the preceding equation and the implicit function theorem, any equation such as $\psi(x) = 0$. We let the reader show that the former rank conditions do not depend on the coordinate system on U.

EXAMPLE 3.3 If f is a section of $\mathscr{E}$ over X, then $f(X)$ is closed fibered submanifold of $\mathscr{E}$ of dimension n.

DEFINITION 3.4 If $\Phi: \mathscr{E} \to \mathscr{E}'$ is an X-morphism and f' is a section of $\mathscr{E}'$ over X, then we define $\ker_{f'} \Phi$ as the subset of $\mathscr{E}$ made by the points $\ker_{f'} \Phi = \{(x, y) \in \mathscr{E} \mid \Phi(x, y) = f'(x)\}$. We say that $\ker_{f'} \Phi$ is the "*kernel of* Φ *with respect to* f'".

We may ask for conditions on Φ and f' that could make $\ker_{f'} \Phi$ a fibered submanifold of $\mathscr{E}$.

In fact $\ker_{f'} \Phi$ must be a submanifold of $\mathscr{E}$. From the implicit function theorem it follows that Φ must have constant rank. However we need also that $\pi|_{\mathscr{R}}$ must be surjective.

Now, if $\ker_{f'} \Phi = \varnothing$ there is nothing to prove. If $\ker_{f'} \Phi \neq \varnothing$ there exists at least one point $(x_0, y_0) \in \ker_{f'} \Phi \subset \mathscr{E}$ and we can use local coordinates on a neighbourhood of that point in $\mathscr{E}$. $\ker_{f'} \Phi$ will then be defined locally by the equation $\Phi(x, y) = f'(x)$. Imagine there exists a local function $\psi(x, y')$ such that $\psi(x, \Phi(x, y)) \equiv 0$, then if $\psi(x, f'(x)) \not\equiv 0$, we could deduce an equation connecting the x^i alone, contradicting the surjectivity of $\pi|_{\mathscr{R}}$. Finally we thus need $f'(X) \subset \operatorname{im} \Phi$.

From this and the implicit function theorem we obtain the following propositions:

PROPOSITION 3.5 If $\Phi : \mathscr{E} \to \mathscr{E}'$ is a morphism of constant rank, then the image of Φ, denoted by im Φ, is a fibered submanifold of $\mathscr{E}'$.

PROPOSITION 3.6 However, if f' is a section of $\mathscr{E}'$ over X such that $f'(X) \subset \operatorname{im} \Phi$, then $\ker_{f'} \Phi$ is a fibered submanifold of $\mathscr{E}$.

PROPOSITION 3.7 If $\mathscr{R}$ is a fibered submanifold of $\mathscr{E}$, then, for each point of $\mathscr{R}$, there exists a neighbourhood $\mathscr{U}$ of this point in $\mathscr{E}$, a fibered manifold $\pi' : \mathscr{E}' \to U$ with $U = \pi(\mathscr{U})$, a section f' of $\mathscr{E}'$ over U and a morphism $\Phi : \mathscr{U} \to \mathscr{E}'$ of constant rank over id_U, such that $\mathscr{R} \cap \mathscr{U} = \ker_{f'} \Phi$.

REMARK 3.8 We shall make an extensive use of the last proposition in order to associate an operator, at least locally, with any given system. In fact, if $(x_0, y_0) \in \mathscr{R} \subset \mathscr{E}$, then we can take a coordinate neighbourhood $\mathscr{U}$ of this point in $\mathscr{E}$, such that $\mathscr{R} \cap \mathscr{U}$ is described by the equation $\Phi^\tau(x, y) = 0$ with $\tau = 1, \ldots, \operatorname{codim} \mathscr{R}$ and $\Phi^\tau(x_0, y_0) = 0$. Setting $U = \pi(\mathscr{U})$, we can introduce the trivial fibered manifold $\mathscr{E}' = U \times \mathbb{R}^{\operatorname{codim} \mathscr{R}}$ over U, with local coordinates (x^i, y^τ), the "*zero*" section of this bundle and the morphism $\Phi : \mathscr{U} \to \mathscr{E}' : (x, y) \to (x, y' = \Phi(x, y))$.

REMARK 3.9 It is sometimes difficult to check the hypothesis of constant rank. For example it is possible to define the same fibered submanifold $\mathscr{R}$ of a fibered manifold $\mathscr{E}$ by means of two morphisms, one Φ' with constant rank "*on $\mathscr{R}$ but not on $\mathscr{E}$*", the other one Φ'' with constant rank "*on $\mathscr{E}$*".

In both cases im Φ' and im Φ'' exist but Φ'' only gives rise to an exact sequence as in the proposition.

EXAMPLE 3.10 $X = \mathbb{R}$, $\mathscr{E} = \mathscr{E}' = \mathscr{E}'' = \mathbb{R} \times \mathbb{R}^2 = \mathbb{R}^3$. The sections of $\mathscr{E}'$ and $\mathscr{E}''$ to be considered are the zero sections. With evident notations $\mathscr{R} = \ker_0 \Phi' = \ker_0 \Phi''$. We use local coordinates (x, y, z) for $\mathscr{E}$, (x', y', z') for $\mathscr{E}'$, (x', y'', z'') for $\mathscr{E}''$.

$$\Phi' \begin{cases} z(y + x) = z' \\ y + x = y' \\ x = x' \end{cases} \qquad \operatorname{rank} T(\Phi') = \operatorname{rank} \begin{bmatrix} y + x & z & z \\ 0 & 1 & 1 \\ 0 & 0 & 1 \end{bmatrix}$$

$$\operatorname{rank} T(\Phi')|_{\mathscr{R}} = 2, \qquad \operatorname{rank} T(\Phi')|_{\mathscr{E} - \mathscr{R}} = 3$$

$$\Phi'' \begin{cases} y + x = z' \\ y + x = y' \\ x = x' \end{cases} \qquad \operatorname{rank} T(\Phi'') = \operatorname{rank} \begin{bmatrix} 0 & 1 & 1 \\ 0 & 1 & 1 \\ 0 & 0 & 1 \end{bmatrix} = 2$$

$\mathscr{R} = \{(x, y, z) \in \mathbb{R}^3 \mid y + x = 0\}$ is a fibered submanifold of $\mathbb{R}^3$. The reader will find himself im Φ'. As for Φ'' we have im $\Phi'' = \{(x, y, z) \in \mathbb{R}^3 \mid z - y = 0\}$.

4 Vector bundles

Let $\pi : E \to X$ and $\pi' : E' \to X'$ be vector bundles.

DEFINITION 4.1 A map $\Phi : E \to E'$ fibered over $\varphi : X \to X'$ is called a "*homomorphism of vector bundles*" if the restriction Φ_x of Φ to any fiber E_x is a homomorphism of vector spaces. The same definition holds for monomorphisms, epimorphisms and isomorphisms.

In particular if $\pi : E \to X$ and $\pi' : E' \to X$ are vector bundles over the same base manifold X, an X-morphism $\Phi : E \to E'$ is called an X-homomorphism if Φ is a homomorphism and a similar definition holds for X-monomorphisms, X-epimorphisms and X-isomorphisms.

An X-morphism will be called simply a morphism if there can be no confusion.

Using local coordinates (x, v) for E and (x, v') for E', we have

$$\Phi : E \to E' : (x, v) \to (x, v' = \Phi(x, v))$$

with $\Phi(x, v) \equiv A(x)v$ where $[A(x)]$ is a $m \times m'$ matrix if $m = \dim E$, $m' = \dim E'$.

If Φ is an isomorphism, then $[A(x)]$ is a square matrix of maximum rank.

The kernel of a homomorphism is always considered with respect to the zero section of E' and is denoted simply $\ker \Phi$. In general it is not a vector bundle because it may not satisfy the local triviality property.

EXAMPLE 4.2 $X = [0, 1] \subset \mathbb{R}$.

$$\Phi : X \times \mathbb{R} \to X \times \mathbb{R} : (x, u) \to (x, xu).$$

However, when Φ is an epimorphism, then $\operatorname{rank} \Phi_x = \dim E' = m'$, $\forall x \in X$ and $\ker \Phi$ is the vector bundle $\ker \Phi = \bigcup_{x \in X} \ker \Phi_x$. We have $\dim \ker \Phi = \dim E - \dim E'$.

Now, when $\Phi : E \to E'$ is a morphism of vector bundles we can define on E' an equivalence relation, saying that $(x, v'_1) \sim (x, v'_2)$ if there exists $(x, v) \in E$ such that $v'_2 - v'_1 = \Phi(x, v)$.

DEFINITION 4.3 The quotient of E' by this equivalence relation is called the cokernel of Φ and is denoted $\operatorname{coker} \Phi$.

The reader will show, as an exercise, that $\operatorname{coker} \Phi$ is a vector bundle when Φ is a monomorphism and that, in this case, we have

$$\dim E' = \dim E + \dim \operatorname{coker} \Phi.$$

Moreover $\operatorname{im} \Phi$ is then a vector subbundle of E' and we have

$$\dim E = \dim \operatorname{im} \Phi.$$

Let now E, E', E'' be vector bundles over X and $\Phi : E \to E'$, $\Psi : E' \to E''$ be homomorphisms such that $\Psi \circ \Phi = 0$.

DEFINITION 4.4 The sequence of vector bundles over X:

$$E \xrightarrow{\Phi} E' \xrightarrow{\Psi} E''$$

is said to be "*exact*", if:

1) $\operatorname{im} \Phi = \ker \Psi$ as subsets of E'.
2) The sequence of vector spaces:

$$E_x \xrightarrow{\Phi_x} E'_x \xrightarrow{\Psi_x} E''_x$$

is exact, $\forall x \in X$.

REMARK 4.5 In the case of vector bundles, the two preceding conditions are equivalent. However we shall extend this definition to arbitrary fibered manifolds, under its present form.

EXAMPLE 4.6 If Φ is a monomorphism and Ψ is an epimorphism such that $\Psi \circ \Phi = 0$, any exact sequence:

$$0 \longrightarrow E \xrightarrow{\Phi} E' \xrightarrow{\Psi} E'' \longrightarrow 0$$

will be called a "*short exact sequence*".

Let V_r be a vector space over $\mathbb{R}$ and $\Phi_r : V_{r-1} \to V_r$ be a map, $\forall r$.

DEFINITION 4.7 We say that

$$\cdots \longrightarrow V_{r-1} \xrightarrow{\Phi_r} V_r \xrightarrow{\Phi_{r+1}} V_{r+1} \longrightarrow \cdots$$

is a sequence if $\Phi_{r+1} \circ \Phi_r = 0$, $\forall r$.

It follows that in general $\operatorname{im} \Phi_r \subset \ker \Phi_{r+1}$.

DEFINITION 4.8 We define $B_r = \operatorname{im} \Phi_r$, $Z_r = \ker \Phi_{r+1}$ and we introduce $H_r = Z_r/B_r$.

We shall now consider a "*finite length sequence*":

$$0 \longrightarrow V_0 \xrightarrow{\Phi_1} V_1 \xrightarrow{\Phi_2} \cdots \xrightarrow{\Phi_n} V_n \longrightarrow 0.$$

We have by definition:

$$\dim Z_r = \dim B_r + \dim H_r$$

and

$$\dim V_r = \dim Z_r + \dim B_{r+1}.$$

It follows that $\dim V_r = \dim B_r + \dim B_{r+1} + \dim H_r$ and we obtain the important formula:

$$\sum_{r=0}^{n} (-1)^r \dim V_r = \sum_{r=0}^{n} (-1)^r \dim H_r.$$

DEFINITION 4.9 It is called the "*Euler–Poincaré*" formula.

REMARK 4.10 If the above sequence is exact, we obtain the useful formula:

$$\sum_{r=0}^{n} (-1)^r \dim V_r = 0.$$

5 Fibered operations

Let $\pi : \mathscr{E} \to X$ and $\pi' : \mathscr{E}' \to X$ be two fibered manifolds over the same base manifold X.

DEFINITION 5.1 We call "*fibered product*" of $\mathscr{E}$ and $\mathscr{E}'$ over X, the fibered manifold over X consisting of all pairs of points of $\mathscr{E}$ and $\mathscr{E}'$ having the same projection on X. It will be denoted by $\mathscr{E} \times_X \mathscr{E}'$.

A system of local coordinates will be (x, y, y').

It is possible to endow $\mathscr{E} \times_X \mathscr{E}'$ with two other projections, one onto $\mathscr{E} : (x, y, y') \to (x, y)$ and the other onto $\mathscr{E}' : (x, y, y') \to (x, y')$ that make it also fibered manifold over $\mathscr{E}$ and $\mathscr{E}'$ respectively. We have the commutative diagram:

$$\begin{array}{ccc} \mathscr{E} \times_X \mathscr{E}' & \longrightarrow & \mathscr{E}' \\ \downarrow & & \downarrow \pi' \\ \mathscr{E} & \xrightarrow{\pi} & X \end{array}$$

Now let $\varphi : X \to X'$ be a map and $\pi' : \mathscr{E}' \to X'$ be a fibered manifold over X'. We denote by $\varphi^{-1}(\mathscr{E}')$ the subset of $X \times \mathscr{E}'$ consisting of all pairs of points $(x, (x', y'))$ such that $x' = \varphi(x) = \pi(x', y')$.

The reader will easily show that this set, with the projection onto X induced by the projection $X \times \mathscr{E}' \to X$ onto the first factor, is a fibered manifold over X.

DEFINITION 5.2 We call $\varphi^{-1}(\mathscr{E}')$ the "*reciprocal image*" of $\mathscr{E}'$ by φ.

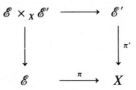

We have the commutative diagram:

$$\begin{array}{ccc} \varphi^{-1}(\mathscr{E}') & \longrightarrow & \mathscr{E}' \\ \downarrow & & \downarrow{\scriptstyle \pi'} \\ X & \xrightarrow{\ \varphi\ } & X' \end{array}$$

If U' is an open subset of X' and U an open subset of X such that $U' = \varphi(U)$, we have:

$$\varphi^{-1}(\mathscr{E}'|_{U'}) = \varphi^{-1}(\mathscr{E}')|_U$$

REMARK 5.3 If X is a submanifold of X' and ι the injection of X into X', then we can identify X with its image in X' by ι and we have:

$$\mathscr{E}'|_X = \iota^{-1}(\mathscr{E}') = \pi'^{-1}(X)$$

This shows the consistency of our notations. As a matter of fact we have also the relations:

$$\mathscr{E} \times_X \mathscr{E}' = \pi^{-1}(\mathscr{E}') = \pi'^{-1}(\mathscr{E}).$$

PROPOSITION 5.4 If $\Phi : \mathscr{E} \to \mathscr{E}'$ is a morphism over $\varphi : X \to X'$, then there exists a X-morphism Ψ and a morphism Ψ' over φ such that the following diagram is commutative:

$$\begin{array}{ccccc} & & \overset{\Phi}{\frown\!\!\!\longrightarrow} & & \\ \mathscr{E} & \xrightarrow{\ \Psi\ } & \varphi^{-1}(\mathscr{E}') & \xrightarrow{\ \Psi'\ } & \mathscr{E}' \\ \downarrow & & \downarrow & & \downarrow{\scriptstyle \pi'} \\ X & = & X & \xrightarrow{\ \varphi\ } & X' \end{array}$$

Proof We define $\Psi : \mathscr{E} \to X \times \mathscr{E}'$ using local coordinates by $\Psi : (x, y) \to (x, (\varphi(x), \Phi(x, y)))$. It follows from the definition of $\varphi^{-1}(\mathscr{E}')$ that ψ factors through $\Psi : \mathscr{E} \to \varphi^{-1}(\mathscr{E}')$: $(x, y) \to (x, \Phi(x, y))$. We define $\Psi' : X \times \mathscr{E}' \to \mathscr{E}'$ as the projection onto the second factor: $(x, (x', y')) \to (x', y')$. We can restrict Ψ' to $\varphi^{-1}(\mathscr{E}') \subset X \times \mathscr{E}'$ and get a morphism denoted also by $\Psi' : \varphi^{-1}(\mathscr{E}') \to \mathscr{E}'$.

It is easy to check that $\Psi' \circ \Psi = \Phi$ because

$$\Psi' \circ \Psi : (x, y) \to (x, (\varphi(x), \Phi(x, y))) \to (\varphi(x), \Phi(x, y))$$

C.Q.F.D.

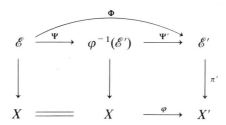

REMARK 5.5 When there will be no confusion we shall also denote by $\Phi: \mathscr{E} \to \varphi^{-1}(\mathscr{E}')$ the morphism induced by Φ.

EXAMPLE 5.6 If $\mathscr{E} = X \times \mathbb{R}$ and $\mathscr{E}' = X' \times \mathbb{R}$ then $\varphi^{-1}(\mathscr{E}') = \mathscr{E}$.

6 Vertical bundles

Let $\pi: \mathscr{E} \to X$ be a fibered manifold and $T(\mathscr{E})$ be the tangent bundle of $\mathscr{E}$. We have the commutative diagram where $T(\pi)$ is defined in a natural way:

$$\begin{array}{ccc} T(\mathscr{E}) & \xrightarrow{T(\pi)} & T(X) \\ \downarrow & & \downarrow \\ \mathscr{E} & \xrightarrow{\pi} & X \end{array}$$

Using the preceding proposition and the remark following it, we have a morphism $T(\pi): T(\mathscr{E}) \to \pi^{-1}(T(X))$ over $\mathrm{id}_{\mathscr{E}}$.

Now π and $T(\pi)$ have constant rank because π is an epimorphism. Moreover, as π and the projection $T(\mathscr{E}) \to \mathscr{E}$ are epimorphisms, the later diagram shows that $T(\pi)$ is also an epimorphism.

Using local coordinates $(x^1, \ldots, x^n; y^1, \ldots, y^m; u^1, \ldots, u^n; v^1, \ldots, v^m)$ or simply $(x, y; u, v)$ for $T(\mathscr{E})$ over a coordinate domain $\mathscr{U} \subset \mathscr{E}$, then the morphism $T(\pi): T(\mathscr{E}) \to \pi^{-1}(T(X)): (x, y; u, v) \to (x, y; u)$ is easily seen to be also an epimorphism.

Finally $T(\mathscr{E})$ is a vector bundle over $\mathscr{E}$ and, as $T(X)$ is a bundle over X, $\pi^{-1}(T(X))$ is also a vector bundle over $\mathscr{E}$.

We define $V(\mathscr{E}) = \ker T(\pi)$ by the short exact sequence of vector bundles over $\mathscr{E}$:

$$0 \longrightarrow V(\mathscr{E}) \longrightarrow T(\mathscr{E}) \xrightarrow{T(\pi)} \pi^{-1}(T(X)) \longrightarrow 0$$

$V(\mathscr{E})$ is the subbundle of $T(\mathscr{E})$ consisting of all vectors tangent to the fibers, that is to say with $u = 0$ and local coordinates for $V(\mathscr{E})$ are $(x, y; 0, v)$.

DEFINITION 6.1 $V(\mathscr{E})$ is called the "*vertical bundle*" of $\mathscr{E}$ and will be denoted by E when there can be no confusion.

EXAMPLE 6.2 For the bundle $\mathrm{id}_X: X \to X$, we define $V(X) = 0$.

EXAMPLE 6.3 Let X be a manifold and $\omega: (x) \to (x, \omega_{ij}(x))$ be a section of $S_2 T^*$. By taking the inverse of the matrix $[\omega_{ij}(x)]$ when it is possible, we define a section $\omega^{-1}: (x) \to (x, \omega^{ij}(x))$ of $S_2 T$. We then define the Christoffel symbols as a field over X of geometric objects defined by:

$$\gamma^i_{jk}(x) = \tfrac{1}{2}\omega^{il}(x)(\partial_j \omega_{kl}(x) + \partial_k \omega_{jl}(x) - \partial_l \omega_{jk}(x))$$

The reader can easily check by a straightforward computation that $\gamma: (x) \to (x, \gamma^i_{jk}(x))$ is a section of an affine bundle $\pi: \mathscr{E} \to X$, with coordinates y^i_{jk} and transition functions:

$$\bar{y}^l_{\sigma\tau} = \frac{\partial x^j}{\partial \bar{x}^\sigma}\frac{\partial x^k}{\partial \bar{x}^\tau}\frac{\partial \bar{x}^l}{\partial x^i} y^i_{jk} + \frac{\partial x^j}{\partial \bar{x}^\sigma}\frac{\partial x^k}{\partial \bar{x}^\tau}\frac{\partial^2 \bar{x}^l}{\partial x^j \partial x^k}$$

The vertical bundle of $\mathscr{E}$ has local coordinates $(x, y^i_{jk}; 0, v^i_{jk})$ and transition function:

$$\bar{v}^l_{\sigma\tau} = \frac{\partial x^j}{\partial \bar{x}^\sigma}\frac{\partial x^k}{\partial \bar{x}^\tau}\frac{\partial \bar{x}^l}{\partial x^i} v^i_{jk}$$

It follows that $\mathscr{E}$ is modelled on the vector bundle $S_2 T^* \otimes T$ and that we have $V(\mathscr{E}) = \pi^{-1}(S^2 T^* \otimes T)$.

More generally we have the proposition:

PROPOSITION 6.4 If $\pi: \mathscr{E} \to X$ is an affine bundle over X, modelled on a vector bundle $E \to X$, then we can identify $\pi^{-1}(E)$ and $V(\mathscr{E})$ as vector bundles over $\mathscr{E}$.

Proof It is similar to that of example 6.3 and left to the reader as an easy exercise. In fact if the transition functions of $\mathscr{E}$ are $\bar{x} = \varphi(x)$, $\bar{y} = A(x)y + B(x)$ then it follows that $\bar{v} = A(x)v$ are among the transition functions for $V(\mathscr{E})$ and E, by definition. C.Q.F.D.

Now let $\Phi: \mathscr{E} \to \mathscr{E}'$ be an X-morphism and let $T(\Phi): T(\mathscr{E}) \to T(\mathscr{E}')$ be the corresponding morphism over Φ. If $(x, y') \in \mathscr{E}'$ is the image of $(x, y) \in \mathscr{E}$ by Φ, then $T_{(x,y)}(\Phi): T_{(x,y)}(\mathscr{E}) \to T_{(x,y')}(\mathscr{E}')$ is a map between vector spaces.

It is then easy to construct the following commutative and exact diagram:

$$\begin{array}{ccccccccc}
0 & \longrightarrow & V(\mathscr{E}) & \longrightarrow & T(\mathscr{E}) & \xrightarrow{T(\pi)} & \pi^{-1}(T(X)) & \longrightarrow & 0 \\
 & & \downarrow{\scriptstyle V(\Phi)} & & \downarrow{\scriptstyle T(\Phi)} & & \downarrow & & \\
0 & \longrightarrow & V(\mathscr{E}') & \longrightarrow & T(\mathscr{E}') & \xrightarrow{T(\pi')} & \pi'^{-1}(T(X)) & \longrightarrow & 0
\end{array}$$

where $V(\Phi)$ is the homomorphism of vector bundles induced by $T(\Phi)$ and fibered over Φ.

7 Exact sequences

Let $\pi : \mathscr{E} \to X$, $\pi' : \mathscr{E}' \to X$, $\pi'' : \mathscr{E}'' \to X$ be fibered manifolds over X and $\Phi : \mathscr{E} \to \mathscr{E}'$, $\Psi : \mathscr{E}' \to \mathscr{E}''$ be X-morphisms.

DEFINITION 7.1 We say that the sequence of fibered manifolds over X:

$$\mathscr{E} \xrightarrow{\Phi} \mathscr{E}' \xrightarrow{\Psi} \mathscr{E}''$$

is exact if there exists a section f'' of $\mathscr{E}''$ such that:

1) $\operatorname{im} \Phi = \ker_{f''} \Psi$ as subsets of $\mathscr{E}'$.
2) the sequence of vector spaces:

$$V_{(x,y)}(\mathscr{E}) \xrightarrow{V(\Phi)} V_{(x,y')}(\mathscr{E}') \xrightarrow{V(\Psi)} V_{(x,y'')}(\mathscr{E}'')$$

is exact $\forall (x, y) \in \mathscr{E}$, with $y' = \Phi(x, y)$ and $y'' = \Psi(x, y')$.

REMARK 7.2 Using local coordinates, we have the identities:

1) $\Psi \circ \Phi = f'' \circ \pi \Rightarrow \Psi(x, \Phi(x, y)) \equiv f''(x), \forall (x, y) \in \mathscr{E}$

2) $V(\Psi) \circ V(\Phi) = 0 \Rightarrow \dfrac{\partial \Psi}{\partial y'}(x, \Phi(x, y)) \dfrac{\partial \Phi}{\partial y}(x, y) \equiv 0, \forall (x, y) \in \mathscr{E}$

where, for simplicity, we have omitted the index.

EXAMPLE 7.3 Let $\Phi : \mathscr{E} \to \mathscr{E}'$ be a morphism of constant rank and f' a section of $\mathscr{E}'$ such that $f'(X) \subset \operatorname{im} \Phi$. Then we have the following commutative and exact diagram:

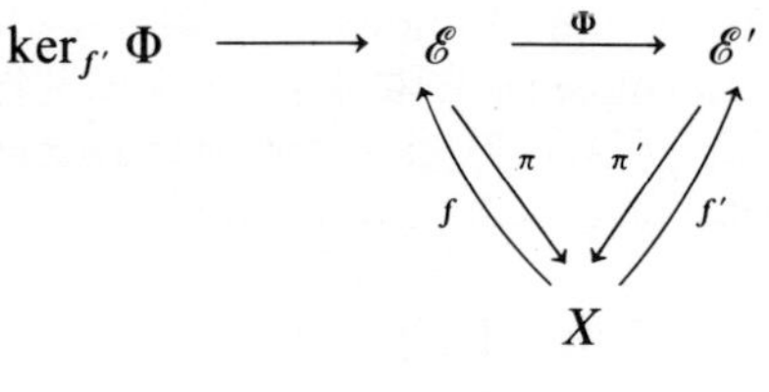

Any section f of $\mathscr{E}$ such that $\Phi \circ f = f'$ is a section of the fibered manifold $\ker_{f'} \Phi$ over X.

DEFINTION 7.4 A morphism $\Phi : \mathscr{E} \to \mathscr{E}'$ is called a "*monomorphism*" if Φ is an injective immersion. We then say that the sequence $0 \to \mathscr{E} \xrightarrow{\Phi} \mathscr{E}'$ is exact.

DEFINITION 7.5 A morphism $\Phi : \mathscr{E} \to \mathscr{E}'$ is called an "*epimorphism*" if Φ is a surjective submersion. We then say that the sequence $\mathscr{E} \xrightarrow{\Phi} \mathscr{E}' \to 0$ is exact.

EXAMPLE 7.6 From the preceding example, we obtain the short exact sequence:

$$0 \longrightarrow \ker_{f'} \Phi \longrightarrow \mathscr{E} \xrightarrow{\ \Phi\ } \operatorname{im} \Phi \longrightarrow 0$$

8 Normal bundles

Let $\varphi : X \to X'$ be a regular map. According to proposition 5.4 we have the following commutative and exact diagram:

$$\begin{array}{ccccc} & & \overset{T(\varphi)}{\frown} & & \\ T(X) & \longrightarrow & \varphi^{-1}(T(X')) & \longrightarrow & T(X') \\ \downarrow{\scriptstyle \pi} & & \downarrow & & \downarrow{\scriptstyle \pi'} \\ X & = & X & \xrightarrow{\ \varphi\ } & X' \end{array}$$

By definition $T(\varphi)$ is an injective map because φ is regular. We then define the vector bundle $N(\varphi) = \operatorname{coker}(T(\varphi))$ as the cokernel of the induced monomorphism $T(\varphi) : T(X) \to \varphi^{-1}(T(X'))$. We have the short exact sequence of vector bundles over X:

$$0 \longrightarrow T(X) \xrightarrow{\ T(\varphi)\ } \varphi^{-1}(T(X')) \longrightarrow N(\varphi) \longrightarrow 0$$

DEFINITION 8.1 $N(\varphi)$ is called the "*normal bundle*" of φ.

More generally, let $\pi : \mathscr{E} \to X$, $\pi' : \mathscr{E}' \to X$ be fibered manifolds over X and $\Phi : \mathscr{E} \to \mathscr{E}'$ be an X-morphism.

We have the following commutative diagram, with exact lines:

$$\begin{array}{ccccccccc} 0 & \longrightarrow & V(\mathscr{E}) & \longrightarrow & T(\mathscr{E}) & \longrightarrow & \pi^{-1}(T(X)) & \longrightarrow & 0 \\ & & \downarrow & \searrow^{V(\Phi)} & \downarrow & \searrow^{T(\Phi)} & \downarrow & & \\ 0 & \longrightarrow & \Phi^{-1}(V(\mathscr{E}')) & \longrightarrow & \Phi^{-1}(T(\mathscr{E}')) & \longrightarrow & \Phi^{-1}(\pi'^{-1}(T(X))) & \longrightarrow & 0 \\ & & \downarrow & \swarrow & \downarrow & \swarrow & \downarrow & & \\ 0 & \longrightarrow & V(\mathscr{E}') & \longrightarrow & T(\mathscr{E}') & \longrightarrow & \pi'^{-1}(T(X)) & \longrightarrow & 0 \end{array}$$

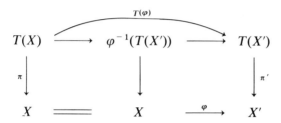

From the fact that $\Phi^{-1} \circ \pi'^{-1} = (\pi' \circ \Phi)^{-1} = \pi^{-1}$, we deduce the important proposition:

PROPOSITION 8.2 Let $\pi : \mathscr{E} \to X$, $\pi' : \mathscr{E}' \to X$ and $\Phi : \mathscr{E} \to \mathscr{E}'$ be an X-monomorphism. Then we have the commutative and exact diagram of vector bundles over $\mathscr{E}$:

$$
\begin{array}{ccccccccc}
 & & 0 & & 0 & & 0 & & \\
 & & \downarrow & & \downarrow & & \downarrow & & \\
0 & \longrightarrow & V(\mathscr{E}) & \longrightarrow & T(\mathscr{E}) & \xrightarrow{T(\pi)} & \pi^{-1}(T(X)) & \longrightarrow & 0 \\
 & & \downarrow & & \downarrow {\scriptstyle T(\Phi)} & & \| & & \\
0 & \longrightarrow & \Phi^{-1}(V(\mathscr{E}')) & \longrightarrow & \Phi^{-1}(T(\mathscr{E}')) & \longrightarrow & \pi^{-1}(T(X)) & \longrightarrow & 0 \\
 & & \downarrow & & \downarrow & & \downarrow & & \\
0 & \longrightarrow & N(\Phi) & = & N(\Phi) & \longrightarrow & 0 & & \\
 & & \downarrow & & \downarrow & & & & \\
 & & 0 & & 0 & & & &
\end{array}
$$

where isomorphic vector bundles have been identified.

DEFINITION 8.3 $N(\Phi)$ defined by the later diagram is called the "*normal bundle*" of the monomorphism Φ.

We invite the reader, as an exercise, to determine the different arrows of this diagram by using local coordinates.

The following proposition, which is an easy consequence of the latter one, will be used in the second part of this book:

PROPOSITION 8.4 If f is a section of the fibered manifold $\pi : \mathscr{E} \to X$, then we can identify $f^{-1}(V(\mathscr{E}))$ and $N(f)$ as vector bundles over X.

Proof As π is a surjective submersion and $\pi \circ f = id_X$ it follows that f is regular and that we can construct $N(f)$ according to proposition 8.2. Now we can consider $f : X \to \mathscr{E}$ as a X-monomorphism of fibered manifolds

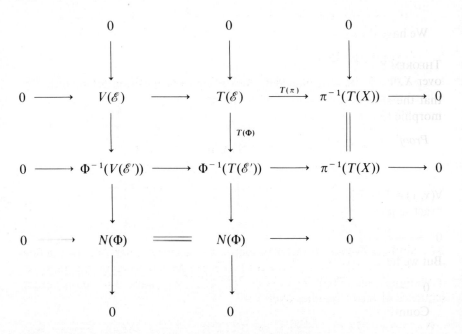

over X. The proposition follows from the fact that, by definition, $V(X) = 0$ in the former diagram and that we identify isomorphic vector bundles over X.
C.Q.F.D.

EXAMPLE 8.5 Let $\pi: \mathscr{E} \to X$ be an affine bundle modelled on a vector bundle E over X. Then we have the isomorphisms:

$$N(f) \approx f^{-1}(V(\mathscr{E})) \approx f^{-1}(\pi^{-1}(E)) = E$$

We have the following theorem, also used in the second part of this book:

THEOREM 8.6 Let $\pi: \mathscr{E} \to X$, $\pi': \mathscr{E}' \to X$, $\pi'': \mathscr{E}'' \to X$ be fibered manifolds over X, $\Phi: \mathscr{E} \to \mathscr{E}'$ a monomorphism and $\Psi: \mathscr{E}' \to \mathscr{E}''$ an epimorphism such that the short sequence $0 \to \mathscr{E} \xrightarrow{\Phi} \mathscr{E}' \xrightarrow{\Psi} \mathscr{E}'' \to 0$ is exact. Then $N(\Phi)$ is isomorphic to the reciprocal image by π of a vector bundle over X.

Proof We have the short exact sequence of vector spaces:

$$0 \longrightarrow V_{(x,y)}(\mathscr{E}) \xrightarrow{V(\Phi)} V_{(x,y')}(\mathscr{E}') \xrightarrow{V(\Psi)} V_{(x,y'')}(\mathscr{E}'') \longrightarrow 0$$

$\forall (x, y) \in \mathscr{E}$ with $y' \neq \Phi(x, y)$ and $y'' = \Psi(x, y')$. We deduce from it the short exact sequence of vector bundles over $\mathscr{E}$:

$$0 \longrightarrow V(\mathscr{E}) \xrightarrow{V(\Phi)} \Phi^{-1}(V(\mathscr{E}')) \longrightarrow (\Psi \circ \Phi)^{-1}(V(\mathscr{E}'')) \longrightarrow 0$$

But we have also the short exact sequence of vector bundles over $\mathscr{E}$:

$$0 \longrightarrow V(\mathscr{E}) \xrightarrow{V(\Phi)} \Phi^{-1}(V(\mathscr{E}')) \longrightarrow N(\Phi) \longrightarrow 0.$$

Counting the dimensions, we have an isomorphism

$$N(\Phi) \approx (\Psi \circ \Phi)^{-1}(V(\mathscr{E}''))$$

However, by definition, there must exist a section f'' of $\mathscr{E}''$ such that $\operatorname{im} \Phi = \ker_{f''} \Psi$ and $\Psi \circ \Phi = f'' \circ \pi$. It follows that

$$\begin{aligned} N(\Phi) &\approx (f'' \circ \pi)^{-1}(V(\mathscr{E}'')) \\ &\approx \pi^{-1}(f''^{-1}(V(\mathscr{E}''))) \\ &\approx \pi^{-1}(N(f'')) \end{aligned}$$

using proposition 8.4. C.Q.F.D.

EXAMPLE 8.7 As Ψ is an epimorphism it has constant rank and $f''(X) \subset \mathscr{E}'' = \operatorname{im} \Psi$. Then we can take for Φ the monomorphism $\iota: \ker_{f''} \Phi \to \mathscr{E}'$.

We shall now consider this theorem when dealing with affine bundles. For this, let $\pi: \mathscr{E} \to X$ and $\pi': \mathscr{E}' \to X$ be affine bundles modelled on the vector bundles $E \to X$ and $E' \to X$. An X-morphism $\Phi : \mathscr{E} \to \mathscr{E}'$ which respects the affine structure of each fiber will be called an X-homomorphism and will have local expression $\Phi : (x, y) \to (x, y' = A(x)y + B(x))$ over a trivialising open set $U \subset X$, where A is a $m \times m'$ matrix with $\dim E = m$, $\dim E' = m'$.

As rank $T(\Phi) = n + \text{rank}\,[A(x)]$, it follows that Φ is of constant rank if and only if the homomorphism $V(\Phi) : E \to E'$ has constant rank.

PROPOSITION 8.8 If $\Phi : \mathscr{E} \to \mathscr{E}'$ is a homomorphism of affine bundles over X of constant rank, then there exists an affine bundle coker Φ and an epimorphism of affine bundles $\Psi : \mathscr{E}' \to \text{coker}\ \Phi$ such that the sequence $\mathscr{E} \xrightarrow{\Phi} \mathscr{E}' \xrightarrow{\Psi} \text{coker}\ \Phi \to 0$ is exact. Moreover we can identify coker Φ with the cokernel of $V(\Phi) : E \to E'$.

Proof The constancy of the rank of Φ implies that $V(\Phi)$ has constant rank and that dim coker $V_x(\Phi)$ is independent of $x \in X$. It follows that coker $V(\Phi) = \bigcup_{x \in X} \text{coker}\ V_x(\Phi)$ is a vector bundle over X.

Now we can define an equivalence relation on $\mathscr{E}'_x$, using local coordinates (x, y') for $\mathscr{E}'$ as follows:

$$(x, y'_1) \sim (x, y'_2) \Leftrightarrow y'_2 - y'_1 = A(x)v \quad \text{with} \quad (x, v) \in E.$$

We denote by coker Φ_x the set of equivalence classes and call Ψ_x the natural projection $\mathscr{E}'_x \to \text{coker}\ \Phi_x$. It is easy to see that coker Φ_x is an affine space modelled on coker $V_x(\Phi)$ and that the linear map associated to Ψ_x is the projection of E'_x onto coker $V_x(\Phi)$.

Moreover, if $y'_1 = A(x)y_1 + B(x)$ and $y'_2 = A(x)y_2 + B(x)$ then $y'_2 - y'_1 = A(x)(y_2 - y_1)$ and it follows that $\Psi_x \circ \Phi_x(\mathscr{E}_x)$ is in one equivalence class only. This fact allows us to identify coker Φ_x and coker $V_x(\Phi)$, and thus also coker $\Phi = \bigcup_{x \in X} \text{coker}\ \Phi_x$ with coker $V(\Phi)$.

This gives to coker Φ the structure of an affine bundle modelled on coker $V(\Phi)$. The resulting map $\Psi : \mathscr{E}' \to \text{coker}\ \Phi$ is a homomorphism of affine bundles and its associated map $V(\Psi)$ is the natural projection $E' \to \text{coker}\ V(\Phi)$.

Now, by definition, we have the exact sequence of vector bundles over X:

$$E \xrightarrow{V(\Phi)} E' \xrightarrow{V(\Psi)} \text{coker}\ V(\Phi) \longrightarrow 0.$$

From the isomporphism $V(\mathscr{E}) \approx \pi^{-1}(E)$ we obtain the isomorphisms $V_{(x, y)}(\mathscr{E}) \approx E_x$ and it follows that the sequence:

$$\mathscr{E} \xrightarrow{\Phi} \mathscr{E}' \xrightarrow{\Psi} \text{coker}\ \Phi \longrightarrow 0$$

is also exact. C.Q.F.D.

REMARK 8.9 Practically, we choose an open covering $\{U_\alpha\}$ of X trivialising both $\mathscr{E}$ and $\mathscr{E}'$, and such that we can select on each open set U of this covering, a submatrix of $[A(x)]$ of maximal rank. Using Cramer's rules, we then define a surjective map $E' \mid U \to U \times \mathbb{R}^{m''}$ with local coordinates (x, v'') on $U \times \mathbb{R}^{m''}$ by the formulas $v'' = \alpha(x)v'$ in such a way that $\alpha(x) \cdot A(x) = 0$ that is to say $[\alpha(x)] \cdot [A(x)] = 0$ in matrix notation. Finally we obtain the map $\mathscr{E}' \mid U \to U \times \mathbb{R}^{m''} : y'' = \alpha(x)y' - \alpha(x) \cdot B(x)$ with local coordinates (x, y'') on $U \times \mathbb{R}^{m''}$ and the kernel is to be taken with respect to the zero section of coker $V(\Phi)$ over U. The last step is to patch together these local results.

REMARK 8.10 If $E \dashrightarrow \mathscr{E} \to X$ is an affine bundle over X, modelled on the vector bundle E, using local coordinates (x, y) on $\mathscr{E}$ we shall define $(x, y_2) - (x, y_1) = (x, y_2 - y_1) \in E$ for any pair of points of $\mathscr{E}$.

Similarly, if E is a vector bundle, using local coordinates (x, v) on E, we shall define

$$(x, v_1) + (x, v_2) = (x, v_1 + v_2) \in E.$$

We now have the following proposition:

PROPOSITION 8.11 A sequence of affine bundles over X:

$$\mathscr{E} \xrightarrow{\ \Phi\ } \mathscr{E}' \xrightarrow{\ \Psi\ } \mathscr{E}''$$

is exact if and only if its associated sequence:

$$E \xrightarrow{\ V(\Phi)\ } E' \xrightarrow{\ V(\Psi)\ } E''$$

of vector bundles over X, is exact.

Proof Using local coordinates, we have: $y' = A(x)y + B(x)$ and $y'' = \alpha(x)y' + \beta(x)$ with $\alpha(x)A(x) = 0$ and $\alpha(x) \cdot B(x) + \beta(x) = f''(x)$. We thus have the commutative diagram:

$$\begin{array}{ccccc}
E & \xrightarrow{\ V(\Phi)\ } & E' & \xrightarrow{\ V(\Psi)\ } & E'' \\
\vdots\downarrow & & \vdots\downarrow & & \vdots\downarrow \\
\mathscr{E} & \xrightarrow{\ \Phi\ } & \mathscr{E}' & \xrightarrow{\ \Psi\ } & \mathscr{E}'' \\
\downarrow & & \downarrow & & \uparrow\downarrow f'' \\
X & = & X & = & X
\end{array}$$

N.C. It is evident because, according to proposition 6.4 we have the isomorphism $V(\mathscr{E}) \approx \pi^{-1}(E)$.

S.C. Take $(x, y') \in \mathscr{E}'$ and $(x, y) \in \mathscr{E}$ with $\Psi(x, y') = f''(x)$. It follows that $\Psi(x, y' - \Phi(x, y)) = 0$ because $\Psi(x, \Phi(x, y)) = f''(x)$ and thus $V(\Psi)(y' - \Phi(x, y)) = 0$. There $\exists (x, v) \in E$ with $y' - \Phi(x, y) = V(\Phi)(v)$ and finally $y' = \Phi(x, y + v)$. C.Q.F.D.

9 Jet bundles

Let f and g be two sections of a fibered manifold $\pi : \mathscr{E} \to X$ and $x \in \operatorname{dom} f \cap \operatorname{dom} g$.

DEFINITION 9.1 For any integer $q \geq 0$, we say that f and g are "*q-equivalent at x*" if:

$$f^k(x) = g^k(x) \quad \text{and} \quad \partial_\mu f^k(x) = \partial_\mu g^k(x) \qquad \forall 1 \leq |\mu| \leq q.$$

The reader will show easily that this definition does not depend on the coordinate system adopted on a neighbourhood of x. It follows that we have an equivalence relation on the set of sections f of $\mathscr{E}$ such that $x \in \operatorname{dom} f$.

DEFINITION 9.2 The equivalence class of a section f is called the "*q-jet of f at x*" and is denoted by $j_q(f)(x)$.

The point $x \in X$ is called the "*source*" of the jet and the point $f(x) \in \mathscr{E}$ is called the "*target*" of the jet.

REMARK 9.3 If the reader wants an intuitive definition, he has just to consider the Taylor's development of f at x up to the order q.

We can introduce the set $J_q(\mathscr{E})_x$ of all the q-jets at x of sections of $\mathscr{E}$ over a neighbourhood of x and set $J_q(\mathscr{E}) = \bigcup_{x \in X} J_q(\mathscr{E})_x$.

DEFINITION 9.4 We say that $J_q(\mathscr{E})$ is the "*bundle of q-jets of* $\mathscr{E}$".

We denote by $\pi_q^{q+r} : J_{q+r}(\mathscr{E}) \to J_q(\mathscr{E})$ the map sending a $(q + r)$-jet onto the q-jet that it determines and we identify $J_0(\mathscr{E})$ and $\mathscr{E}$.

REMARK 9.5 The corresponding intuitive idea is just to truncate at the order q the Taylor development initially considered up to the order $q + r$.

DEFINITION 9.6 The projection π_0^q is called "*target projection*". Similarly the projection $\pi : J_q(\mathscr{E}) \to X$ is called "*source projection*".

PROPOSITION 9.7 $J_q(\mathscr{E})$ can be considered as a fibered manifold both over X or $\mathscr{E}$.

Proof The two projections have already been defined. Now, with respect to convenient fibered charts of $\mathscr{E}$ with domains $\mathscr{U}$ and $\bar{\mathscr{U}}$ projecting respectively onto the open sets U and $\bar{U}$ of X, the transition functions take the following form:

$$\begin{cases} \bar{y}^l = \psi^l(x^i, y^k) \\ \bar{x}^j = \varphi^j(x^i) \end{cases}$$

Let a section of $\mathscr{E}$ over $U \cap \bar{U}$ have local expressions $f^k(x^i)$ with the local coordinates (x^i, y^k) and $\bar{f}^l(\bar{x}^j)$ with the local coordinates $(\bar{x}^j, \bar{y}^l)$. By definition we have the identity:

$$\bar{f}^l(\varphi^j(x^i)) \equiv \psi^l(x^i, f^k(x^i)), \qquad \forall x \in U \wedge \bar{U}.$$

A straightforward computation then shows that:

$$\frac{\partial^q \bar{f}^l}{\partial \bar{x}^{j_1} \cdots \partial \bar{x}^{j_q}} = \frac{\partial x^{i_1}}{\partial \bar{x}^{j_1}} \cdots \frac{\partial x^{i_q}}{\partial \bar{x}^{j_q}} \frac{\partial \bar{y}^l}{\partial y^k} \frac{\partial^q f^k}{\partial x^{i_1} \cdots \partial x^{i_q}}$$

+ terms involving the partial derivatives of the f^k of order less than q.

Introducing the multi-index notation $\mu = (\mu_1, \ldots, \mu_n)$ and defining the "*length*" of μ by $|\mu| = \mu_1 + \cdots + \mu_n$, we obtain:

$$\frac{\partial^{|\nu|} \bar{f}^l(\varphi(x))}{(\partial \bar{x})^\nu} = \chi^l_\nu(x, f^k(x), \partial_\mu f^k(x))$$

with $k, l = 1, \ldots, m$ and $1 \leq |\mu| \leq |\nu|$.

Let $s(n, m, q)$ be the total number of derivatives of order $\leq q$ and ≥ 1. We can construct the fibered manifold $J_q(\mathscr{E})$ by patching together coordinate domains $\mathscr{U} \times \mathbb{R}^s$, using local coordinates p^k_μ for $\mathbb{R}^s$ with $1 \leq |\mu| \leq q$ and transition functions $\bar{p}^l_\nu = \chi^l_\nu(x^i, y^k, p^k_\mu)$ on $(\mathscr{U} \cap \bar{\mathscr{U}}) \times \mathbb{R}^s$.

The reader will check easily that we can apply to this situation the theorem 1.18. We thus obtain a fibered manifold over X, using the projection $\mathscr{U} \times \mathbb{R}^s \to U : (x, y, p) \to (x)$ or a bundle over $\mathscr{E}$, using the projection $\mathscr{U} \times \mathbb{R}^s \to \mathscr{U} : (x, y, p) \to (x, y)$.

It is finally well known that the number of p of order q is equal to $m \cdot [(q + n - 1)!/q!(n - 1)!]$ and we get easily the formula $\dim J_q(\mathscr{E}) = n + m[(q + n)!/q!n!]$.

C.Q.F.D.

In particular, if Y is a copy of X, we call $I_q(X \times Y) \subset J_q(X \times Y)$ the bundle of q-jets of all invertible maps from X to Y and it is defined by the single condition $\det [p^k_i] \neq 0$.

We shall, in general, consider only the connected component of the identity map, that is to say the component containing the points (x, y, p) with $p_i^k = \delta_i^k$, where δ_i^k is the Kronecker symbol.

DEFINITION 9.8 If $E \to X$ is a vector bundle over X, $\pi : \mathscr{E} \to X$ is a fibered manifold over X and $P \to \mathscr{E}$ is a vector bundle over $\mathscr{E}$, then the tensor product of P and the reciprocal image $\pi^{-1}(E)$ of E over $\mathscr{E}$ will be simply denoted by $E \otimes P$ when there will be no confusion.

PROPOSITION 9.9 $J_q(\mathscr{E})$ is an affine bundle over $J_{q-1}(\mathscr{E})$, modelled on the vector bundle $S_q T^* \otimes V(\mathscr{E})$.

Proof The proposition follows easily from the definitions of $T^* = T^*(X)$ and $V(\mathscr{E})$, using the fact that the transition functions of $J_q(\mathscr{E})$ are such that:

$$\bar{p}^l_{j_1 \cdots j_q} = \frac{\partial x^{i_1}}{\partial \bar{x}^{j_1}} \cdots \frac{\partial x^{i_q}}{\partial \bar{x}^{j_q}} \frac{\partial \bar{y}^l}{\partial y^k} p^k_{i_1 \cdots i_q}$$

$$+ \text{ terms involving only } x, y \text{ and the } p^k_\mu \text{ with } |\mu| \leq q - 1.$$

with $p^k_{i_1 \cdots i_q}$ symmetric with respect to any transposition of the indices.

C.Q.F.D.

We shall now describe the vector bundle $T(J_q(\mathscr{E}))$ over $J_q(\mathscr{E})$ using local coordinates $(x^i, y^k, p^k_\mu; u^i, v^k, v^k_\mu)$ with $1 \leq |\mu| \leq q$ or $(x^i, y^k_\mu; u^i, v^k_\mu)$ with $0 \leq |\mu| \leq q$ and $v^k_0 = v^k$, or simply $(x, y, p; u, v)$.

Among the transition functions we have:

$$\begin{cases} \bar{u}^j = \dfrac{\partial \varphi^j}{\partial x^i}(x) u^i \\[2ex] \bar{v}^l = \dfrac{\partial \psi^l(x, y)}{\partial x^i} u^i + \dfrac{\partial \psi^l(x, y)}{\partial y^k} v^k \\[2ex] \bar{v}^l_\nu = \dfrac{\partial \chi^l_\nu(x, y, p)}{\partial x^i} u^i + \dfrac{\partial \chi^l_\nu(x, y, p)}{\partial y^k} v^k + \dfrac{\partial \chi^l_\nu(x, y, p)}{\partial p^k_\mu} v^k_\mu \end{cases}$$

It follows that we can use the same local coordinates for the vertical bundle $V(J_q(\mathscr{E}))$ over $J_q(\mathscr{E})$ but we have to set $u^i = 0$.

In fact we prefer to adopt the local coordinates $(x, y, p; v)$ for $V(J_q(\mathscr{E}))$ with the following transition functions:

$$\begin{cases} \bar{v}^l = \dfrac{\partial \psi^l(x, y)}{\partial y^k} v^k \\[2ex] \bar{v}^l_\nu = \dfrac{\partial \chi^l_\nu}{\partial y^k}(x, y, p) v^k + \dfrac{\partial \chi^l_\nu(x, y, p)}{\partial p^k_\mu} v^k_\mu \end{cases}$$

Then, if we set equal to zero all the components except those of order q, we have $\bar{v}^l_\nu = (\partial \chi^l_\nu/\partial p^k_\mu)(x, y, p)v^k_\mu$ with $|\mu| = |\nu| = q$, and we obtain the following useful proposition.

PROPOSITION 9.10 We have the short exact sequence of vector bundles over $J_q(\mathscr{E})$:

$$0 \longrightarrow S_qT^* \otimes V(\mathscr{E}) \xrightarrow{\varepsilon_q} V(J_q(\mathscr{E})) \xrightarrow{V(\pi^q_{q-1})} (\pi^q_{q-1})^{-1}(V(J_{q-1}(\mathscr{E}))) \longrightarrow 0$$

Proof It suffices to define the monomorphism ε_q locally as follows:

$$\varepsilon_q : (x, y, p; v^k_\mu)|\mu| = q \longrightarrow (x, y, p; v^l_\nu = 0 \quad \text{if} \quad |\nu| < q, v^k_\mu)|\mu| = q.$$

C.Q.F.D.

The proof of the following proposition is similar and left to the reader:

PROPOSITION 9.11 We have the short exact sequence of vector bundles over $J_q(\mathscr{E})$:

$$0 \longrightarrow S_qT^* \otimes V(\mathscr{E}) \xrightarrow{\varepsilon_q} T(J_q(\mathscr{E})) \xrightarrow{T(\pi^q_{q-1})} (\pi^q_{q-1})^{-1}(T(J_{q-1}(\mathscr{E}))) \longrightarrow 0$$

We will need the following technical lemma:

LEMMA 9.12 There exists a natural isomorphism $V(J_q(\mathscr{E})) \approx J_q(V(\mathscr{E}))$ of vector bundles over $J_q(\mathscr{E})$.

Proof We shall first prove this lemma for $q = 1$, using local coordinates $(x, y, p; v^k, v^k_i)$ for $V(J_1(\mathscr{E}))$, $(x, y; v^k)$ for $V(\mathscr{E})$ and $(x, y, p; v^k, v^k_{,i})$ for $J_1(V(\mathscr{E}))$.

We have the following transition function with respect to convenient coordinate domains:

$$V(\mathscr{E})\begin{cases} \bar{x}^j = \varphi^j(x^i) \\ \bar{y}^l = \psi^l(x^i, y^k) \\ \bar{v}^l = \dfrac{\partial \psi^l}{\partial y^k}(x, y)v^k \end{cases}$$

It follows that we have for $J_1(V(\mathscr{E}))$ the transition function:

$$\frac{\partial \varphi^j}{\partial x^i}(x)\bar{v}^l_{,j} = \frac{\partial^2 \psi^l(x, y)}{\partial x^i \partial y^k} v^k + \frac{\partial^2 \psi^l}{\partial y^k \partial y^r} p^r_i v^k + \frac{\partial \psi^l}{\partial y^k}(x, y)v^k_{,i}$$

Now we have for $J_1(\mathscr{E})$ the transition function:

$$\frac{\partial \varphi^j}{\partial x^i}(x)\bar{p}^l_j = \frac{\partial \psi^l}{\partial x^i}(x, y) + \frac{\partial \psi^l}{\partial y^r}(x, y)p^r_i$$

and it follows that for $V(J_1(\mathscr{E}))$ we have:

$$\frac{\partial \varphi^j}{\partial x^i}(x)\bar{v}^l_j = \frac{\partial^2 \psi^l(x, y)}{\partial x^i \partial y^k} v^k + \frac{\partial^2 \psi^l}{\partial y^k \partial y^r}(x, y)p^r_i v^k + \frac{\partial \psi^l(x, y)}{\partial y^r} v^r_i$$

and the natural isomorphism is obtained by taking $v^k_{,i} = v^k_i$. It is easy to see that this result does not depend on the coordinate system.

We shall now use an induction on q. Let us suppose that we have found the desired isomorphism $V(J_{q-1}(\mathscr{E})) \approx J_{q-1}(V(\mathscr{E}))$ by setting $v^k_{,\mu} = v^k_\mu$ $\forall 1 \leq |\mu| \leq q - 1$. By definition we have the transition function:

$$\begin{cases} \bar{p}^l_\nu = \chi^l_\nu(x, y, p) \\ \bar{v}^l_\nu = \dfrac{\partial \chi^l_\nu}{\partial y^k}(x, y, p)v^k + \dfrac{\partial \chi^l_\nu}{\partial p^k_\mu}(x, y, p)v^k_\mu \end{cases}$$

for $V(J_{q-1}(\mathscr{E}))$ and the transition function for $J_{q-1}(V(\mathscr{E}))$ are:

$$\bar{p}^l_\nu = \chi^l_\nu(x, y, p)$$

$$\bar{v}^l_{,\nu} = \frac{\partial \chi^l_\nu}{\partial y^k}(x, y, p)v^k + \frac{\partial \chi^l_\nu}{\partial p^k_\mu}(x, y, p)v^k_{,\mu}$$

It follows that for $V(J_q(\mathscr{E}))$ we have

$$\frac{\partial \varphi^j}{\partial x^i}(x)\bar{p}^l_{\nu+1_j} = \frac{\partial \chi^l_\nu}{\partial x^i}(x, y, p) + \frac{\partial \chi^l_\nu}{\partial y^k}(x, y, p)p^k_i + \frac{\partial \chi^l_\nu}{\partial p^k_\mu}(x, y, p)p^k_{\mu+1_i}$$

and

$$\begin{aligned} \frac{\partial \varphi^j}{\partial x^i}(x)\bar{v}^l_{\nu+1_j} = {} & \frac{\partial^2 \chi^l_\nu}{\partial x^i \partial y^r} v^r + \frac{\partial^2 \chi^l_\nu}{\partial x^i \partial p^r_\lambda} v^r_\lambda + \frac{\partial^2 \chi^l_\nu}{\partial y^k \partial y^r_\lambda} p^k_i v^r_\lambda + \frac{\partial \chi^l_\nu}{\partial y^k} v^k_i \\ & + \frac{\partial^2 \chi^l_\nu}{\partial y^r \partial p^k_\mu} p^k_{\mu+1_i} v^r + \frac{\partial^2 \chi^l_\nu}{\partial p^r_\lambda \partial p^k_\mu} p^k_{\mu+1_i} v^r_\lambda + \frac{\partial \chi^l_\nu}{\partial p^k_\mu} v^k_{\mu+1_i}. \end{aligned}$$

Similarly, for $J_q(V(\mathscr{E}))$ we have:

$$\begin{aligned} \frac{\partial \varphi^j}{\partial x^i}(x)\bar{v}^l_{,\nu+1_j} = {} & \frac{\partial^2 \chi^l_\nu}{\partial x^i \partial y^k} v^k + \frac{\partial^2 \chi^l_\nu}{\partial x^i \partial p^k_\mu} v^k_{,\mu} + \frac{\partial^2 \chi^l_\nu}{\partial y^k_\lambda \partial y^r} p^r_i v^k_{,\lambda} + \frac{\partial \chi^l_\nu}{\partial y^k} v^k_{,i} \\ & + \frac{\partial^2 \chi^l_\nu}{\partial y^r \partial p^k_\mu} p^k_{\mu+1_i} v^r + \frac{\partial^2 \chi^l_\nu}{\partial p^r_\lambda \partial p^k_\mu} p^r_{\lambda+1_i} v^k_{,\mu} + \frac{\partial \chi^l_\nu}{\partial p^k_\mu} v^k_{,\mu+1_i} \end{aligned}$$

and we deduce the isomorphism $V(J_q(\mathscr{E})) \approx J_q(V(\mathscr{E}))$ by setting $v^k_{,\mu} = v^k_\mu$ for $1 \leq |\mu| \leq q$. This concludes the proof of the lemma. C.Q.F.D.

Subsequently we shall identify $V(J_q(\mathscr{E}))$ and $J_q(V(\mathscr{E}))$ as vector bundles over $J_q(\mathscr{E})$ and we have the short exact sequence, using a slightly abusive notation:

$$0 \longrightarrow S_q T^* \otimes E \longrightarrow J_q(E) \longrightarrow J_{q-1}(E) \longrightarrow 0$$

where we have to consider the vector bundles involved as vector bundles over $J_q(\mathscr{E})$.

We end this chapter with a very important but technical proposition:

PROPOSITION 9.13 If we consider $J_{r+s}(\mathscr{E})$, $J_r(J_s(\mathscr{E}))$ and $J_{r+1}(J_{s-1}(\mathscr{E}))$ as fibered submanifolds of $J_1^{r+s}(\mathscr{E})$ in a natural way, we have the relation:

$$\forall r, s \geq 0 \qquad J_{r+s}(\mathscr{E}) = J_r(J_s(\mathscr{E})) \cap J_{r+1}(J_{s-1}(\mathscr{E}))$$

REMARK 9.14 $J_1^{r+s}(\mathscr{E})$ is written for

$$\underbrace{J_1(J_1(\ldots(J_1(\mathscr{E})).}_{(r+s)\text{ times}}$$

Proof We shall define the former fibered submanifolds of $J_1^{r+s}(\mathscr{E})$ using local coordinates by means of the following trick:

We adopt for $J_1(\mathscr{E})$ local coordinates (x, y^k_i) where $i = 0, 1, \ldots, n$ and $y^k_0 = y^k$, $y^k_i = p^k_i$ if $1 \leq i \leq n$. Now we have $\dim J_1(\mathscr{E})_x = m(n+1)$, $\forall x \in X$ and $\dim J_1^r(\mathscr{E})_x = m(n+1)^r$, $\forall x \in X$.

We define $J_r(\mathscr{E})$ as a submanifold of $J_1^r(\mathscr{E})$ by means of the equation:

$$y^k_{i_1 \cdots i_\alpha \cdots i_\beta \cdots i_r} - y^k_{i_1 \cdots i_\beta \cdots i_\alpha \cdots i_r} = 0$$

where it must be understood that $y^k_{0 \cdots 0} = y^k$.

In fact if there are some zeros among the indices the former equations allow us to push them on the left and to keep the corresponding non-zero indices as usual indices for the p^k_μ with $|\mu| \leq r$. Moreover if there is no zero, the former equations just tell us that the coordinates on $J_r(\mathscr{E})$ can be taken as the corresponding p^k_μ with $|\mu| = r$.

Using the same trick, we can define $J_r(J_s(\mathscr{E}))$ as a fibered submanifold of $J_1^{r+s}(\mathscr{E})$ defined by the equation:

$$\begin{cases} y^k_{i_1 \cdots i_{\alpha_2} \cdots i_{\alpha_1} \cdots i_r i_{r+1} \cdots i_{r+s}} - y^k_{i_1 \cdots i_{\alpha_1} \cdots i_{\alpha_2} \cdots i_r i_{r+1} \cdots i_{r+s}} = 0 \\ y^k_{i_1 \cdots i_r i_{r+1} \cdots i_{\beta_2} \cdots i_{\beta_1} \cdots i_{r+s}} - y^k_{i_1 \cdots i_r i_{r+1} \cdots i_{\beta_1} \cdots i_{\beta_2} \cdots i_{r+s}} = 0 \end{cases}$$

with $1 \leq \alpha_1, \alpha_2 \leq r$ and $r + 1 \leq \beta_1, \beta_2 \leq r + s$.

We can also define $J_{r+1}(J_{s-1}(\mathscr{E}))$ as a fibered submanifold of $J_1^{r+s}(\mathscr{E})$ by means of similar equations with $1 \leq \alpha_1, \alpha_2 \leq r+1$ and $r+2 \leq \beta_1, \beta_2 \leq r+s$.

As a consequence, we have for $J_r(J_s(\mathscr{E})) \cap J_{r+1}(J_{s-1}(\mathscr{E}))$:

$$\begin{cases} y^k_{i_1 \cdots i_\alpha \cdots i_r i_\beta \cdots i_{r+1} \cdots i_{r+s}} - y^k_{i_1 \cdots i_\alpha \cdots i_r i_{r+1} \cdots i_\beta \cdots i_{r+s}} = 0 \\ y^k_{i_1 \cdots i_\beta \cdots i_r i_\alpha \cdots i_{r+1} \cdots i_{r+s}} - y^k_{i_1 \cdots i_\alpha \cdots i_r i_\beta \cdots i_{r+1} \cdots i_{r+s}} = 0 \\ y^k_{i_1 \cdots i_\beta \cdots i_r i_{r+1} \cdots i_\alpha \cdots i_{r+s}} - y^k_{i_1 \cdots i_\beta \cdots i_r i_\alpha \cdots i_{r+1} \cdots i_{r+s}} = 0 \end{cases}$$

By addition we obtain:

$$y^k_{i_1 \cdots i_\beta \cdots i_r i_{r+1} \cdots i_\alpha \cdots i_{r+s}} - y^k_{i_1 \cdots i_\alpha \cdots i_r i_{r+1} \cdots i_\beta \cdots i_{r+s}} = 0.$$

This concludes the proof because we can exhibit $J_{r+s}(\mathscr{E})$ as a fibered submanifold of $J_1^{r+s}(\mathscr{E})$ defined by the equation

$$y^k_{i_1 \cdots i_\beta \cdots i_\alpha \cdots i_{r+s}} - y^k_{i_1 \cdots i_\alpha \cdots i_\beta \cdots i_{r+s}} = 0. \qquad \text{C.Q.F.D.}$$

EXAMPLE 9.15 $m = 1, \quad n = 1, \quad r = 1, \quad s = 2, \quad \mathscr{E} = \mathbb{R} \times \mathbb{R}$

$$J_1(\mathscr{E}) : (x, y_0, y_1) \to (x, y, p_1)$$

$$J_1^3(\mathscr{E}) : (x, y_{000}, y_{001}, y_{010}, y_{100}, y_{011}, y_{101}, y_{110}, y_{111})$$

$$J_1(J_2(\mathscr{E})) \subset J_1^3(\mathscr{E}) \begin{cases} y_{010} - y_{001} = 0 \\ y_{110} - y_{101} = 0 \end{cases}$$

$$J_2(J_1(\mathscr{E})) \subset J_1^3(\mathscr{E}) \begin{cases} y_{100} - y_{010} = 0 \\ y_{101} - y_{011} = 0 \end{cases}$$

$$J_1(J_2(\mathscr{E})) \cap J_2(J_1(\mathscr{E})) \begin{cases} y_{000} & \to y \\ y_{100} = y_{010} = y_{001} & \to p_1 \\ y_{110} = y_{101} = y_{011} & \to p_{11} \\ y_{111} & \to p_{111} \end{cases}$$

$$J_3(\mathscr{E}) : (x, y, p_1, p_{11}, p_{111}).$$

CHAPTER 2

1 Differential operators

Let $\mathscr{E}$ and $\mathscr{E}'$ be two fiber manifolds over X with $\dim \mathscr{E}_x = m$ and $\dim \mathscr{E}'_x = m'$.

DEFINITION 1.1 A non-linear differential operator of order q from $\mathscr{E}$ to $\mathscr{E}'$ is a map $\Phi \circ j_q : \mathscr{E} \to \mathscr{E}'$ between the sets of germs of sections of $\mathscr{E}$ and $\mathscr{E}'$, where $\Phi : J_q(\mathscr{E}) \to \mathscr{E}'$ is an X-morphism.

Writing $\mathscr{D} = \Phi \circ j_q$ we have the commutative diagram:

$$\begin{array}{ccc} J_q(\mathscr{E}) & \xrightarrow{\ \Phi\ } & \mathscr{E}' \\ {\scriptstyle j_q}\uparrow & & \| \\ \mathscr{E} & \xrightarrow{\ \mathscr{D}\ } & \mathscr{E}' \end{array}$$

If we choose a fiber chart for $J_q(\mathscr{E})$ over an open set $U \subset X$ and local coordinates (x, y, p), this means that, to any local section f of $\mathscr{E}$ over U, there corresponds the local section f' of $\mathscr{E}'$ over U such that:

$$f' = \Phi \circ j_q(f)$$

with

$$f'(x) = \Phi \circ j_q(f)(x) = \Phi(x^i, f^k(x), \partial_\mu f^k(x)) \qquad 1 \leq |\mu| \leq q$$

when Φ is defined itself by

$$\Phi : (x, y, p) \to (x, \Phi(x, y, p)).$$

REMARK 1.2 For simplicity we shall often write $\mathscr{D}$ and its corresponding morphism as follows, using local coordinates:

$$(x, y) \to (x, y' = \Phi(x^i, y^k, \partial_\mu y^k))$$

If we derive the members of the last equality with respect to x^i, we obtain:

$$\partial_i f'(x) = \frac{\partial \Phi}{\partial x^i}(j_q(f)(x)) + \frac{\partial \Phi}{\partial y^k}(j_q(f)(x))\partial_i f^k(x) + \frac{\partial \Phi}{\partial p^k_\mu}(j_q(f)(x))\partial_{\mu+1_i} f^k(x)$$

It follows that there exists a unique morphism

$$\rho_r(\Phi) : J_{q+r}(\mathscr{E}) \to J_r(\mathscr{E}') \quad \text{such that} \quad j_r(f') = \rho_r(\Phi) \circ j_{q+r}(f).$$

DEFINITION 1.3 The "*r-prolongation*" $\rho_r(\Phi): J_{q+r}(\mathscr{E}) \to J_r(\mathscr{E}')$ of Φ is this unique morphism such that $\rho_r(\Phi) \circ j_{q+r} = j_r \circ \mathscr{D}$. Moreover we have the commutative diagram:

$$\begin{array}{ccc} J_{q+r}(\mathscr{E}) & \xrightarrow{\rho_r(\Phi)} & J_r(\mathscr{E}') \\ \uparrow{\scriptstyle j_{q+r}} & & \uparrow{\scriptstyle j_r} \\ \mathscr{E} & \xrightarrow{\mathscr{D}} & \mathscr{E}' \end{array}$$

We use to set $\rho_0(\Phi) = \Phi$.

REMARK 1.4 In actual practice, in order to get the local formulas for the prolongations of Φ, we can use the q-jet of a section instead of the jet coordinates, then differentiate as many times as it is necessary and finally substitute the jet-coordinates again.

However the prolongations are made easy by means of a new kind of derivative.

DEFINITION 1.5 If Φ is a function defined on the open set $\mathscr{U}_q \subset J_q(\mathscr{E})$, we call "*formal derivative*" of Φ with respect to x^i, the function $d\Phi/dx^i$ defined on the open set $(\pi_q^{q+1})^{-1}(\mathscr{U}_q) \subset J_{q+1}(\mathscr{E})$ by the formula:

$$\frac{d\Phi}{dx^i} = \frac{\partial \Phi}{\partial x^i}(x, y, p) + \frac{\partial \Phi}{\partial y^k}(x, y, p)p_i^k + \frac{\partial \Phi}{\partial p_\mu^k}(x, y, p)p_{\mu+1_i}^k$$

We call $\partial\Phi/\partial x^i$ the "*usual derivative*", or simply the derivative of Φ with respect to x^i when there is no confusion.

In order to use this definition, we just need to point out that $\Phi \in C^\infty(J_q(\mathscr{E}))$ can be considered as an X-morphism $\Phi: J_q(\mathscr{E})) \to X \times \mathbb{R}$. In particular, if $\Phi: J_q(\mathscr{E}) \to \mathscr{E}'$ is defined in local coordinates by

$$\Phi: (x, y, p) \to (x, y' = \Phi(x, y, p)),$$

then we have

$$p_i' = \frac{d\Phi}{dx^i}.$$

In this way we obtain the prolongation of Φ, up to any order, by differentiating Φ formally, as many times as necessary.

The reader will easily check the following properties of formal differentiation:

1 $$\frac{d(\Phi_1 + \Phi_2)}{dx^i} = \frac{d\Phi_1}{dx^i} + \frac{d\Phi_2}{dx^i}$$

2 $$\frac{d(\Phi_1 \cdot \Phi_2)}{dx^i} = \Phi_1 \cdot \frac{d\Phi_2}{dx^i} + \Phi_2 \cdot \frac{d\Phi_1}{dx^i}$$

3 $$\frac{d}{dx^i}\left(\frac{d\Phi}{dx^j}\right) = \frac{d}{dx^j}\left(\frac{d\Phi}{dx^i}\right)$$

Let $(x, y, p) \to (\bar{x} = \varphi(x), \bar{y} = \psi(x, y), \bar{p} = \chi(x, y, p))$ be the more general change of coordinates on $J_q(\mathscr{E})$. If we introduce $\bar{\Phi}(\bar{x}, \bar{y}, \bar{p})$ such that $\bar{\Phi}(\varphi(x), \psi(x, y), \chi(x, y, p)) \equiv \Phi(x, y, p)$ then we have the following proposition:

Proposition 1.6

$$\frac{d\bar{\Phi}}{d\bar{x}^j} \cdot \frac{\partial \varphi^j(x)}{\partial x^i} = \frac{d\Phi}{dx^i}$$

Proof From the definition of $\bar{\Phi}$ we have:

$$\frac{\partial\bar{\Phi}}{\partial\bar{x}^j}\frac{\partial\varphi^j}{\partial x^i} + \frac{\partial\bar{\Phi}}{\partial\bar{y}^l}\frac{\partial\psi^l}{\partial x^i} + \frac{\partial\bar{\Phi}}{\partial\bar{p}^l_\nu}\frac{\partial\chi^l_\nu}{\partial x^i} = \frac{\partial\Phi}{\partial x^i} \qquad \times 1$$

$$\frac{\partial\bar{\Phi}}{\partial\bar{y}^l}\frac{\partial\psi^l}{\partial y^k} + \frac{\partial\bar{\Phi}}{\partial\bar{p}^l_\nu}\frac{\partial\chi^l_\nu}{\partial y^k} = \frac{\partial\Phi}{\partial y^k} \qquad \times p^k_i$$

$$\frac{\partial\bar{\Phi}}{\partial\bar{p}^l_\nu}\frac{\partial\chi^l_\nu}{\partial p^k_\mu} = \frac{\partial\Phi}{\partial p^k_\mu} \qquad \times p^k_{\mu+1_i}$$

and the proof follows from the fact that

$$\bar{p}^l_j\frac{\partial\varphi^j}{\partial x^i} = \frac{d\psi^l}{dx^i} \quad \text{and} \quad \bar{p}^l_{\nu+1_j}\frac{\partial\varphi^j}{\partial x^i} = \frac{d\chi^l_\nu}{dx^i}$$

C.Q.F.D.

Definition 1.7 A non-linear operator $\mathscr{D} = \Phi \circ j_q : \mathscr{E} \to \mathscr{E}'$, is said to be "*regular*" if the morphisms $\rho_r(\Phi) : J_{q+r}(\mathscr{E}) \to J_r(\mathscr{E}')$ have constant rank.

Example 1.8 The operator $j_q : \mathscr{E} \to J_q(\mathscr{E})$ is such that the image of a section f of $\mathscr{E}$ is the section $j_q(f)$ of $J_q(\mathscr{E})$. It is determined by the morphism of constant rank $\mathrm{id}_{J_q(\mathscr{E})} : J_q(\mathscr{E}) \to J_q(\mathscr{E})$. The monomorphisms $\rho_r(\mathrm{id}_{J_q(\mathscr{E})}) : J_{q+r}(\mathscr{E}) \to J_r(J_q(\mathscr{E}))$ have already been described in local coordinates and have constant rank.

DEFINITION 1.9 For any morphism $\Phi : \mathscr{E} \to \mathscr{E}'$ and any integer $r \geq 0$, we define the morphism $J_r(\Phi) : J_r(\mathscr{E}) \to J_r(\mathscr{E}')$ by the commutative diagram:

$$\begin{array}{ccc} J_r(\mathscr{E}) & \xrightarrow{J_r(\Phi)} & J_r(\mathscr{E}') \\ \uparrow{\scriptstyle j_r} & & \uparrow{\scriptstyle j_r} \\ \mathscr{E} & \xrightarrow{\Phi} & \mathscr{E}' \end{array}$$

Let $\Phi : (x, y) \to (x, y' = \Phi(x, y))$, then

$$\rho_1(\Phi) : (x, y, p) \to \left(x, y' = \Phi(x, y), p' = \frac{\partial \Phi(x, y)}{\partial x} + \frac{\partial \Phi(x, y)}{\partial y} p = \frac{d\Phi}{dx}\right)$$

and $\rho_r(\Phi)$ can be computed similarly, using formal differentiations.

EXAMPLE 1.10 $\mathrm{id}_{J_q(\mathscr{E})} = J_q(\mathrm{id}_{\mathscr{E}}) = \rho_q(\mathrm{id}_{\mathscr{E}})$.

The two following propositions are very useful.

PROPOSITION 1.11 If

$$\mathscr{E} \xrightarrow{\Phi} \mathscr{E}' \xrightarrow{\Psi} \mathscr{E}''$$

is an exact sequence of fibered manifold over X with respect to the section f'' of $\mathscr{E}''$ over X, then

$$J_q(\mathscr{E}) \xrightarrow{J_q(\Phi)} J_q(\mathscr{E}') \xrightarrow{J_q(\Psi)} J_q(\mathscr{E}'')$$

is also an exact sequence of fibered manifolds over X, with respect to the section $j_q(f'')$ of $J_q(\mathscr{E}'')$ over X.

Proof From the definition, for any section f of $\mathscr{E}$ over X we have

$$\Psi \circ \Phi \circ f = f''.$$

It follows that

$$j_q(\Psi \circ \Phi \circ f) = J_q(\Psi) \circ j_q(\Phi \circ f) = J_q(\Psi) \circ J_q(\Phi) \circ j_q(f) = j_q(f'')$$

and that $J_q(\Psi) \circ J_q(\Phi) \circ f_q = j_q(f'')$ for any section f_q of $J_q(\mathscr{E})$.

We shall now use an induction on q. As the case $q = 0$ is already satisfied, let us suppose that the proposition has been proved for $q - 1$. Then, from proposition 1.9.9 and 1.6, we have a commutative diagram of affine bundles

$$
\begin{array}{ccccc}
S_q T^* \otimes E & \xrightarrow{V(\Phi)} & S_q T^* \otimes E' & \xrightarrow{V(\Psi)} & S_q T^* \otimes E'' \\
\vdots\downarrow & & \vdots\downarrow & & \vdots\downarrow \\
J_q(\mathscr{E}) & \xrightarrow{J_q(\Phi)} & J_q(\mathscr{E}') & \xrightarrow{J_q(\Psi)} & J_q(\mathscr{E}'') \\
\downarrow \pi^q_{q-1} & & \downarrow \pi^q_{q-1} & & \downarrow \pi^q_{q-1} \\
J_{q-1}(\mathscr{E}) & \xrightarrow{J_{q-1}(\Phi)} & J_{q-1}(\mathscr{E}') & \xrightarrow{J_{q-1}(\Psi)} & J_{q-1}(\mathscr{E}'')
\end{array}
$$

REMARK 1.12 In this diagram and in the sequel, when a map is a tensor product of maps, one of them being the identity, we suppress the identity and write simply $V(\Phi)$ instead of $\mathrm{id}_{S_q T^*} \otimes V(\Phi)$ as in the upper row.

Now the sequence

$$\mathscr{E} \xrightarrow{\Phi} \mathscr{E}' \xrightarrow{\Psi} \mathscr{E}''$$

is exact and from the property 2 of the definition it follows that the sequence of vector spaces

$$V_{(x,y)}(\mathscr{E}) \xrightarrow{V_{(x,y)}(\Phi)} V_{(x,y')}(\mathscr{E}') \xrightarrow{V_{(x,y')}(\Psi)} V_{(x,y'')}(\mathscr{E}'')$$

is exact, $\forall (x, y) \in \mathscr{E}$ with $y' = \Phi(x, y)$ and $y'' = \Psi(x, y')$.

Then the sequence of vector spaces:

$$S_q T^*_x \otimes V_{(x,y)}(\mathscr{E}) \xrightarrow{V_{(x,y)}(\Phi)} S_q T^*_x \otimes V_{(x,y')}(\mathscr{E}') \xrightarrow{V_{(x,y')}(\Psi)} S_q T^*_x \otimes V_{(x,y'')}(\mathscr{E}'')$$

is exact as the reader can check.

It follows that the upper row of the diagram is an exact sequence of vector bundles, over the exact sequence of fibered manifolds represented by the lower row.

As the following diagram chase will be met in the sequel, we will detail it.

Take a point $a \in J_q(\mathscr{E}')$ projecting onto $x \in X$, and such that we have $J_q(\Psi)(a) = j_q(f'')(x)$. Applying π^q_{q-1} it follows that $J_{q-1}(\Psi)(\pi^q_{q-1}(a)) = j_{q-1}(f'')(x)$.

Now, by the induction hypothesis, $\exists b \in J_{q-1}(\mathscr{E})$ projecting onto x, such that $J_{q-1}(\Phi)(b) = \pi^q_{q-1}(a)$. As $J_q(\mathscr{E})$ is an affine bundle over $J_{q-1}(\mathscr{E})$, there exists $c \in J_q(\mathscr{E})$ such that $\pi^q_{q-1}(c) = b$.

Set $a' = J_q(\Phi)(c)$, then $\pi^q_{q-1}(a') = \pi^q_{q-1}(a)$ and there exist $\alpha \in S_q T^* \otimes E'$ such that $a - a' = \alpha$.

But $J_q(\Psi)(a') = J_q(\Psi) \circ J_q(\Phi)(c) = j_q(f'')(x) = J_q(\Psi)(a)$ and it follows that $V(\Psi)(\alpha) = 0$.

Finally, as the upper row is exact, there exists $\gamma \in S_q T^* \otimes E$ such that $V(\Phi)(\gamma) = \alpha$ and we have $J_q(\Phi)(c + \gamma) = a' + \alpha = a$.

The middle row is thus exact and this concludes the proof. C.Q.F.D.

COROLLARY 1.13 If $\Phi : \mathscr{E} \to \mathscr{E}'$ is a monomorphism (epimorphism, isomorphism) then $J_q(\Phi) : J_q(\mathscr{E}) \to J_q(\mathscr{E}')$ is also a monomorphism (epimorphism, isomorphism).

REMARK 1.14 In particular, if f' is a section of $\mathscr{E}'$ over X such that $\operatorname{im} f' \subset \operatorname{im} \Phi$ then $\operatorname{im} j_q(f') \subset \operatorname{im} J_q(\Phi)$ and it follows that we have, when Φ is a morphism of constant rank

$$J_q(\ker_{f'} \Phi) = \ker_{j_q(f')} J_q(\Phi).$$

PROPOSITION 1.15 If we have the morphisms

$$\Phi : J_q(\mathscr{E}) \to \mathscr{E}' \quad \text{and} \quad \Phi' : J_r(\mathscr{E}') \to \mathscr{E}'',$$

then for any integer $s \geq 0$, we have:

$$\rho_s(\Phi' \circ \rho_r(\Phi)) = \rho_s(\Phi') \circ \rho_{r+s}(\Phi).$$

Proof Let $\mathscr{D} = \Phi \circ j_q$ and $\mathscr{D}' = \Phi' \circ j_r$. The operator $\mathscr{D}' \circ \mathscr{D} : \mathscr{E} \to \mathscr{E}''$ has order $q + r$ and is determined by the composite morphism:

$$\Phi' \circ \rho_r(\Phi) : J_{q+r}(\mathscr{E}) \to J_r(\mathscr{E}') \to \mathscr{E}''$$

It follows that

$$\rho_s(\Phi' \circ \rho_r(\Phi)) \circ j_{q+r+s} = j_s \circ \mathscr{D}' \circ \mathscr{D}$$

But also

$$\rho_s(\Phi') \circ \rho_{r+s}(\Phi) \circ j_{q+r+s} = \rho_s(\Phi') \circ j_{r+s} \circ \mathscr{D} = j_s \circ \mathscr{D}' \circ \mathscr{D}$$

and we deduce the proposition from the fact that any operator is determined by a unique morphism. C.Q.F.D.

PROPOSITION 1.16 If $\Phi : J_q(\mathscr{E}) \to \mathscr{E}'$ is a morphism, then the following diagram is commutative:

$$\begin{array}{ccc}
J_{q+r+s}(\mathscr{E}) & \xrightarrow{\rho_{r+s}(\Phi)} & J_{r+s}(\mathscr{E}') \\
\downarrow{\scriptstyle \rho_s(\mathrm{id}_{J_{q+r}(\mathscr{E})})} & \searrow{\scriptstyle \rho_s(\rho_r(\Phi))} & \downarrow{\scriptstyle \rho_s(\mathrm{id}_{J_r(\mathscr{E}')})} \\
J_s(J_{q+r}(\mathscr{E})) & \xrightarrow[J_s(\rho_r(\Phi))]{} & J_s(J_r(\mathscr{E}'))
\end{array}$$

Proof Composing with j_{q+r+s} on the right, we get:

$$
\begin{aligned}
\rho_s(\mathrm{id}_{J_r(\mathscr{E}')}) \circ \rho_{r+s}(\Phi) \circ j_{q+r+s} &= \rho_s(\mathrm{id}_{J_r(\mathscr{E}')}) \circ j_{r+s} \circ \Phi \circ j_q \\
&= j_s \circ j_r \circ \Phi \circ j_q \\
\rho_s(\rho_r(\Phi)) \circ j_{q+r+s} &= j_s \circ \rho_r(\Phi) \circ j_{q+r} \\
&= j_s \circ j_r \circ \Phi \circ j_q \\
J_s(\rho_r(\Phi)) \circ \rho_s(\mathrm{id}_{J_{q+r}(\mathscr{E})}) \circ j_{q+r+s} &= J_s(\rho_r(\Phi)) \circ j_s \circ j_{q+r} \\
&= j_s \circ \rho_r(\Phi) \circ j_{q+r} \\
&= j_s \circ j_r \circ \Phi \circ j_q
\end{aligned}
$$

and the proposition follows from the fact that the operator $j_s \circ j_r \circ \mathscr{D}$ has order $q + r + s$ and is determined by a unique morphism. In fact we have the following diagram between operators:

$$
\begin{array}{ccc}
\mathscr{E} & \xrightarrow{\mathscr{D}} & \mathscr{E}' \\
{\scriptstyle j_{q+r}}\downarrow & \searrow{\scriptstyle j_r \circ \mathscr{D}} & \downarrow{\scriptstyle j_r} \\
J_{q+r}(\mathscr{E}) & \xrightarrow[\rho_r(\Phi)]{} & J_r(\mathscr{E}')
\end{array}
$$

C.Q.F.D.

APPLICATION 1.17 The commutative diagram:

$$
\begin{array}{ccc}
J_{q+r}(\mathscr{E}) & \xrightarrow{\rho_r(\Phi)} & J_r(\mathscr{E}') \\
{\scriptstyle \rho_r(\mathrm{id}_{J_q(\mathscr{E})})}\downarrow & & \| \\
J_r(J_q(\mathscr{E})) & \xrightarrow[J_r(\Phi)]{} & J_r(\mathscr{E}')
\end{array}
$$

allows us to define:

$$\rho_r(\Phi) = J_r(\Phi) \circ \rho_r(\mathrm{id}_{J_q(\mathscr{E})})$$

APPLICATION 1.18 We have the formulas:

$$\pi^r_{r-1} \circ \rho_r(\Phi) \circ j_{q+r} = \pi^r_{r-1} \circ j_r \circ \Phi \circ j_q = j_{r-1} \circ \Phi \circ j_q$$

$$\rho_{r-1}(\Phi) \circ \pi^{q+r}_{q+r-1} \circ j_{q+r} = \rho_{r-1}(\Phi) \circ j_{q+r-1} = j_{r-1} \circ \Phi \circ j_q$$

It follows that the diagram:

$$
\begin{array}{ccc}
J_{q+r}(\mathscr{E}) & \xrightarrow{\rho_r(\Phi)} & J_r(\mathscr{E}') \\
{\scriptstyle \pi^{q+r}_{q+r-1}}\downarrow & & \downarrow{\scriptstyle \pi^r_{r-1}} \\
J_{q+r-1}(\mathscr{E}) & \xrightarrow[\rho_{r-1}(\Phi)]{} & J_{r-1}(\mathscr{E}')
\end{array}
$$

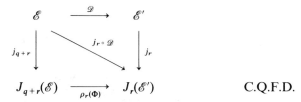

commutes and that $\rho_r(\Phi)$ is a morphism of affine bundles, fibered over $\rho_{r-1}(\Phi)$.

Let $\Phi : (x, y) \to (x, y' = \Phi(x, y))$ be expressed in local coordinates without any index, then

$$p'_\nu = \frac{d^{|\nu|}\Phi}{(dx)^\nu} = p^k_{\mu+\nu} \frac{\partial \Phi}{\partial p^k_\mu}(x, y, p) \quad \text{with } |\mu| = q$$
$$+ \text{ terms of order} < q + |\nu| \text{ in the } p$$

with

$$\mu = (\mu_1, \ldots, \mu_n), \quad \nu = (\nu_1, \ldots, \nu_n) \quad \text{and} \quad \mu + \nu = (\mu_1 + \nu_1, \ldots, \mu_n + \nu_n)$$

and there exists a morphism:

$$\sigma_r(\Phi) : S_{q+r}T^* \otimes E \to S_r T^* \otimes E'$$

fibered over $\rho_r(\Phi)$, the local expression of which is

$$v'_\nu = \frac{\partial \Phi}{\partial p^k_\mu}(x, y, p) v^k_{\mu+\nu} \quad \text{with } |\mu| = q, |\nu| = r$$

Finally we have the commutative and exact diagram:

$$\begin{array}{ccc}
0 & & 0 \\
\downarrow & & \downarrow \\
S_{q+r}T^* \otimes E & \xrightarrow{\sigma_r(\Phi)} & S_r T^* \otimes E' \\
\varepsilon_{q+r} \downarrow & & \downarrow \varepsilon_r \\
J_{q+r}(E) & \xrightarrow{V(\rho_r(\Phi))} & J_r(E') \\
\downarrow & & \downarrow \\
J_{q+r-1}(E) & \xrightarrow{V(\rho_{r-1}(\Phi))} & J_{r-1}(E') \\
\downarrow & & \downarrow \\
0 & & 0
\end{array}$$

In fact, from the last formula, we see that $\sigma_r(\Phi)$ is fibered over Φ.

DEFINITION 1.19 The morphism of vector bundles:

$$\sigma_r(\Phi) : S_{q+r}T^* \otimes E \to S_r T^* \otimes E'$$

which is fibered over $\Phi : J_q(\mathscr{E}) \to \mathscr{E}'$ is called the "*r-symbol*" of Φ and $\sigma_0(\Phi) = \sigma(\Phi)$ is called simply the "*symbol*" of Φ.

Now let E and F be vector bundles over X.

DEFINITION 1.20 A linear (differential) operator of order q from E to F is a map $\Phi \circ j_q : E \to F$ between the sets of germs of sections of E and F, where $\Phi : J_q(E) \to F$ is a vector bundle homomorphism.

Writing $\mathscr{D} = \Phi \circ j_q$, we have the commutative diagram:

$$\begin{array}{ccc} J_q(E) & \xrightarrow{\Phi} & F \\ \uparrow{\scriptstyle j_q} & & \| \\ E & \xrightarrow{\mathscr{D}} & F \end{array}$$

and, to any section ξ of E, there corresponds by $\mathscr{D}$ the section $\eta = \mathscr{D} \cdot \xi$ of F.

If Φ is expressed in local coordinates by:

$$\Phi : (x, u_\mu^k) \to (x, v^l = A_k^{l\mu}(x) u_\mu^k)$$

then

$$\eta^l(x) = A_k^{l\mu}(x) \partial_\mu \xi^k(x)$$

We have used local coordinates: (x^i, u^k) for E, (x^i, u_μ^k) for $J_q(E)$ with $k = 1, \ldots, \dim E, 0 \leq |\mu| \leq q, u_0^k = u^k$ and (x^i, v^l) for F with $l = 1, \ldots, \dim F$.

A section of $J_q(E)$ is denoted $\xi_q : (x) \to (x, \xi_\mu^k(x))$ and we have considered the section $j_q(\xi) : (x) \to (x, \partial_\mu \xi^k(x))$.

EXAMPLE 1.21 Using local coordinates we define the Spencer operator $D : J_q(E) \to T^* \otimes J_{q-1}(E)$ by the formulas:

$$(x, \xi_\mu^k(x)) \quad 0 \leq |\mu| \leq q \to \left(x, \left(\frac{\partial \xi_\mu^k(x)}{\partial x^i} - \xi_{\mu+1_i}^k(x)\right) dx^i\right) \quad 0 \leq |\mu| \leq q-1$$

In fact using local coordinates $\eta_{\mu,i}^k(x)$ for a section of $T^* \otimes J_{q-1}(E)$, we have $\eta_{\mu,i}^k(x) = \partial \xi_\mu^k(x)/\partial x^i - \xi_{\mu+1_i}^k(x)$. It is easy to check that $D \cdot \xi_q = 0$ if and only if $\xi_q = j_q(\xi)$ with $\xi = \pi_0^q \circ \xi_q$.

We have the exact sequence:

$$0 \longrightarrow E \xrightarrow{j_q} J_q(E) \xrightarrow{D} T^* \otimes J_{q-1}(E)$$

We shall use D later on.

We then have a commutative and exact diagram similar to the one of application 1.18:

$$
\begin{array}{ccc}
0 & & 0 \\
\downarrow & & \downarrow \\
S_{q+r}T^* \otimes E & \xrightarrow{\sigma_r(\Phi)} & S_r T^* \otimes F \\
\downarrow{\scriptstyle \varepsilon_{q+r}} & & \downarrow{\scriptstyle \varepsilon_r} \\
J_{q+r}(E) & \xrightarrow{\rho_r(\Phi)} & J_r(F) \\
\downarrow{\scriptstyle \pi^{q+r}_{q+r-1}} & & \downarrow{\scriptstyle \pi^{r}_{r-1}} \\
J_{q+r-1}(E) & \xrightarrow{\rho_{r-1}(\Phi)} & J_{r-1}(F) \\
\downarrow & & \downarrow \\
0 & & 0
\end{array}
$$

and a commutative diagram similar to the one of proposition 1.16:

$$
\begin{array}{ccc}
J_{q+r+s}(E) & \xrightarrow{\rho_{r+s}(\Phi)} & J_{r+s}(F) \\
\downarrow{\scriptstyle \rho_s(\mathrm{id}_{J_{q+r}(E)})} & \searrow{\scriptstyle \rho_s(\rho_r(\Phi))} & \downarrow{\scriptstyle \rho_s(\mathrm{id}_{J_r(F)})} \\
J_s(J_{q+r}(E)) & \xrightarrow[J_s(\rho_r(\Phi))]{} & J_s(J_r(F))
\end{array}
$$

In a similar way we say that $\mathscr{D} = \Phi \circ j_q$ is "*regular*" if $\rho_r(\Phi)$ has constant rank $\forall r \geq 0$.

2 Non-linear systems

Let us come back to the general case.

DEFINITION 2.1 We call a fiber submanifold $\mathscr{R}_q$ of $J_q(\mathscr{E})$ a "*non-linear system of partial differential equations*" of order q on $\mathscr{E}$, or simply non-linear system of order q on $\mathscr{E}$,

If E is a vector bundle, we call linear system of order q on E a "*vector sub-bundle*" R_q of $J_q(E)$.

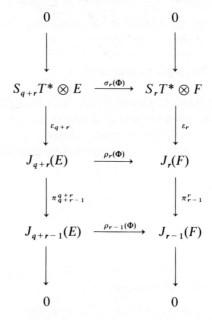

DEFINITION 2.2 A "*solution*" of $\mathcal{R}_q$ is a local section f of $\mathcal{E}$ over an open set $U \subset X$ such that

$$j_q(f)(x) \in \mathcal{R}_q, \qquad \forall x \in U.$$

DEFINITION 2.3 The "*r-prolongation*" of $\mathcal{R}_q$ is the subset

$$\mathcal{R}_{q+r} = J_r(\mathcal{R}_q) \cap J_{q+r}(\mathcal{E})$$

of $J_{q+r}(\mathcal{E})$ where $J_r(\mathcal{R}_q)$ is regarded as a fiber submanifold of $J_r(J_q(\mathcal{E}))$.

REMARK 2.4 $\mathcal{R}_q \to X$ is a fiber manifold by definition but $\mathcal{R}_{q+r} \to X$ is not a fiber manifold in general (see examples). However we have

$$\pi_{q+r}^{q+r+1}(\mathcal{R}_{q+r+1}) \subset \mathcal{R}_{q+r}.$$

DEFINITION 2.5 The system $\mathcal{R}_q \subset J_q(\mathcal{E})$ is said to be "*regular*" if $\mathcal{R}_{q+r}$ is a fiber submanifold of $J_{q+r}(\mathcal{E})$, $\forall r \geq 0$.

DEFINITION 2.6 The "*symbol*" of $\mathcal{R}_q$ is the family

$$G_q = V(\mathcal{R}_q) \cap (S_q T^* \otimes V(\mathcal{E}))$$

of vector spaces over $\mathcal{R}_q$.

As $\mathcal{R}_q$ is a fiber submanifold of $J_q(\mathcal{E})$, we have the following exact and commutative diagram where ι_q is a monomorphism of fiber manifolds over X, and π the projection onto X:

$$\begin{array}{ccccc} 0 & \longrightarrow & \mathcal{R}_q & \xrightarrow{\iota_q} & J_q(\mathcal{E}) \\ & & {\scriptstyle \pi|\mathcal{R}_q = \iota_q \circ \pi} \searrow & & \swarrow {\scriptstyle \pi} \\ & & & X & \end{array}$$

π and $\pi \circ \iota_q$ are epimorphism. Practically we shall identify $\mathcal{R}_q$ and its image by ι_q and we shall write also π instead of $\pi|_{\mathcal{R}_q}$ in order to indicate the projection onto X.

It follows that $\iota_q^{-1}(V(J_q(\mathcal{E}))) = V(J_q(\mathcal{E}))|_{\mathcal{R}_q}$ and we have the exact sequence of vector bundles over $\mathcal{R}_q$:

$$0 \longrightarrow V(\mathcal{R}_q) \xrightarrow{V(\iota_q)} V(J_q(\mathcal{E}))|_{\mathcal{R}_q} \longrightarrow N(\iota_q) \longrightarrow 0$$

and the commutative and exact diagram:

$$
\begin{array}{ccccccccc}
 & & 0 & & 0 & & & & \\
 & & \downarrow & & \downarrow & & & & \\
0 & \longrightarrow & G_q & \longrightarrow & S_qT^* \otimes V(\mathscr{E}) & \longrightarrow & N(\iota_q) & & \\
 & & \downarrow & & \downarrow \varepsilon_q & & \| & & \\
0 & \longrightarrow & V(\mathscr{R}_q) & \longrightarrow & V(J_q(\mathscr{E}))|_{\mathscr{R}_q} & \longrightarrow & N(\iota_q) & \longrightarrow & 0
\end{array}
$$

where $N(\iota_q)$ is the normal bundle of the imbedding of $\mathscr{R}_q$ into $J_q(\mathscr{E})$ that we shall denote F_0 for reasons that will become clear later on.

REMARK 2.7 As G_q is not always a vector bundle over $\mathscr{R}_q$, we should have to consider a family of diagrams over each point $(x, y, p) \in \mathscr{R}_q$. However, in order to simplify the notation we do not in general specify these points.

We shall now define a monomorphism $\Delta_{r,s}: S_{r+s}T^* \to S_rT^* \otimes S_sT^*$ of vector bundles over X, as the composition of the monomorphism

$$S_{r+s}T^* \to \otimes^{r+s} T^* = (\otimes^r T^*) \otimes (\otimes^s T^*)$$

and the homomorphism $(\otimes^r T^*) \otimes (\otimes^s T^*) \to S_rT^* \otimes S_sT^*$ induced by the projections $\otimes^r T^* \to S_rT^*$ and $\otimes^s T^* \to S_sT^*$.

It is easy to check, using local coordinates $(\omega_{\mu+\nu})$ for $S_{r+s}T^*$ and $(\omega_{\mu,\nu})$ for $S_rT^* \otimes S_sT^*$ with $|\mu| = r$ and $|\nu| = s$, that $(\Delta_{r,s}(\omega))_{\mu,\nu} = \omega_{\mu+\nu}$ and this does not depend on the coordinate system.

There is of course a monomorphism also called

$$\Delta_{r,q}: S_{q+r}T^* \otimes E \to S_rT^* \otimes S_qT^* \otimes E$$

induced by $\Delta_{r,q}$ and we can compose it with the morphism

$$S_rT^* \otimes S_qT^* \otimes E \to S_rT^* \otimes F_0$$

induced by the morphism $S_qT^* \otimes E \to F_0$, the kernel of which is G_q.

DEFINITION 2.8 We call the family G_{q+r} of vector spaces over $\mathscr{R}_q$, kernel of the preceding morphism, $S_{q+r}T^* \otimes E \to S_rT^* \otimes F_0$, the "*r-prolongation*" of G_q.

REMARK 2.9 Even if G_q is a vector bundle over $\mathscr{R}_q$, G_{q+r} is not in general a vector bundle over $\mathscr{R}_q$.

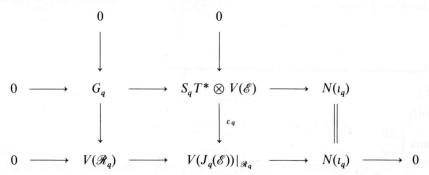

We shall make precise all these definitions, using local coordinates. The reader will find at the end of this Part, problems and examples that illustrate the above abstract definitions.

As $\mathscr{R}_q$ is a fiber submanifold of $J_q(\mathscr{E})$, $\forall(x_0, y_0, p_0) \in \mathscr{R}_q$ there exists a neighbourhood $\mathscr{U}_q$ of this point in $J_q(\mathscr{E})$ such that $\mathscr{R}_q \cap \mathscr{U}_q$ can be described locally by the equation:

$$\mathscr{R}_q \qquad \Phi^\tau(x^i, y^k, p^k_\mu) = 0$$

with

$$\operatorname{rank}\left[\frac{\partial\Phi^\tau}{\partial y^k}(x, y, p), \frac{\partial\Phi^r}{\partial p^k_\mu}(x, y, p)\right] = \operatorname{rank}\left[\frac{\partial\Phi^\tau}{\partial y^k}(x_0, y_0, p_0), \frac{\partial\Phi^\tau}{\partial p^k_\mu}(x_0, y_0, p_0)\right]$$
$$= \operatorname{codim} \mathscr{R}_q$$

and

$$i = 1, \ldots, n; \quad k = 1, \ldots, m; \quad \tau = 1, \ldots, \operatorname{codim} \mathscr{R}_q = m\frac{(q+n)!}{q!n!} - \dim \mathscr{R}_q$$

REMARK 2.10 Facing each set of equations we indicate only the fiber manifold, or the family of vector spaces, and not the restriction or open sets needed in order to use local coordinates.

It follows that $\mathscr{R}_{q+1} \cap (\pi^{q+1}_q)^{-1}(\mathscr{U}_q)$ is defined locally by the equations:

$$\mathscr{R}_{q+1} \qquad \Phi^\tau(x, y, p) = 0, \quad d_i\Phi^\tau \equiv \frac{d\Phi^\tau}{dx^i} = 0$$

with

$$\frac{d\Phi^\tau}{dx^i} = \frac{\partial\Phi^\tau}{\partial x^i}(x, y, p) + \frac{\partial\Phi^\tau}{\partial y^k}(x, y, p)p^k_i + \frac{\partial\Phi^\tau}{\partial p^k_\mu}(x, y, p)p^k_{\mu+1_i}$$

Similarly $\mathscr{R}_{q+r} \cap (\pi^{q+r}_q)^{-1}(\mathscr{U}_q)$ is defined locally by the equations:

$$\mathscr{R}_{q+r} \qquad \Phi^\tau = 0, \quad d_\nu\Phi^\tau = 0 \qquad 1 \le |\nu| \le r$$

In particular we have:

$$d_\nu\Phi^\tau = \sum_{|\mu|=q} \frac{\partial\Phi^\tau}{\partial p^k_\mu}(x, y, p)p^k_{\mu+\nu} + \text{terms of order} < q + |\nu|.$$

REMARK 2.11 From now on we shall suppress the index τ when the context makes clear its use and we shall adopt the local coordinates $(x, y, p; v)$ on $V(J_q(\mathscr{E}))|_{\mathscr{U}_q}$, or simply on $V(J_q(\mathscr{E}))$ when there can be no confusion on the coordinate domain. Moreover we shall write $\mathscr{R}_q$ instead of $\mathscr{R}_q \cap \mathscr{U}_q$ when the context makes clear the coordinate domain.

Using proposition 1.3.7 and shrinking the coordinate domain if necessary, we can find a fibered manifold $\mathscr{E}'$ over U, with local coordinates (x, y'), a map $\Phi : \mathcal{U}_q \to \mathcal{E}' : (x, y, p) \to (x, y' = \Phi(x, y, p))$ where $\Phi(x, y, p)$ is as above, and a section f' of $\mathcal{E}'$ over U, which is the zero section in this coordinate system, such that $\mathcal{R}_q \cap \mathcal{U}_q = \ker_{f''} \Phi$.

Using now theorem 1.8.6 and local coordinates $(x, y'; v')$ for $V(\mathcal{E}')$, we have an isomorphism $N(\iota_q)|(\mathcal{R}_q \cap \mathcal{U}_q) \approx (\Phi \circ \iota_q)^{-1}(V(\mathcal{E}'))$ and we can define the restriction to $\mathcal{R}_q \cap \mathcal{U}_q$ of the epimorphism

$$V(J_q(\mathcal{E}))|_{\mathcal{R}q} \to N(\iota_q)$$

by the local formulas:

$$(x, y, p; v) \to \left(x, y, p; v' = \frac{\partial \Phi}{\partial y^k}(x, y, p)v^k + \frac{\partial \Phi}{\partial p^k_\mu}(x, y, p)v^k_\mu \right)$$

where $(x, y, p) \in \mathcal{R}_q \cap \mathcal{U}_q$. Moreover, as the former map is an epimorphism of vector bundles over $\mathcal{R}_q \cap \mathcal{U}_q$, we have:

$$\operatorname{rank}\left[\frac{\partial \Phi^\tau}{\partial y^k}(x, y, p), \frac{\partial \Phi^\tau}{\partial p^k_\mu}(x, y, p) \right] = \operatorname{codim} \mathcal{R}_q, \quad \forall (x, y, p) \in \mathcal{R}_q \cap \mathcal{U}_q$$

We define $V(\mathcal{R}_q)|(\mathcal{R}_q \cap \mathcal{U}_q)$ by the equation:

$$V(\mathcal{R}_q) \qquad \frac{\partial \Phi^\tau}{\partial y^k}(x, y, p)v^k + \frac{\partial \Phi^\tau}{\partial p^k_\mu}(x, y, p)v^k_\mu = 0$$

and also $G_q|(\mathcal{R}_q \cap \mathcal{U}_q)$ by the equation:

$$G_q \qquad \frac{\partial \Phi^\tau}{\partial p^k_\mu}(x, y, p)v^k_\mu = 0, \quad |\mu| = q$$

The homomorphism $S_{q+r}T^* \otimes E \to S_rT^* \otimes F_0$ of vector bundles over $\mathcal{R}_q$ can be expressed, using local coordinates $(x, y, p; v'_\nu)$ for $S^rT^* \otimes F_0$ by:

$$(x, y, p; v^k_{\mu+\nu}) \to \left(x, y, p; v'_\nu = \frac{\partial \Phi}{\partial p^k_\mu}(x, y, p)v^k_{\mu+\nu} \right)$$

with $k = 1, \ldots, m$ and $|\mu| = q$, $|\nu| = r$.

Finally we can define $G_{q+r}|(\mathcal{R}_q \cap \mathcal{U}_q)$ by:

$$G_{q+r} \qquad \frac{\partial \Phi^\tau}{\partial p^k_\mu}(x, y, p)v^k_{\mu+\nu} = 0, \quad |\mu| = q, |\nu| = r.$$

We recall here that E and F_0 have to be pulled back over $\mathcal{R}_q$.

Definition 2.12 We define the family F_1 of vector spaces over $\mathcal{R}_q$ as the cokernel of the homomorphism $S_{q+1}T^* \otimes E \to T^* \otimes F_0$ of vector bundles over $\mathcal{R}_q$. We have the exact sequence:

$$0 \to G_{q+1} \to S_{q+1}T^* \otimes E \to T^* \otimes F_0 \to F_1 \to 0$$

THEOREM 2.13 When $\mathcal{R}_q \subset J_q(\mathcal{E})$ is a regular system of order q, then for any integer $r \geq 0$, we have an exact sequence of vector bundles over $\mathcal{R}_{q+r}$:

$$0 \to V(\mathcal{R}_{q+r}) \to V(J_{q+r}(\mathcal{E}))|_{\mathcal{R}_{q+r}} \to J_r(N(\iota_q))|_{\mathcal{R}_{q+r}}$$

Proof By definition $\mathcal{R}_{q+r} = J_r(\mathcal{R}_q) \cap J_{q+r}(\mathcal{E})$ and we have the short exact sequence:

$$0 \to V(\mathcal{R}_q) \to V(J_q(\mathcal{E}))|_{\mathcal{R}_q} \to N(\iota_q) \to 0$$

Applying J_r to this sequence and using lemma 1.9.12 we obtain the exact sequence of vector bundles, pulled back over $\mathcal{R}_{q+r}$ by reciprocal image:

$$0 \to V(J_r(\mathcal{R}_q))|_{\mathcal{R}_{q+r}} \to V(J_r(J_q(\mathcal{E})))|_{\mathcal{R}_{q+r}} \to J_r(N(\iota_q))|_{\mathcal{R}_{q+r}} \to 0.$$

Now we have the monomorphism:

$$\rho_r(\mathrm{id}_{J_q(\mathcal{E})}): J_{q+r}(\mathcal{E}) \to J_r(J_q(\mathcal{E}))$$

It gives rise to a monomorphism of vector bundles over $\mathcal{R}_{q+r}$, that is:

$$V(J_{q+r}(\mathcal{E}))|_{\mathcal{R}_{q+r}} \to V(J_r(J_q(\mathcal{E})))|_{\mathcal{R}_{q+r}}.$$

However we have:

$$V(\mathcal{R}_{q+r}) = V(J_r(\mathcal{R}_q))|_{\mathcal{R}_{q+r}} \cap V(J_{q+r}(\mathcal{E}))|_{\mathcal{R}_{q+r}}$$

and the theorem follows from the commutative and exact diagram:

$$\begin{array}{ccccccc}
 & & 0 & & 0 & & \\
 & & \downarrow & & \downarrow & & \\
0 & \longrightarrow & V(\mathcal{R}_{q+r}) & \longrightarrow & V(J_{q+r}(\mathcal{E}))|_{\mathcal{R}_{q+r}} & \longrightarrow & J_r(N(\iota_q))|_{\mathcal{R}_{q+r}} \\
 & & \downarrow & & \downarrow & & \| \\
0 & \longrightarrow & V(J_r(\mathcal{R}_q))|_{\mathcal{R}_{q+r}} & \longrightarrow & V(J_r(J_q(\mathcal{E})))|_{\mathcal{R}_{q+r}} & \longrightarrow & J_r(N(\iota_q))|_{\mathcal{R}_{q+r}}
\end{array}$$

C.Q.F.D.

COROLLARY 2.14 We have the exact sequence $\forall r \geq 0$:

$$0 \to (\pi_q^{q+r})^{-1}(G_{q+r}) \to V(\mathcal{R}_{q+r}) \to (\pi_{q+r-1}^{q+r})^{-1}(V(\mathcal{R}_{q+r-1}))$$

where the family G_{q+r} of vector spaces over $\mathcal{R}_q$ has been pulled back over $\mathcal{R}_{q+r}$ by reciprocal image.

Proof Using proposition 1.9.10 the corollary easily follows from "*diagram chasing*" in the following commutative diagram with exact rows and exact second and third columns.

For the reader we have indicated on the diagram the successive chases that will prove the exactness of the former sequence at $V(\mathcal{R}_{q+r})$.

$$
\begin{array}{ccccccccc}
 & & 0 & & 0 & & 0 & & \\
 & & \downarrow & & \downarrow & & \downarrow & & \\
0 & \longrightarrow & (\pi^{q+r}_{q})^{-1}(G_{q+r}) & \longrightarrow & S^{q+r}T^* \otimes V(\mathscr{E}) & \longrightarrow & S^{r}T^* \otimes N(\iota_q) & & \\
 & & \downarrow & & \downarrow & & \downarrow & & \\
0 & \longrightarrow & V(\mathscr{R}_{q+r}) & \longrightarrow & V(J_{q+r}(\mathscr{E}))|_{\mathscr{R}_{q+r}} & \longrightarrow & J_r(N(\iota_q))|_{\mathscr{R}_{q+r}} & & \\
 & & \downarrow & & \downarrow & & \downarrow & & \\
0 & \longrightarrow & (\pi^{q+r}_{q+r-1})^{-1}(V(\mathscr{R}_{q+r-1})) & \longrightarrow & (\pi^{q+r}_{q+r-1})^{-1}(V(J_{q+r-1}(\mathscr{E})))|_{\mathscr{R}_{q+r}} & \longrightarrow & (\pi^{q+r}_{q+r-1})^{-1}(J_{r-1}(N(\iota_q)))|_{\mathscr{R}_{q+r-1}} & & \\
 & & & & \downarrow & & \downarrow & & \\
 & & & & 0 & & 0 & &
\end{array}
$$

(connecting arrows labelled 1, 2, 3, 4, 5, 6)

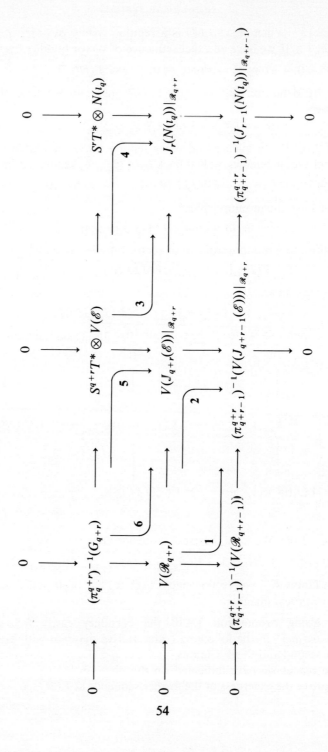

REMARK 2.15 In future we will no longer indicate the chases as this quickly becomes a "routine" for the advised reader. Moreover when there is no confusion the short exact sequence of the corollary will be simply written:

$$0 \to G_{q+r} \to V(\mathscr{R}_{q+r}) \to V(\mathscr{R}_{q+r-1})$$

as a sequence of families of vector spaces over $\mathscr{R}_{q+r}$.

As the reader will discover in the second part of this book, many systems of p.d.e. are defined by means of differential operators.

For this reason we make a separate study of these systems in the pages below.

Let $\pi : \mathscr{E} \to X$, $\pi' : \mathscr{E}' \to X$ be fibered manifolds over X, $\Phi : J_q(\mathscr{E}) \to \mathscr{E}'$ be a morphism of constant rank and f' be a section of $\mathscr{E}'$ over X such that $\operatorname{im} f' \subset \operatorname{im} \Phi$.

As Φ has constant rank, it follows that $\mathscr{R}_q = \ker_{f'} \Phi$ is a fibered submanifold of $J_q(\mathscr{E})$ and can be viewed as a system of order q on $\mathscr{E}$. We define

$$\mathscr{R}_{q+r} = J_r(\mathscr{R}_q) \cap J_{q+r}(\mathscr{E}).$$

PROPOSITION 2.16 $\mathscr{R}_{q+r} = \ker_{j_r(f')} \rho_r(\Phi)$

REMARK 2.17 The reader will see that this is in fact the "*classical*" way to define prolongation.

Proof We have the commutative and exact diagram:

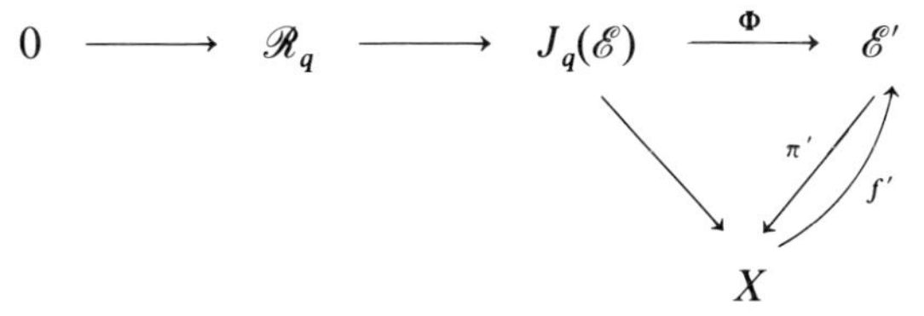

Applying J_r we obtain the commutative and exact diagram:

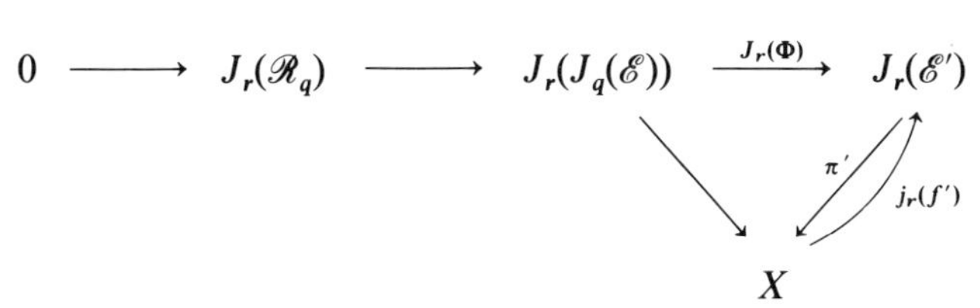

and the proposition follows from a chase in the commutative and exact diagram:

$$
\begin{array}{ccccccc}
 & & & & J_{q+r}(\mathscr{E}) & \xrightarrow{\rho_r(\Phi)} & J_r(\mathscr{E}') \\
 & & & & \downarrow{\scriptstyle \rho_r(\mathrm{id}_{J_q(\mathscr{E})})} & & \| \\
0 & \longrightarrow & J_r(\mathscr{R}_q) & \longrightarrow & J_r(J_q(\mathscr{E})) & \xrightarrow{J_r(\Phi)} & J_r(\mathscr{E}')
\end{array}
$$

C.Q.F.D.

In particular, if Φ is regular, there $\mathscr{R}_{q+r}$ is a fibered manifold over X, $\forall r \geq 0$.

In a way similar to that of proposition 2.16, we have the exact sequences:

$$0 \longrightarrow G_q \longrightarrow S_q T^* \otimes E \xrightarrow{\sigma(\Phi)} E'$$

$$0 \longrightarrow G_{q+r} \longrightarrow S_{q+r} T^* \otimes E \xrightarrow{\sigma_r(\Phi)} S_r T^* \otimes E'$$

where the different vector bundles have been pulled back over $\mathscr{R}_q$ by reciprocal image, while keeping the same notations for the homomorphisms $\sigma(\Phi)$ and $\sigma_r(\Phi)$.

In local coordinates we have:

$\mathscr{R}_q$ $$\Phi^\tau(x, y, p) = f'^\tau(x) \quad \tau = 1, \ldots, \operatorname{codim} \mathscr{R}_q$$

$\mathscr{R}_{q+r}$ $$d_\nu \Phi^\tau = \partial_\nu f'(x), \quad 0 \leq |\nu| \leq r$$

G_q $$\frac{\partial \Phi^\tau}{\partial p^k_\mu}(x, y, p) v^k_\mu = 0, \quad |\mu| = q, (x, y, p) \in \mathscr{R}_q$$

G_{q+r} $$\frac{\partial \Phi^\tau}{\partial p^k_\mu}(x, y, p) v^k_{\mu+\nu} = 0, \quad |\mu| = q, |\nu| = r, (x, y, p) \in \mathscr{R}_q$$

REMARK 2.18 We note that the symbol G_q of the kernel $\mathscr{R}_q$ of Φ is the kernel of the symbol $\sigma(\Phi)$ of Φ.

REMARK 2.19 The definition of G_{q+r} is coherent with the one already given. In fact, we can always suppose that Φ is an epimorphism, because otherwise we have just to define $\mathscr{E}' = \operatorname{im} \Phi$. Then of course $\operatorname{im} f' \subset \operatorname{im} \Phi$ for any section f' of $\mathscr{E}'$ and using theorem 1.8.6 we have $(f' \circ \pi)^{-1}(V(\mathscr{E}')) \approx N(\iota_q)$.

REMARK 2.20 When Φ is an epimorphism we have the isomorphisms

$$N(\iota_q) \approx (f' \circ \pi)^{-1}(V(\mathscr{E}')) \approx \pi^{-1}(N(f'))$$

of vector bundles over $\mathscr{R}_q$. This topological restriction on $\mathscr{R}_q$ is not in general satisfied. This is the reason why we shall give results that do not depend on the fact that $\mathscr{R}_q$ can be considered as the kernel of a morphism $\Phi : J_q(\mathscr{E}) \to \mathscr{E}'$ with respect to a given section f' of $\mathscr{E}'$. However, this situation will be that of the systems considered in the second part of this book and we shall indicate how this property can be introduced in order to simplify the proofs of many results that hold in the general case.

3 Formal properties

Let $\mathscr{R}_q \subset J_q(\mathscr{E})$ be a regular system of order q on $\mathscr{E}$. From the diagram of corollary 2.14 it follows that the morphism $\pi^{q+r+1}_{q+r} : \mathscr{R}_{q+r+1} \to \mathscr{R}_{q+r}$ induced by the epimorphism $\pi^{q+r+1}_{q+r} : J_{q+r+1}(\mathscr{E}) \to J_{q+r}(\mathscr{E})$ is in general neither a monomorphism nor an epimorphism.

EXAMPLE 3.1 Let E be a vector bundle over X and $D : J_1(E) \to T^* \otimes E$ be the Spencer operator already defined. Set $D = \Phi \circ j_1$ with $\Phi : J_1(J_1(E)) \to T^* \otimes E$ and define $R'_{r+1} = \ker \rho_r(\Phi)$.

Using local coordinates (x, u) on E, we have

R'_1 $\qquad u^k_{,i} - u^k_i = 0$

R'_2 $\qquad u^k_{,ij} - u^k_{i,j} = 0 \qquad u^k_{,i} - u^k_i = 0$

and

$$J_2(E) = \pi^2_1(R'_2) \subset R'_1 \subset J_1(J_1(E)).$$

We are thus led to the following definition:

DEFINITION 3.2 We say that a regular system $\mathscr{R}_q \subset J_q(\mathscr{E})$ of order q on $\mathscr{E}$ is "*formally integrable*" if the morphism $\pi^{q+r+1}_{q+r} : \mathscr{R}_{q+r+1} \to \mathscr{R}_{q+r}$ is an epimorphism, $\forall r \geq 0$.

Such a property for a system of p.d.e. is very useful in order to construct a "*formal solution*" of $\mathscr{R}_q$. In fact, when it is fulfilled, we can choose a point $(x_0, y_0, p_0) \in \mathscr{R}_q$ and construct locally, on a neighbourhood of x_0, the Taylor development at x_0, using at each step the equations of $\mathscr{R}_{q+r}$ containing effectively some jet-coordinates of order $q + r$. Moreover, from the properties of the formal differentiations used for the prolongations of $\mathscr{R}_q$, it follows that these equations of order $q + r$ are linear with respect to the p^k_μ with $|\mu| = q + r$. Using the implicit function theorem and Gramer's rules we can parametrise $\mathscr{R}_{q+r}$, expressing some jet-coordinates, called "*principal*", locally as functions of the x and of the other jet-coordinates, called "*parametric*". The possibility of using, at each step, the parametrisation of

$\mathscr{R}_{q+r}$ in order to parametrise $\mathscr{R}_{q+r+1}$, while keeping the same parametric and principal jet-coordinates of order $\leq q+r$, is given by the following proposition:

PROPOSITION 3.3 If $\mathscr{R}_q \subset J_q(\mathscr{E})$ is formally integrable, then G_{q+r} is a vector bundle over $\mathscr{R}_q$, $\forall r > 0$.

Proof As $\mathscr{R}_q$ is formally integrable, $\mathscr{R}_{q+r}$ is a fibered manifold $\forall r \geq 0$ and $\pi^{q+r+1}_{q+r} : \mathscr{R}_{q+r+1} \to \mathscr{R}_{q+r}$ has constant rank because it is an epimorphism. From corollary 2.14 we have the short exact sequences:

$$0 \to G_{q+r+1} \to V(\mathscr{R}_{q+r+1}) \to V(\mathscr{R}_{q+r}) \to 0$$

But G_{q+r+1} depends only on $\pi^{q+r+1}_q(\mathscr{R}_{q+r+1}) = \mathscr{R}_q$ and the family of vector spaces G_{q+r+1} over $\mathscr{R}_q$ is such that

$$\dim G_{q+r+1} = \dim \mathscr{R}_{q+r+1} - \dim \mathscr{R}_{q+r} = \text{cst.}$$

It follows that G_{q+r} is a vector bundle over $\mathscr{R}_q$, $\forall r > 0$. C.Q.F.D.

We shall look at the former formal series in the analytic case, that is to say when X is an analytic manifold and $\mathscr{E}$ an analytic fibered manifold over X.

DEFINITION 3.4 We call analytic system of p.d.e. an analytic fibered submanifold $\mathscr{R}_q \subset J_q(\mathscr{E})$.

PROPOSITION 3.5 In the analytic case, if the formal series constructed above is convergent, then it is a solution of $\mathscr{R}_q$.

Proof Set

$$f^k(x) = y_0^k + \sum_{|\mu|=1}^{\infty} \frac{1}{\mu!} p^k_{0\mu}(x - x_0)^\mu$$

for the limit of the formal series, $\forall x \in U$, sufficiently small neighbourhood of x_0 in X. By definition

$$\mu! = (\mu_1)! \cdots (\mu_n)! \quad \text{and} \quad (x - x_0)^\mu = (x^1 - x_0^1)^{\mu_1} \cdots (x^n - x_0^n)^{\mu_n}$$

On U we define:

$$\Phi(x, f^k(x), \partial_\mu f^k(x)) \equiv \Phi(j_q(f)(x))$$

and we check that:

$$\frac{\partial}{\partial x^i}(\Phi(j_q(f)(x))) = \frac{\partial \Phi}{\partial x^i}(j_q(f)(x)) + \frac{\partial \Phi}{\partial y^k}(j_q(f)(x))\partial_i f^k(x)$$

$$+ \frac{\partial \Phi}{\partial p^k_\mu}(j_q(f)(x))\partial_{\mu+1_i} f^k(x)$$

$$= \frac{d\Phi}{dx^i}(j_{q+1}(f)(x)).$$

It follows that:

$$\Phi(j_q(f)(x)) = \Phi(j_q(f)(x_0)) + \sum_{|\nu|=1}^{\infty} \frac{(x-x_0)^\nu}{\nu!} d_\nu \Phi(j_{q+|\nu|}(f)(x_0))$$

$$= 0$$

because an analytic function that has all its derivatives equal to zero, when they are evaluated at a point x_0, is identically equal to zero on a sufficiently small neighbourhood of x_0. C.Q.F.D.

REMARK 3.6 In the analytic case, it can be shown (20) that any formal development constructed as above is convergent on a convenient neighbourhood of x_0, but this fact is outside of the scope of this book. However, in C^∞ case, we shall give later on an example, found by H. Lewy, of a formally integrable system such that it is possible to construct formal series that are not convergent.

Let us come back to the C^∞ case.

As we have already seen, a system of p.d.e. is not in general formally integrable.

We shall even exhibit some examples of linear systems such that $\pi^{q+1}_q : \mathcal{R}_{q+1} \to \mathcal{R}_q$ is an epimorphism, when $\pi^{q+2}_{q+1} : \mathcal{R}_{q+2} \to \mathcal{R}_{q+1}$ is not an epimorphism.

It seems also very difficult to determine $\bar{\mathcal{R}}_q = \pi^\infty_q(\mathcal{R}_\infty) \subset J_q(\mathscr{E})$ for a given system $\mathcal{R}_q \subset J_q(\mathscr{E})$. We only know that, in general, $\bar{\mathcal{R}}_q \subset \mathcal{R}_q$ with $\bar{\mathcal{R}}_q \neq \mathcal{R}_q, \forall_q$.

EXAMPLE 3.7 For the Spencer linear differential operator D, we have seen that $\bar{R}'_1 = \pi^\infty_1(R'_\infty) = \pi^2_1(R'_2) \neq R'_1$.

It is thus essential to have a concrete "*criterion*" that allows us to be able to check the formal integrability of any system of p.d.e. by means of only a finite number of operations.

We shall now give such a criterion for nonlinear systems of p.d.e. The idea will be to use only the surjectivity of the morphism $\pi_q^{q+1}: \mathscr{R}_{q+1} \to \mathscr{R}_q$ and an algebraic property of the symbol G_q of $\mathscr{R}_q$ and its prolongations G_{q+r}.

DEFINITION 3.8 In the sequel we shall define: for $r, s \geq 0$:

$$\mathscr{R}_{q+r}^{(s)} = \pi_{q+r}^{q+r+s}(\mathscr{R}_{q+r+s})$$

$\mathscr{R}_{q+r}^{(s)}$ is not in general a fibered submanifold of $J_{q+r}(\mathscr{E})$.

LEMMA 3.9 If $\mathscr{R}'$ and $\mathscr{R}''$ are fibered submanifolds of $\mathscr{E}$ and if $\mathscr{R} = \mathscr{R}' \cap \mathscr{R}''$ is also a fibered submanifold of $\mathscr{E}$, then $J_q(\mathscr{R}) = J_q(\mathscr{R}') \cap J_q(\mathscr{R}'')$, $\forall q \geq 0$.

Proof If $\mathscr{U}$ is an open set of $\mathscr{E}$, then we have $J_q(\mathscr{U}) = (\pi_0^q)^{-1}(\mathscr{U}) \subset J_q(\mathscr{E})$ from the property of the operation "J_q".

We have also $J_q(\mathscr{R}) \subseteq J_q(\mathscr{R}') \cap J_q(\mathscr{R}'')$.

Now from proposition 1.3.7, it is possible to find, for any point of $\mathscr{R}$, a neighbourhood $\mathscr{U}$ of that point in $\mathscr{E}$ projecting onto $U \subset X$, fibered manifolds $\pi': \mathscr{E}' \to U$, $\pi'': \mathscr{E}'' \to U$, morphisms $\Phi': \mathscr{U} \to \mathscr{E}'$, $\Phi'': \mathscr{U} \to \mathscr{E}''$ and sections f' of $\mathscr{E}'$, f'' of $\mathscr{E}''$ such that:

$$\mathscr{R}' \cap \mathscr{U} = \ker_{f'} \Phi', \qquad \mathscr{R}'' \cap \mathscr{U} = \ker_{f''} \Phi''$$

We can introduce the section $(f', f''): (x) \to (x, f'(x), f''(x))$ of $\mathscr{E}' \times_U \mathscr{E}''$ and the morphism $\Phi: \mathscr{U} \to \mathscr{E}' \times_U \mathscr{E}'': (x, y) \to (x, \Phi'(x, y), \Phi''(x, y))$ such that $\mathscr{R} \cap \mathscr{U} = \ker_{(f', f'')} \Phi$.

$\mathscr{R} \cap \mathscr{U}$ is defined locally by the equation:

$$\mathscr{R} \qquad \Phi'(x, y) = f'(x), \qquad \Phi''(x, y) = f''(x)$$

Then $J_q(\mathscr{R} \cap \mathscr{U}) = J_q(\mathscr{R}) \cap (\pi_0^q)^{-1}(\mathscr{U})$ is defined locally by the equation:

$$J_q(\mathscr{R}) \qquad d_\nu \Phi' = \partial_\nu f'(x), \qquad d_\nu \Phi'' = \partial_\nu f''(x) \qquad 0 \leq |\nu| \leq q$$

It follows that:

$$J_q(\mathscr{R} \cap \mathscr{U}) = J_q(\mathscr{R}' \cap \mathscr{U}) \cap J_q(\mathscr{R}'' \cap \mathscr{U})$$

and

$$J_q(\mathscr{R}) \cap (\pi_0^q)^{-1}(\mathscr{U}) = (J_q(\mathscr{R}') \cap J_q(\mathscr{R}'')) \cap (\pi_0^q)^{-1}(\mathscr{U}).$$

As $\mathscr{U}$ is open in $\mathscr{E}$, $(\pi_0^q)^{-1}(\mathscr{U})$ is open in $J_q(\mathscr{E})$ and it follows that

$$J_q(\mathscr{R}) = J_q(\mathscr{R}') \cap J_q(\mathscr{R}'').$$

REMARK 3.10 Practically we can say that the prolongation of the intersection of two systems of p.d.e. is the intersection of the prolongations. We notice however that the set of equations of the intersection of two systems is formed from the union of the set of equations of these latter two systems.

LEMMA 3.11 If $\mathscr{R}$ is a fibered submanifold of $\mathscr{E}$ and if we look at $J_{r+s}(\mathscr{E})$ and $J_r(J_s(\mathscr{R}))$ as fibered submanifolds of $J_r(J_s(\mathscr{E}))$, then we have:

$$J_{r+s}(\mathscr{R}) = J_r(J_s(\mathscr{R})) \cap J_{r+s}(\mathscr{E})$$

Proof Using standard notation we set $\mathscr{R} \cap \mathscr{U} = \ker_{f'} \Phi$. Applying $J_r \circ J_s$ and J_{r+s} to the exact sequence.

$$0 \longrightarrow \mathscr{R} \cap \mathscr{U} \longrightarrow \mathscr{U} \xrightarrow{\Phi} \mathscr{E}'$$

we obtain the commutative and exact diagram:

$$\begin{array}{ccccccc}
0 & \longrightarrow & J_{r+s}(\mathscr{R} \cap \mathscr{U}) & \longrightarrow & J_{r+s}(\mathscr{U}) & \xrightarrow{\rho_{r+s}(\Phi)} & J_{r+s}(\mathscr{E}') \\
 & & \downarrow & & \downarrow{\scriptstyle \rho_r(\mathrm{id}_{J_s(\mathscr{U})})} & & \downarrow{\scriptstyle \rho_r(\mathrm{id}_{J_s(\mathscr{E}')})} \\
0 & \longrightarrow & J_r(J_s(\mathscr{R} \cap \mathscr{U})) & \longrightarrow & J_r(J_s(\mathscr{U})) & \xrightarrow{J_r(\rho_s(\Phi))} & J_r(J_s(\mathscr{E}'))
\end{array}$$

where the morphism on the left is induced by $\rho_r(\mathrm{id}_{J_s(\mathscr{U})})$.

It follows that:

$$J_{r+s}(\mathscr{R} \cap \mathscr{U}) = J_r(J_s(\mathscr{R} \cap \mathscr{U})) \cap J_{r+s}(\mathscr{U}).$$

But

$$J_r(J_s(\mathscr{R} \cap \mathscr{U})) = J_r(J_s(\mathscr{R}) \cap (\pi_0^s)^{-1}(\mathscr{U})) = J_r(J_s(\mathscr{R})) \cap (\pi_0^{r+s})^{-1}(\mathscr{U})$$

and of course

$$J_{r+s}(\mathscr{R}) \subseteq J_r(J_s(\mathscr{R})) \cap J_{r+s}(\mathscr{E}).$$

Then

$$J_{r+s}(\mathscr{R}) \cap (\pi_0^{r+s})^{-1}(\mathscr{U}) = (J_r(J_s(\mathscr{R})) \cap J_{r+s}(\mathscr{E})) \cap (\pi_0^{r+s})^{-1}(\mathscr{U}).$$

and the lemma follows from the fact that $(\pi_0^{r+s})^{-1}(\mathscr{U})$ is open in $J_{r+s}(\mathscr{E})$ because $\mathscr{U}$ is open in $\mathscr{E}$. C.Q.F.D.

We deduce from the two preceding lemmas the following proposition:

PROPOSITION 3.12 If $\mathscr{R}_q \subset J_q(\mathscr{E})$ is a system of order q on $\mathscr{E}$ and if $\mathscr{R}_{q+r}$ is a fibered submanifold of $J_{q+r}(\mathscr{E})$, then $(\mathscr{R}_{q+r})_{+s} = \mathscr{R}_{q+r+s}$, $\forall s \geq 0$.

Proof By definition:

$$\mathscr{R}_{q+r} = J_r(\mathscr{R}_q) \cap J_{q+r}(\mathscr{E})$$

and

$$(\mathscr{R}_{q+r})_{+s} = J_s(\mathscr{R}_{q+r}) \cap J_{q+r+s}(\mathscr{E})$$

It follows from the lemmas above that:

$$\begin{aligned}
(\mathscr{R}_{q+r})_{+s} &= J_s(J_r(\mathscr{R}_q)) \cap J_s(J_{q+r}(\mathscr{E})) \cap J_{q+r+s}(\mathscr{E}) \\
&= J_s(J_r(\mathscr{R}_q)) \cap J_{q+r+s}(\mathscr{E}) \\
&= J_s(J_r(\mathscr{R}_q)) \cap J_{r+s}(J_q(\mathscr{E})) \cap J_{q+r+s}(\mathscr{E}) \\
&= J_{r+s}(\mathscr{R}_q) \cap J_{q+r+s}(\mathscr{E}) \\
&= \mathscr{R}_{q+r+s}.
\end{aligned}$$

C.Q.F.D.

REMARK 3.13 When $\mathscr{R}_q = \ker_{f'} \Phi$, we have the following commutative and exact diagrams:

$$\begin{array}{ccccccc}
0 & \longrightarrow & \mathscr{R}_{q+r} & \longrightarrow & J_{q+r}(\mathscr{E}) & \xrightarrow{\rho_r(\Phi)} & J_r(\mathscr{E}') \\
 & & \downarrow & & \downarrow{\scriptstyle \rho_r(\mathrm{id}_{J_q(\mathscr{E})})} & & \Vert \\
0 & \longrightarrow & J_r(\mathscr{R}_q) & \longrightarrow & J_r(J_q(\mathscr{E})) & \xrightarrow{J_r(\Phi)} & J_r(\mathscr{E}')
\end{array}$$

$$\begin{array}{ccccccc}
0 & \longrightarrow & (\mathscr{R}_{q+r})_{+s} & \longrightarrow & J_{q+r+s}(\mathscr{E}) & \xrightarrow{\rho_s(\rho_r(\Phi))} & J_s(J_r(\mathscr{E}')) \\
 & & \downarrow & & \downarrow{\scriptstyle \rho_s(\mathrm{id}_{J_{q+r}(\mathscr{E})})} & & \Vert \\
0 & \longrightarrow & J_s(\mathscr{R}_{q+r}) & \longrightarrow & J_s(J_{q+r}(\mathscr{E})) & \xrightarrow{J_s(\rho_r(\Phi))} & J_s(J_r(\mathscr{E}'))
\end{array}$$

But we have also the commutative and exact diagram:

$$\begin{array}{ccccccc}
0 & \longrightarrow & \mathscr{R}_{q+r+s} & \longrightarrow & J_{q+r+s}(\mathscr{E}) & \xrightarrow{\rho_{r+s}(\Phi)} & J_{r+s}(\mathscr{E}') \\
 & & \downarrow & & \downarrow{\scriptstyle \rho_{r+s}(\mathrm{id}_{J_q(\mathscr{E})})} & & \Vert \\
0 & \longrightarrow & J_{r+s}(\mathscr{R}_q) & \longrightarrow & J_{r+s}(J_q(\mathscr{E})) & \xrightarrow{J_{r+s}(\Phi)} & J_{r+s}(\mathscr{E}')
\end{array}$$

and from proposition 1.16 we deduce the commutative and exact diagram:

$$\begin{array}{ccccccc}
0 & \longrightarrow & \mathscr{R}_{q+r+s} & \longrightarrow & J_{q+r+s}(\mathscr{E}) & \xrightarrow{\rho_{r+s}(\Phi)} & J_{r+s}(\mathscr{E}') \\
 & & \downarrow & & \Vert & & \downarrow{\scriptstyle \rho_s(\mathrm{id}_{J_r(\mathscr{E}')})} \\
0 & \longrightarrow & (\mathscr{R}_{q+r})_{+s} & \longrightarrow & J_{q+r+s}(\mathscr{E}) & \xrightarrow{\rho_s(\rho_r(\Phi))} & J_s(J_r(\mathscr{E}'))
\end{array}$$

We conclude that $(\mathscr{R}_{q+r})_{+s} = \mathscr{R}_{q+r+s}$ by means of an easy diagram chase, using the fact that $\rho_s(\mathrm{id}_{J_r(\mathscr{E}')})$ is a monomorphism.

Of course we may notice that:

$$(\mathscr{R}_{q+r})_{+s} = \ker_{j_s(j_r(f'))} \rho_s(\rho_r(\Phi)) \quad \text{and} \quad \mathscr{R}_{q+r+s} = \ker_{j_{r+s}(f')} \rho_{r+s}(\Phi).$$

Proposition 3.14 If G_{q+1} is a vector bundle over $\mathscr{R}_q$ and if the map $\pi_q^{q+1}: \mathscr{R}_{q+1} \to \mathscr{R}_q$ is surjective, then $\mathscr{R}_{q+1}$ is a fibered submanifold of $J_{q+1}(\mathscr{E})$.

Proof With standard notations, we can write locally $\mathscr{R}_q \cap \mathscr{U}_q = \ker_{f'} \Phi$ and we have the following commutative and exact diagram, where $\mathscr{U}_{q+1}$ stands for $(\pi_q^{q+1})^{-1}(\mathscr{U}_q)$ and is open in $J_{q+1}(\mathscr{E})$:

$$\begin{array}{ccccccc}
0 & \longrightarrow & G_{q+1}|(\mathscr{R}_q \cap \mathscr{U}_q) & \longrightarrow & S_{q+1}T^* \otimes E & \xrightarrow{\sigma_1(\Phi)} & T^* \otimes E' \\
 & & \downarrow & & \downarrow & & \downarrow \\
0 & \longrightarrow & \mathscr{R}_{q+1} \cap \mathscr{U}_{q+1} & \longrightarrow & \mathscr{U}_{q+1} & \xrightarrow{\rho_1(\Phi)} & J_1(\mathscr{E}') \\
 & & \downarrow & & \downarrow \pi_q^{q+1} & & \downarrow \pi_0^1 \\
0 & \longrightarrow & \mathscr{R}_q \cap \mathscr{U}_q & \longrightarrow & \mathscr{U}_q & \xrightarrow{\Phi} & \mathscr{E}'
\end{array}$$

We notice that $\mathscr{U}_{q+1}$ is an affine bundle over $\mathscr{U}_q$ and that $J_1(\mathscr{E}')$ is also an affine bundle over $\mathscr{E}'$.

As G_{q+1} is a vector bundle over $\mathscr{R}_q$ it follows that $\sigma_1(\Phi)$ has constant rank. As Φ has also constant rank because $\mathscr{R}_q$ is a fibered submanifold of $J_q(\mathscr{E})$ by hypothesis, using the result quoted just before proposition 1.8.8 and proposition 1.8.11, we deduce that $\rho_1(\Phi)$ has constant rank when it is restricted to $\mathscr{U}_{q+1}|(\mathscr{R}_q \cap \mathscr{U}_q) = (\pi_q^{q+1})^{-1}(\mathscr{R}_q \cap \mathscr{U}_q)$.

As $\mathscr{U}_{q+1}$ is open in $J_{q+1}(\mathscr{E})$ it follows that $\mathscr{R}_{q+1}$ is a fibered submanifold of $J_{q+1}(\mathscr{E})$. Moreover we have the relation:

$$\dim G_{q+1} = \dim \mathscr{R}_{q+1} - \dim \mathscr{R}_q. \qquad \text{C.Q.F.P.}$$

We shall now detail the proof, using local coordinates. We have successively:

$$\mathscr{R}_q \qquad \Phi^\tau(x, y, p) = 0$$

$$\mathscr{R}_{q+1} \qquad \frac{\partial \Phi^\tau}{\partial x^i}(x, y, p) + \frac{\partial \Phi^\tau}{\partial y^k}(x, y, p)p_i^k + \frac{\partial \Phi^\tau}{\partial p_\mu^k}(x, y, p)p_{\mu+1_i}^k = 0$$

$$G_{q+1} \qquad \frac{\partial \Phi^\tau}{\partial p_\mu^k}(x, y, p)v_{\mu+1_i}^k = 0 \quad \text{with } (x, y, p) \in \mathscr{R}_q \quad \text{and} \quad |\mu| = q$$

It follows that the Jacobian matrix of $\rho_1(\Phi)$ has the following form, with $(x, y, p) \in \mathscr{R}_q$:

p of order $q+1$	p of order $\leq q$	y	
$\dfrac{\partial \Phi^\tau}{\partial p^k_\mu}(x, y, p) \quad \lvert\mu\rvert = q$	×	×	$\dfrac{d\Phi^\tau}{dx^i}$
0	$\dfrac{\partial \Phi^\tau}{\partial p^k_\mu}(x, y, p)$	$\dfrac{\partial \Phi^\tau}{\partial y^k}(x, y, p)$	Φ^τ

In order to parametrise $\mathscr{R}_{q+1}$, we can parametrise first $\mathscr{R}_q$, then use Gramer's rules for the inhomogeneous linear system with respect to the $p^k_{\mu+1_i}$ with $\lvert\mu\rvert = q$.

The fact that the map $\pi^{q+1}_q : \mathscr{R}_{q+1} \to \mathscr{R}_q$ is surjective will not bring any new equations of order q after the elimination of the $p^k_{\mu+1_i}$ with $\lvert\mu\rvert = q$, using linear combinations such as

$$A^i_\tau(x, y, p)\,\frac{d\Phi^\tau}{dx^i}.$$

From elementary linear algebra, it then follows that the rank of the total matrix is the sum of the rank of the matrix $[(\partial\Phi^\tau/\partial p^k_\mu)(x, y, p)]$ on the left upper sight, that is to say

$$m \cdot \frac{(n+q)!}{(q+1)!(n-1)!} - \dim G_{q+1} \quad \text{when } (x, y, p) \in \mathscr{R}_q$$

plus the rank of the Jacobian matrix of Φ.

The following technical lemma is very useful:

Lemma 3.15 If U is a convex neighbourhood of the origin in $\mathbb{R}^n$ and if $f : U \to \mathbb{R}$ is such that $f \in C^\infty(U)$ and $f(0) = 0$ then

$$f(x) = \sum_{i=1}^{n} x^i \cdot \int_0^1 \frac{\partial f(tx)}{\partial x^i}\,dt$$

Proof

$$\frac{d}{dt}(f(tx)) = \sum_{i=1}^{n} x^i \frac{\partial f}{\partial x^i}(tx)$$

and

$$\sum_{i=1}^{n} \int_0^1 \frac{d}{dt} f(tx)\,dt = [f(tx)]_0^1 = f(x) - f(0) = f(x).$$

C.Q.F.D.

Using the implicit function theorem and this lemma the reader will easily show, as an exercise, that the linear combination written above and containing no $p^k_{\mu+1_i}$ with $|\mu| = q$ are such that:

$$A^i_\tau(x, y, p)\frac{d\Phi^\tau}{dx^i} \equiv B_\tau(x, y, p)\cdot\Phi^\tau$$

The existence of such identities is in general the only way to check the surjectivity of the map $\pi^{q+1}_q : \mathscr{R}_{q+1} \to \mathscr{R}_q$.

The total number of the above identities that must be verified is equal to:

$$n\cdot\dim F_0 - m\frac{(n+q)!}{(q+1)!(n-1)!} + \dim G_{q+1} = \dim F_1$$

and we note, from its definition, that F_1 is a vector bundle over $\mathscr{R}_q$ if and only if G_{q+1} is a vector bundle over $\mathscr{R}_q$.

Combining the preceding proposition and the proposition 3.3, we obtain:

THEOREM 3.16 A system $\mathscr{R}_q \subset J_q(\mathscr{E})$ is formally integrable if and only if G_{q+r+1} is a vector bundle over $\mathscr{R}_q$ and the maps $\pi^{q+r+1}_{q+r} : \mathscr{R}_{q+r+1} \to \mathscr{R}_{q+r}$ are surjective, $\forall r \geq 0$.

Proof The reader will detail the easy induction on r that must be used.

C.Q.F.D.

We shall now look at the solution of a non-linear system of p.d.e.

PROPOSITION 3.17 If $\mathscr{R}_q \subset J_q(\mathscr{E})$ is a system of order q on $\mathscr{E}$ and if $\mathscr{R}_{q+1}$ is a fibered submanifold of $J_{q+1}(\mathscr{E})$, then a section f of $\mathscr{E}$ over an open set $U \subset X$ is a solution of $\mathscr{R}_q$ if and only if it is a solution of $\mathscr{R}_{q+1}$.

Proof Using standard notation, we may assume, without loss of generality, that $\mathscr{U}_q$ projects onto U. As f is a solution of $\mathscr{R}_q$ we have $f(U) \subset \mathscr{R}_q \cap \mathscr{U}_q$, that is to say:

$$\Phi^\tau(j_q(f)(x)) = 0, \qquad \forall x \in U.$$

It follows that

$$\frac{d\Phi^\tau}{dx^i}(j_{q+1}(f)(x)) \equiv \partial_i(\Phi^\tau(j_q(f)(x)) = 0, \qquad \forall x \in U$$

and f is also a solution of $\mathscr{R}_{q+1}$.

Now if f is a solution of $\mathscr{R}_{q+1}$, then we must have $\Phi^\tau(j_q(f)(x)) = 0\ \forall x \in U$ and f is surely a solution of $\mathscr{R}_q$. C.Q.F.D.

4 Criterion theorem

We shall now state one of the main theorems in this chapter. It will be the formal counterpart of the local and technical study of the map $\pi_q^{q+1}: \mathscr{R}_{q+1} \to \mathscr{R}_q$ that we have already seen.

THEOREM 4.1 If $\mathscr{R}_q \subset J_q(\mathscr{E})$ is a system of order q on $\mathscr{E}$ and if F_1 is a vector bundle over $\mathscr{R}_q$, then there exists a section κ of this vector bundle such that we have the exact sequence:

$$\mathscr{R}_{q+1} \xrightarrow{\pi_q^{q+1}} \mathscr{R}_q \underset{0}{\overset{\kappa}{\rightrightarrows}} F_1$$

where 0 is written for the zero section.

REMARK 4.2 As we have already noticed, F_1 is defined by the exact sequence:

$$0 \to G_{q+1} \to S_{q+1}T^* \otimes E \to T^* \otimes F_0 \to F_1 \to 0$$

and is a vector bundle over $\mathscr{R}_q$ if and only if G_{q+1} is a vector bundle over $\mathscr{R}_q$. For reasons that will be clear later on, we have denoted $N(\iota_q)$ by F_0 as a vector bundle over $\mathscr{R}_q$.

Proof By definition we have the short exact sequence of vector bundles over $\mathscr{R}_q$:

$$0 \to V(\mathscr{R}_q) \to V(J_q(\mathscr{E}))|_{\mathscr{R}_q} \to F_0 \to 0$$

and we deduce from it the short exact sequence:

$$0 \to T^* \otimes V(\mathscr{R}_q) \to T^* \otimes V(J_q(\mathscr{E})) \to T^* \otimes F_0 \to 0.$$

Moreover we know that $J_1(\mathscr{R}_q)$ is an affine bundle over $\mathscr{R}_q$ and that $J_1(J_q(\mathscr{E}))$ is also an affine bundle over $J_q(\mathscr{E})$. It follows that we have the following commutative and exact three dimensional diagram.

For simplicity we indicate on this diagram (top of page 67) neither the maps nor the zeros.

The proof follows from a diagram chase that we will now detail.

Take $a \in \mathscr{R}_q$. As $\pi_q^{q+1}: J_{q+1}(\mathscr{E}) \to J_q(\mathscr{E})$ is an epimorphism, there exists $b \in J_{q+1}(\mathscr{E})$ such that $\pi_q^{q+1}(b) = \iota_q(a)$. Let $c = \rho_1(\mathrm{id}_{J_q(\mathscr{E})})(b) \in J_1(J_q(\mathscr{E}))$. We have a natural projection $\pi_0^1: J_1(J_q(\mathscr{E})) \to J_q(\mathscr{E})$ and $\pi_0^1(c) = \iota_q(a)$.

Let $\beta \in J_1(\mathscr{R}_q)$ such that $\pi_0^1(\beta) = a$ and set $\gamma = J_1(\iota_q)(\beta) \in J_1(J_q(\mathscr{E}))$.

We have

$$\pi_0^1(\gamma) = \pi_0^1 \circ J_1(\iota_q)(\beta) = \iota_q \circ \pi_0^1(\beta) = \iota_q(a) = \pi_0^1(c)$$

As $J_1(J_q(\mathscr{E}))$ is an affine bundle over $J_q(\mathscr{E})$ modelled on the vector bundle $T^* \otimes V(J_q(\mathscr{E}))$ and $a \in \mathscr{R}_q$, we can identify $\mathscr{R}_q$ and its image by ι_q in $J_q(\mathscr{E})$

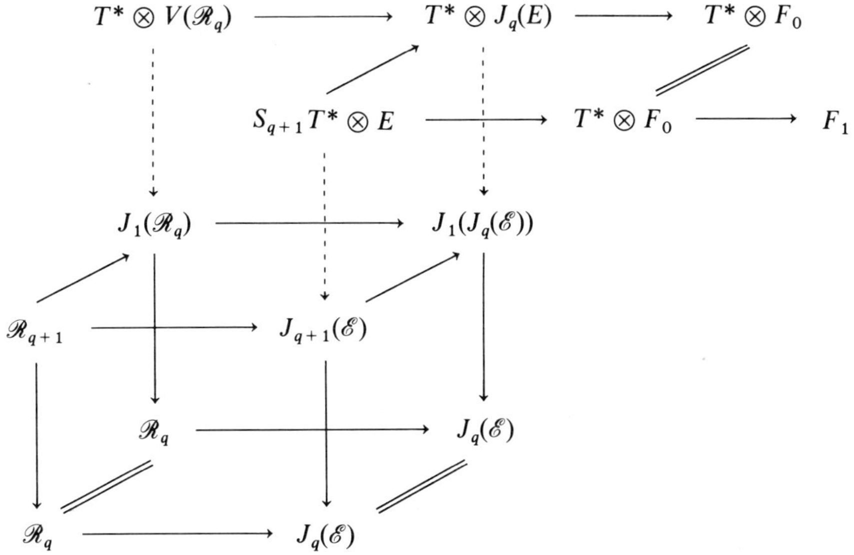

and there exists $a' = c - \gamma \in T^* \otimes V(J_q(\mathscr{E}))$. We project a' onto $T^* \otimes F_0$, then onto F_1 in order to get $a'' \in F_1$, which is of course over $a \in \mathscr{R}_q$.

The reader will easily check that a'' does not depend on the choice of b or β, because of the two successive projections.

We set $a'' = \kappa(a)$ and we can consider κ as a section of F_1 over $\mathscr{R}_q$.

As $\mathscr{R}_{q+1} = J_1(\mathscr{R}_q) \cap J_{q+1}(\mathscr{E})$ by definition, the image by κ of an element $a \in \mathscr{R}_q^{(1)} = \pi_q^{q+1}(\mathscr{R}_{q+1}) \subset \mathscr{R}_q$ corresponds to the zero section of F_1 over $\mathscr{R}_q$.

Conversely, if $a \in \mathscr{R}_q$ is such that $\kappa(a)$ is the image of a by the zero section of F_1 over $\mathscr{R}_q$, this means that a' is the sum of the image of $\beta' \in T^* \otimes V(\mathscr{R}_q)$ by the monomorphism $T^* \otimes V(\mathscr{R}_q) \to T^* \otimes V(J_q(\mathscr{E}))$ and of $b' \in S_{q+1}T^* \otimes E$ by the morphism $S_{q+1}T^* \otimes V(\mathscr{E}) \to T^* \otimes V(J_q(\mathscr{E}))$ where β' and b' are over a and the last morphism is just $(\mathrm{id}_{T^*} \otimes \varepsilon_q) \circ \Delta_{1,q}$. It follows that:

$$\rho_1(\mathrm{id}_{J_q(\mathscr{E})})(b - b') = J_1(\iota_q)(\beta + \beta')$$

because the two former morphisms are induced by $J_1(\iota_q)$ and $\rho_1(\mathrm{id}_{J_q(\mathscr{E})})$ respectively. Of course, from the definitions we have: $\pi_q^{q+1}(b - b') = \iota_q(a)$ and $\pi_0^1(\beta + \beta') = a$ and it follows that there exists $\alpha \in \mathscr{R}_{q+1}$ such that $\iota_{q+1}(\alpha) = b - b'$ and $\pi_q^{q+1}(\alpha) = a$.

This concludes the proof of the theorem. C.Q.F.D.

The transcription of this theorem, using local coordinates, has already been done. In fact if $\pi_q^{q+1} : \mathscr{R}_{q+1} \to \mathscr{R}_q$ is surjective, then it is an epimorphism whenever G_{q+1} or F_1 is a vector bundle over $\mathscr{R}_q$. Then κ must be the zero section and this means that the dim F_1 former identities must be satisfied.

EXAMPLE 4.3 $n = 3, m = 1, q = 2$, trivial bundles over $X = \mathbb{R}^3$.

$$\mathscr{R}_2 \begin{cases} \Phi^4 \equiv p_{33} & = 0 \\ \Phi^3 \equiv p_{23} - \frac{1}{2}(p_{13})^2 & = 0 \\ \Phi^2 \equiv p_{22} - p_{11}(p_{13})^2 & = 0 \\ \Phi^1 \equiv p_{12} - p_{11} \cdot p_{13} & = 0 \end{cases}$$

$\mathscr{R}_2$ is already parametrised and $\dim \mathscr{R}_2 = 9$.

$$\begin{cases} x^1, x^2, x^3, y, p_1, p_2, p_3, p_{11}, p_{13} \text{ are parametric coordinates for } \mathscr{R}_2 \\ \qquad p_{12}, p_{22}, p_{23}, p_{33} \text{ are principal coordinates for } \mathscr{R}_2. \end{cases}$$

$$G_2 \begin{cases} v_{33} & = 0 \\ v_{23} - p_{13}v_{13} & = 0 \\ v_{22} - 2p_{11}p_{13}v_{13} - (p_{13})^2 v_{11} & = 0 \\ v_{12} - p_{11}v_{13} - p_{13}v_{11} & = 0 \end{cases}$$

We have $\dim G_2 = 2$ and G_2 is a vector bundle over $\mathscr{R}_2$. It is easy to check that $\dim G_3 = 2$ and that G_3 is also a vector bundle over $\mathscr{R}_2$.

We write the dimensions (of the fibers) in a circle and get

$$\begin{array}{ccccccccc} 0 \to & G_3 & \to & S_3 T^* \otimes E & \to & T^* \otimes F_0 & \to & F_1 & \to 0 \\ & (2) & & (10) & & (12) & & (4) & \end{array}$$

We check that $2 - 10 + 12 - 4 = 0$ and we have 4 identities:

$$\begin{cases} \dfrac{d\Phi^3}{dx^3} - \dfrac{d\Phi^4}{dx^2} + p_{13}\dfrac{d\Phi^4}{dx^1} \equiv 0, \\[2ex] \dfrac{d\Phi^2}{dx^3} - \dfrac{d\Phi^3}{dx^2} - p_{13}\dfrac{d\Phi^3}{dx^1} + 2p_{11}p_{13}\dfrac{d\Phi^4}{dx^1} \equiv 0 \\[2ex] \dfrac{d\Phi^1}{dx^3} - \dfrac{d\Phi^3}{dx^1} + p_{11}\dfrac{d\Phi^4}{dx^1} \equiv 0, \\[2ex] \dfrac{d\Phi^1}{dx^2} - \dfrac{d\Phi^2}{dx^1} + p_{11}\dfrac{d\Phi^3}{dx^1} + p_{13}\dfrac{d\Phi^1}{dx^1} \equiv 0 \end{cases}$$

In particular it follows that $\pi_2^3 : \mathscr{R}_3 \to \mathscr{R}_2$ is an epimorphism and that $\dim \mathscr{R}_3 = \dim G_3 + \dim \mathscr{R}_2 = 11$.

The reader is strongly advised to do this exercise with $\Phi^2 = p_{22} - p_{12}p_{13}$ in order to see that we must consider G_2 and its prolongations as family of vector spaces over $\mathscr{R}_2$.

REMARK 4.4 The proof of the latter theorem is simplified when $\mathscr{R}_q$ is defined by a differential operator $\mathscr{D} = \Phi \circ j_q$, where $\Phi : J_q(\mathscr{E}) \to \mathscr{E}'$ is a morphism of constant rank. Without any loss of generality we shall suppose that Φ is an epimorphism.

In this case we have the following commutative and exact diagram:

$$\begin{array}{ccccccccc}
 & & & & S_{q+1}T^* \otimes E & \xrightarrow{\sigma_1(\Phi)} & T^* \otimes E' & \longrightarrow F_1 \longrightarrow 0 \\
 & & & & \vdots\downarrow & & \vdots\downarrow & \\
0 & \longrightarrow & \mathscr{R}_{q+1} & \xrightarrow{\iota_{q+1}} & J_{q+1}(\mathscr{E}) & \xrightarrow{\rho_1(\Phi)} & J_1(\mathscr{E}') & \\
 & & \downarrow{\scriptstyle \pi^{q+1}_q} & & \downarrow{\scriptstyle \pi^{q+1}_q} & & \downarrow{\scriptstyle \pi^1_0} & \\
0 & \longrightarrow & \mathscr{R}_q & \xrightarrow{\iota_q} & J_q(\mathscr{E}) & \xrightarrow{\Phi} & \mathscr{E}' &
\end{array}$$

where $J_{q+1}(\mathscr{E})$ is an affine bundle over $J_q(\mathscr{E})$ modelled on the vector bundle $S_{q+1}T^* \otimes E$ and $J_1(\mathscr{E}')$ is an affine bundle over $\mathscr{E}'$ modelled on the vector bundle $T^* \otimes E'$. We can pull back the latter vector bundles over $\mathscr{R}_q$ for our convenience.

In this case, as Φ is an epimorphism, we have the isomorphism

$$T^* \otimes E' \approx T^* \otimes F_0$$

and we notice that $\rho_1(\Phi)$ has not necessarily constant rank, even if G_{q+1} or F_1 are vector bundles over $\mathscr{R}_q$.

Let $a \in \mathscr{R}_q$ projecting onto $x \in X$. We have $\Phi(\iota_q(a)) = f'(x)$ if $\mathscr{R}_q$ is the kernel of Φ with respect to the section f' of $\mathscr{E}'$.

Let $b \in J_{q+1}(\mathscr{E})$ be such that $\pi^{q+1}_q(b) = \iota_q(a)$ and set $c = \rho_1(\Phi)(b) \in J_1(\mathscr{E}')$. Then we have $\pi^1_0(c) = \pi^1_0 \circ \rho_1(\Phi)(b) = \Phi \circ \pi^{q+1}_q(b) = \Phi \circ \iota_q(a) = f'(x)$ and b, c project onto x as a does.

It follows that there exists $\gamma \in T^* \otimes V(\mathscr{E}')$, over a and such that $\gamma = c - j_1(f')(x)$ because $\pi^1_0(j_1(f')(x)) = f'(x) = \pi^1_0(c)$.

If a' is the projection of γ onto F_1, we define $\kappa : \mathscr{R}_q \to F_1$ by $\kappa(a) = a'$.

As a' is over a like γ, then κ can be considered as a section of the vector bundle F_1 over $\mathscr{R}_q$.

This section does not depend on the choice of b. In fact if $b' \in J_{q+1}(\mathscr{E})$ is such that $\pi^{q+1}_q(b') = i_q(a)$, then there exists $\beta \in S_{q+1}T^* \otimes E$ which is over a and such that $\beta = b - b'$. If $c' = \rho_1(\Phi)(b')$ we have $c' - c = \sigma_1(\Phi)(\beta) \in T^* \otimes E'$. Then c' and c have the same projection a' onto F_1.

Now if there exists $\alpha \in \mathscr{R}_{q+1}$ such that $a = \pi^{q+1}_q(\alpha)$ then we can take $b = \iota_{q+1}(\alpha)$ and it follows that $c = j_1(f')(x)$ and that a' is the image of a by the zero section of F_1 over $\mathscr{R}_q$.

Conversely, if $a \in \mathscr{R}_q$ is such that a' is its image by the zero section of F_1 over $\mathscr{R}_q$, then we must have $\gamma = \sigma_1(\Phi)(\beta)$ for some $\beta \in S_{q+1}T^* \otimes E$ over a, with $\gamma = c - j_1(f')(x)$ and $c = \rho_1(\Phi)(b)$ for some $b \in J_{q+1}(\mathscr{E})$ with $\pi^{q+1}_q(b) = \iota_q(a)$.

It follows that $\rho_1(\Phi)(b - \beta) = c - \gamma = j_1(f')(x)$ and $\pi^{q+1}_q(b - \beta) = \pi^{q+1}_q(b) = \iota_q(a)$. Then there exists $\alpha \in \mathscr{R}_{q+1}$ such that $\iota_{q+1}(\alpha) = b - \beta$.

We have $\iota_q(\pi_q^{q+1}(\alpha)) = \iota_q(a)$ and $\pi_q^{q+1}(\alpha) = a$ because ι_q is a monomorphism and the proof is now complete. C.Q.F.D.

Using similar methods, and the results of Chapter 3 we shall now prove the following theorem that will be a key trick in the study of the formal integrability of a system.

THEOREM 4.5 Let $\mathscr{R}_q \subset J_q(\mathscr{E})$ be a system of order q on $\mathscr{E}$ such that $\mathscr{R}_{q+1}$ is a fibered submanifold of $J_{q+1}(\mathscr{E})$. If G_{q+1} is a vector bundle over $\mathscr{R}_q$ and if G_q is 2-acyclic, then $\mathscr{R}_{q+1}^{(1)} = (\mathscr{R}_q^{(1)})_{+1}$.

Proof As G_{q+1} is a vector bundle over $\mathscr{R}_q$, using proposition 3.14, it follows that $\pi_q^{q+1}: \mathscr{R}_{q+1} \to \mathscr{R}_q$ is a morphism of constant rank and that $\mathscr{R}_{q+1}$ is an affine bundle over $\mathscr{R}_q^{(1)} = \pi_q^{q+1}(\mathscr{R}_{q+1}) \subset \mathscr{R}_q$, modelled on the restriction of G_{q+1} to $\mathscr{R}_q^{(1)}$. Moreover we have

$$\dim G_{q+1} = \dim \mathscr{R}_{q+1} - \dim \mathscr{R}_1^{(1)}.$$

We shall need the following important lemma:

LEMMA 4.6 Under the hypothesis of the theorem, we have the exact sequence of vector bundles over $\mathscr{R}_q$:

$$0 \to G_{q+2} \to S_{q+2}T^* \otimes E \to S_2 T^* \otimes F_0 \to T^* \otimes F_1$$

Proof We shall prove that such a sequence holds when the last morphisms are respectively the second prolongation of the morphisms $S_q T^* \otimes E \to F_0$ and the first prolongation of the epimorphism $T^* \otimes F_0 \to F_1$ already defined.

In fact the lemma follows from a diagram chase in the following commutative diagram which has exact columns because G_q is 2-acyclic and exact rows, except perhaps the upper one:

$$\begin{array}{ccccccccc}
 & & 0 & & 0 & & 0 & & \\
 & & \downarrow & & \downarrow & & \downarrow & & \\
0 \to & & G_{q+2} & \to & S_{q+2}T^* \otimes E & \to & S_2 T^* \otimes F_0 & \to T^* \otimes F_1 & \\
 & & \downarrow \delta & & \downarrow \delta & & \downarrow \delta & \| & \\
0 \to & & T^* \otimes G_{q+1} & \to & T^* \otimes S_{q+1}T^* \otimes E & \to & T^* \otimes T^* \otimes F_0 & \to T^* \otimes F_1 & \to 0 \\
 & & \downarrow \delta & & \downarrow \delta & & \downarrow \delta & & \\
0 \to & & \Lambda^2 T^* \otimes G_q & \to & \Lambda^2 T^* \otimes S_q T^* \otimes E & \to & \Lambda^2 T^* \otimes F_0 & & \\
 & & \downarrow \delta & & \downarrow \delta & & \downarrow & & \\
0 \to & & \Lambda^3 T^* \otimes S_{q-1}T^* \otimes E & = & \Lambda^3 T^* \otimes S_{q-1} T^* \otimes E & \to & 0 & &
\end{array}$$

C.Q.F.D.

REMARK 4.7 As F_1 is a vector bundle over $\mathscr{R}_q$, then $J_1(F_1)$ is a vector bundle over $J_1(\mathscr{R}_q)$ and we have the exact sequence of vector bundles over $\mathscr{R}_{q+1}$:

$$0 \to T^* \otimes F_1 \to J_1(F_1)|\mathscr{R}_{q+1} \to (\pi_q^{q+1})^{-1}(F_1) \to 0.$$

Now, using proposition 1.16, we get the commutative three-dimensional diagram:

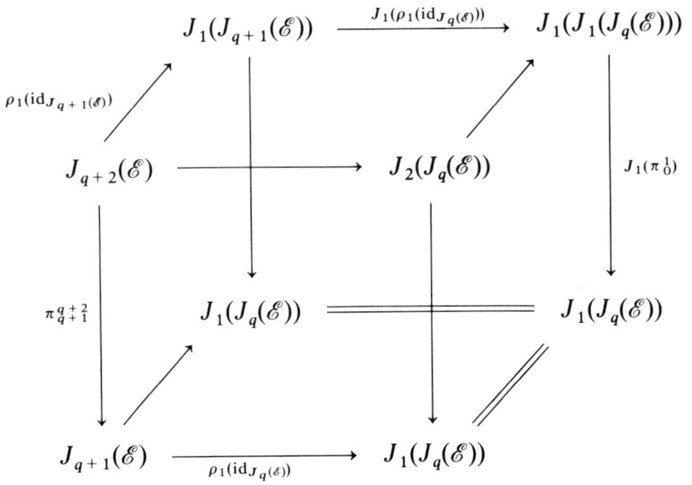

As the morphism $\pi_q^{q+1} : \mathscr{R}_{q+1} \to \mathscr{R}_q$ has constant rank, from the preceding diagram we deduce:

$$\begin{aligned}\mathscr{R}_{q+1}^{(1)} = \pi_{q+1}^{q+2}(\mathscr{R}_{q+2}) &= \pi_{q+1}^{q+2}(J_1(\mathscr{R}_{q+1}) \cap J_{q+2}(\mathscr{E}))\\ &\subseteq J_1(\pi_q^{q+1})(J_1(\mathscr{R}_{q+1})) \cap J_{q+1}(\mathscr{E})\\ &= J_1(\mathscr{R}_q^{(1)}) \cap J_{q+1}(\mathscr{E})\end{aligned}$$

and

$$\mathscr{R}_{q+1}^{(1)} \subseteq (\mathscr{R}_q^{(1)})_{+1}.$$

From the last diagram and the remark following the preceding lemma, we get the two commutative and exact diagrams of vector bundles pulled back over $\mathscr{R}_{q+1}$:

$$\begin{array}{ccccccc}
0 \to & S_2T^* \otimes V(\mathscr{R}_q) & \to & S_2T^* \otimes V(J_q(\mathscr{E})) & \to & S_2T^* \otimes F_0 & \to 0\\
& \downarrow{\scriptstyle \varepsilon_1 \circ \delta} & & \downarrow{\scriptstyle \varepsilon_1 \circ \delta} & & \downarrow{\scriptstyle \varepsilon_1 \circ \delta} & \\
0 \to & J_1(T^* \otimes V(\mathscr{R}_q))|_{\mathscr{R}_{q+1}} & \to & J_1(T^* \otimes V(J_q(\mathscr{E})))|_{\mathscr{R}_{q+1}} & \to & J_1(T^* \otimes F_0)|_{\mathscr{R}_{q+1}} & \to 0
\end{array}$$

$$\begin{array}{ccccc}
S_{q+2}T^* \otimes E & \longrightarrow & S_2 T^* \otimes F_0 & \longrightarrow & T^* \otimes F_1 \\
\downarrow{\scriptstyle \varepsilon_1 \circ \delta} & & \downarrow{\scriptstyle \varepsilon_1 \circ \delta} & & \downarrow{\scriptstyle \varepsilon_1 \circ \delta} \\
J_1(S_{q+1}T^* \otimes E)|_{\mathscr{R}_{q+1}} & \longrightarrow & J_1(T^* \otimes F_0)|_{\mathscr{R}_{q+1}} & \longrightarrow & J_1(F_1)|_{\mathscr{R}_{q+1}}
\end{array}$$

As in the proof of theorem 4.1, there exists a section κ_1 of the vector bundle $T^* \otimes F_1$ over $\mathscr{R}_{q+1}$ such that we have the exact sequence:

$$\mathscr{R}_{q+2} \xrightarrow{\pi^{q+2}_{q+1}} \mathscr{R}_{q+1} \underset{0}{\overset{\kappa_1}{\rightrightarrows}} T^* \otimes F_1$$

For this we have to use the same diagram chasing in the following commutative and exact diagram where the upper part is exact because of the lemma and is pulled back over $\mathscr{R}_{q+1}$:

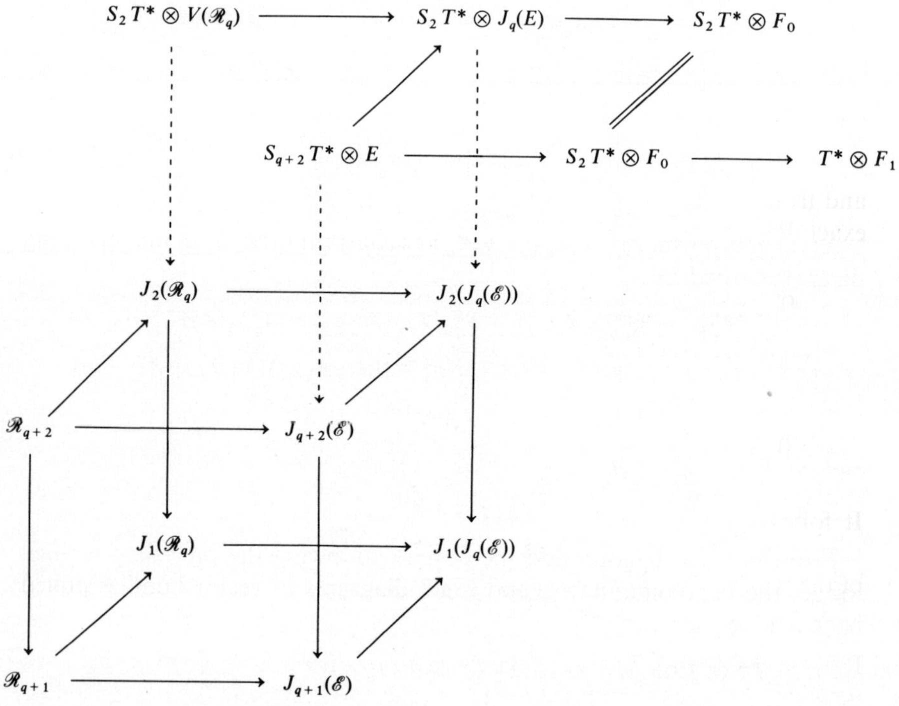

Finally we have the exact sequence:

$$0 \longrightarrow \mathscr{R}^{(1)}_{q+1} \longrightarrow \mathscr{R}_{q+1} \underset{0}{\overset{\kappa_1}{\rightrightarrows}} T^* \otimes F_1$$

But from theorem 4.1 we have the exact sequence:

$$0 \longrightarrow \mathscr{R}_q^{(1)} \longrightarrow \mathscr{R}_q \underset{0}{\overset{\kappa}{\rightrightarrows}} F_1$$

from which follows the exact sequence:

$$0 \longrightarrow J_1(\mathscr{R}_q^{(1)}) \longrightarrow J_1(\mathscr{R}_q) \underset{0}{\overset{J_1(\kappa)}{\rightrightarrows}} J_1(F_1)$$

because if $0 : \mathscr{R}_q \to F_1$ is the zero section of F_1 over $\mathscr{R}_q$, then

$$J_1(0) : J_1(\mathscr{R}_q) \to J_1(F_1)$$

is also the zero section of $J_1(F_1)$ over $J_1(\mathscr{R}_q)$.

By definition we also have:

$$\begin{aligned}(\mathscr{R}_q^{(1)})_{+1} = J_1(\mathscr{R}_q^{(1)}) \cap J_{q+1}(\mathscr{E}) &= J_1(\mathscr{R}_q^{(1)}) \cap J_1(\mathscr{R}_q) \cap J_{q+1}(\mathscr{E}) \\ &= J_1(\mathscr{R}_q^{(1)}) \cap \mathscr{R}_{q+1}\end{aligned}$$

Restricting to $\mathscr{R}_{q+1} \subset J_1(\mathscr{R}_q)$ the last exact sequence, we obtain the new exact sequence:

$$0 \longrightarrow (\mathscr{R}_q^{(1)})_{+1} \longrightarrow \mathscr{R}_{q+1} \underset{0}{\overset{J_1(\kappa)}{\rightrightarrows}} J_1(F_1)|_{\mathscr{R}_{q+1}}$$

and from the former diagrams, we deduce the following commutative and exact diagram:

$$\begin{array}{ccccccc}
0 & \longrightarrow & \mathscr{R}_{q+1}^{(1)} & \longrightarrow & \mathscr{R}_{q+1} & \underset{0}{\overset{\kappa_1}{\rightrightarrows}} & T^* \otimes F_1 \\
 & & \downarrow & & \| & & \downarrow \varepsilon_1 \\
0 & \longrightarrow & (\mathscr{R}_q^{(1)})_{+1} & \longrightarrow & \mathscr{R}_{q+1} & \underset{0}{\overset{J_1(\kappa)}{\rightrightarrows}} & J_1(F_1)|_{\mathscr{R}_{q+1}}
\end{array}$$

It follows that $\mathscr{R}_{q+1}^{(1)} = (\mathscr{R}_q^{(1)})_{+1}$ because $\varepsilon_1 : T^* \otimes F_1 \to J_1(F_1)|_{\mathscr{R}_{q+1}}$ is a monomorphism of vector bundles over $\mathscr{R}_{q+1}$ and this concludes the proof of the theorem. C.Q.F.D.

REMARK 4.8 The former proof can be simplified when $\mathscr{R}_q \subset J_q(\mathscr{E})$ is defined as the kernel of an epimorphism $\Phi : J_q(\mathscr{E}) \to \mathscr{E}'$ with respect to a section f' of $\mathscr{E}'$ over X.

We leave the reader to do the chase in the following three-dimensional diagram which is commutative and exact, because we have already noticed that $F_0 \approx E'$:

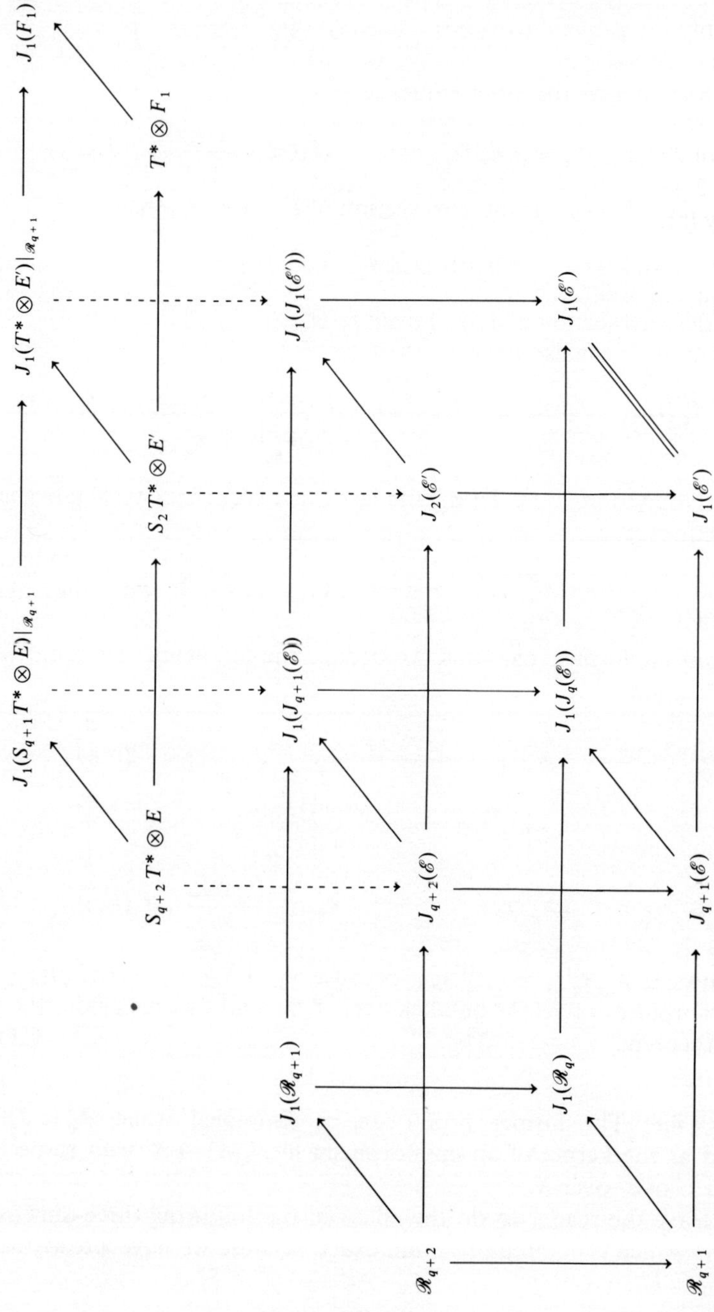

$J_1(S_{q+1}T^* \otimes E)|_{\mathscr{R}_{q+1}}$
$J_1(T^* \otimes E')|_{\mathscr{R}_{q+1}}$
$J_1(F_1)|_{\mathscr{R}_{q+1}}$
$S_{q+2}T^* \otimes E$
$S_2T^* \otimes E'$
$T^* \otimes F_1$
$J_1(\mathscr{R}_{q+1})$
$J_1(J_{q+1}(\mathscr{E}))$
$J_1(J_1(\mathscr{E}'))$
$J_{q+2}(\mathscr{E})$
$J_2(\mathscr{E}')$
$J_1(\mathscr{R}_q)$
$J_1(J_q(\mathscr{E}))$
$J_1(\mathscr{E}')$
$\mathscr{R}_{q+2}$
$\mathscr{R}_{q+1}$
$J_{q+1}(\mathscr{E})$
$J_1(\mathscr{E}')$

The same proof can be used with local coordinates on $\mathcal{R}_q \cap \mathcal{U}_q$ where $\mathcal{U}_q$ is open in $J_q(\mathscr{E})$. However we have to say that a proof of the latter theorem using only local coordinates and classical computations seems impossible because of the complexity of the symbol sequences.

COROLLARY 4.9 "*Criterion of formal integrability*" Let $\mathcal{R}_q \subset J_q(\mathscr{E})$ be a system of order q on $\mathscr{E}$ such that $\mathcal{R}_{q+r}$ is a fibered submanifold of $J_{q+r}(\mathscr{E})$. If G_q is 2-acyclic and if the map $\pi_q^{q+1}: \mathcal{R}_{q+1} \to \mathcal{R}_q$ is surjective, then $\mathcal{R}_q$ is formally integrable.

Proof Using corollary 2.14 and the fact that $\pi_q^{q+1}: \mathcal{R}_{q+1} \to \mathcal{R}_q$ is an epimorphism, it follows that G_{q+1} is a vector bundle over $\mathcal{R}_q$ such that

$$\dim G_{q+1} = \dim \mathcal{R}_{q+1} - \dim \mathcal{R}_q.$$

Now, from proposition 3.1.16, G_{q+r} is also a vector bundle over $\mathcal{R}_q$, $\forall r \geq 1$, and is also 2-acyclic according to the definition of acyclicity.
Using the former theorem we obtain:

$$\mathcal{R}^{(1)}_{q+1} = (\mathcal{R}^{(1)}_q)_{+1} = (\mathcal{R}_q)_{+1} = \mathcal{R}_{q+1}.$$

and the map $\pi_{q+1}^{q+2}: \mathcal{R}_{q+2} \to \mathcal{R}_{q+1}$ is surjective. It is an epimorphism because G_{q+2} is a vector bundle over $\mathcal{R}_q$ and using proposition 3.14, $\mathcal{R}_{q+2}$ is a fibered submanifold of $J_{q+2}(\mathscr{E})$ such that

$$\dim G_{q+2} = \dim \mathcal{R}_{q+2} - \dim \mathcal{R}_{q+1}.$$

We are now in position to apply theorem 4.5 to $\mathcal{R}_{q+1}$ and the corollary follows easily by an inductive use of this theorem. It follows that

$$\pi_{q+r}^{q+r+1}: \mathcal{R}_{q+r+1} \to \mathcal{R}_{q+r}$$

is an epimorphism and

$$\dim G_{q+r+1} = \dim \mathcal{R}_{q+r+1} - \dim \mathcal{R}_{q+r}. \qquad \text{C.Q.F.D.}$$

Using proposition 3.4.1, we can always prolong a system of p.d.e., sufficiently enough for its symbol to become involutive.

DEFINITION 4.10 A system $\mathcal{R}_q \subset J_q(\mathscr{E})$ of order q on $\mathscr{E}$ is said to be "*involutive*" if it is formally integrable and if its symbol G_q is involutive.
It is sometimes very useful to have the similar corollary:

COROLLARY 4.11 "*Criterion of involutiveness*" Let $\mathcal{R}_q \subset J_q(\mathscr{E})$ be a system of order of an $\mathscr{E}$ such that $\mathcal{R}_{q+r}$ is a fibered submanifold of $J_{q+r}(\mathscr{E})$. If G_q is involutive and if the map $\pi_q^{q+1}: \mathcal{R}_{q+1} \to \mathrm{R}_q$ is surjective, then $\mathcal{R}_q$ is involutive.

Such a criterion was formulated in the first place by M. Janet (19) in his doctoral thesis, dating back to 1921.

However his author did not give a precise link between the involutiveness of the symbol and the formal integrability of the system, which he called "*passivity*". Moreover this result was depending largely on the system of coordinates.

The criterion was formulated precisely by D. Quillen (39) and H. Goldschmidt (12) for linear systems of p.d.e. and by H. Goldschmidt in 1967 (14) for non-linear systems.

His proof, based on the use of Spencer's family of vector spaces, to be introduced later on, cannot be related to the use of local coordinates. For this reason we gave a quite different proof following closely the "*operational*" method of M. Janet.

We shall see below how it is possible to adapt the methods of this author within the formal theory of linear systems of p.d.e.

For a faithful but up to date study of the work of M. Janet, the reader can refer to the paper (47) and the book (48) of J. M. Thomas or to the excellent book of J. F. Ritt (43).

5 Prolongation theorem

In a book written by M. Janet, one can read the following sentence:

"Etant donné un systeme e.d.p. quelconque, on peut par dérivations, éliminations, résolutions d'équations, "*en nonbre fini*", ou bien constater que le système est impossible ou bien le ramener à un système passif équivalent" (21).

Later on, he adds:

"Dans ce genre de raisonnements, il est sous-entendu que l'on se place au voisinage d'un système de valeurs des variables indépendantes et des dérivées, pour lequel toutes les résolutions successives supposées dans le texte sont possibles, conformément à la théorie générale des fonctions implicites" (21).

We shall study this result, within the formal theory of non-linear systems of p.d.e., using theorem 4.5.

First of all we have the proposition:

PROPOSITION 5.1 Let $\mathscr{R}_q \subset J_q(\mathscr{E})$ be a regular system of order q on $\mathscr{E}$. If G_{q+1} is a vector bundle over $\mathscr{R}_q$ and if G_q is 2-acyclic, then $\mathscr{R}^{(1)}_{q+r} = (\mathscr{R}^{(1)}_q)_{+r}$.

Proof We shall use an induction on r.

- The case $r = 1$ is just that of theorem 4.5.
- Imagine now that $\mathscr{R}^{(1)}_{q+r} = (\mathscr{R}^{(1)}_q)_{+r}$.

As G_{q+1} is a vector bundle over $\mathscr{R}_q$ and G_q is 2-acyclic, we know from proposition 3.1.16 that G_{q+r} is a vector bundle over $\mathscr{R}_q$, $\forall r \geq 1$.

It follows that $\pi^{q+r+1}_{q+r} : \mathscr{R}_{q+r+1} \to \mathscr{R}_{q+r}$ is a morphism of constant rank and $\mathscr{R}^{(1)}_{q+r}$ is a fibered submanifold of $J_{q+r}(\mathscr{E})$.

Using proposition 3.12, we have:

$$(\mathscr{R}^{(1)}_q)_{+r+1} = ((\mathscr{R}^{(1)}_q)_{+r})_{+1} = (\mathscr{R}^{(1)}_{q+r})_{+1}$$

But G_{q+r+1} is a vector bundle over $\mathscr{R}_q$ and its pull back $(\pi^{q+r}_q)^{-1}(G_{q+r+1})$ is also a vector bundle over $\mathscr{R}_{q+r}$ which is also 2-acyclic because G_q is 2-acyclic.

Using theorem 4.5, we obtain:

$$(\mathscr{R}^{(1)}_q)_{+r+1} = (\mathscr{R}^{(1)}_{q+r})_{+1} = \mathscr{R}^{(1)}_{q+r+1}. \qquad \text{C.Q.F.D.}$$

We shall now restrict ourselves to the situation to be found in the second part of this book, that is to say we shall only consider regular systems $\mathscr{R}_q \subset J_q(\mathscr{E})$ such that the morphisms $\pi^{q+r+s}_{q+r} : \mathscr{R}_{q+r+s} \to \mathscr{R}_{q+r}$ have constant rank $\forall r, s \geq 0$.

It follows that $\mathscr{R}^{(s)}_{q+r} = \pi^{q+r+s}_{q+r}(\mathscr{R}_{q+r+s})$ is a fibered submanifold of $J_{q+r}(\mathscr{E})$.

DEFINITON 5.2 When this is the case $\mathscr{R}_q$ is said to be "*sufficiently regular*":

For r fixed we have a chain of subsets:

$$J_{q+r}(\mathscr{E}) \supset \mathscr{R}_{q+r} \supset \mathscr{R}^{(1)}_{q+r} \supset \cdots \supset \mathscr{R}^{(s)}_{q+r} \supset \cdots.$$

Counting the dimension we have:

$$\dim J_{q+r}(\mathscr{E}) \geq \dim \mathscr{R}_{q+r} \geq \dim \mathscr{R}^{(1)}_{q+r} \geq \cdots \geq \dim \mathscr{R}^{(s)}_{q+r} \geq \cdots$$

and there must exist an integer $s(r)$ such that the chain becomes stationary, that is to say:

$$\mathscr{R}^{(s(r))}_{q+r} = \mathscr{R}^{(s(r)+1)}_{q+r} = \cdots = \bar{\mathscr{R}}_{q+r}$$

Unfortunately, if we know that such a finite number exists, it is practically impossible to compute it and this kind of argument is not very useful for applications.

On the other hand we know that G_{q+r} becomes involutive when r is big enough, and this fact is easy to check.

Without loss of generality let us suppose from now on that G_q is involutive.

By definition $(\mathscr{R}_q)_{+r} = \mathscr{R}_{q+r} \subset J_r(\mathscr{R}_q)$. We have the commutative diagram:

$$\begin{array}{ccc}
J_{q+r+s}(\mathscr{E}) & \xrightarrow{\rho_r(\mathrm{id}_{J_{q+s}(\mathscr{E})})} & J_r(J_{q+s}(\mathscr{E})) \\
\downarrow{\scriptstyle \pi^{q+r+s}_{q+r}} & & \downarrow{\scriptstyle J_r(\pi^{q+s}_q)} \\
J_{q+r}(\mathscr{E}) & \xrightarrow{\rho_r(\mathrm{id}_{J_q(\mathscr{E})})} & J_r(J_q(\mathscr{E}))
\end{array}$$

From which we deduce the commutative diagram:

$$\begin{array}{ccc}
\mathscr{R}_{q+r+s} & \longrightarrow & J_r(\mathscr{R}_{q+s}) \\
\downarrow{\scriptstyle \pi^{q+r+s}_{q+r}} & & \downarrow{\scriptstyle J_r(\pi^{q+s}_q)} \\
\mathscr{R}_{q+r} & \longrightarrow & J_r(\mathscr{R}_q)
\end{array}$$

As π^{q+r+s}_{q+r} is a morphism of constant rank we have

$$\mathscr{R}^{(s)}_{q+r} \subseteq J_r(\mathscr{R}^{(s)}_q)$$

and it follows that

$$\mathscr{R}^{(s)}_{q+r} \subseteq J_r(\mathscr{R}^{(s)}_q) \cap J_{q+r}(\mathscr{E}) = (\mathscr{R}^{(s)}_q)_{+r}$$

Moreover, as

$$\pi^{q+r+1}_{q+r} \circ \pi^{q+r+s+1}_{q+r+1} = \pi^{q+r+s+1}_{q+r}$$

we have

$$\pi^{q+r+1}_{q+r}(\mathscr{R}^{(s)}_{q+r+1}) = \mathscr{R}^{(s+1)}_{q+r}$$

and the commutative diagram:

$$\begin{array}{ccccc}
0 & \longrightarrow & \mathscr{R}^{(s)}_{q+r+1} & \longrightarrow & J_1(\mathscr{R}^{(s)}_{q+r}) \\
 & & \downarrow & & \downarrow \\
0 & \longrightarrow & \mathscr{R}^{(s+1)}_{q+r} & \longrightarrow & \mathscr{R}^{(s)}_{q+r} \\
 & & \downarrow & & \downarrow \\
 & & 0 & & 0
\end{array}$$

DEFINITION 5.3 We define $G^{(s)}_{q+r}$ as the symbol of $\mathscr{R}^{(s)}_{q+r}$.

LEMMA 5.4 $\delta(G^{(s)}_{q+r+1}) \subset T^* \otimes G^{(s)}_{q+r}$.

Proof From the last diagram we get a monomorphism

$$0 \to G^{(s)}_{q+r+1} \to T^* \otimes G^{(s)}_{q+r}$$

and the lemma follows from the fact that this monomorphism is induced by the monomorphism

$$\rho_1(\mathrm{id}_{J_{q+r}(\mathscr{E})}) : J_{q+r+1}(\mathscr{E}) \to J_1(J_{q+r}(\mathscr{E}))$$

and that $G^{(s)}_{q+r+1}$ is a vector bundle over $\mathscr{R}^{(s)}_{q+r+1}$ because

$$\dim G^{(s)}_{q+r+1} = \dim \mathscr{R}^{(s)}_{q+r+1} - \dim \mathscr{R}^{(s+1)}_{q+r}$$

accordingly to the short exact sequence:

$$0 \longrightarrow G^{(s)}_{q+r+1} \longrightarrow V(\mathscr{R}^{(s)}_{q+r+1}) \xrightarrow{V(\pi^{q+r+1}_{q+r})} V(\mathscr{R}^{(s+1)}_{q+r}) \longrightarrow 0$$

C.Q.F.D.

To end this chapter we shall now study the systems $\mathscr{R}^{(s)}_{q+r} \subset J_{q+r}(\mathscr{E})$.

PROPOSITION 5.5 $\mathscr{R}_q$ and $\mathscr{R}^{(s)}_{q+r}$ have the same solution $\forall r, s \geq 0$.

Proof From proposition 3.17, every solution of $\mathscr{R}_q$ is a solution of $\mathscr{R}_{q+r+s}$ and thus of $\mathscr{R}^{(s)}_{q+r}$. Conversely every solution of $\mathscr{R}^{(s)}_{q+r}$ is a solution of $\mathscr{R}_{q+r}$ because $\mathscr{R}^{(s)}_{q+r} \subset \mathscr{R}_{q+r}$ and thus is a solution of $\mathscr{R}_q$. C.Q.F.D.

As did M. Janet, we will now construct inductively a chain of systems

$$\mathscr{R}_q, \mathscr{R}^{(1)}_{q_1}, \ldots, \mathscr{R}^{(s)}_{q_s}, \ldots$$

with involutive symbols and such that

$$(\mathscr{R}^{(s)}_{q_s})_{+r} = \mathscr{R}^{(s)}_{q_s+r} \qquad \forall r, s \geq 0.$$

We shall detail the inductive process.

Step 1 G_q is involutive by hypothesis.

- If $\mathscr{R}^{(1)}_q = \mathscr{R}_q$, then $\mathscr{R}_q$ is involutive because of the criterion.
- If $\mathscr{R}^{(1)}_q \neq \mathscr{R}_q$ and "G_{q+1} *is a vector bundle over* $\mathscr{R}_q$", then from proposition 5.1, we have $(\mathscr{R}^{(1)}_q)_{+r} = \mathscr{R}^{(1)}_{q+r}$. However $G^{(1)}_q$ is not in general involutive. We have to prolong $\mathscr{R}^{(1)}_q$ a finite number of times, say $q_1 - q$, in order to get a system $\mathscr{R}^{(1)}_{q_1}$ such that $G^{(1)}_{q_1}$ is involutive and such that:

$$(\mathscr{R}^{(1)}_{q_1})_{+r} = (\mathscr{R}^{(1)}_q)_{q_1-q+r} = \mathscr{R}^{(1)}_{q_1+r}.$$

Step s Imagine that we have constructed inductively a system $\mathscr{R}^{(s)}_{q_s}$ of order q_s on $\mathscr{E}$, such that $G^{(s)}_{q_s}$ is involutive and $(\mathscr{R}^{(s)}_{q_s})_{+r} = \mathscr{R}^{(s)}_{q_s+r}$.

Step $s+1$ We have

$$\pi^{q_s+1}_{q_s}((\mathscr{R}^{(s)}_{q_s})_{+1}) = \pi^{q_s+1}_{q_s}(\mathscr{R}^{(s)}_{q_s+1}) = \mathscr{R}^{(s+1)}_{q_s}$$

- If $\mathscr{R}^{(s+1)}_{q_s} = \mathscr{R}^{(s)}_{q_s}$, then $\mathscr{R}^{(s)}_{q_s}$ is involutive because of the criterion.
- If $\mathscr{R}^{(s+1)}_{q_s} \neq \mathscr{R}^{(s)}_{q_s}$ and if "$(G^{(s)}_{q_s})_{+1}$ *is a vector bundle over* $\mathscr{R}^{(s)}_{q_s}$", from proposition 5.1, we have

$$(\mathscr{R}^{(s+1)}_{q_s})_{+r} = ((\mathscr{R}^{(s)}_{q_s})^{(1)})_{+r} = ((\mathscr{R}^{(s)}_{q_s})_{+r})^{(1)} = (\mathscr{R}^{(s)}_{q_s+r})^{(1)} = \mathscr{R}^{(s+1)}_{q_s+r}$$

REMARK 5.6 We notice that the fact that $G^{(s)}_{q_s+1}$ is a vector bundle over $\mathscr{R}^{(s)}_{q_s+1}$ means that $(G^{(s)}_{q_s})_{+1}$ is a vector bundle over $\mathscr{R}^{(s+1)}_{q_s} \subset \mathscr{R}^{(s)}_{q_s}$ and not necessarily over $\mathscr{R}^{(s)}_{q_s}$ as this must be in order to be able to use proposition 5.1.

However $G^{(s+1)}_{q_s}$ is not in general involutive. We have to prolong $\mathscr{R}^{(s+1)}_{q_s}$ a finite number of times say $q_{s+1} - q_s$, in order to get a system $\mathscr{R}^{(s+1)}_{q_{s+1}}$ such that $G^{(s+1)}_{q_{s+1}}$ is involutive and such that:

$$(\mathscr{R}^{(s+1)}_{q_{s+1}})_{+r} = (\mathscr{R}^{(s+1)}_{q_s})_{q_{s+1}-q_s+r} = \mathscr{R}^{(s+1)}_{q_{s+1}+r}.$$

As this will be the case in the second part of this book, we shall now suppose that $G^{(s)}_{q+r}$ is the pull back over $\mathscr{R}^{(s)}_{q+r}$ by the composition

$$\mathscr{R}^{(s)}_{q+r} \to \mathscr{R}_{q+r} \to \mathscr{R}_q$$

of a vector bundle over $\mathscr{R}_q$ also denoted by $G^{(s)}_{q+r}$ and that this operation is compatible with the prolongation, that is to say the pull back of a prolongation is the prolongation of the pull back.

The reader will check easily that this hypothesis allows us to satisfy the requirements for the inductive process to work.

The next theorem will show that this process must stop after a finite number of steps.

THEOREM 5.7 There exist integers $s \geq 0$ and $q_s \geq q$ such that $\mathscr{R}^{(s)}_{q_s} \subset J_{q_s}(\mathscr{E})$ is an involutive system with the same solution as $\mathscr{R}_q$ and such that:

$$(\mathscr{R}^{(s)}_{q_s})_{+r} = \mathscr{R}^{(s)}_{q_s+r}, \qquad \forall r \geq 0.$$

Proof As the argument of the proof is purely an algebraic one, for later use and in order to simplify it, we shall prove this theorem for linear systems. The reader can adapt the proof to the non-linear case, using the former assumption on the symbols.

THEOREM 5.8 Let E be a vector bundle over X and $R_q \subset J_q(E)$ be a linear system of order q on E. If R_q is sufficiently regular, then there exist integers

$s \geq 0$ and $q_s \geq q$ such that $R^{(s)}_{q_s} \subset J_{q_s}(E)$ is an involutive system with the same solutions as R_q and such that

$$(R^{(s)}_{q_s})_{+r} = R^{(s)}_{q_s+r}, \qquad \forall r \geq 0.$$

Proof Under the hypothesis of the theorem, the symbol $G^{(s)}_{q+r}$ of $R^{(s)}_{q+r}$ is a vector bundle over X.

We can construct $R_q, R^{(1)}_{q_1}, \ldots, R^{(s)}_{q_s}, \ldots,$ as above and the theorem will be proved if we can show that there exists an integer s_0 such that

$$G^{(s)}_{q+r} = G^{(s_0)}_{q+r}, \qquad \forall s \geq s_0, \forall r \geq 0.$$

In fact, in the inductive process, the symbols $G^{(s)}_{q_s} = G^{(s_0)}_{q_s}$ are involutive and it follows that $q_s = q_{s_0}$. But at this time we have $R^{(s_0)}_{q_{s_0}} \supset R^{(s_0+1)}_{q_{s_0}} \supset \cdots$ and $\dim R^{(s_0)}_{q_{s_0}} \geq \dim R^{(s_0+1)}_{q_{s_0}} \geq \cdots$. Finally there exists $s \geq s_0$ such that $R^{(s)}_{q_s} = R^{(s)}_{q_{s_0}} = R^{(s+1)}_{q_{s_0}} = \cdots$ and the process must stop after a finite number of steps.

We see that the linear system $R^{(s)}_{q_s}$ thus obtained satisfies the conditions of the theorem.

It thus remains only to demonstrate the proposition:

PROPOSITION 5.9 $\exists s_0 \geq 0, \quad G^{(s)}_{q+r} = G^{(s_0)}_{q+r}, \quad \forall s \geq s_0, \quad \forall r \geq 0.$

Proof From proposition 1.6.4 we can identify E and $V(E)$. Then $G^{(s)}_{q+r}$ is, $\forall r, s \geq 0$, a vector sub-bundle of $S_{q+r}T^* \otimes E$, where the tensor product is to be considered over X.

Moreover from lemma 5.4, we have the commutative diagram:

$$\begin{array}{ccc} G^{(s+1)}_{q+r+1} & \xrightarrow{\ \delta\ } & T^* \otimes G^{(s+1)}_{q+r} \\ \downarrow & & \downarrow \\ G^{(s)}_{q+r+1} & \xrightarrow{\ \delta\ } & T^* \otimes G^{(s)}_{q+r} \end{array}$$

Let now $x \in X$, U be a neighbourhood of x in X trivialising E and (x^i, v^k_μ) be local coordinates for $J_q(E)$ over U, with $0 \leq |\mu| \leq q$ and $v^k_0 = v^k$.

We define locally $G^{(s)}_{q+r}$ by a linear system such as:

$$A^{\tau\mu}_k(x)v^k_\mu = 0 \qquad |\mu| = q + r, k = 1, \ldots, \dim E.$$

For x fixed we can consider the vector space over $\mathbb{R}$ generated by the polynomials in $\chi_1, \ldots, \chi_n$ with value in $\mathbb{R}^m$ and indexed by τ:

$$\{A^{\tau\mu}_k(x)\chi_\mu\} \in \mathbb{R}^m[\chi] = \mathbb{R}^m[\chi_1, \ldots, \chi_n]$$

$(G^{(s)}_{q+r})_{+1}$ is defined locally by the linear system $A^{\tau\mu}_k(x)v^k_{\mu+1_i} = 0$ with $|\mu| = q + r$.

It follows that the vector space associated to $(G^{(s)}_{q+r})_{+1}$ is generated by the products by χ_i of the polynomials associated to $G^{(s)}_{q+r}$.

Moreover, from the last diagram it follows that $\delta(G^{(s)}_{q+r+1}) \subset T^* \otimes G^{(s)}_{q+r}$ and thus $(G^{(s)}_{q+r})_{+1} \supset G^{(s)}_{q+r+1}$ by means of proposition 3.1.14.

Using elementary linear algebra we see that the vector space associated to $(G^{(s)}_{q+r})_{+1}$ is a vector subspace of the vector space associated to $G^{(s)}_{q+r+1}$.

For s fixed, we call $M^{(s)}_{q+r}$ the $\mathbb{R}[\chi]$-submodule of the $\mathbb{R}[\chi]$-module $\mathbb{R}^m[\chi]$ generated by the polynomials associated as above to $G^{(s)}_q, \ldots, G^{(s)}_{q+r}$. The action of $\mathbb{R}[\chi]$ on $\mathbb{R}^m[\chi]$ is obtained from the multiplication law of the ring $\mathbb{R}[\chi]$ using the fact that $\mathbb{R}^m[\chi] = (\mathbb{R}[\chi])^m$ and that $\mathbb{R}^m$ is an $\mathbb{R}$-module in a natural way. It is easy to see that this definition does not depend on the coordinate system on X about x or on the particular equations defining the symbols.

Moreover $M^{(s)}_{q+r_1} \subseteq M^{(s)}_{q+r_2}$ whenever $r_1 \leq r_2$.

We shall need some results from commutative algebra that will also be used later on (59, 60).

Let A be a ring and M an A-module. We define the $A[\chi]$-module $M[\chi]$ from the sums with a finite number of terms such as $\sum_\mu m_\mu \chi_\mu$ by setting $\chi_i \cdot (\sum_\mu m_\mu \chi_\mu) = \sum_\mu m_\mu \cdot \chi_{\mu+1_i}$ with $m_\mu \in M$ and $m_\mu \neq 0$ only for a finite number of multi indices $\mu = (\mu_1, \ldots, \mu_n)$.

Proposition 5.10 For a module, the following properties are equivalent.

1 Every non empty set of sub-modules has a maximal element.
2 Every ascending chain of submodules is stationary.
3 Every submodule is finitely generated.

Definition 5.11 An A-module satisfying one of those properties is called a "*Noetherian A-module*".

Proof We shorten the proof as it can be found in many textbooks.

1 → 2 There exists a maximal element among the elements of the chain.
2 → 1 If not one could construct inductively an increasing chain of infinite length.
1 → 3 The set of finitely generated submodules of a given module $N \subset M$ is not empty because it contains at least the zero submodule. Thus it has a maximal element, say N_0. If $N_0 \neq N$, consider $N_0 + A \cdot m$ with $m \in N$, $m \notin N_0$. This module is finitely generated and strictly contains N_0, contrary to the hypothesis. Finally $N = N_0$ is finitely generated.
3 → 1 Let $N_1 \subseteq N_2 \subseteq \cdots$ be an increasing chain of submodules of M. Then $N = \bigcup_{r=1}^{\infty} N_r$ is a submodule of M, which is thus

finitely generated, say by $m_1, \ldots, m_\alpha, \ldots, m_a$. If $m_\alpha \in N_{r(\alpha)}$, set $s = \sup_{1 \le \alpha \le a} r(\alpha)$. Then $m_\alpha \in N_s$ and we have the equality $N = N_s$ that shows that the chain is stationary. C.Q.F.D.

PROPOSITION 5.12 If M is a Noetherian A-module, then $M[\chi]$ is a Noetherian $A[\chi]$-module.

Proof We shall first give the proof in the case of one indeterminate χ.

Any element of $M[\chi]$ can be written as $m_r\chi^r + \cdots + m_0$ and its multiplication by χ gives the element $m_r\chi^{r+1} + \cdots + m_0\chi$ of $M[\chi]$.

If N is a submodule of $M[\chi]$, we denote by $I_r(N)$ the set of $m \in M$ such that there exists a polynomial $m\chi^r + m_{r-1}\chi^{r-1} + \cdots + m_0 \in N$. It is easy to see that $I_r(N)$ is a submodule of M and that we have the increasing chain of submodules:

$$I_0(N) \subseteq I_1(N) \subseteq \cdots.$$

Moreover, if $L \subseteq N$ are two submodules of M then

$$I_r(L) \subseteq I_r(N), \qquad \forall r \ge 0.$$

The proof of the proposition follows from the next lemma:

LEMMA 5.13 If $L \subseteq N$ are submodules of $M[\chi]$ and if $I_r(L) = I_r(N)$, $\forall r \ge 0$, then $L = N$.

Proof If $\Phi = m_r\chi^r + m_{r-1}\chi^{r-1} + \cdots + m_0 \in N$, then $m_r \in I_r(N) = I_r(L)$. Then we can find a polynomial $\psi_0 = m_r\chi^r + m'_{r-1}\chi^{r-1} + \cdots + m'_0 \in L$ and $\Phi - \psi_0 \in N$ but is of degree $r - 1$ in χ. Repeating this process in order to reduce the degree, we successively find out $\psi_0, \psi_1, \ldots, \psi_r \in L$ such that $\Phi - \psi_0 - \cdots - \psi_r = 0$ that is to say $\Phi = \psi_0 + \cdots + \psi_r \in L$. C.Q.F.D.

Now let $N_1 \subseteq N_2 \subseteq \cdots$ be an increasing chain of $\mathbb{R}[\chi]$-submodules of $M[\chi]$. We can consider the double indexed set $I_r(N_i)$, with $r \ge 0$ and $i \ge 1$, of A-submodules of M.

As M is Noetherian this set has a maximum element, say $I_s(N_j)$.

- If $r \ge s$, $i \ge j$ then $I_s(N_j) \subseteq I_s(N_i) \subseteq I_r(N_i)$ and it follows that $I_r(N_i) = I_s(N_j)$.
- If $0 \le r < s$, then $I_r(N_1) \subseteq I_r(N_2) \subseteq \cdots$ and there exists $j(r)$ such that $I_r(N_i) = I_r(N_{j(r)})$ $\forall i \ge j(r)$. We set $k = \sup(j, j(0), \ldots, j(s-1))$ and take $i \ge k$.

Then if $0 \le r < s$ we have $I_r(N_i) = I_r(N_{j(r)}) = I_r(N_k)$ and if $s \le r$ we have $I_r(N_i) = I_s(N_i) = I_r(N_k)$. It follows that $I_r(N_i) = I_r(N_k)$, $\forall r \ge 0$.

Finally, as $N_k \subseteq N_i$, from the lemma above we find that $N_k = N_i$, $\forall i \geq k$ and from the proposition 5.10 we deduce that $M[\chi]$ is a Noetherian $A[\chi]$-module.

By induction on the number of indeterminates, we successively find out that $M[\chi_1]$ is a Noetherian $A[\chi_1]$-module, that $M[\chi_1][\chi_2] = M[\chi_1, \chi_2]$ is a Noetherian $A[\chi_1][\chi_2] = A[\chi_1, \chi_2]$-module and finally that $M[\chi_1, \ldots, \chi_n]$ is a Noetherian $A[\chi_1, \ldots, \chi_n]$-module.

This concludes the proof of the proposition 5.12. C.Q.F.D.

COUNTER EXAMPLE 5.14 A polynomial ring with an infinite number of indeterminates is not Noetherian. We can consider a strictly increasing chain such as $(\chi_1), (\chi_1, \chi_2), \ldots$.

We finally come back to the proof of proposition 5.9.

As $\mathbb{R}^m$ is a vector space over $\mathbb{R}$ it is of course a Noetherian $\mathbb{R}$-module. From the last proposition $\mathbb{R}^m[\chi_1, \ldots, \chi_n]$ is a Noetherian $\mathbb{R}[\chi_1, \ldots, \chi_n]$-module.

It follows that the increasing chain $M_q^{(s)} \subseteq M_{q+1}^{(s)} \subseteq \cdots$ is stationary and we call $M^{(s)}$ its maximum element which is a finitely generated submodule of $\mathbb{R}^m[\chi]$.

Now, as $G_{q+r}^{(s)} \supseteq G_{q+r}^{(s+1)}$, we have $M_{q+r}^{(s)} \subseteq M_{q+r}^{(s+1)}$, $\forall s \geq 0$, and the increasing chain $M = M^{(0)} \subseteq M^{(1)} \subseteq \cdots \subseteq M^{(s)} \subseteq \cdots$ is stationary, that is to say, there exists an integer s_0 such that $M^{(s)} = M^{(s_0)}$, $\forall s \geq s_0$.

This implies $G_{q+r}^{(s)} = G_{q+r}^{(s_0)}$, $\forall s \geq s_0$ and $r \geq 0$.

Finally this equality is true $\forall x \in X$, because the $G_{q+r}^{(s)}$ are, by hypothesis, vector bundles over X, $\forall r, s \geq 0$. C.Q.F.D.

CHAPTER 3

1 Spencer cohomology

Let $\mathcal{R}_q \subset J_q(\mathcal{E})$ be a system of order q on $\mathcal{E}$. We define the symbol G_q of $\mathcal{R}_q$ as the family of vector spaces over $\mathcal{R}_q$:

$$G_q \qquad\qquad V(\mathcal{R}_q) \cap (S_q T^* \otimes V(\mathcal{E}))$$

We have the exact sequence over $\mathcal{R}_q$ with $E = V(\mathcal{E})$ and $F_0 = N(\iota_q)$:

$$0 \longrightarrow G_q \longrightarrow S_q T^* \otimes E \xrightarrow{\ \sigma\ } F_0$$

DEFINITION 1.1 We then define the "*r-prolongation*" G_{q+r} of G_q as the family of vector spaces over $\mathcal{R}_q$ that makes commutative and exact the following diagram:

$$\begin{array}{ccccccc}
0 \longrightarrow & G_{q+r} & \longrightarrow & S_{q+r}T^* \otimes E & \xrightarrow{\ \sigma_r\ } & & S_rT^* \otimes F_0 \\
 & & & & \searrow^{\Delta r,\, q} & \nearrow_{\sigma} & \\
 & & & & & S_rT^* \otimes S_qT^* \otimes E &
\end{array}$$

We have already seen how to determine G_{q+r} using local coordinates, and at the same time σ_r with $\sigma_0 = \sigma$.

We shall now look for the intrinsic properties of the families G_{q+r} of vector spaces over $\mathcal{R}_q$.

We strongly advise the reader to keep in mind some easy concrete cases before going ahead through the next formal developments.

DEFINITION 1.2 We call δ the monomorphism:

$$\delta : S_{r+1}T^* \to T^* \otimes S_rT^*$$

which is the composition of the canonical injection

$$S_{r+1}T^* \to \otimes^{r+1} T^* = T^* \otimes (\otimes^r T^*)$$

with the homomorphism

$$T^* \otimes (\otimes^r T^*) \to T^* \otimes S_rT^*$$

induced by the canonical projection $\otimes^r T^* \to S_rT^*$.

Using local coordinates, with $\omega = (\omega_\mu) = (v_\mu)$ an element of $S_{r+1}T^*$, we get:

$$\delta\omega = dx^i\delta_i\omega \quad \text{with} \quad (\delta_i\omega)_\mu = \omega_{\mu+1_i} = v_{\mu+1_i}.$$

We may extend δ to a morphism:

$$\delta : \Lambda^p T^* \otimes S_{r+1}T^* \to \Lambda^{p+1}T^* \otimes S_r T^*$$

by means of the formula:

$$\delta\omega = dx^i \wedge \delta_i\omega \quad \text{with} \quad (\delta_i\omega)_\mu = \omega_{\mu+1_i}, |\mu| = r$$

using the local coordinates

$$\omega_\mu = v_{\mu,I}\, dx^I$$

and considering a section of $\Lambda^p T^* \otimes S_{r+1}T^*$ as a differential p-form with value in $S_{r+1}T^*$. We have introduced the classical notations

$$dx^I = dx^{i_1} \wedge \cdots \wedge dx^{i_p}$$

with

$$I = (i_1, \ldots, i_p) \quad \text{and} \quad i_1 < i_2 < \cdots < i_p.$$

PROPOSITION 1.3 $\delta^2 = \delta \circ \delta = 0$.

Proof
$$\begin{aligned}(\delta^2\omega)_\mu = (\delta(\delta\omega))_\mu &= dx^i \wedge (\delta\omega)_{\mu+1_i}\\ &= dx^i \wedge dx^j \wedge \omega_{\mu+1_i+1_j}\\ &= v_{\mu+1_i+1_j,I}\, dx^i \wedge dx^j \wedge dx^I\\ &= v_{\mu+1_j+1_i,I}\, dx^j \wedge dx^i \wedge dx^I\\ &= -v_{\mu+1_i+1_j,I}\, dx^i \wedge dx^j \wedge dx^I\\ &= 0\end{aligned}$$
C.Q.F.D.

Setting $S_{r-n}T^* = 0$ if $r - n < 0$ we obtain the following sequence:

$$0 \longrightarrow S_r T^* \xrightarrow{\delta} T^* \otimes S_{r-1}T^* \xrightarrow{\delta} \cdots \xrightarrow{\delta} \Lambda^n T^* \otimes S_{r-n}T^* \longrightarrow 0$$

PROPOSITION 1.4 This sequence is exact.

Proof We shall work over a fixed point $x \in X$, choosing local coordinates and using an induction on n.

If $n = 1$, we just need to prove that the sequence:

$$0 \longrightarrow S_{r+1}T^* \xrightarrow{\delta} T^* \otimes S_r T^* \longrightarrow 0$$

is exact $\forall r \geq 0$. This is a trivial consequence of the fact that $(\delta\omega)_\mu = \omega_{\mu+1}\, dx^1$ with $|\mu| = r$.

Let us suppose that we have already proved that the latter sequences were exact in $(n-1)$ variables. We shall prove that they are also exact in n variables.

We can decompose $\omega_\mu = v_{\mu, I}\, dx^I$ into two parts, one of them ω'_μ not containing dx^n and the other one ω''_μ containing dx^n.

According to the study for the case $n = 1$, there exists

$$\tau \in \Lambda^{p-1}T^* \otimes S_{r-p+1}T^* \quad \text{such that} \quad dx^n \wedge \delta_n \tau = \omega''.$$

If we consider

$$\begin{aligned}\bar{\omega} &= \omega - \delta\tau \\ &= \omega - \sum_{i=1}^{n-1} dx^i \wedge \delta_i \tau - dx^n \wedge \delta_n \tau\end{aligned}$$

we have thus

$$\begin{aligned}\bar{\omega}_\mu &= \omega'_\mu - \sum_{i=1}^{n-1} dx^i \wedge (\delta_i \tau)_\mu \\ &= \omega'_\mu - \sum_{i=1}^{n-1} dx^i \wedge \tau_{\mu+1_i}\end{aligned}$$

and this fact proves that $\bar{\omega}_\mu$ does not contain dx^n. Moreover:

$$\delta\bar{\omega} = \delta\omega - \delta^2\tau = 0 \quad \text{and} \quad \delta_n \bar{\omega} = 0.$$

It follows that $\bar{\omega}_\mu = 0$ if $\mu_n \neq 0$ and $\bar{\omega}$, involving only $(n-1)$ variables, is such that $\delta\bar{\omega} = 0$. From the induction hypothesis, $\exists \sigma$ such that $\bar{\omega} = \delta\sigma$ and we finally get

$$\omega = \bar{\omega} + \delta\tau = \delta(\sigma + \tau). \qquad \text{C.Q.F.D.}$$

If now E is a vector bundle over X, similar definitions allow one to extend the map

$$\delta : \Lambda^p T^* \otimes S_{r+1} T^* \to \Lambda^{p+1} T^* \otimes S_r T^*$$

to a map also denoted by

$$\delta : \Lambda^p T^* \otimes S_{r+1} T^* \otimes E \to \Lambda^{p+1} T^* \otimes S_r T^* \otimes E.$$

We set

$$\omega = (\omega^k_\mu) \quad \text{with} \quad \omega^k_\mu = v^k_{\mu, I}\, dx^I, k = 1, \ldots, \dim E$$

and we define

$$(\delta\omega)^k_\mu = dx^i \wedge (\delta_i \omega)^k_\mu = dx^i \wedge \omega^k_{\mu+1_i}.$$

We have the similar result:

PROPOSITION 1.5 The following sequence is exact $\forall r \geq 0$.

$$0 \longrightarrow S_r T^* \otimes E \xrightarrow{\delta} T^* \otimes S_{r-1} T^* \otimes E \xrightarrow{\delta} \cdots \xrightarrow{\delta} \Lambda^n T^* \otimes S_{r-n} T^* \otimes E \longrightarrow 0.$$

From the commutative diagram:

$$\begin{array}{ccc} S_{q+r+1} T^* & \xrightarrow{\Delta r+1,q} & S_{r+1} T^* \otimes S_q T^* \\ \delta \downarrow & & \downarrow \delta \\ T^* \otimes S_{q+r} T^* & \xrightarrow{\Delta r,q} & T^* \otimes S_r T^* \otimes S_q T^* \end{array}$$

we easily deduce the commutative diagram:

$$\begin{array}{ccc} \Lambda^p T^* \otimes S_{q+r+1} T^* & \xrightarrow{\Delta r+1,q} & \Lambda^p T^* \otimes S_{r+1} T^* \otimes S_q T^* \\ \delta \downarrow & & \downarrow \delta \\ \Lambda^{p+1} T^* \otimes S_{q+r} T^* & \xrightarrow{\Delta r,q} & \Lambda^{p+1} T^* \otimes S_r T^* \otimes S_q T^* \end{array}$$

The latter diagram gives rise to the two following ones:

$$\begin{array}{ccc} \Lambda^p T^* \otimes S_{q+r+1} T^* \otimes E & \xrightarrow{\Delta r+1,q} & \Lambda^p T^* \otimes S_{r+1} T^* \otimes S_q T^* \otimes E \\ \delta \downarrow & & \downarrow \delta \\ \Lambda^{p+1} T^* \otimes S_{q+r} T^* \otimes E & \xrightarrow{\Delta r,q} & \Lambda^{p+1} T^* \otimes S_r T^* \otimes S_q T^* \otimes E \end{array}$$

$$\begin{array}{ccc} \Lambda^p T^* \otimes S_{r+1} T^* \otimes S_q T^* \otimes E & \xrightarrow{\sigma} & \Lambda^p T^* \otimes S_{r+1} T^* \otimes F_0 \\ \delta \downarrow & & \downarrow \delta \\ \Lambda^{p+1} T^* \otimes S_r T^* \otimes S_q T^* \otimes E & \xrightarrow{\sigma} & \Lambda^{p+1} T^* \otimes S_r T^* \otimes F_0 \end{array}$$

We invite the reader to specify the different maps involved by means of local coordinates.

Combining the two preceding diagrams we get the commutative and exact diagram of families of vector spaces over $\mathscr{R}_q$:

$$\begin{array}{ccccccc} 0 \to \Lambda^p T^* \otimes G_{q+r+1} & \to & \Lambda^p T^* \otimes S_{q+r+1} T^* \otimes E & \xrightarrow{\sigma_{r+1}} & \Lambda^p T^* \otimes S_{r+1} T^* \otimes F_0 \\ \downarrow \delta & & \downarrow \delta & & \downarrow \delta \\ 0 \to \Lambda^{p+1} T^* \otimes G_{q+r} & \to & \Lambda^{p+1} T^* \otimes S_{q+r} T^* \otimes E & \xrightarrow{\sigma_r} & \Lambda^{p+1} T^* \otimes S_r T^* \otimes F_0 \end{array}$$

The induced map δ on the left is still such that $\delta^2 = 0$ and we may consider the sequence:

$$0 \longrightarrow G_{q+r} \xrightarrow{\delta} T^* \otimes G_{q+r-1} \xrightarrow{\delta} \cdots \xrightarrow{\delta} \Lambda^n T^* \otimes G_{q+r-n} \longrightarrow 0$$

with

$$G_{q+r-p} = \begin{cases} S_{q+r-p} T^* \otimes E & \text{if} \quad r < p \\ 0 & \text{if} \quad q + r - p < 0 \end{cases}$$

DEFINITION 1.6 This sequence is called the "*δ-sequence*".

REMARK 1.7 It is often useful to write it as follows:

$$0 \longrightarrow G_{q+r} \xrightarrow{\delta} T^* \otimes G_{q+r-1} \xrightarrow{\delta} \cdots \xrightarrow{\delta} \Lambda^r T^* \otimes G_q \xrightarrow{\delta} \delta(\Lambda^r T^* \otimes G_q) \longrightarrow 0$$

REMARK 1.8 We recall that $\dim \Lambda^s T^* \otimes_{(x, y, p)} G_{q+r-s}$ may depend on the point $(x, y, p) \in \mathscr{R}_q$ at which the δ-sequence is "*evaluated*".

DEFINITION 1.9 We denote by $H^p_{q+r-p} = H^p_{q+r-p}(G_q)$ the cohomology at $\Lambda^p T^* \otimes G_{q+r-p}$ of the δ-sequence, that is to say the quotient of $\ker \delta \subset \Lambda^p T^* \otimes G_{q+r-p}$ by $\operatorname{im} \delta \subset \Lambda^p T^* \otimes G_{q+r-p}$.

DEFINITION 1.10 G_q is said to be "*s-acyclic*" if

$$H^p_{q+r} = 0 \qquad \forall 0 \leq p \leq s, \qquad \forall r \geq 0.$$

DEFINITION 1.11 G_q is said to be "*involutive*" if

$$H^p_{q+r} = 0 \qquad \forall 0 \leq p \leq n, \qquad \forall r \geq 0$$

that is to say if it is n-acyclic.

REMARK 1.12 G_q is involutive if and only if the δ-sequences written as in the above remark 1.7 are exact.

REMARK 1.13 If G_q is s-acyclic or involutive G_{q+r} is also s-acyclic or involutive. The above definitions are thus consistent because we know that G_{q+r} depends only on G_q.

We have the important proposition:

PROPOSITION 1.14 G_q is always 1-acyclic.

Proof This can be deduced easily from a chase in the following commutative diagram where the three rows and the two left columns are exact. We let the reader do the chase himself. It follows that we always have:

$$H^0_{q+r} = 0, \qquad H^1_{q+r} = 0 \qquad \forall r \geq 0.$$

$$\begin{array}{ccccccc}
 & 0 & & 0 & & 0 \\
 & \downarrow & & \downarrow & & \downarrow \\
0 \to & G_{q+r} & \to & S_{q+r}T^* \otimes E & \xrightarrow{\sigma_r} & S_r T^* \otimes F_0 \\
 & \delta\downarrow & & \delta\downarrow & & \downarrow\delta \\
0 \to & T^* \otimes G_{q+r-1} & \to & T^* \otimes S_{q+r-1}T^* \otimes E & \xrightarrow{\sigma_{r-1}} & T^* \otimes S_{r-1}T^* \otimes F_0 \\
 & \delta\downarrow & & \delta\downarrow & & \downarrow\delta \\
0 \to & \Lambda^2 T^* \otimes G_{q+r-2} & \to & \Lambda^2 T^* \otimes S_{q+r-2}T^* \otimes E & \xrightarrow{\sigma_{r-2}} & \Lambda^2 T^* \otimes S_{r-2}T^* \otimes F_0
\end{array}$$

C.Q.F.D.

As the family G_q is defined over $\mathscr{R}_q$, the family G_{q+r} is also defined over $\mathscr{R}_q$ and we may look for the s-prolongation $(G_{q+r})_{+s}$ of G_{q+r} as a family of vector spaces over $\mathscr{R}_q$.

PROPOSITION 1.15 $(G_{q+r})_{+s} = (G_q)_{+(r+s)} = G_{q+r+s}$

Proof It can be deduced at once, by induction, from the preceding proposition, using the commutative and exact diagram:

$$\begin{array}{ccccccc}
0 & \longrightarrow & G_{q+r+s} & \xrightarrow{\delta} & T^* \otimes G_{q+r+s-1} & \xrightarrow{\delta} & \Lambda^2 T^* \otimes G_{q+r+s-2} \\
 & & \vdots & & \| & & \| \\
0 & \longrightarrow & (G_{q+r})_{+s} & \xrightarrow{\delta} & T^* \otimes (G_{q+r})_{+(s-1)} & \xrightarrow{\delta} & \Lambda^2 T^* \otimes (G_{q+r})_{+(s-2)}
\end{array}$$

and the fact that the proposition is trivial for $s = 0$. C.Q.F.D.

The 2-acyclicity of the symbol G_q is used through the following important proposition.

PROPOSITION 1.16 If G_q is 2-acyclic and if G_{q+1} is a vector bundle over $\mathscr{R}_q$, then G_{q+r} is a vector bundle over $\mathscr{R}_q$, $\forall r \geq 1$.

Proof By definition we have the exact sequence:

$$T^* \otimes G_{q+1} \to \Lambda^2 T^* \otimes G_q \to \Lambda^3 T^* \otimes S_{q-1} T^* \otimes E$$

It follows that we have the exact sequence:

$$0 \longrightarrow \delta(T^* \otimes G_{q+1}) \longrightarrow \Lambda^2 T^* \otimes S_q T^* \otimes E \xrightarrow{\sigma \oplus \delta} \Lambda^2 T^* \otimes F_0 \oplus \Lambda^3 T^* \otimes S_{q-1} T^* \otimes E$$

Thus the function $(x, y, p) \to \dim \delta(T^* \otimes G_{q+1})_{(x,y,p)}$ admits a minimum which is reached at every point $(x, y, p) \in \mathscr{R}_q$ where certain determinants do not cancel.

Now the exact sequence:

$$0 \longrightarrow G_{q+2} \xrightarrow{\delta} T^* \otimes G_{q+1} \xrightarrow{\delta} \delta(T^* \otimes G_{q+1}) \longrightarrow 0$$

gives:

$$\dim (G_{q+2})_{(x,y,p)} + \dim (\delta(T^* \otimes G_{q+1}))_{(x,y,p)} = n \dim (G_{q+1})_{(x,y,p)}$$

But, from the exact sequence:

$$0 \longrightarrow G_{q+2} \longrightarrow S_{q+2} T^* \otimes E \xrightarrow{\sigma_2} S_2 T^* \otimes F_0$$

we deduce that the function:

$$(x, y, p) \to \dim (G_{q+2})_{(x,y,p)}$$

admits a minimum which is reached at every point $(x, y, p) \in \mathscr{R}_q$ where certain determinants do not cancel.

According to the above dimension formula and the fact that G_{q+1} is a vector bundle over $\mathscr{R}_q$, it follows that G_{q+2} and $\delta(T^* \otimes G_{q+1})$ are also vector bundles over $\mathscr{R}_q$. We now need the following technical lemma:

LEMMA 1.17 If

$$E \xrightarrow{\Phi} F \xrightarrow{\Psi} G$$

is an exact sequence of vector bundles over X, then $K = \ker \Phi$ and $Q = \operatorname{coker} \Psi$ are vector bundles over X.

Proof By definition we have the exact sequence:

$$0 \longrightarrow K \longrightarrow E \xrightarrow{\Phi} F \xrightarrow{\Psi} G \longrightarrow Q \longrightarrow 0$$

Counting the dimensions we obtain:

$$\dim K_x + \dim Q_x = \dim E_x - \dim F_x + \dim G_x = \text{cst}$$

The function $x \to \dim K_x + \dim Q_x$ is constant. However the functions $x \to \dim K_x$ and $x \to \dim Q_x$ each have a minimum which is reached at every point $x \in X$ where some determinants do not cancel, as $\dim K_x$ and $\dim Q_x$ depend on the rank of some matrices.

It follows that $\dim K_x = \text{cst}$ and $\dim Q_x = \text{cst}$ and this proves the lemma.

C.Q.F.D.

Coming back to the proof of the last proposition, we have the exact sequences $\forall r \geq 0$:

$$0 \longrightarrow G_{q+r+3} \xrightarrow{\delta} T^* \otimes G_{q+r+2} \xrightarrow{\delta} \Lambda^2 T^* \otimes G_{q+r+1} \longrightarrow \Lambda^3 T^* \otimes S_{q+r} T^* \otimes E$$

According to the last lemma, it follows that G_{q+r+3} is a vector bundle over $\mathscr{R}_q$, whenever G_{q+r+2} and G_{q+r+1} are vector bundles over $\mathscr{R}_q$.

An easy induction on r achieves the proof. C.Q.F.D.

2 Involutive symbols

As the acyclicity or involutiveness of the symbol G_q of a system $\mathscr{R}_q \subset J_q(\mathscr{E})$ gives the key to many formal properties of $\mathscr{R}_q$ itself, it is therefore fundamental to check these properties by means of a finite number of operations only.

However, if the definitions we gave above are nice from a formal point of view, it is not easy to look at them in concrete cases. Nevertheless, we notice that there are not many systems, whose symbols are acyclic but not involutive. The examples stated in this book are among the few such cases known.

We shall indicate a "*test*" for checking the involutiveness of a symbol and will describe it carefully as it is a difficult topic, not very well treated in the literature. We shall detail the computations and give some tricks to help the reader in dealing with concrete problems.

When dealing with families of vector spaces over $\mathscr{R}_q$ defined on a neighbourhood of a given point, we shall adopt local coordinates (x, y, p) on a neighbourhood $\mathscr{U}_q \subset J_q(\mathscr{E})$ of this point, and we shall indicate the effects of a change of coordinates when this becomes necessary.

In the sequel, we choose a fixed point $(x, y, p) \in \mathscr{R}_q$ and consider the neighbourhood $U = \pi(\mathscr{U}_q)$ of x in X.

In order to simplify the notations, we shall not indicate this point and thus write G_{q+r} instead of $(G_{q+r})_{(x,y,p)}$ when the context is clear. For the same reason, an element $(x, y, p; v_\mu^k) \in G_{q+r}$ will be described simply by its local coordinates v_μ^k with $k = 1, \ldots, m$, $|\mu| = q + r$, $r \geq 0$.

DEFINITION 2.1 We define

$$(G_{q+r})^i = \{\omega \in G_{q+r} | \delta_1\omega = 0, \ldots, \delta_i\omega = 0\}$$

with $i = 1, \ldots, n$ and $(G_{q+r})^n = 0$ in a given coordinate system.

DEFINITION 2.2 With respect to the same coordinate system, we call the v_μ^k such that $\mu_1 \neq 0$ "*components of class 1*" and more generally the v_μ^k such that $\mu_1 = 0, \ldots, \mu_{i-1} = 0$, $\mu_i \neq 0$ are "*components of class i*".

It follows that $(G_{q+r})^i$ is formed from the elements of G_{q+r} with zero components of class $1, \ldots, i$.

With respect to the same system of coordinates, we have, for $i = 1, \ldots, n$ the exact sequences:

$$0 \longrightarrow (G_{q+r})^i \longrightarrow (G_{q+r})^{i-1} \xrightarrow{\delta_i} (G_{q+r-1})^{i-1}$$

with

$$(G_{q+r})^0 = G_{q+r} \forall r \geq 0 \quad \text{and} \quad G_{q-1} = S_{q-1}T^* \otimes E$$

For any $r \geq 0$ we have the inclusions:

$$G_{q+r} = (G_{q+r})^0 \supseteq (G_{q+r})^1 \supseteq \cdots \supseteq (G_{q+r})^{n-1} \supseteq (G_{q+r})^n = 0$$

We set:

$$\alpha_{q+r}^i = \dim (G_{q+r})^{i-1} - \dim (G_{q+r})^i$$

and we get

$$\alpha_{q+r}^i \leq \dim (G_{q+r-1})^{i-1}$$

Adding the relation we obtain:

$$\begin{cases} \dim (G_{q+r})^{n-1} = \alpha_{q+r}^n \\ \dim (G_{q+r})^{i-1} = \alpha_{q+r}^i + \cdots + \alpha_{q+r}^n \\ \dim G_{q+r} = \alpha_{q+r}^1 + \cdots + \alpha_{q+r}^n \end{cases}$$

and the important inequality:

$$\alpha_{q+r}^1 + \cdots + \alpha_{q+r}^n \leq \alpha_{q+r-1}^1 + 2 \cdot \alpha_{q+r-1}^2 + \cdots + n \cdot \alpha_{q+r-1}^n$$

We shall now compare G_q and G_{q+1} We have:

$$\dim (G_q)^{i-1} = \alpha_q^i + \cdots + \alpha_q^n$$
$$\dim G_{q+1} \leq \alpha_q^1 + 2 \cdot \alpha_q^2 + \cdots + n \cdot \alpha_q^n$$

IMPORTANT REMARK 2.3 The integers α_{q+r}^{i} have been considered in 1920 by M. Janet when $m = 1$. The integers α_{q+r}^{i} depend both on the point $(x, y, p) \in \mathscr{R}_q$ and on the system of coordinates on $U \subset X$. We can denote them as $\alpha_{q+r}^{i}(x, y, p)$.

The next very useful example has to be kept in mind.

IMPORTANT EXAMPLE 2.4 $n = 5, m = 1, q = 2$

1

$$G_2\begin{cases} v_{45} - v_{13} = 0 \\ v_{35} - v_{12} = 0 \\ v_{33} - v_{24} = 0 \\ v_{25} - v_{11} = 0 \\ v_{23} - v_{14} = 0 \\ v_{22} - v_{13} = 0 \end{cases} \qquad \dim G_2 = 9$$

$$\begin{array}{rl} \{v_{11}, v_{12}, v_{13}, v_{14}, v_{15}\} & \text{class } 1 \\ \{v_{22}, v_{23}, v_{24}, v_{25}\} & \text{class } 2 \\ \{v_{33}, v_{34}, v_{35}\} & \text{class } 3 \\ \{v_{44}, v_{45}\} & \text{class } 4 \\ \{v_{55}\} & \text{class } 5 \end{array}$$

$$\alpha_2^1 = 5, \quad \alpha_2^2 = 1, \quad \alpha_2^3 = 1, \quad \alpha_2^4 = 1, \quad \alpha_2^5 - 1$$

$$13 = \dim G_3 < 5 + 2 \times 1 + 3 \times 1 + 4 \times 1 + 5 \times 1 = 19$$

2 If we effect the permutation

$$\begin{pmatrix} 1 & 2 & 3 & 4 & 5 \\ 1 & 3 & 2 & 4 & 5 \end{pmatrix}$$

we get in the new coordinate system:

$$G_2\begin{cases} v_{45} - v_{12} = 0 \\ v_{35} - v_{11} = 0 \\ v_{34} - v_{22} = 0 \\ v_{33} - v_{12} = 0 \\ v_{25} - v_{13} = 0 \\ v_{23} - v_{14} = 0 \end{cases} \qquad \dim G_2 = 9$$

$$\alpha_2^1 = 5, \quad \alpha_2^2 = 2, \quad \alpha_2^3 = 0, \quad \alpha_2^4 = 1, \quad \alpha_2^5 = 1$$

$$13 = \dim G_3 < 5 + 2 \times 2 + 3 \times 0 + 4 \times 1 + 5 \times 1 = 18$$

3 If we effect the permutation

$$\begin{pmatrix} 1 & 2 & 3 & 4 & 5 \\ 5 & 4 & 3 & 2 & 1 \end{pmatrix}$$

with respect to case **1**, we get in the new coordinate system:

$$G_2 \begin{cases} v_{55} - v_{14} = 0 \\ v_{45} - v_{13} = 0 \\ v_{44} - v_{12} = 0 \\ v_{35} - v_{12} = 0 \\ v_{34} - v_{25} = 0 \\ v_{33} - v_{24} = 0 \end{cases} \qquad \dim G_2 = 9$$

$$\alpha_2^1 = 5 \geq \alpha_2^2 = 4 \geq \alpha_2^3 = 0 \geq \alpha_2^4 = 0 \geq \alpha_2^5 = 0$$

$$13 = \dim G_3 = 5 + 2 \times 4 + 3 \times 0 + 4 \times 0 + 5 \times 0$$

Looking back to that example, we may in the general case consider the following equations.

$$G_q \qquad A_k^{\tau\mu}(x, y, p)v_\mu^k = 0 \qquad |\mu| = q$$

$$(G_q)^i \qquad \begin{cases} A_k^{\tau\mu}(x, y, p)v_\mu^k = 0 \\ v_{1i_2\ldots i_q}^k = 0 \\ \cdots\cdots\cdots\cdots \\ v_{ii_2\ldots i_q}^k = 0 \end{cases}$$

These linear systems for $i = 1, \ldots, n$ make it easy to compute successively $\alpha_q^1, \ldots, \alpha_q^n$ in the given coordinate system for $U \subset X$, the point $(x, y, p) \in \mathcal{R}_q$ being fixed.

We shall now define at the same point $(x, y, p) \in \mathcal{R}_q$ other integers $\bar{\alpha}_q^i$, similar to the α_q^i but independent of the coordinate system on $U \subset X$.

Let a_j^i be n^2 arbitrary constants, that is to say chosen in such a way that they satisfy certain inequalities that we will specify.

Let us consider the following systems of linear equations:

$$\bar{G}_q \equiv G_q \qquad A_k^{\tau\mu}(x, y, p)v_\mu^k = 0 \qquad |\mu| = q$$

$$(\bar{G}_q)^j \qquad \begin{cases} A_k^{\tau\mu}(x, y, p)v_\mu^k = 0 \\ a_1^{i_1}v_{i_1i_2\ldots i_q}^k = 0 \\ \cdots\cdots\cdots\cdots \\ a_j^{i_1}v_{i_1i_2\ldots i_q}^k = 0 \end{cases}$$

where we choose the a_j^i in such a way as to obtain minimum dimension for $(\bar{G}_q)^i$, $i = 1, \ldots, n$. The a_j^i must be such that some determinants do not cancel. For example $\det [a_j^i] \neq 0$.

For such a_j^i we set:

$$\bar{v}^k_{j_1 \dots j_q} = a^{i_1}_{j_1} \dots a^{i_q}_{j_q} v^k_{i_1 \dots i_q}$$

With evident notations we get:

$$\bar{G}_q \qquad B_l^{\tau\nu}(x, y, p)\bar{v}_\nu^l = 0 \qquad |\nu| = q$$

$$(\bar{G}_q)^j \qquad \begin{cases} B_l^{\tau\nu}(x, y, p)\bar{v}_\nu^l = 0 \\ \bar{v}^l_{1 j_2 \dots j_q} = 0 \\ \dots\dots\dots\dots\dots\dots \\ \bar{v}^l_{j j_2 \dots j_q} = 0 \end{cases}$$

DEFINITION 2.5 The numbers $\bar{\alpha}_q^i(x, y, p)$ are called the "*characters*" of the symbol G_q at $(x, y, p) \in \mathscr{R}_q$.

We still have

$$\dim G_{q+r} \leq \bar{\alpha}_q^1 + 2 \cdot \bar{\alpha}_q^1 + \cdots + n \cdot \bar{\alpha}_q^n$$

and

$$\dim G_q = \bar{\alpha}_q^1 + \cdots + \bar{\alpha}_q^n$$

The preceding method allows one to find a coordinate system on U that makes the integers:

$$\alpha_q^1, \alpha_q^1 + \alpha_q^2, \dots, \alpha_q^1 + \cdots + \alpha_q^{n-1}.$$

successively maximum at $(x, y, p) \in \mathscr{R}_q$.

DEFINTION 2.6 Such a coordinate system on U is said to be "*δ-regular*" at $x \in U$.

For $i = 1, \dots, n - 1$ we have thus:

$$\alpha_q^1 + \cdots + \alpha_q^i \leq \bar{\alpha}_q^1 + \cdots + \bar{\alpha}_q^i.$$

DEFINITION 2.7 We shall say that a coordinate system for $\mathscr{U}_q \subset J_q(\mathscr{E})$, neighbourhood of a point $(x, y, p) \in \mathscr{R}_q$, is δ-regular at this point if the corresponding coordinate system for $U = \pi(\mathscr{U}_q)$, neighbourhood of $x \in X$, is δ-regular at x.

In the sequel we shall generally study the symbol G_q of $\mathscr{R}_q$ at a point $(x, y, p) \in \mathscr{R}_q$ by choosing a system of coordinates δ-regular at this point, and denoting by $\alpha_q^i = \alpha_q^i(x, y, p)$ the corresponding characters.

We shall now study the relations between the characters.

PROPOSITION 2.8 We have the inequalities:

$$0 \leq \alpha_q^i \leq m \cdot \binom{q+n-i-1}{n-i}$$

and

$$\alpha_q^1 \geq \alpha_q^2 \geq \cdots \geq \alpha_q^n$$

Proof With respect to any coordinate system, we have the exact sequences:

$$0 \longrightarrow (G_q)^i \longrightarrow (G_q)^{i-1} \xrightarrow{\delta_i} (S_{q-1}T^*)^{i-1} \otimes E$$

and thus the relations:

$$0 \leq \alpha_q^i = \dim (G_q)^{i-1} - \dim (G_q)^i \leq m \dim (S^{q-1}T^*)^{i-1}$$

with

$$\dim (S_{q-1}T^*)^{i-1} = \frac{(q+n-i-1)!}{(q-1)!(n-i)!}$$

These relations are also satisfied by the characters.

Coming back to former notations, we define

$$\operatorname{rank} (G_q)^i = m\binom{q+n-1}{n-1} - \dim (G_q)^i$$

and we consider the new family of vector spaces:

$$(\tilde{G}_q)^i \qquad \left\{\begin{array}{ll} A_k^{\tau\mu}(x, y, p)v_\mu^k & = 0 \\ a_1^{i_1}v_{i_1 i_2 \ldots i_q}^k & = 0 \\ \cdots\cdots\cdots\cdots\cdots & \\ a_{i-1}^{i_1}v_{i_1 i_2 \ldots i_q}^k & = 0 \\ a_{i+1}^{i_1}v_{i_1 i_2 \ldots i_q}^k & = 0 \end{array}\right. \quad \text{(first four lines: } (\bar{G}_q)^{i-1})$$

Elementary linear algebra shows that:

$$\operatorname{rank} (\bar{G}_q)^{i+1} \leq \operatorname{rank} (\bar{G}_q)^{i-1} + [\operatorname{rank} (\bar{G}_q)^i - \operatorname{rank} (\bar{G}_q)^{i-1}] + [\operatorname{rank} (\tilde{G}_q)^i - \operatorname{rank} (\bar{G}_q)^{i-1}]$$

and thus:

$$\operatorname{rank} (\bar{G}_q)^{i+1} - \operatorname{rank} (\bar{G}_q)^i \leq \operatorname{rank} (\tilde{G}_q)^i - \operatorname{rank} (\bar{G}_q)^{i-1}$$

But the a_j^i have been chosen such that the numbers rank $(\bar{G}_q)^i$ are maximum. It follows that:

$$\text{rank}\,(\tilde{G}_q)^i \leq \text{rank}\,(\bar{G}_q)^i$$

and

$$\text{rank}\,(\bar{G}_q)^{i+1} - \text{rank}\,(\bar{G}_q)^i \leq \text{rank}\,(\bar{G}_q)^i - \text{rank}\,(\bar{G}_q)^{i-1}.$$

Finally

$$\dim\,(\bar{G}_q)^i - \dim\,(\bar{G}_q)^{i+1} \leq \dim\,(\bar{G}_q)^{i-1} - \dim\,(\bar{G}_q)^i$$

or

$$\bar{\alpha}_q^{i+1} \leq \bar{\alpha}_q^i \qquad \forall (x, y, p) \in \mathscr{R}_q \qquad \text{C.Q.F.D.}$$

EXAMPLE 2.9 In the case **3** above, the coordinate system is δ-regular because $\alpha_2^1 = 5$, its maximum value and $\alpha_2^1 + \alpha_2^2 = 5 + 4 = 9 = \dim G_2$, its maximum value too. The five integers of case **3** are the characters of G_2.

There is however another way to look at a δ-regular coordinate system which is more useful. We shall now describe it.

When the coordinate system is δ-regular at a point $(x, y, p) \in \mathscr{R}_q$, then the numbers $\alpha_q^1, \alpha_q^1 + \alpha_q^2, \ldots, \alpha_q^1 + \cdots + \alpha_q^{n-1}$ are maximum. But $\dim G_q = \alpha_q^1 + \cdots + \alpha_q^n$ and the numbers $\alpha_q^n, \alpha_q^n + \alpha_q^{n-1}, \ldots, \alpha_q^n + \cdots + \alpha_q^2$ are minimum. Let us define the integers

$$\beta_q^i = m\binom{q + n - i - 1}{n - i} - \alpha_q^i.$$

The δ-regular system makes the numbers $\beta_q^n, \beta_q^n + \beta_q^{n-1}, \ldots, \beta_q^n + \cdots + \beta_q^2$ maximum and $\beta_q^1 + \cdots + \beta_q^n = \text{rank}\, G_q$.

With such a δ-regular coordinate system at a point $(x, y, p) \in \mathscr{R}_q$, we can solve the equations of G_q successively at this point in the following way:

class n The equations are solved with respect to β_q^n different components v_μ^k of class n. We substitute into the other ones, that will contain no components of class n. The equations obtained are called "*equations of class n*".

class i The remaining equations, that contain only components of class $\leq i$, are solved with respect to β_q^i different components of class i. The corresponding equations are called "*equations of class i*". We substitute into the other ones that will contain no components of class i.

class 1 The remaining β_q^1 equations, that contain only components of class 1 are solved with respect to β_q^1 different components of class 1. These equations are called "*equations of class 1*".

We thus obtain, for $i = 1, \ldots, n$ some components of class i, called "*principal*", as linear combinations of the other components of class $\leq i$, called "*parametric*".

DEFINITION 2.10 If v_μ^k, $|\mu| = q$ is a principal component of order q, we call $v_{\mu+\nu}^k$, $|\mu + \nu| = q + r$ a principal component of order $q + r$. The other components of order $q + r$ are called parametric.

Practically, we shall write the equations of G_q of class i, in the following form, using matrix notation:

$$v'^\tau \equiv \boxed{\begin{matrix}\text{principal}\\ \text{component}\\ \text{of class } i\end{matrix}} + A(x, y, p) \cdot \begin{pmatrix}\text{parametric}\\ \text{components}\\ \text{of class } \leq i\end{pmatrix} = 0$$

and we shall say that this equation τ is "*solved*" with respect to the principal corresponding component, which will be framed for clarity.

DEFINITION 2.11 $x^1, \ldots, x^i$ are called "*multiplicative variables*" for any equation of class i and $x^{i+1}, \ldots, x^n$ are called "*non-multiplicative variables*" for the same equation.

IMPORTANT TRICK 2.12 It is very useful to write out all these variables on a board $P(G_q)$, writing only i for x^i when it is a multiplicative variable and using a dot when it is a non-multiplicative variable.

EXAMPLE 2.13 In the above case **3**, it is easy to check that the given equations are in a "*solved form*".

$$G_2\begin{cases} v'^6 \equiv \boxed{v_{55}} - v_{14} = 0 & 1\ 2\ 3\ 4\ 5 \\ v'^5 \equiv \boxed{v_{45}} - v_{13} = 0 & 1\ 2\ 3\ 4\ \cdot \\ v'^4 \equiv \boxed{v_{44}} - v_{12} = 0 & 1\ 2\ 3\ 4\ \cdot \\ v'^3 \equiv \boxed{v_{35}} - v_{12} = 0 & 1\ 2\ 3\ \cdot\ \cdot \\ v'^2 \equiv \boxed{v_{34}} - v_{25} = 0 & 1\ 2\ 3\ \cdot\ \cdot \\ v'^1 \equiv \boxed{v_{33}} - v_{24} = 0 & 1\ 2\ 3\ \cdot\ \cdot \end{cases} \quad P(G_2)$$

REMARK 2.14 If $\mathcal{R}_q$ is defined by local equations, then:

$$\mathcal{R}_q \qquad \Phi^\tau(x, y, p) = 0 \qquad 1 \leq \operatorname{ord} p \leq q$$

$$G_q \qquad \frac{\partial \Phi^\tau}{\partial p_\mu^k}(x, y, p) v_\mu^k = 0 \qquad |\mu| = q$$

Using the implicit function theorem, we can solve some of the equations of $\mathscr{R}_q$ with respect to the p^k_μ with the same indices as the principal v^k_μ. We define the board $P(\mathscr{R}_q)$ by adding as many dotted rows to the board $P(G_q)$ as the number β^0_q of unconnected equations of order $\leq q - 1$.

The following proposition shows the importance of the P-board.

PROPOSITION 2.15 If we prolong each of the solved equations of G_q by using only the corresponding multiplicative variables, we get

$$\beta^1_q + 2 \cdot \beta^2_q + \cdots + n \cdot \beta^n_q$$

independent equations.

Proof The prolonged equations obtained by using x^i contain $v^k_{\mu+1_i}$ instead of v^k_μ. These equations do not contain any component of order $q + 1$ and of class $> i$. They are solved with respect to $\beta^n_q + \cdots + \beta^i_q$ different components of class i and order $q + 1$ and are thus independent between themselves and from the other equations of order $q + 1$ obtained with multiplicative variables other than x^i. C.Q.F.D.

An easy induction, which is left to the reader, shows that we have the formula:

$$\beta^1_q + 2 \cdot \beta^2_q + \cdots + n \cdot \beta^n_q = m\binom{q+n}{n-1} - (\alpha^1_q + 2 \cdot \alpha^2_q + \cdots + n \cdot \alpha^n_q)$$

DEFINITION 2.16 We say that G_q is "*involutive*" at a point $(x, y, p) \in \mathscr{R}_q$ if, at this point, we have the relation:

$$\dim G_{q+1} = \alpha^1_q + 2 \cdot \alpha^2_q + \cdots + n \cdot \alpha^n_q$$

We shall now show the link between this operational definition and the formal definition by means of the δ-sequences.

PROPOSITION 2.17 G_q is involutive at a point $(x, y, p) \in \mathscr{R}_q$ if and only if there exists on a neighbourhood $\mathscr{U}_q \subset J_q(\mathscr{E})$ of this point, a system of coordinates such that we have, at this point, the short exact sequences:

$$0 \longrightarrow (G_{q+1})^i \longrightarrow (G_{q+1})^{i-1} \xrightarrow{\delta_i} (G_q)^{i-1} \longrightarrow 0$$

Proof

N.C. By definition there exists a coordinate system which is δ-regular at $(x, y, p) \in \mathscr{R}_q$ and such that:

$$\dim (G_q)^i = \alpha^{i+1}_q + \cdots + \alpha^n_q$$

Moreover we have the relations:

$$\dim (G_{q+1})^{i-1} - \dim (G_{q+1})^{i} \leq \dim (G_q)^{i-1}$$

Adding these relations, we get:

$$\dim G_{q+1} \leq \alpha_q^1 + 2 \cdot \alpha_q^2 + \cdots + n \cdot \alpha_q^n$$

As we have in fact an equality, we must have:

$$\dim (G_{q+1})^{i-1} - \dim (G_{q+1})^{i} = \dim (G_q)^{i-1}.$$

S.C. A coordinate system which is δ-regular at $(x, y, p) \in \mathscr{R}_q$ is such that, the numbers $\dim (G_q)^i$ are minimum. It follows that

$$\dim (G_q)^i \geq \alpha_q^{i+1} + \cdots + \alpha_q^n$$

and

$$\dim (G_{q+1})^{i-1} - \dim (G_{q+1})^{i} = \dim (G_q)^{i-1}$$

Adding we get

$$\dim G_{q+1} \geq \alpha_q^1 + 2 \cdot \alpha_q^2 + \cdots + n \cdot \alpha_q^n.$$

As the α_q^i are the characters of G_q at $(x, y, p) \in \mathscr{R}_q$, we have also

$$\dim G_{q+1} \leq \alpha_q^1 + 2 \cdot \alpha_q^2 + \cdots + m \cdot \alpha_q^n.$$

Thus

$$\dim G_{q+1} = \alpha_q^1 + 2 \cdot \alpha_q^2 + \cdots + n \cdot \alpha_q^n$$

and

$$\dim (G_q)^i = \alpha_q^{i+1} + \cdots + \alpha_q^n.$$

Finally any coordinate system in which the sequences of the proposition are exact at a point $(x, y, p) \in \mathscr{R}_q$ is δ-regular at this point. C.Q.F.D.

Definition 2.18 G_q is said to be "*involutive*" if it is involutive at any point $(x, y, p) \in \mathscr{R}_q$.

We have

$$\dim (G_{q+1})_{(x,y,p)} = \alpha_q^1(x, y, p) + \cdots + n \cdot \alpha_q^n(x, y, p) \qquad \forall (x, y, p) \in \mathscr{R}_q$$

Remark 2.19 We have to notice that the dimensions of the vector spaces $(G_q)^i_{(x,y,p)}$ can vary on $\mathscr{R}_q$, even if the preceding relation holds.

Lemma 2.20 If E is a vector bundle over X, then $S_q T^* \otimes E$ is involutive $\forall q \geq 1$.

Proof If dim $E = m$, we have

$$\alpha_q^i = m \cdot \binom{q+n-i-1}{n-i}$$

and

$$\dim S_{q+1}T^* \otimes E = m\binom{q+n}{n-1} = \alpha_q^1 + 2 \cdot \alpha_q^2 + \cdots + n \cdot \alpha_q^n$$

It follows that we have, in any coordinate system, the short exact sequences:

$$0 \longrightarrow (S_{q+1}T^* \otimes E)^i \longrightarrow (S_{q+1}T^* \otimes E)^{i-1} \xrightarrow{\delta_i} (S_q T^* \otimes E)^{i-1} \longrightarrow 0$$

EXAMPLE 2.21 $n = 2, m = 1, q = 1, \mathscr{E}$ trivial bundle over $\mathbb{R}^2$.

$\mathscr{R}_1$ $\qquad \Phi(x, y, p) \equiv (p_2)^2 + (p_1)^2 - (x^1 p_1 + x^2 p_2) + y = 0$

$\mathscr{R}_1$ is a subfibered manifold of $J_1(\mathscr{E})$ because

$$\frac{\partial \Phi}{\partial y}(x, y, p) = 1$$

G_1 $\qquad (2p_2 - x^2)v_2 + (2p_1 - x^1)v_1 = 0.$

Let $\mathscr{R}'_1 \subset \mathscr{R}_1$ defined by

$$y - \frac{(x^1)^2 + (x^2)^2}{4} = 0, \quad p_1 - \frac{x^1}{2} = 0, \quad p_2 - \frac{x^2}{2} = 0.$$

We have $\dim G_1|_{\mathscr{R}'_1} = 2$ and G_1 is involutive at any point of $\mathscr{R}'_1$, because

$$G_1|_{\mathscr{R}'_1} \equiv T^* \otimes E (\neq G'_1)$$

Now

$$\dim G_1|_{\mathscr{R}_1 - \mathscr{R}'_1} = 1$$

and it is easy to see that the characters are

$$\alpha_1^1 = 1 \geq \alpha_1^2 = 0,$$

but they were $\alpha_1^1 = 1 \geq \alpha_1^2 = 1$ on $\mathscr{R}'_1$.

The reader will check easily that G_1 is involutive on $\mathscr{R}_1$ though the characters change on $\mathscr{R}_1$.

Coming back to proposition 2.15, we see that, if G_q is involutive, all the prolonged equations obtained by means of the non-multiplicative variables must be linear combinations of the prolonged equations obtained by means of the multiplicative variables. The checking of this property will be called the "*δ-test*".

EXAMPLE 2.22 In the example following the definition of the board $P(G_q)$ we check easily the identities:

$$\begin{aligned}
&v_5'^5 - v_4'^6 + v_1'^3 - v_1'^4 \equiv 0\\
&v_5'^4 - v_4'^5 - v_1'^2 \equiv 0\\
&v_5'^3 - v_3'^6 - v_1'^2 \equiv 0\\
&v_5'^2 - v_3'^5 + v_2'^6 - v_1'^1 \equiv 0\\
&v_5'^1 - v_3'^3 + v_2'^5 \equiv 0\\
&v_4'^3 - v_3'^5 - v_1'^1 \equiv 0\\
&v_4'^2 - v_3'^4 + v_2'^5 \equiv 0\\
&v_4'^1 - v_3'^2 + v_2'^4 - v_2'^3 \equiv 0
\end{aligned}$$

LEMMA 2.23 For $i = 1, \ldots, n$ we have the commutative diagrams:

$$\begin{array}{ccc}
(S_{q+r+1}T^* \otimes E)^{i-1} & \xrightarrow{\sigma_{r+1}} & (S_{r+1}T^* \otimes F_0)^{i-1}\\
\downarrow{\scriptstyle \delta_i} & & \downarrow{\scriptstyle \delta_i}\\
(S_{q+r}T^* \otimes E)^{i-1} & \xrightarrow{\sigma_r} & (S_rT^* \otimes F_0)^{i-1}
\end{array}$$

Proof Using local coordinates, we have:

σ_r $$v'^{\tau}_{\nu} = A^{\tau\mu}_k(x, y, p)v^k_{\mu+\nu} \qquad |\mu| = q, |\nu| = r$$

σ_{r+1} $$v'^{\tau}_{\nu+1_i} = A^{\tau\mu}_k(x, y, p)v^k_{\mu+\nu+1_i}$$

As

$$\mu + (\nu + 1_i) = (\mu + \nu) + 1_i$$

we have

$$\sigma_r \circ \delta_i = \delta_i \circ \sigma_{r+1}.$$

Now we notice that

$$(\mu + \nu)_j = \mu_j + \nu_j > 0 \quad \text{whenever } \nu_j > 0.$$

It follows that any component v'^{τ}_{ν} of class j is expressed as a linear combination of components $v^k_{\mu+\nu}$ of class $\leq j$ and we can restrict the maps σ_r and σ_{r+1} as in the above diagram for $j = 1, \ldots, i - 1$. C.Q.F.D.

We set:

$$(\sigma_r(S_{q+r}T^* \otimes E))^i = \sigma_r(S_{q+r}T^* \otimes E) \cap (S_rT^* \otimes F_0)^i$$

Proposition 2.24 For $r \geq 0$, the maps:

$$\delta_i : (G_{q+r+1})^{i-1} \to (G_{q+r})^{i-1}$$

are surjective if and only if the maps:

$$\sigma_{r+1} : (S_{q+r+1}T^* \otimes E)^{i-1} \to (\sigma_{r+1}(S_{q+r+1}T^* \otimes E))^{i-1}$$

are surjective.

Proof Using lemma 2.20 and the preceding lemma, we have the commutative and exact diagram of families of vector spaces over $\mathscr{R}_q$.

$$\begin{array}{ccccccccc}
 & & 0 & & 0 & & 0 & & \\
 & & \downarrow & & \downarrow & & \downarrow & & \\
0 & \to & (G_{q+r+1})^i & \to & (S_{q+r+1}T^* \otimes E)^i & \xrightarrow{\sigma_{r+1}} & (\sigma_{r+1}((S_{q+r+1}T^* \otimes E)^{i-1}))^i & & \\
 & & \downarrow & & \downarrow & & \downarrow & & \\
0 & \to & (G_{q+r+1})^{i-1} & \to & (S_{q+r+1}T^* \otimes E)^{i-1} & \xrightarrow{\sigma_{r+1}} & \sigma_{r+1}((S_{q+r+1}T^* \otimes E)^{i-1}) & \to & 0 \\
 & & \downarrow \delta_i & & \downarrow \delta_i & & \downarrow \delta_i & & \\
0 & \to & (G_{q+r})^{i-1} & \to & (S_{q+r}T^* \otimes E))^{i-1} & \xrightarrow{\sigma_r} & \sigma_r((S_{q+r}T^* \otimes E)^{i-1}) & \to & 0 \\
 & & & & \downarrow & & \downarrow & & \\
 & & & & 0 & & 0 & &
\end{array}$$

N.C. If the maps

$$\delta_i : (G_{q+r+1})^{i-1} \to (G_{q+r})^{i-1}$$

are surjective a diagram chasing shows that:

$$\begin{aligned}
\sigma_{r+1}((S_{q+r+1}T^* \otimes E)^i) &= (\sigma_{r+1}((S_{q+r+1}T^* \otimes E)^{i-1}))^i \\
&= ((\sigma_{r+1}((S_{q+r+1}T^* \otimes E)^{i-2}))^{i-1})^i \\
&= (\sigma_{r+1}((S_{q+r+1}T^* \otimes E)^{i-2}))^i \\
&= \cdots\cdots\cdots\cdots\cdots\cdots\cdots\cdots \\
&= (\sigma_{r+1}(S_{q+r+1}T^* \otimes E))^i
\end{aligned}$$

S.C. If the maps

$$\sigma_{r+1} : (S_{q+r+1}T^* \otimes E)^{i-1} \to (\sigma_{r+1}(S_{q+r+1}T^* \otimes E))^{i-1}$$

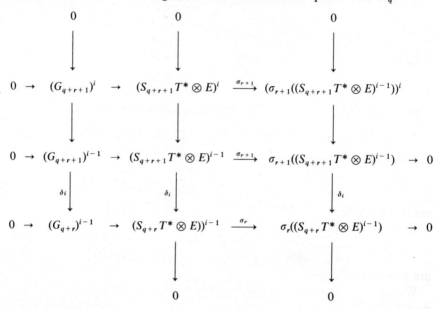

are surjective, using the diagram we obtain:

$$\begin{aligned}\sigma_{r+1}((S_{q+r+1}T^* \otimes E)^i) &\subset (\sigma_{r+1}((S_{q+r+1}T^* \otimes E)^{i-1}))^i \\ &\subset (\sigma_{r+1}(S_{q+r+1}T^* \otimes E))^i \\ &= \sigma_{r+1}((S_{q+r+1}T^* \otimes E)^i)\end{aligned}$$

and the surjectivity of the maps

$$\delta : (G_{q+r+1})^{i-1} \to (G_{q+r})^{i-1}$$

can be deduced at once from an easy chase. C.Q.F.D.

REMARK 2.25 When $r = 0$, the preceding proposition brings out the δ-test.

We shall use the last proposition in order to get the next theorem.

THEOREM 2.26 If G_q is involutive, then G_{q+r} is involutive $\forall r \geq 0$.

Proof According to proposition 2.17 it suffices to prove that the maps

$$\delta_i : (G_{q+r+1})^{i-1} \to (G_{q+r})^{i-1}$$

are surjective when the maps

$$\delta_i : (G_{q+1})^{i-1} \to (G_q)^{i-1}$$

are surjective.

Using the last proposition it is equivalent to show that the maps

$$\sigma_{r+1} : (S_{q+r+1}T^* \otimes E)^{i-1} \to (\sigma_{r+1}(S_{q+r+1}T^* \otimes E))^{i-1}$$

are surjective for $i = 1, \ldots, n$.

We shall now prove, using an induction on i, that, if the former maps σ_r are surjective for $i = 1, \ldots, n$, then the maps σ_{r+1} are also surjective for $i = 1, \ldots, n$.

This fact will simply follow from a chase in the commutative and exact diagram:

$$\begin{array}{ccccccccc}
0 \longrightarrow & (S_{q+r+1}T^* \otimes E)^i & \longrightarrow & (S_{q+r+1}T^* \otimes E)^{i-1} & \xrightarrow{\delta_i} & (S_{q+r}T^* \otimes E)^{i-1} & \longrightarrow 0 \\
& \downarrow \sigma_{r+1} & & \downarrow \sigma_{r+1} & & \downarrow \sigma_r & \\
0 \longrightarrow & (\sigma_{r+1}(S_{q+r+1}T^* \otimes E))^i & \longrightarrow & (\sigma_{r+1}(S_{q+r+1}T^* \otimes E))^{i-1} & \xrightarrow{\delta_i} & (\sigma_r(S_{q+r}T^* \otimes E))^{i-1} & \longrightarrow 0 \\
& & & & & \downarrow & \\
& & & & & 0 &
\end{array}$$

We have just in fact to start the induction, taking $i = n$ and to notice that the left column is made of zeros in that case. C.Q.F.D.

COROLLARY 2.27 If G_q is involutive, we have the exact sequences of families of vector spaces over $\mathcal{R}_q$.

$$0 \to G_{q+r} \xrightarrow{\delta} T^* \otimes G_{q+r-1} \xrightarrow{\delta} \cdots \xrightarrow{\delta} \Lambda^r T^* \otimes G_q \xrightarrow{\delta} \delta(\Lambda^r T^* \otimes G_q) \to 0$$

and

$$H^p_{q+r} = 0 \qquad \forall 0 \le p \le n, \forall r \ge 0.$$

Proof According to the preceding theorem, the maps:

$$\delta_i : (G_{q+r+1})^{i-1} \to (G_{q+r})^{i-1}$$

are surjective for $i = 1, \ldots, n$. The corollary can be deduced from this fact, using an induction on n similar to that of proposition 1.4. We leave the details to the reader. C.Q.F.D.

REMARK 2.28 The operational and formal definitions of involutiveness are thus coherent.

THEOREM 2.29 If G_q is involutive at $(x_0, y_0, p_0) \in \mathcal{R}_q$ and if G_{q+1} is a vector bundle over $\mathcal{R}_q$, then G_q is also involutive on a sufficiently small neighbourhood of (x_0, y_0, p_0) in $\mathcal{R}_q$.

Proof As G_q is involutive at $(x_0, y_0, p_0) \in \mathcal{R}_q$, we can choose on a convenient neighbourhood of this point in $J_q(\mathscr{E})$ a system of coordinates which is δ-regular at this point and, for $i = 1, \ldots, n$ we have the exact sequences:

$$0 \longrightarrow (G_q)^{i-1} \longrightarrow (S_q T^* \otimes E)^{i-1} \xrightarrow{\sigma} F_0$$

Taking into account the different determinants involved in the study of the rank of σ, we shall have for any $(x, y, p) \in \mathcal{R}_q$ sufficiently close to $(x_0, y_0, p_0) \in \mathcal{R}_q$

$$\dim (G_q)^{i-1}_{(x,y,p)} \le \dim (G_q)^{i-1}_{(x_0,y_0,p_0)}$$

or

$$\alpha^i_q(x, y, p) + \cdots + \alpha^n_q(x, y, p) \le \alpha^i_q(x_0, y_0, p_0) + \cdots + \alpha^n_q(x_0, y_0, p_0)$$

Adding these inequalities we get:

$$\alpha^1_q(x, y, p) + \cdots + n \cdot \alpha^n_q(x, y, p) \le \alpha^1_q(x_0, y_0, p_0) + \cdots + n \cdot \alpha^n_q(x_0, y_0, p_0)$$

But we know that:

$$\dim G_{q+1} \leq \alpha_q^1 + 2\alpha_q^2 + \cdots + n \cdot \alpha_q^n \qquad \forall (x, y, p) \in \mathcal{R}_q$$

and, as G_{q+1} is a vector bundle over $\mathcal{R}_q$, we have

$$\dim (G_{q+1})_{(x,y,p)} = \dim (G_{q+1})_{(x_0,y_0,p_0)}$$

Moreover, as G_q is involutive at (x_0, y_0, p_0), we have:

$$\dim (G_{q+1})_{(x_0,y_0,p_0)} = \alpha_q^1(x_0, y_0, p_0) + \cdots + n \cdot \alpha_q^n(x_0, y_0, p_0)$$

It follows that

$$\dim (G_{q+1})_{(x,y,p)} = \alpha_q^1(x, y, p) + \cdots + n \cdot \alpha_q^n(x, y, p)$$

G_q is thus involutive at $(x, y, p) \in \mathcal{R}_q$ and the system of coordinates is also δ-regular at this point. Finally:

$$\alpha_q^i(x, y, p) = \alpha_q^i(x_0, y_0, p_0) \qquad \text{C.Q.F.D.}$$

COROLLARY 2.30 If G_q is involutive and if G_{q+1} is a vector bundle over $\mathcal{R}_q$, then G_{q+r} is a vector bundle over $\mathcal{R}_q$, $\forall r \geq 0$.

Proof A direct and formal proof has already been given in proposition 1.16. The proof we present here will allow us to compute the numbers $\dim G_{q+r}$ by means of the characters.

First of all we have the inequalities:

$$\binom{q+n-i}{n-i} \geq \alpha_q^i(x, y, p) + \cdots + \alpha_q^n(x, y, p)$$
$$\geq \bar{\alpha}_q^i(x, y, p) + \cdots + \bar{\alpha}_q^n(x, y, p) \geq 0.$$

Let $(x_0, y_0, p_0) \in \mathcal{R}_q$ be a point which is a minimum of the function

$$(x, y, p) \to \bar{\alpha}_q^i(x, y, p) + \cdots + \bar{\alpha}_q^n(x, y, p).$$

As G_q is involutive, we can find a system of coordinates δ-regular at this point and we have:

$$\alpha_q^i(x_0, y_0, p_0) + \cdots + \alpha_q^n(x_0, y_0, p_0) = \bar{\alpha}_q^i(x_0, y_0, p_0) + \cdots + \bar{\alpha}_q^n(x_0, y_0, p_0)$$

Moreover

$$\dim (G_q)^{i-1}_{(x,y,p)} = \dim (G_q)^{i-1}_{(x_0,y_0,p_0)}$$

at any point $(x, y, p) \in \mathcal{R}_q$ such that certain determinants do not cancel. As we have:

$$\bar{\alpha}_q^i(x, y, p) + \cdots + \bar{\alpha}_q^n(x, y, p) \geq \bar{\alpha}_q^i(x_0, y_0, p_0) + \cdots + \bar{\alpha}_q^n(x_0, y_0, p_0)$$

the function

$$(x, y, p) \to \bar{\alpha}_q^i(x, y, p) + \cdots + \bar{\alpha}_q^n(x, y, p)$$

reaches its minimum value

$$\bar{\alpha}_q^i(x_0, y_0, p_0) + \cdots + \bar{\alpha}_q^n(x_0, y_0, p_0)$$

at any point of $\mathscr{R}_q$ where certain determinants do not cancel.

Taking $1 \leq i \leq n$, we can find a point $(x_0, y_0, p_0) \in \mathscr{R}_q$ that minimises separately all the sums

$$\bar{\alpha}_q^i(x, y, p) + \cdots + \bar{\alpha}_q^n(x, y, p)$$

and we shall adopt a system of coordinates δ-regular at this point. Then

$$\bar{\alpha}_q^i(x, y, p) + \cdots + \bar{\alpha}_q^n(x, y, p) \geq \bar{\alpha}_q^i(x_0, y_0, p_0) + \cdots + \bar{\alpha}_q^n(x_0, y_0, p_0).$$

Adding these inequalities and using the fact that G_q is involutive we get:

$$\dim (G_{q+1})_{(x,y,p)} = \bar{\alpha}_q^1(x, y, p) + \cdots + \bar{\alpha}_q^n(x, y, p) \geq \bar{\alpha}_q^1(x_0, y_0, p_0) + \cdots + \bar{\alpha}_q^n(x_0, y_0, p_0) = \dim (G_{q+1})_{(x_0, y_0, p_0)}$$

As G_{q+1} is a vector bundle over $\mathscr{R}_q$, we have:

$$\dim (G_{q+1})_{(x,y,p)} = \dim (G_{q+1})_{(x_0, y_0, p_0)} \qquad \forall (x, y, p) \in \mathscr{R}_q$$

It follows that

$$\bar{\alpha}_q^1(x, y, p) + \cdots + \bar{\alpha}_q^n(x_0, y_0, p_0) = \bar{\alpha}_q^1(x_0, y_0, p_0) + \cdots + \bar{\alpha}_q^n(x_0, y_0, p_0)$$

and

$$\bar{\alpha}_q^i(x, y, p) = \bar{\alpha}_q^i(x_0, y_0, p_0) = \alpha_q^i(x_0, y_0, p_0) = \alpha_q^i$$

Combining proposition 2.16 and theorem 2.26, we have:

$$\alpha_{q+r+1}^i = \alpha_{q+r}^i + \cdots + \alpha_{q+r}^n \qquad \forall r \geq 0$$

and we obtain the useful formula:

$$\dim G_{q+r} = \alpha_q^1 + \cdots + \binom{r+i-1}{i-1}\alpha_q^i + \cdots + \binom{r+n-1}{n-1}\alpha_q^n$$

The number $\dim G_{q+r}$ is expressed by a polynomial in r of degree $\leq (n-1)$ and G_{q+r} is of course a vector bundle over $\mathscr{R}_q$, $\forall r \geq 0$. C.Q.F.D.

As we have seen in the last chapter that it is convenient to deal with involutive symbols, it would be very useful for us to show that G_{q+r} becomes involutive when r is big enough. In order to prove this fact, we now need a digression that will show how to transform the study of $\mathscr{R}_q$ into that of a first order system.

3 Lowered systems

Let $\mathscr{R}_q \subset J_q(\mathscr{E})$ be a system of order q on $\mathscr{E}$ and $\hat{\mathscr{R}}_1$ the image of $\mathscr{R}_q$ by the monomorphism

$$\rho_1(\mathrm{id}_{J_{q-1}(\mathscr{E})}) : J_q(\mathscr{E}) \to J_1(J_{q-1}(\mathscr{E})).$$

We set $\hat{\mathscr{E}} = J_{q-1}(\mathscr{E})$.

THEOREM 3.1 $\hat{\mathscr{R}}_1 \subset J_1(\hat{\mathscr{E}})$ is a system of order 1 on $\hat{\mathscr{E}}$ and $j_{q-1} : \mathscr{E} \to \hat{\mathscr{E}}$ establishes a one to one and onto correspondence between the set of solutions of $\mathscr{R}_q$ and that of $\hat{\mathscr{R}}_1$. Moreover, if $\mathscr{R}_{q+r}$ is identified with its image by the monomorphism

$$\rho_{r+1}(\mathrm{id}_{\hat{\mathscr{E}}}) : J_{q+r}(\mathscr{E}) \to J_{r+1}(\hat{\mathscr{E}})$$

then we have the identifications:

$$\mathscr{R}_{q+r} = \hat{\mathscr{R}}_{r+1}, G_{q+r} = \hat{G}_{r+1} \qquad \forall r \geq 0$$

Proof We have

$$\mathscr{R}_q \subset J_q(\mathscr{E}) \subset J_1(J_{q-1}(\mathscr{E})) = J_1(\hat{\mathscr{E}}).$$

1 As $\mathscr{R}_q$ is a fibered submanifold of $J_q(\mathscr{E})$ and $J_q(\mathscr{E})$ a fibered submanifold of $J_1(J_{q-1}(\mathscr{E}))$, then, $\mathscr{R}_q$ is also a fibered submanifold of $J_1(J_{q-1}(\mathscr{E}))$ and $\hat{\mathscr{R}}_1 \subset J_1(\hat{\mathscr{E}})$ is a system of order 1 on $\hat{\mathscr{E}}$.

Using local coordinates, we have:

$$\mathscr{R}_q \qquad \Phi^\tau(x, y, p) = 0 \qquad 1 \leq \operatorname{ord} p \leq q$$

Now we define

$$\rho_1(\mathrm{id}_{\hat{\mathscr{E}}}) : J_q(\mathscr{E}) \to J_1(\hat{\mathscr{E}})$$

by:

$$\hat{y}^k_\mu = y^k_\mu, \hat{p}^k_{\mu,i} = p^k_{\mu+1_i} \qquad 0 \leq |\mu| \leq q-1$$

setting $\hat{y}^k_0 = y^k_0 = y^k$, and we have:

$$\hat{\mathscr{R}}_1 \qquad \begin{cases} \hat{\Phi}^\tau(x, \hat{y}^k_\mu, \hat{p}^k_{\mu,i}) = 0 & 0 \leq |\mu| \leq q-1 \\ \hat{p}^k_{\mu,i} - \hat{y}^k_{\mu+1_i} = 0 & 0 \leq |\mu| \leq q-2 \\ \hat{p}^k_{\mu+1_{i,j}} - \hat{p}^k_{\mu+1_{j,i}} = 0 & |\mu| = q-2 \end{cases}$$

Now let f be a section of $\mathscr{E}$ solution of $\mathscr{R}_q$. Then $\hat{f} = j_{q-1}(f)$ is a section of $\hat{\mathscr{E}}$, but we have

$$j_1(j_{q-1}(f)) \equiv j_1(\hat{f}) \equiv \rho_1(id_{\hat{\mathscr{E}}}) \circ j_q(f)$$

and the relations

$$\partial_i(\partial_\mu f^k(x)) \equiv \partial_{\mu+1_i} f^k(x)$$

because

$$f : (x) \to (x, f^k(x))$$

is a section of $\mathscr{E}$. Moreover, by hypothesis,

$$\Phi^\tau(x, f^k(x), \partial_\mu f^k(x)) \equiv 0, \qquad \forall x \in U \subset X.$$

It follows that $\hat{f}$ is a solution of $\hat{\mathscr{R}}_1$ because we have:

$$\hat{\Phi}^\tau(x, y^k_\mu, p^k_{\mu+1_i}) \equiv \Phi^\tau(x, y, p) \quad \text{with } 0 \leq |\mu| \leq q - 1.$$

Conversely, let

$$f_{q-1} : (x) \to (x, f^k(x), f^k_\mu(x)) \quad \text{with } 1 \leq |\mu| \leq q - 1$$

or

$$\hat{f} : (x) \to (x, \hat{f}^k_\mu(x)) \quad \text{with } 0 \leq |\mu| \leq q - 1$$

be a section of $\hat{\mathscr{E}}$, solution of $\hat{\mathscr{R}}_1$. We have

$$\partial_i \hat{f}^k_\mu(x) = \hat{f}^k_{\mu+1_i}(x) \equiv 0 \quad \text{for } 0 \leq |\mu| \leq q - 2.$$

Thus

$$\hat{f}^k_\mu(x) = \partial_\mu \hat{f}^k(x) = \partial_\mu f^k(x) \quad \text{for } 0 \leq |\mu| \leq q - 1$$

and there exists a section f of $\mathscr{E}$ such that $\hat{f} = j_{q-1}(f)$. Finally we have $\Phi^\tau(x, f^k(x), f^k_\mu(x)) \equiv 0$ with $1 \leq |\mu| \leq q$ and f is a solution of $\mathscr{R}_q$.

2 By definition we have:

$$\mathscr{R}_{q+r} = J_r(\mathscr{R}_q) \cap J_{q+r}(\mathscr{E})$$
$$\hat{\mathscr{R}}_{r+1} = J_r(\hat{\mathscr{R}}_1) \cap J_{r+1}(\hat{\mathscr{E}})$$

and

$$\hat{\mathscr{R}}_{r+1} = J_r(\mathscr{R}_q) \cap J_{r+1}(J_{q-1}(\mathscr{E}))$$

According to proposition 1.9.13, we have:

$$J_{q+r}(\mathscr{E}) = J_r(J_q(\mathscr{E})) \cap J_{r+1}(J_{q-1}(\mathscr{E}))$$

It follows that:

$$\begin{aligned}\hat{\mathscr{R}}_{r+1} &= J_r(\mathscr{R}_q) \cap J_r(J_q(\mathscr{E})) \cap J_{r+1}(J_{q-1}(\mathscr{E}))\\ &= J_r(\mathscr{R}_q) \cap J_{q+r}(\mathscr{E}) = \mathscr{R}_{q+r}\end{aligned}$$

3 The last identification is more difficult to establish. First of all we need the following technical lemma:

LEMMA 3.2 We have the commutative diagram over $J_{q+r}(\mathscr{E})$:

$$\begin{array}{ccccc} S_{q+r}T^* \otimes E & \xrightarrow{\Delta r+1,\, q-1} & S_{r+1}T^* \otimes S_{q-1}T^* \otimes E & \xrightarrow{\varepsilon_{q-1}} & S_{r+1}T^* \otimes \hat{E} \\ \Big\downarrow{\scriptstyle \varepsilon_{q+r}} & & & & \Big\downarrow{\scriptstyle \varepsilon_{r+1}} \\ V(J_{q+r}(\mathscr{E})) & & \xrightarrow{V(\rho_{r+1}(\mathrm{id}_{\hat{E}}))} & & V(J_{r+1}(\hat{\mathscr{E}}))|_{J_{q+r}(\mathscr{E})} \end{array}$$

Proof Using proposition 2.1.16 we have the commutative diagram:

$$\begin{array}{ccc} J_{q+r}(\mathscr{E}) & \xrightarrow{\rho_{r+1}(\mathrm{id}_{\hat{E}})} & J_{r+1}(\hat{\mathscr{E}}) \\ \Big\downarrow{\scriptstyle \pi^{q+r}_{q+r-1}} & & \Big\downarrow{\scriptstyle \pi^{r+1}_{r}} \\ J_{q+r-1}(\mathscr{E}) & \xrightarrow{\rho_r(\mathrm{id}_{\hat{E}})} & J_r(\hat{\mathscr{E}}) \end{array}$$

and thus the exact sequence of vector bundles over $J_{q+r}(\mathscr{E})$

$$0 \rightarrow S_{q+r}T^* \otimes E \rightarrow S_{r+1}T^* \otimes \hat{E}$$

Local coordinates for the first bundle are $v^k_{\mu+\nu}$ with $|\mu + \nu| = q + r$, and for the second $\bar{v}^k_{\mu,\nu}$ with $|\mu| = q - 1, 0 \leq |\nu| \leq r + 1$. The reader will prove the lemma for himself as an exercise, using the map in the upper part of the diagram:

$$\bar{v}^k_{\mu,\nu} = v^k_{\mu+\nu}.$$

C.Q.F.D.

REMARK 3.3 In the upper row of the diagram, we have only to introduce the transition laws of $J_{q-1}(\mathscr{E})$ and we can pull back this exact sequence over $\mathscr{R}_q = \hat{\mathscr{R}}_1$ and obtain the composition:

$$S_{q+r}T^* \otimes E \xrightarrow{\Delta r+1,\, q-1} S_{r+1}T^* \otimes S_{q-1}T^* \otimes E \xrightarrow{\varepsilon_{q-1}} S_{r+1}T^* \otimes \hat{E}$$

We shall use the above diagram first of all for $r = 0$ and obtain the following two commutative and exact diagrams:

$$
\begin{array}{ccccccccc}
 & & 0 & & 0 & & 0 & & \\
 & & \downarrow & & \downarrow & & \downarrow & & \\
0 & \longrightarrow & V(\mathscr{R}_q) & \longrightarrow & V(J_q(\mathscr{E}))|_{\mathscr{R}_q} & \longrightarrow & N(\iota_q) & \longrightarrow & 0 \\
 & & \| & & \downarrow & & \downarrow & & \\
0 & \longrightarrow & V(\hat{\mathscr{R}}_1) & \longrightarrow & V(J_1(\hat{\mathscr{E}}))|_{\hat{\mathscr{R}}_1} & \longrightarrow & N(\hat{\iota}_1) & \longrightarrow & 0 \\
 & & \downarrow & & \downarrow & & \downarrow & & \\
 & & 0 & \longrightarrow & A & = & B & \longrightarrow & 0 \\
 & & & & \downarrow & & \downarrow & & \\
 & & & & 0 & & 0 & &
\end{array}
$$

$$
\begin{array}{ccccccccc}
 & & 0 & & 0 & & 0 & & \\
 & & \downarrow & & \downarrow & & \downarrow & & \\
0 & \longrightarrow & S_qT^* \otimes E & \longrightarrow & T^* \otimes \hat{E} & \longrightarrow & C & \longrightarrow & 0 \\
 & & \downarrow{\scriptstyle \varepsilon_q} & & \downarrow{\scriptstyle \varepsilon_1} & & \| & & \\
0 & \longrightarrow & V(J_q(\mathscr{E}))|_{\mathscr{R}_q} & \longrightarrow & V(J_1(\hat{\mathscr{E}}))|_{\hat{\mathscr{R}}_1} & \longrightarrow & A & \longrightarrow & 0 \\
 & & \downarrow{\scriptstyle V(\pi^q_{q-1})} & & \downarrow{\scriptstyle V(\pi^1_0)} & & \downarrow & & \\
0 & \longrightarrow & (\pi^q_{q-1})^{-1}V(J_{q-1}(\mathscr{E}))|_{\mathscr{R}_q} & = & (\pi^1_0)^{-1}V(\hat{\mathscr{E}})|_{\hat{\mathscr{R}}_1} & \longrightarrow & 0 & & \\
 & & \downarrow & & \downarrow & & & & \\
 & & 0 & & 0 & & & &
\end{array}
$$

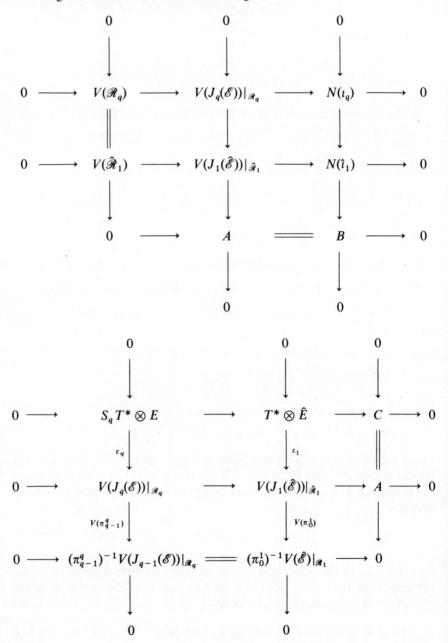

Finally, using the isomorphisms $A \approx B \approx C$ we obtain the commutative and exact diagram:

$$
\begin{array}{ccccccccc}
 & & 0 & & 0 & & 0 & & \\
 & & \downarrow & & \downarrow & & \downarrow & & \\
0 & \longrightarrow & G_q & \longrightarrow & S_qT^* \otimes E & \xrightarrow{\sigma} & F_0 & & \\
 & & \Vert & & \downarrow & & \downarrow & & \\
0 & \longrightarrow & \hat{G}_1 & \longrightarrow & T^* \otimes \hat{E} & \xrightarrow{\hat{\sigma}} & \hat{F}_0 & & \\
 & & \downarrow & & \downarrow & & \downarrow & & \\
 & & 0 & \longrightarrow & C & = & B & \longrightarrow & 0 \\
 & & & & \downarrow & & \downarrow & & \\
 & & & & 0 & & 0 & &
\end{array}
$$

We thus obtain our first identification $G_q = \hat{G}_1$ as families of vector spaces over $\mathscr{R}_q = \hat{\mathscr{R}}_1$.

Now by definition G_{q+r} is the kernel of the composition:

$$S_{q+r}T^* \otimes E \xrightarrow{\Delta r,q} S_rT^* \otimes S_qT^* \otimes E \xrightarrow{\sigma} S_rT^* \otimes F_0$$

and $\hat{G}_{r+1}$ is the kernel of the composition:

$$S_{r+1}T^* \otimes \hat{E} \xrightarrow{\Delta r,1} S_rT^* \otimes T^* \otimes \hat{E} \xrightarrow{\hat{\sigma}} S_rT^* \otimes \hat{F}_0$$

We have the commutative diagram:

$$
\begin{array}{ccc}
S_{q+r}T^* \otimes E & \xrightarrow{\Delta r,q} & S_rT^* \otimes S_qT^* \otimes E \\
{\scriptstyle \Delta r+1,q-1}\downarrow & & \downarrow{\scriptstyle \delta} \\
S_{r+1}T^* \otimes S_{q-1}T^* \otimes E & \xrightarrow{\delta} & S_rT^* \otimes T^* \otimes S_{q-1}T^* \otimes E
\end{array}
$$

$$
\begin{array}{ccc}
v^k_{\mu+\nu+1_i} & \longrightarrow & v^k_{\nu,\mu+1_i} \\
\downarrow & & \downarrow \\
v^k_{\nu+1_i,\mu} & \longrightarrow & v^k_{\nu,i,\mu}
\end{array}
$$

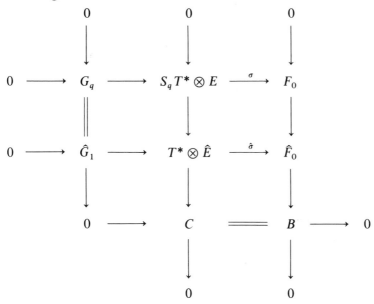

and the commutative diagram:

$$\begin{array}{ccc}
S_{r+1}T^* \otimes S_{q-1}T^* \otimes E & \xrightarrow{\delta} & S_rT^* \otimes T^* \otimes S_{q-1}T^* \otimes E \\
\Big\downarrow{\scriptstyle \varepsilon_{q-1}} & & \Big\downarrow{\scriptstyle \varepsilon_{q-1}} \\
S_{r+1}T^* \otimes \hat{E} & \xrightarrow{\delta} & S_rT^* \otimes T^* \otimes \hat{E}
\end{array}$$

from which we easily deduce the commutative diagram:

$$\begin{array}{ccccc}
 & & 0 & & 0 \\
 & & \big\downarrow & & \big\downarrow \\
0 & \longrightarrow & S_{q+r}T^* \otimes E & \xrightarrow{\Delta r, q} & S_rT^* \otimes S_qT^* \otimes E \\
 & & \big\downarrow & & \big\downarrow \\
0 & \longrightarrow & S_{r+1}T^* \otimes \hat{E} & \xrightarrow{\Delta r, 1} & S_rT^* \otimes T^* \otimes \hat{E} \\
 & & \big\downarrow & & \big\downarrow \\
0 & \longrightarrow & Q & \longrightarrow & S_rT^* \otimes C \\
 & & \big\downarrow & & \big\downarrow \\
 & & 0 & & 0
\end{array}$$

This diagram is also easily seen to be exact, because

$$J_{q+r}(\mathscr{E}) = J_r(J_q(\mathscr{E})) \cap J_{r+1}(J_{q-1}(\mathscr{E})) \subset J_r(J_1(J_{q-1}(\mathscr{E})))$$

Using the commutative diagram:

$$\begin{array}{ccc}
S_qT^* \otimes E & \xrightarrow{\sigma} & F_0 \\
\big\downarrow & & \big\downarrow \\
T^* \otimes \hat{E} & \xrightarrow{\hat{\sigma}} & \hat{F}_0
\end{array}$$

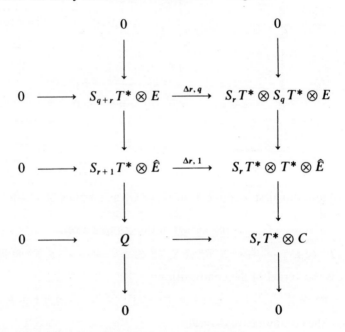

we obtain the commutative and exact diagram:

$$
\begin{array}{ccccccc}
 & & 0 & & 0 & & \\
 & & \downarrow & & \downarrow & & \\
 & & S_r T^* \otimes S_q T^* \otimes E & \xrightarrow{\sigma} & S_r T^* \otimes F_0 & & \\
 & & \downarrow & & \downarrow & & \\
 & & S_r T^* \otimes T^* \otimes \hat{E} & \xrightarrow{\hat{\sigma}} & S_r T^* \otimes \hat{F}_0 & & \\
 & & \downarrow & & \downarrow & & \\
0 & \longrightarrow & S_r T^* \otimes C & = & S_r T^* \otimes B & \longrightarrow & 0 \\
 & & \downarrow & & \downarrow & & \\
 & & 0 & & 0 & &
\end{array}
$$

Finally, we have the commutative and exact diagram:

$$
\begin{array}{ccccccc}
 & & 0 & & 0 & & 0 \\
 & & \downarrow & & \downarrow & & \downarrow \\
0 & \longrightarrow & G_{q+r} & \longrightarrow & S_{q+r} T^* \otimes E & \xrightarrow{\sigma_r} & S_r T^* \otimes F_0 \\
 & & \| & & \downarrow & & \downarrow \\
0 & \longrightarrow & \hat{G}_{r+1} & \longrightarrow & S_{r+1} T^* \otimes \hat{E} & \xrightarrow{\hat{\sigma}_r} & S_r T^* \otimes \hat{F}_0 \\
 & & \downarrow & & \downarrow & & \downarrow \\
 & & 0 & \longrightarrow & Q & \longrightarrow & S_r T^* \otimes B \\
 & & & & \downarrow & & \downarrow \\
 & & & & 0 & & 0
\end{array}
$$

and the desired identification $G_{q+r} = \hat{G}_{r+1}$.

This proves the theorem. C.Q.F.D.

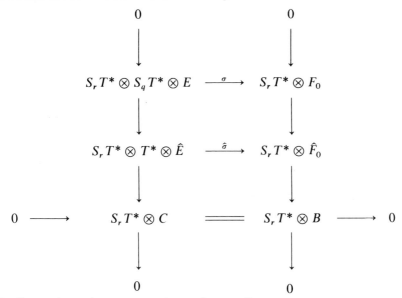

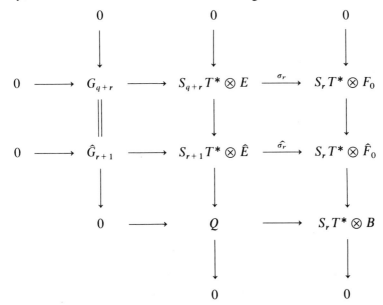

THEOREM 3.4 G_q is involutive if and only if $\hat{G}_1$ is involutive.

Proof The image of $(S_qT^* \otimes E)^i \subset S_qT^* \otimes E$ by the monomorphism $S_qT^* \otimes E \to T^* \otimes \hat{E}$ belongs to $(T^* \otimes \hat{E})^i$. In fact, using local coordinates, we have:

$$\begin{cases} \hat{v}^k_{\mu,j} = v^k_{\mu+1_j} & |\mu| = q-1 \\ \hat{v}^k_{\mu,j} = 0 & |\mu| \leq q-1 \end{cases}$$

If $j = 1, \ldots, i$, then $v^k_{\mu+1_j}$ will be of class $1, \ldots, i$ and thus equal to zero. Using an induction on i, we obtain the commutative and exact diagram:

$$\begin{array}{ccccccccc}
 & & 0 & & 0 & & 0 & & \\
 & & \downarrow & & \downarrow & & \downarrow & & \\
0 & \longrightarrow & (S_qT^* \otimes E)^i & \longrightarrow & (T^* \otimes \hat{E})^i & \longrightarrow & (C)^i & \longrightarrow & 0 \\
 & & \downarrow & & \downarrow & & \downarrow & & \\
0 & \longrightarrow & (S_qT^* \otimes E)^{i-1} & \longrightarrow & (T^* \otimes \hat{E})^{i-1} & \longrightarrow & (C)^{i-1} & \longrightarrow & 0 \\
 & & \downarrow{\scriptstyle \delta_i} & & \downarrow{\scriptstyle \delta_i} & & & & \\
0 & \longrightarrow & (S_{q-1}T^* \otimes E)^{i-1} & \xrightarrow{\varepsilon_{q-1}} & \hat{E} & & & &
\end{array}$$

It follows that we have monomorphisms:

$$0 \to (C)^i \to C$$

and a commutative and exact diagram:

$$\begin{array}{ccccccc}
 & & 0 & & 0 & & 0 \\
 & & \downarrow & & \downarrow & & \downarrow \\
0 & \longrightarrow & (G_q)^i & \longrightarrow & (S_qT^* \otimes E)^i & \xrightarrow{\sigma} & F_0 \\
 & & \| & & \downarrow & & \downarrow \\
0 & \longrightarrow & (\hat{G}_1)^i & \longrightarrow & (T^* \otimes \hat{E})^i & \xrightarrow{\hat{\sigma}} & \hat{F}_0 \\
 & & \downarrow & & \downarrow & & \downarrow \\
 & & 0 & \longrightarrow & (C)^i & \longrightarrow & B \\
 & & & & \downarrow & & \downarrow \\
 & & & & 0 & & 0
\end{array}$$

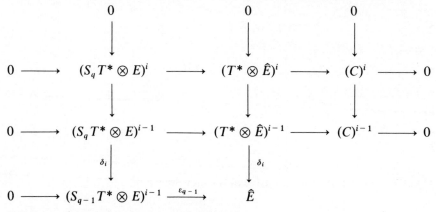

Finally $\dim (G_q)^i = \dim (\hat{G}_1)^i$ and $\dim G_{q+r} = \dim \hat{G}_{r+1}$. Any system of coordinates which is δ-regular for G_q is also δ-regular for $\hat{G}_1$ and conversely. Moreover we have $\alpha_q^i = \hat{\alpha}_1^i$ and this proves the theorem.

C.Q.F.D.

COROLLARY 3.5 We can identify the two sequences:

$$0 \longrightarrow G_{q+r} \xrightarrow{\delta} T^* \otimes G_{q+r-1} \xrightarrow{\delta} \cdots \xrightarrow{\delta} \Lambda^r T^* \otimes G_q \xrightarrow{\delta} \delta(\Lambda^r T^* \otimes G_q) \longrightarrow 0$$

$$0 \longrightarrow \hat{G}_{r+1} \xrightarrow{\delta} T^* \otimes \hat{G}_r \xrightarrow{\delta} \cdots \xrightarrow{\delta} \Lambda^r T^* \otimes \hat{G}_1 \xrightarrow{\delta} \delta(\Lambda^r T^* \otimes \hat{G}_1) \longrightarrow 0$$

Proof It is left to the reader to use the following commutative and exact diagram:

$$\begin{array}{ccccc}
0 \longrightarrow & S_{q+r}T^* \otimes E & \xrightarrow{\delta} & T^* \otimes S_{q+r-1}T^* \otimes E \\
& \downarrow {\scriptstyle \Delta r+1,\, q-1} & & \downarrow {\scriptstyle \Delta r,\, q-1} \\
0 \longrightarrow & S_{r+1}T^* \otimes S_{q-1}T^* \otimes E & \xrightarrow{\delta} & T^* \otimes S_r T^* \otimes S_{q-1}T^* \otimes E \\
& \downarrow {\scriptstyle \varepsilon_{q-1}} & & \downarrow {\scriptstyle \varepsilon_{q-1}} \\
0 \longrightarrow & S_{r+1}T^* \otimes \hat{E} & \xrightarrow{\delta} & T^* \otimes S_r T^* \otimes \hat{E}
\end{array}$$

C.Q.F.D.

4 Prolongation theorem

We shall now consider a system $\mathscr{R}_1 \subset J_1(\mathscr{E})$ of order one on $\mathscr{E}$ and study the prolongations G_q of its symbol G_1.

Using local coordinates, we have:

$$G_1 \qquad A_k^{\tau i}(x, y, p)v_i^k = 0$$

$$G_q \qquad A_k^{\tau i}(x, y, p)v_{\mu+1_i}^k = 0 \qquad |\mu| = q - 1$$

We denote by $\omega \in S_q T^* \otimes V(\mathscr{E})$ the element with components v_μ^k at the point $(x, y, p) \in \mathscr{R}_1$ fixed in the sequel.

For k fixed, $1 \leq k \leq m$, we introduce the formal polynomial $\omega^k \in \mathbb{R}[X] = \mathbb{R}[X^1, \ldots, X^n]$:

$$\omega^k = \sum_{|\mu|=q} v_\mu^k \frac{X^\mu}{\mu!} = \sum_{\mu_1+\cdots+\mu_n=q} v_{\mu_1,\ldots,\mu_n}^k \frac{(X^1)^{\mu_1}\cdots(X^n)^{\mu_n}}{\mu_1!\cdots\mu_n!}$$

For simplicity we shall sometimes suppress k and write:

$$\omega = \sum_{|\mu|=q} v_\mu \frac{X^\mu}{\mu!}$$

$$G_1 \qquad A^i(x, y, p)v_i = 0$$

$$G_q \qquad A^i(x, y, p)v_{\mu+1_i} = 0 \qquad |\mu| = q - 1$$

using matrix notation.

It is easy to check that the formal polynomial corresponding to the element $\delta_i\omega \in G_{q-1}$ is just the partial derivative of ω with respect to X^i and is an homogeneous polynomial of degree $q - 1$.

As $\mathbb{R}[X] = (R[X^2, \ldots, X^n])[X^1]$ we may define the homogeneous polynomial of degree r, $\omega_r \in \mathbb{R}^m[X^2, \ldots, X^n]$ for $\mu_1 = q - r$ fixed:

$$\omega_r = \sum_{\mu_2 + \cdots + \mu_n = r} v_{\mu_1, \mu_2, \ldots, \mu_n} \frac{(X^2)^{\mu_2} \cdots (X^n)^{\mu_n}}{\mu_2! \cdots \mu_n!}$$

and we set

$$\omega = \sum_{r=0}^{q} \omega_r \frac{(X^1)^{q-r}}{(q-r)!}$$

We can identify $(S_q T^*)^1$ with the vector space of homogeneous polynomials of degree q in $X^2, \ldots, X^n$.

Using these notations, we bring the equations of G_q into the following form:

$$\omega \in G_q \Leftrightarrow A^i(x, y, p)v_{\mu+1_i} \frac{X^\mu}{\mu!} = 0$$

with $|\mu| = q - 1$, or into the equivalent form:

$$\omega \in G_q \Leftrightarrow A^i(x, y, p) \frac{\partial \omega}{\partial X^i} = 0$$

If we look for the powers of X^1, an easy computation shows that we have:

$$\omega \in G_q \Leftrightarrow \sum_{r=0}^{q-1} \left(\sum_{i=2}^{n} A^i(x, y, p) \frac{\partial \omega_{r+1}}{\partial X^i} + A^1(x, y, p)\omega_r \right) \frac{(X^1)^{q-r-1}}{(q-r-1)!} = 0$$

We shall later on use the equivalent formulation:

$$\omega \in G_q \Leftrightarrow \sum_{i=2}^{n} A^i(x, y, p) \frac{\partial \omega_{r+1}}{\partial X^i} + A^1(x, y, p)\omega_r = 0 \qquad \forall r = 0, \ldots, q - 1$$

We have already shown in an example that the sequences over $\mathscr{R}_1$:

$$0 \longrightarrow G_{r+1} \xrightarrow{\delta} T^* \otimes G_r \xrightarrow{\delta} \cdots \xrightarrow{\delta} \Lambda^r T^* \otimes G_1 \xrightarrow{\delta} \delta(\Lambda^r T^* \otimes G_1) \longrightarrow 0$$

may be exact at any point of $\mathscr{R}_1$, even if the dimensions of the different vector spaces were changing.

The following theorem is thus fundamental in order to study any system of p.d.e.:

PROLONGATION THEOREM 4.1 If G_q is the symbol of a system $\mathscr{R}_q \subset J_q(\mathscr{E})$, then there exists an integer $q'(n, m, q) \geq q$ such that $G_{q'}$ is involutive.

Proof We shall follow Sweeny (46) and use the preceding theorems to bring the proof to that of the following theorem:

THEOREM 4.2 If G_1 is the symbol of a system $\mathscr{R}_1 \subset J_1(\mathscr{E})$, there exists an integer $q(n, m) \geq 1$ such that G_q is involutive.

Proof This uses an induction on n. For $n = 0, 1$ the proof is trivial. Let us suppose that the proof has been given for $n - 1$ and set:

$$q_0 = q(n - 1, m)$$

$$q_1 = m\binom{q_0 + n}{n - 1} + q_0 + 3$$

In the sequel we fix the point $(x, y, p) \in \mathscr{R}_1$ and we choose a coordinate system that maximizes

$$\sum_{q=q_0+1}^{q_1} \dim \delta_1^{q_1 - q}(G_{q_1})$$

and such that the applications:

$$(G_{q+1})^{i-1} \xrightarrow{\delta_i} (G_q)^{i-1}$$

are surjective for $1 < i \leq n$ and $q \geq q_0$, using the induction hypothesis.

LEMMA 4.3 $(G_q)^1 \subseteq \delta_1^{q_1 - q}(G_{q_1}) \subseteq G_q$

$$\forall q = q_0 + 1, \ldots, q_1$$

Proof We shall work towards a contradiction. Let us imagine that there exists $q_0 + 1 \leq q \leq q_1$ and $\omega \in (G_q)^1$ with $\omega \notin \delta_1^{q_1 - q}(G_q)$. Then there exists $\tau \in (G_{q_1})^1$ such that $\omega = \delta_2^{q_1 - q}(\tau)$, and we can choose $\tau_1, \ldots, \tau_N \in G_{q_1}$ such that $\delta_1^{q_1 - q}(\tau_1), \ldots, \delta_1^{q_1 - q}(\tau_N)$ are a basis of $\delta_1^{q_1 - q}(G_{q_1})$.

If we consider the new system of coordinates $\bar{x}^2 = x^2 - tx^1$, $\bar{x}^i = x^i$ for $i \neq 2$, we may define $\bar{X}^2 = X^2 - tX^1$, $\bar{X}^i = X^i$ for $i \neq 2$.

As

$$\frac{\partial}{\partial \bar{X}^1} = \frac{\partial}{\partial X^1} + t\frac{\partial}{\partial X^2}, \frac{\partial}{\partial \bar{X}^i} = \frac{\partial}{\partial X^i} \qquad \text{for } i \neq 1,$$

and

$$\omega = v_\mu \frac{X^\mu}{\mu!} = \bar{v}_\mu \frac{\bar{X}^\mu}{\mu!},$$

we have

$$\bar{\delta}_1 = \delta_1 + t\delta_2.$$

For small t, the elements $\bar{\delta}_1^{q_1-q}(\tau_1), \ldots, \bar{\delta}_1^{q_1-q}(\tau_N)$ are linearly independent and generate a subspace of G_q tending to $\delta_1^{q_1-q}(G_{q_1})$ when $t \to 0$.

As $\omega \notin \delta_1^{q_1-q}(G_{q_1})$, ω must be linearly independent of

$$\bar{\delta}_1^{q_1-q}(\tau_1), \ldots, \bar{\delta}_1^{q_1-q}(\tau_N)$$

for t small enough. This means that

$$\dim \bar{\delta}_1^{q_1-q}(G_{q_1}) > \dim \delta_1^{q_1-q}(G_{q_1})$$

because

$$t^{q_1-q}\omega = (\bar{\delta}_1 - \delta_1)^{q_1-q}(\tau) = \bar{\delta}_1^{q_1-q}(\tau) \quad \text{and} \quad \tau \in (G_{q_1})^1.$$

As the study of the different dimensions involves matrices depending on t, the rank of which cannot diminue when t becomes $\neq 0$ but small enough, we have also

$$\dim \bar{\delta}_1^{q_1-q}(G_{q_1}) \geq \dim \delta_1^{q_1-q}(G_{q_1}), \qquad \forall q = q_0 + 1, \ldots, q_1,$$

and this gives a contradiction. C.Q.F.D.

Now let $\omega \in G_q$ with $q_0 + 1 \leq q < q_1$. We can extend the scalar product on $\mathbb{R}^m$ to a scalar product on $\mathbb{R}^m[X]$, setting:

$$\left\langle v_\mu \frac{X^\mu}{\mu!}, v'_\mu \frac{X^\mu}{\mu!} \right\rangle = \sum_{|\mu|=q} \langle v_\mu, v'_\mu \rangle \frac{1}{\mu!}$$

Let τ_{q_0+1} be the orthogonal projection onto $(G_{q_0+1})^1$ of ω_{q_0+1}, homogeneous polynomial of degree $q_0 + 1$ in $X^2, \ldots, X^n$. Using the lemma, there exists $\tau \in G_{q+1}$ such that $\tau_{q_0+1} = \delta_1^{q-q_0}(\tau)$.

Setting

$$\tau = \sum_{r=0}^{q+1} \tau_r \frac{(X^1)^{q+1-r}}{(q+1-r)!}$$

and deriving with respect to X^1, we obtain:

$$\delta_1 \tau = \frac{\partial \tau}{\partial X^1} = \sum_{r=0}^{q} \tau_r \frac{(X^1)^{q-r}}{(q-r)!}$$

Thus

$$(\delta_1 \tau)_r = \tau_r \qquad \text{for } r = 0, \ldots, q \qquad \text{with } \tau_r = 0 \qquad \text{for } r = 0, \ldots, q_0$$

and

$$(\omega - \delta_1 \tau)_{q_0+1} = \omega_{q_0+1} - \tau_{q_0+1}.$$

If we set $\bar{\omega} = \omega - \delta_1 \tau$, then $\bar{\omega}_{q_0+1}$ is orthogonal to $(G_{q_0+1})^1$. Using the above procedure we can reduce step by step ω to an element

$$\bar{\omega} = \sum_{r=0}^{q} \bar{\omega}_r \frac{(X^1)^{q-r}}{(q-r)!} \quad \text{such that} \quad \bar{\omega}_r = \omega_r \quad \text{for } r = 0, \ldots, q_0$$

and $\bar{\omega}_r$ is orthogonal to $(G_r)^1$ for $r = q_0 + 1, \ldots, q$. We define:

$$\bar{G}_q = \{\bar{\omega} \mid \omega \in G_q\} \qquad \forall q_0 + 1 \leq q \leq q_1 - 1.$$

LEMMA 4.4 $\forall q_0 + 1 \leq q \leq q_1 - 2$ we have the exact sequences:

$$0 \longrightarrow \bar{G}_{q+1} \xrightarrow{\delta_1} \bar{G}_q$$

Proof Let

$$\bar{\omega} \in \bar{G}_{q+1} \subseteq G_{q+1}.$$

As

$$(\delta_1 \bar{\omega})_r = \bar{\omega}_r \qquad \text{for } r = 0, \ldots, q$$

we have

$$\delta_1(\bar{G}_{q+1}) \subset \bar{G}_q.$$

Now we have, because $\bar{G}_{q+1} \subset G_{q+1}$:

$$\sum_{i=2}^{n} A^i(x, y, p) \frac{\partial \bar{\omega}_{r+1}}{\partial X^i} + A^1(x, y, p)\bar{\omega}_r = 0 \qquad \forall r = 0, \ldots, q.$$

If $\bar{\omega}_r = 0$ for $r = 0, \ldots, q_0$ then:

$$\sum_{i=2}^{n} A^i(x, y, p) \frac{\partial \bar{\omega}_{q_0+1}}{\partial X^i} = 0 \Rightarrow \bar{\omega}_{q_0+1} \in (G_{q_0+1})^1$$

But this is impossible unless $\bar{\omega}_{q_0+1} = 0$ as $\bar{\omega}_{q_0+1}$ is orthogonal to $(G_{q_0+1})^1$ by definition.

Similarly we obtain $\bar{\omega}_r = 0$ for $r = q_0 + 1, \ldots, q + 1$ and thus $\bar{\omega} = 0$. It follows that $\bar{\omega}$ is uniquely determined by $\bar{\omega}_0 = \omega_0, \ldots, \bar{\omega}_{q_0} = \omega_{q_0}$.

As $q \geq q_0 + 1$ the above sequences are exact. C.Q.F.D.

According to the lemma we have a chain of injections:

$$0 \longrightarrow \bar{G}_{q_1 - 1} \xrightarrow{\delta_1} \bar{G}_{q_1 - 2} \xrightarrow{\delta_1} \cdots \xrightarrow{\delta_1} \bar{G}_{q_0 + 1}$$

and we deduce

$$\dim \bar{G}_{q_0 + 1} \geq \dim \bar{G}_{q_1 - 1} + (q_1 - q_0 - 2)$$

if no one of the maps δ_1 involved is surjective. But

$$q_1 - q_0 - 2 = m\binom{q_0 + n}{n - 1} + 1 > \dim \bar{G}_{q_0 + 1}$$

and we should deduce $\dim \bar{G}_{q_1 - 1} < 0$ which is impossible.

Finally there must exist $q_0 < q < q_1$ such that $\delta_1 : \bar{G}_{q+1} \to \bar{G}_q$ is surjective, and

$$G_q = \bar{G}_q + \delta_1(G_{q+1}) = \delta_1(\bar{G}_{q+1}) + \delta_1(G_{q+1}) = \delta_1(G_{q+1})$$

This proves that the sequences:

$$(G_{q+1})^{i-1} \xrightarrow{\delta_i} (G_q)^{i-1} \longrightarrow 0$$

are exact $\forall i = 1, \ldots, n$ and G_q is involutive. Of course G_{q_1} is also involutive because $q_1 > q$,

By construction q_1 depends only on (n, m) and not on the point $(x, y, p) \in \mathscr{R}_1$. This proves the theorem. C.Q.F.D.

REMARK 4.5 The number $q'(n, m, q)$ can be determined by the following properties:

$$\begin{cases} q'(0, m, 1) = 0 \\ q'(n, m, 1) = m\binom{q_0 + n}{n - 1} + q_0 + 1 & \text{with } q_0 = q'(n - 1, m, 1) \\ q'(n, m, q) = q'(n, \hat{m}, 1) & \text{with } \hat{m} = m\binom{q + n - 1}{n} \end{cases}$$

However, in general, systems of p.d.e. are such that their symbols become involutive for a smaller number of prolongations.

An important case is the following.

DEFINITION 4.6 A system $\mathscr{R}_q \subset J_q(\mathscr{E})$ is said to be of "*finite type*" if $\exists r \geq 0, G_{q+r} = 0$.

PROPOSITION 4.7 If G_q is a finite type and involutive, then we must have $G_q = 0$.

Proof According to proposition 2.17, we may introduce the characters α_q^i at any point of $\mathscr{R}_q$ and obtain:

$$\dim G_{q+r} = \alpha_q^1 + \cdots + \binom{r+i-1}{i-1}\alpha_q^i + \cdots + \binom{r+n-1}{n-1}\alpha_q^n$$

Thus

$$G_{q+r} = 0 \Rightarrow \dim G_{q+r} = 0 \Rightarrow \alpha_q^i = 0, i = 1, \ldots, n$$

and

$$\dim G_q = \alpha_q^1 + \cdots + \alpha_q^n = 0 \Rightarrow G_q = 0. \qquad \text{C.Q.F.D.}$$

Many systems in mathematical physics are of this type.

5 Complements

For any solved equation τ of class i, we denote the corresponding principal components of class i by V_μ^k and we keep v_ν^l for the parametric components with $|\mu| = |\nu| = q$.

For this equation we have:

1 $\mu_1 = 0, \ldots, \mu_{i-1} = 0, \mu_i > 0$ in V_μ^k.
2 The v_ν^l appearing are of class $\leq i$ and those of class i such that $l > k$.

Choosing the total ordering already mentioned for the total set of components, we may write:

$$V_\mu^k > v_\nu^l$$

and thus

$$V_{\mu+1_j}^k > v_{\nu+1_j}^l \qquad \forall j = 1, \ldots, n.$$

However we have to notice that $V_{\mu+1_j}^k$ is a principal component of order $q+1$ but that $v_{\nu+1_j}^l$ may also be a principal component of order $q+1$. The reader will easily check this fact in the examples.

PROPOSITION 5.1 If V_μ^k is a principal component of class i, then $\forall j > i$, $V_{\mu-1_i+1_j}^k$ is a principal component of class $\geq i$.

Proof The notation $V_{\mu-1_i+1_j}^k$ has a meaning because V_μ^k is of class i and $\mu_i > 0$.

We set $\mu - 1_i + 1_j = \bar{\mu}$. We have

$$\bar{\mu}_1 = \mu_1 = 0, \ldots, \quad \bar{\mu}_{i-1} = \mu_{i-1} = 0, \quad \bar{\mu}_i = \mu_i - 1 \geq 0, \quad \bar{\mu}_j = \mu_j + 1.$$

x^j is a non-multiplicative variable for the equation solved with respect to V_μ^k of class i because $j > i$. The prolongation of this equation with respect to x^j is:

$$V_{\mu+1_j}^k - a_l^\nu(x, y, p)v_{\nu+1_j}^l = 0$$

As G_q is involutive, according to proposition 2.15, this equation is a linear combination of the ones obtained by using the prolongations of the solved equations with respect to their corresponding multiplicative variables.

The higher component of this equation, which is $V_{\mu+1_j}^k$, is equal to the highest component that may be found among the equations entering into the linear combination. As these later ones are principal, it follows that $V_{\mu+1_j}^k$ is a principal component of order $q + 1$ and that these exist $V_{\bar{\mu}}^k$ of class $\bar{i}$ such that:

$$\bar{\mu} + 1_{\bar{j}} = \mu + 1_j \qquad \text{with } \bar{j} \leq \bar{i}$$

We shall study the different cases:

- If $\bar{i} < i$, we have $\bar{j} \leq \bar{i} < i < j$ and $\bar{\mu}_{\bar{j}} + 1 = \mu_{\bar{j}} = 0$ which is impossible.
- If $\bar{i} = i$, we have $\bar{j} \leq i < j$ and if $\bar{j} < i$ one cannot have $1 = 0$. Thus $\bar{\mu}_i + 1 = \mu_i$ and $\bar{j} = i$.
- If $i < \bar{i}$ then $\bar{\mu}_i = 0$ and $\mu_i = 0$ unless $\bar{j} = i$ and $\mu_i = 1$.

It follows that $\bar{j} = i \leq \bar{i}$ and $V_{\bar{\mu}+1_i}^k = V_{\mu+1_j}^k$. This shows that $V_{\bar{\mu}}^k$ is a principal component with $\bar{\mu} = \mu - 1_i + 1_j$. C.Q.F.D.

PROPOSITION 5.2 If there exists an equation of class i, there exists also an equation of class $i + 1$. Equivalently one can say that the classes of the solved equations of an involutive symbol are an increasing chain of consecutive integers. C.Q.F.D.

Proof Let V_μ^k be a principal component of class i such that μ_i is minimum. Using the last proposition, there exists a principal component $V_{\bar{\mu}}^k$ with $\bar{\mu}_i = \mu_i - 1 < \mu_i$ which is impossible unless $\bar{\mu}_i = 0$ and $\mu_i = 1$.

If we take $j = i + 1$, we have:

$$\bar{\mu} + 1_i = \mu + 1_j \quad \text{with} \quad \bar{\mu}_i + 1 = \mu_i = 1$$

thus

$$\bar{\mu}_i = 0 \quad \text{and} \quad \bar{\mu}_{i+1} = \mu_{i+1} + 1 > 0.$$

Finally $V_{\bar{\mu}}^k$ is of class $i + 1$. C.Q.F.D.

PROPOSITION 5.3 The indices μ_i of the principal components of class i are an increasing chain of consecutive integers starting from 1.

Proof The proof of this proposition is similar to that of the preceding one and left to the reader. C.Q.F.D.

We leave the reader to show that these properties make it possible to construct all the sets of principal components corresponding to involutive symbols, when n, m and q are given.

Problems

Show that convenient restrictions of the following non-linear systems are involutive.
$\mathscr{E}$ is a trivial bundle over X.

1) $n = 2, m = 1, q = 2$

$$\mathscr{R}_2 \begin{cases} p_{22} - \frac{1}{3}(p_{11})^3 = 0 \\ p_{12} - \frac{1}{2}(p_{11})^2 = 0 \end{cases}$$

2) $n = 3, m = 1, q = 2$

$$\mathscr{R}_2 \begin{cases} p_{33} = 0 \\ (x^3 + p_{23})p_{13} - p_1 = 0 \end{cases}$$

3) $n = 3, m = 1, q = 2$

$$\mathscr{R}_2 \begin{cases} p_{33} = 0 \\ p_{23} + \frac{1}{2}(p_{13})^2 = 0 \\ p_{22} + 2p_{12}p_{13} + p_{11}(p_{13})^2 = 0 \end{cases}$$

4) $n = 3, m = 1, q = 2$

$$\mathscr{R}_2 \begin{cases} p_{33} = 0 \\ x^3 p_{23} + p_3 - p_2 = 0 \\ p_{22} - p_{23} - x^3 p_{13} = 0 \end{cases}$$

5) $n = 3, m = 1, q = 2$

$$\mathscr{R}_2 \begin{cases} p_{11} = 0 \\ p_{12} - \frac{1}{2}(p_{13})^2 = 0 \\ p_{22} - p_{13}p_{23} = 0 \\ p_{23} - p_{13}p_{33} = 0 \end{cases}$$

(Hint: find a δ-regular coordinate system)

6) $n = 3, m = 1, q = 2$

$$\mathscr{R}_2 \begin{cases} p_{11} - p_{33}(p_{22}p_{33} + (p_{23})^2) = 0 \\ p_{12} + p_{22}p_{33} + \frac{1}{2}(p_{23})^2 = 0 \\ p_{13} + p_{23}p_{33} = 0 \end{cases}$$

(Hint: show that G_2 is a vector bundle on $\mathscr{R}_2$ whenever $p_{22}p_{33} \neq 0$.)

7) $n = 3, m = 1, q = 2$

$$\mathscr{R}_2\begin{cases} p_{23} - \frac{1}{2}(p_{22})^2 = 0 \\ p_{33} - \frac{1}{3}(p_{22})^3 = 0 \\ p_{13} - p_{12}p_{22} + \frac{1}{2}(x^2)^2p_{22} + x^2x^3p_{23} + \frac{1}{2}(x^3)^2p_{33} - x^2p_2 - x^3p_3 + y = 0 \end{cases}$$

8) $n = 2, m = 1, x^2 \neq 0$

$$\mathscr{R}_1\begin{cases} p_2^1 = 0 \\ y^2p_1^1 - x^2 = 0 \end{cases} \quad \text{with} \begin{vmatrix} p_1^1 & p_2^1 \\ p_1^2 & p_2^2 \end{vmatrix} \neq 0$$

Check that G_1 is involutive but that $\mathscr{R}_1^{(1)} \neq \mathscr{R}_1$. Show that $\mathscr{R}_1^{(1)}$ is involutive.

9) Solved Problem

Let X be a manifold with $\dim X = 2$ and $Y = \mathbb{R}^3$. We introduce the bundles $\mathscr{E} = X \times Y$ and $\mathscr{E}' = S^2T^*$ and we call ω a section of $\mathscr{E}'$.

Let $\Phi: J_1(X \times Y) \to S_2T^*$ be a morphism such that $\omega(X) \subset \operatorname{im} \Phi$. We define in local coordinates $\mathscr{R}_1 = \ker_\omega \Phi$ by the three equations:

$$\mathscr{R}_1\begin{cases} \Phi^3(x, y, p) \equiv (p_2^1)^2 + (p_2^2)^2 + (p_2^3)^2 = \omega^3(x) \\ \Phi^2(x, y, p) \equiv p_1^1p_2^1 + p_1^2p_2^2 + p_1^3p_2^3 = \omega^2(x) \\ \Phi^1(x, y, p) \equiv (p_1^1)^2 + (p_1^2)^2 + (p_1^3)^2 = \omega^1(x) \end{cases}$$

with

$$\omega^1(x) = \omega_{11}(x), \quad \omega^2(x) = \omega_{12}(x) = \omega_{21}(x), \quad \omega^3(x) = \omega_{22}(x).$$

We have:

$$G_1\begin{cases} p_2^1v_2^1 + p_2^2v_2^2 + p_2^3v_2^3 = 0 \\ p_1^1v_2^1 + p_2^1v_1^1 + p_1^2v_2^2 + p_2^2v_1^2 + p_1^3v_2^3 + p_2^3v_1^3 = 0 \\ p_1^1v_1^1 + p_1^2v_1^2 + p_1^3v_1^3 = 0 \end{cases}$$

We let the reader check that $\sigma(\Phi)$ and $V(\Phi)$ are epimorphisms over Φ if and only if the rank of the matrix

$$\begin{bmatrix} p_2^1 & p_2^2 & p_2^3 \\ p_1^1 & p_1^2 & p_1^3 \end{bmatrix}$$

is equal to 2.

However we notice that the sum of the squares of the 2×2 determinants involved is equal to:

$$\Delta(x) \equiv \omega^1(x)\omega^3(x) - (\omega^2(x))^2$$

It follows that $\mathscr{R}_1$ is a fibered submanifold of $J_1(X \times Y)$ if and only if $\Delta(x) \neq 0$. This property of the section ω will be assumed in the sequel. We have the commutative and exact diagram:

$$\begin{array}{ccccccccc} 0 & \longrightarrow & \mathscr{R}_1 & \longrightarrow & J_1(X \times Y) & \xrightarrow{\Phi} & S^2T^* & \longrightarrow & 0 \\ & & & & & \searrow & \swarrow\!\!\nearrow \omega & & \\ & & & & & & X & & \end{array}$$

The reader will check that we have also the short exact sequence:

$$0 \longrightarrow G_2 \longrightarrow S_2T^* \otimes V(X \times Y) \xrightarrow{\sigma_1(\Phi)} T^* \otimes S_2T^* \longrightarrow 0$$

of vector bundles over $\mathscr{R}_1$.

As $J_2(X \times Y)$ and $J_1(S_2T^*)$ are affine bundles over $J_1(X \times Y)$ and S_2T^*, from the commutative and exact diagram:

$$\begin{array}{ccccccc} 0 & \longrightarrow & \mathscr{R}_2 & \longrightarrow & J_2(X \times Y) & \xrightarrow{\rho_1(\Phi)} & J_1(S_2T^*) \\ & & \downarrow & & \downarrow & & \downarrow \\ 0 & \longrightarrow & \mathscr{R}_1 & \longrightarrow & J_1(X \times Y) & \xrightarrow{\Phi} & S_2T^* \end{array}$$

we deduce that $\rho_1(\Phi)$ is an epimorphism, that $\mathscr{R}_2$ is a fibered submanifold of $J_2(X \times Y)$ and that $\pi_1^2 : \mathscr{R}_2 \to \mathscr{R}_1$ is an epimorphism.

Now we obtain:

$$G_2 \begin{cases} p_1^1 v_{22}^1 + p_1^2 v_{22}^2 + p_1^3 v_{22}^3 = 0 \\ p_2^1 v_{22}^1 + p_2^2 v_{22}^2 + p_2^3 v_{22}^3 = 0 \\ p_1^1 v_{12}^1 + p_1^2 v_{12}^2 + p_1^3 v_{12}^3 = 0 \\ p_2^1 v_{12}^1 + p_2^2 v_{12}^2 + p_2^3 v_{12}^3 = 0 \\ p_1^1 v_{11}^1 + p_1^2 v_{11}^2 + p_1^3 v_{11}^3 = 0 \\ p_2^1 v_{11}^1 + p_2^2 v_{11}^2 + p_2^3 v_{11}^3 = 0 \end{cases}$$

and

$$\dim (G_2)^2 = 0 < \dim (G_2)^1 = 1 < \dim (G_2)^0 = \dim G_2 = 3$$
$$\dim (G_3)^2 = 0 < \dim (G_3)^1 = 1 < \dim (G_3)^0 = \dim G_3 = 4$$

The exactness of the sequences:

$$0 \longrightarrow (G_3)^1 \longrightarrow G_3 \xrightarrow{\delta_1} G_2 \longrightarrow 0$$
$$0 \longrightarrow (G_3)^1 \xrightarrow{\delta_2} (G_2)^1 \longrightarrow 0$$

then show that G_2 is involutive and that G_{2+r} is a vector bundle over $\mathscr{R}_1$, $\forall r \geq 0$, the basis

$$\left\{\frac{\partial}{\partial x^1}, \frac{\partial}{\partial x^2}\right\}$$

being δ-regular.

$\mathscr{R}_2$ is a non-linear system with involutive symbol. However we notice that $\mathscr{R}_2^{(1)}$ is different from $\mathscr{R}_2$ and thus $\mathscr{R}_3$ or $\mathscr{R}_2^{(1)}$ are not "*a priori*" fibered submanifolds of $J_3(X \times Y)$ or $J_2(X \times Y)$ respectively. In local coordinates $\mathscr{R}_2^{(1)}$ is defined by the equations:

$$\mathscr{R}_2^{(1)} \begin{cases} \Phi^1 = \omega^1(x), \quad \Phi^2 = \omega^2(x), \quad \Phi^3 = \omega^3(x) \\ p^1_{11}\cdot p^1_{22} + p^2_{11}\cdot p^2_{22} + p^3_{11}\cdot p^3_{22} - (p^1_{12})^2 - (p^2_{12})^2 - (p^3_{12})^2 \\ \qquad = -\tfrac{1}{2}(\partial_{22}\omega^1(x) - 2\partial_{12}\omega^2(x) + \partial_{11}\omega^3(x)) \\ p^1_2\cdot p^1_{22} + p^2_2\cdot p^2_{22} + p^3_2\cdot p^3_{22} = \tfrac{1}{2}\partial_2\omega^3(x) \\ p^1_1\cdot p^1_{22} + p^2_1\cdot p^2_{22} + p^3_1\cdot p^3_{22} = \partial_2\omega^2(x) - \tfrac{1}{2}\partial_1\omega^3(x) \\ p^1_2\cdot p^1_{12} + p^2_2\cdot p^2_{12} + p^3_2\cdot p^3_{12} = \tfrac{1}{2}\partial_1\omega^3(x) \\ p^1_1\cdot p^1_{12} + p^2_1\cdot p^2_{12} + p^3_1\cdot p^3_{12} = \tfrac{1}{2}\partial_2\omega^1(x) \\ p^1_2\cdot p^1_{11} + p^2_2\cdot p^2_{11} + p^3_2\cdot p^3_{11} = \partial_1\omega^2(x) - \tfrac{1}{2}\partial_2\omega^1(x) \\ p^1_1\cdot p^1_{11} + {}^2_1\cdot p^2_{11} + p^3_1\cdot p^3_{11} = \tfrac{1}{2}\partial_1\omega^1(x) \end{cases}$$

The symbol $G_2^{(1)}$ of $\mathscr{R}_2^{(1)}$ is defined by the 6 equations of G_2 and by the supplementary equation:

$$p^1_{11}v^1_{22} + p^2_{11}\cdot v^2_{22} + p^3_{11}\cdot v^3_{22} + p^1_{22}\cdot v^1_{11} + p^2_{22}\cdot v^2_{11} + p^3_{22}\cdot v^3_{11} \\ - 2(p^1_{12}\cdot v^1_{12} + p^2_{12}\cdot v^2_{12} + p^3_{12}\cdot v^3_{12}) = 0$$

Finally we have $\dim G_2^{(1)} = 2$ if one of the following determinants is non-zero:

$$\begin{vmatrix} p^1_{11} & p^2_{11} & p^3_{11} \\ p^1_2 & p^2_2 & p^3_2 \\ p^1_1 & p^2_1 & p^3_1 \end{vmatrix}, \quad \begin{vmatrix} p^1_{12} & p^2_{12} & p^3_{12} \\ p^1_2 & p^2_2 & p^3_2 \\ p^1_1 & p^2_1 & p^3_1 \end{vmatrix}, \quad \begin{vmatrix} p^1_{22} & p^2_{22} & p^3_{22} \\ p^1_2 & p^2_2 & p^3_2 \\ p^1_1 & p^2_1 & p^3_1 \end{vmatrix}$$

First of all it is easy to check that these determinants do not vanish identically on $\mathscr{R}_2^{(1)}$. As the map $\mathscr{R}_2^{(1)} \to \mathscr{R}_1$ is surjective, it follows that the restriction of $\mathscr{R}_2^{(1)}$ to any open set of $J_2(X \times Y)$ where one of the above determinants is different from zero, is a fibered submanifold. This restriction will be assumed in the sequel.

$\mathscr{R}_2^{(1)}$ is thus a second order non-linear system and we shall study its symbol $G_2^{(1)}$.

Changing the coordinates, we may suppose, without any loss of generality, that the first of the latter determinants is non-zero.

First we have $\dim (G_2^{(1)})^1 = 0$. Then it is possible to show that:

$$\dim G_3^{(1)} = \dim (G_2^{(1)})_{+1} = 2 \text{ on } \mathscr{R}_2^{(1)}.$$

It follows that $G_2^{(1)}$ is also involutive.

A straightforward but tedious computation shows that

$$\mathscr{R}_2^{(2)} = \pi_2^3(\mathscr{R}_3^{(1)}) = \pi_2^3((\mathscr{R}_2^{(1)})_{+1}) = \mathscr{R}_2^{(1)}$$

and thus $\mathscr{R}_2^{(1)}$ is involutive.

Remark We shall show that, under the preceding hypothesis, the restriction of $\rho_r(\Phi)$ is an epimorphism $\forall r \geq 0$. We use the fact that $\mathscr{R}_{3+r}$ is an affine bundle over $\mathscr{R}_{2+r}^{(1)}$, modelled on G_{3+r}. We have

$$\dim \mathscr{R}_{3+r} = \dim G_{3+r} + \dim \mathscr{R}_{2+r}^{(1)}$$

But

$$\dim G_{3+r} = 3 + (r+1) = r+4$$

because G_2 is involutive.

Moreover

$$\begin{aligned}\dim \mathscr{R}^{(1)}_{2+r} &= \dim (\mathscr{R}^{(1)}_2)_{+r} \\ &= \dim \mathscr{R}^{(1)}_2 + \dim G^{(1)}_3 + \cdots + \dim G^{(1)}_{2+r}\end{aligned}$$

and

$$\begin{aligned}\dim \mathscr{R}^{(1)}_2 &= \dim \mathscr{R}_1 + \dim G^{(1)}_2 \\ &= 6 + 2 = 8\end{aligned}$$

It follows that

$$\dim \mathscr{R}^{(1)}_{2+r} = 2r + 6 + 2 = 2r + 8$$

and

$$\dim \mathscr{R}_{3+r} = (r+4) + (2r+8) = 3r + 12$$

Finally:

$$\begin{cases}\dim J_{3+r}(X \times Y) = 3\binom{5+r}{2} = \dfrac{r^2 + 9r + 20}{2} \cdot 3 \\[2ex] \dim J_{2+r}(S^2 T^*) = 3\binom{4+r}{2} = \dfrac{r^2 + 7r + 12}{2} \cdot 3\end{cases}$$

and we check

$$\dim \mathscr{R}_{3+r} = \dim J_{3+r}(X \times Y) - \dim J_{2+r}(S_2 T^*) \qquad \text{C.Q.F.D.}$$

10) Problem

Let $\mathscr{E}$, $\mathscr{E}'$ be two fibered manifolds over X, f' a section of $\mathscr{E}'$ and $\Phi : J_q(\mathscr{E}) \to \mathscr{E}'$ an epimorphism. Show that, if $\sigma_r(\Phi)$ is an epimorphism $\forall r > 0$, then $\rho_r(\Phi)$ is an epimorphism $\forall r \geq 0$ and $\mathscr{R}_q = \ker_{f'} \Phi \subset J_q(\mathscr{E})$ a formally integrable system.

Application Let $\mathscr{E}$ be an affine bundle modelled on $S_2 T^* \otimes T$, $\gamma : (x) \to (x, \gamma^i_{jk}(x))$ a section of $\mathscr{E}$ which has the following transition functions:

$$\begin{cases}\gamma^i_{jk}(x) = \dfrac{\partial \bar{x}^q}{\partial x^j} \cdot \dfrac{\partial \bar{x}^r}{\partial x^k} \cdot \dfrac{\partial x^i}{\partial \bar{x}^p} \bar{\gamma}^p_{qr}(\varphi(x)) + \dfrac{\partial x^i}{\partial \bar{x}^r} \dfrac{\partial^2 x^r}{\partial x^j \partial x^k} \\[2ex] \bar{x} = \varphi(x)\end{cases}$$

We consider the operator $\mathscr{D} : \mathscr{E} \to \otimes^2 T^*$ of order 1, defined in local coordinates by:

$$\partial_i \gamma^i_{jk}(x) - \partial_k \gamma^i_{ji}(x) + \gamma^i_{ir}(x)\gamma^r_{jk}(x) - \gamma^i_{kr}(x)\gamma^r_{ji}(x) = \omega_{jk}(x)$$

where $\omega : (x) \to (x, \omega_{jk}(x))$ is a section of $\otimes^2 T^*$ and the implicit summation is done on i.

Show that $\mathscr{D}$ is formally integrable.

11) Solved Problem

Let X be a manifold with dim $X = n$ and Y a copy of X. We call $I_q(X \times Y) \subset J_q(X \times Y)$ the bundle of q-jets of inversibles maps from X to Y by identifying a map with its graph. $I_q(X \times Y)$ is an open fibered submanifold of $J_q(X \times Y)$ defined by det $[p_i^k] \neq 0$.

Let $\mathscr{E}$ be the bundle over X considered in the preceding problem, γ a section of $\mathscr{E}$ and χ a section of T^*.

We consider the system $\mathscr{R}_2 \subset I_2(X \times Y)$ defined in local coordinates by eliminating the n functions $\chi_i(x)$ between the equations:

$$\mathscr{R}_2 \qquad \frac{\partial x^i}{\partial y^p}\frac{\partial^2 y^p}{\partial x^j \partial x^k} + \frac{\partial y^q}{\partial x^j}\frac{\partial y^r}{\partial x^k}\frac{\partial x^i}{\partial y^p}\gamma^p_{qr}(y) = \gamma^i_{jk}(x) + \delta^i_j\chi_k(x) + \delta^i_k\chi_j(x)$$

The form of this system does not depend on the coordinate system. Moreover, setting $i = j$ and summing over i, we get:

$$\frac{\partial x^i}{\partial y^r}\frac{\partial^2 y^r}{\partial x^i \partial x^k} + \frac{\partial y^r}{\partial x^k}\gamma^p_{pr}(y) = \gamma^i_{ik}(x) + (n+1)\chi_k(x)$$

This equation allows one to do the elimination easily. In particular the symbol G_2 of $\mathscr{R}_2$ is defined by:

$$G_2 \qquad \frac{\partial x^i}{\partial y^r} v^r_{jk} - \frac{1}{(n+1)}(\delta^i_j v^r_{sk} + \delta^i_k v^r_{sj})\frac{\partial x^s}{\partial y^r} = 0$$

It is then easy to check that $\mathscr{R}_2$ is an affine bundle over $I_1(X \times Y)$ modelled on G_2.

Setting

$$\bar{v}^i_{jk} = \frac{\partial x^i}{\partial y^r} v^r_{jk}$$

we define $\bar{G}_2$ which is isomorphic to G_2 by the equations:

$$\bar{G}_2 \qquad \bar{v}^i_{jk} - \frac{1}{(n+1)}(\delta^i_j \bar{v}^r_{rk} + \delta^i_k \bar{v}^r_{rj}) = 0$$

It is easy to check that the 1-prolongation G_3 of G_2 is isomorphic to the 1-prolongation $\bar{G}_3$ of $\bar{G}_2$:

$$\bar{G}_3 \qquad \bar{v}^i_{jkl} - \frac{1}{(n+1)}(\delta^i_j v^r_{rkl} + \delta^i_k \bar{v}^r_{rjl}) = 0$$

Setting $i = l$ and summing on i, we get:

$$\bar{v}^r_{jkr} - \frac{1}{(n+1)}(\bar{v}^r_{rkj} + \bar{v}^r_{rjk}) = \frac{n-1}{n+1}\bar{v}^r_{rjk} = 0$$

It follows that $\bar{v}^i_{jkl} = 0$ and $G_3 = \bar{G}_3 = 0$.

According to proposition 1.14, G_2 is always 0 and 1-acyclic. Thus, in order to show that G_2 is 2-acyclic it suffices to show that the sequence:

$$0 \rightarrow \Lambda^2 T^* \otimes G_2 \rightarrow \Lambda^3 T^* \otimes T^* \otimes T$$

is exact, or equivalently that the same sequence with $\bar{G}_2$ instead of G_2 is exact.

Let $v^i_{jk:\alpha\beta}$ with $\alpha < \beta$ be local coordinates for the fiber of $\Lambda^2 T^* \otimes G_2$.

We have to look for the solution of the linear system:

$$\bar{v}^i_{j\alpha:\beta\gamma} + \bar{v}^i_{j\beta:\gamma\alpha} + \bar{v}^i_{j\gamma:\alpha\beta} = 0$$

In particular we have:

$$\delta^i_j(\bar{v}^r_{r\alpha:\beta\gamma} + \bar{v}^r_{r\beta:\gamma\alpha} + \bar{v}^r_{r\gamma:\alpha\beta}) + \delta^i_\alpha \bar{v}^r_{rj:\beta\gamma} + \delta^i_\beta \bar{v}^r_{rj:\gamma\alpha} + \delta^i_\gamma \bar{v}^r_{rj:\alpha\beta} = 0$$

Setting $i = j$ and summing on i, we get:

$$\bar{v}^r_{r\alpha:\beta\gamma} + \bar{v}^r_{r\beta:\gamma\alpha} + \bar{v}^r_{r\gamma:\alpha\beta} = 0$$

and thus

$$\delta^i_\alpha \bar{v}^r_{rj:\beta\gamma} + \delta^i_\beta \bar{v}^r_{rj:\gamma\alpha} + \delta^i_\gamma \bar{v}^r_{rj:\alpha\beta} = 0$$

Setting $i = \alpha$ and summing on i, we get:

$$n\bar{v}^r_{rj:\beta\gamma} + \bar{v}^r_{rj:\gamma\beta} + \bar{v}^r_{rj:\gamma\beta} = 0$$

that is to say

$$(n - 2)\bar{v}^r_{rj:\beta\gamma} = 0$$

For $n \neq 2$ we obtain $\bar{v}^r_{rj:\beta\gamma} = 0$ and $\bar{v}^i_{j\alpha:\beta\gamma} = 0$. It follows that G_2 is 2-acyclic and not involutive.

Let us now introduce the section ρ of $T \otimes T^* \otimes \Lambda^2 T^*$:

$$\rho^i_{jkl} = \partial_l \gamma^i_{jk}(x) - \partial_k \gamma^i_{jl}(x) + \gamma^r_{jk}(x)\gamma^i_{rl}(x) - \gamma^r_{jl}(x)\gamma^i_{rk}(x)$$

The reader will check for himself that the elimination of the jet-components of order 3 in the equations of $\mathscr{R}_3$ leads to the equations:

$$\frac{\partial x^i}{\partial y^p}\frac{\partial y^q}{\partial x^j}\frac{\partial y^r}{\partial x^k}\frac{\partial y^s}{\partial x^l}\rho^p_{qrs}(y) = \rho^i_{jkl}(x) + \begin{cases} \delta^i_j(\chi_{kl} - \chi_{lk}) \\ \delta^i_k(\chi_{jl}) \\ -\delta^i_l(\chi_{jk}) \end{cases}$$

with

$$\chi_{kl} = \partial_l \chi_k - \chi_k \chi_l$$

Setting

$$\rho_{jk}(x) = \rho^i_{jki}(x),$$

we get:

$$\frac{\partial y^q}{\partial x^j}\frac{\partial y^r}{\partial x^k}\rho_{qr}(y) = \rho_{jk}(x) + \chi_{kj} - n\chi_{jk}$$

Finally, setting:

$$v^i_{jkl}(x) = \rho^i_{jkl}(x) - \frac{\delta^i_j}{n+1}(\rho_{kl}(x) - \rho_{lk}(x))$$
$$- \frac{\delta^i_k}{n^2-1}(n\rho_{jl}(x) + \rho_{lj}(x)) + \frac{\delta^i_l}{n^2-1}(n\rho_{jk}(x) + \rho_{kj}(x))$$

we obtain the equations:

$$\frac{\partial x^i}{\partial y^p}\frac{\partial y^q}{\partial x^j}\frac{\partial y^r}{\partial x^k}\frac{\partial y^s}{\partial x^l}v^p_{qrs}(y) = v^i_{jkl}(x)$$

In order that the map $\pi^3_2 : \mathscr{R}_3 \to \mathscr{R}_2$ be surjective it is thus necessary and sufficient that $v^i_{jkl}(x) = 0$.

The criterion of involutiveness then shows that, under this condition, $\mathscr{R}_3$ is involutive.

12) Solved Problem

Let X be a manifold with $\dim X = n$ and Y a copy of X. We keep the same notations as in the preceding problem.

To any section ω of S_2T^* we consider the non-linear system $\mathscr{R}_1 \subset I_1(X \times Y)$ obtained in local coordinates by eliminating the function $a(x)$ between the equations:

$$\omega_{kl}(y)\frac{\partial y^k}{\partial x^i}\frac{\partial y^l}{\partial x^j} = a(x)\omega_{ij}(x)$$

Calling $\Delta(x)$ the determinant of the symmetric matrix $[\omega_{ij}(x)]$ we have:

$$\left(\frac{\partial(y^1,\dots,y^n)}{\partial(x^1,\dots,x^n)}\right)^2 \Delta(y) = a^n(x)\Delta(x)$$

This equation allows one to eliminate $a(x)$ easily and to obtain:

$$\mathscr{R}_1 \qquad \frac{\omega_{kl}(y)}{(\Delta(y))^{1/n}}\left(\frac{\partial(y^1,\dots,y^n)}{\partial(x^1,\dots,x^n)}\right)^{-2/n}\frac{\partial y^k}{\partial x^i}\frac{\partial y^l}{\partial x^j} = \frac{\omega_{ij}(x)}{(\Delta(x))^{1/n}}$$

The symbol G_1 of $\mathscr{R}_1$ is defined by the equations:

$$G_1 \qquad \omega_{kl}(y)\left[-\frac{2}{n}\frac{\partial y^k}{\partial x^i}\frac{\partial y^l}{\partial x^j}\frac{\partial x^s}{\partial y^r}v^r_s + \frac{\partial y^k}{\partial x^i}v^l_j + \frac{\partial y^l}{\partial x^j}v^k_i\right] = 0$$

Setting

$$\bar{v}^i_j = \frac{\partial x^i}{\partial y^r}v^r_j$$

we obtain:

$$\bar{G}_1 \qquad -\frac{2}{n}\omega_{ij}(x)\bar{v}^s_s + \omega_{ir}(x)\bar{v}^r_j + \omega_{jr}(x)\bar{v}^r_i = 0$$

and we may restrict X in order that $\mathscr{R}_1$ be a bundle.

We now introduce:

$$\gamma^i_{jk}(x) = \tfrac{1}{2}\omega^{il}(x)(\partial_j \omega_{kl}(x) + \partial_k \omega_{jl}(x) - \partial_l \omega_{jk}(x))$$

with $\omega^{il}(x)\omega_{lj}(x) = \delta^i_j$ and obtain:

$$\frac{\partial x^i}{\partial y^r}\frac{\partial^2 y^r}{\partial x^j \partial x^k} + \frac{\partial y^q}{\partial x^j}\frac{\partial y^r}{\partial x^k}\frac{\partial x^i}{\partial y^p}\gamma^p_{qr}(y) = \gamma^i_{jk}(x) + \delta^i_k\partial_j\alpha(x) + \delta^i_j\partial_k\alpha(x) - \omega_{jk}(x)\omega^{ir}(x)\partial_r\alpha(x)$$

with $a(x) = e^{2\alpha(x)}$ as $a(x) \neq 0$.

Setting $i = j$ and summing on i we get:

$$\frac{\partial x^i}{\partial y^r}\frac{\partial^2 y^r}{\partial x^i \partial x^k} + \frac{\partial y^r}{\partial x^k}\gamma^s_{sr}(y) = \gamma^i_{ik}(x) + n\partial_k\alpha(x)$$

Remark γ^i_{jk} is a section of the affine bundle introduced in the preceding problem. Similarly to the preceding problem we may set

$$\bar{v}^i_{jk} = \frac{\partial x^i}{\partial y^r} v^r_{jk}$$

and obtain:

$\bar{G}_2$
$$\bar{v}^i_{jk} - \frac{1}{n}[\delta^i_k \bar{v}^r_{rj} + \delta^i_j \bar{v}^r_{rk} - \omega_{jk}(x)\omega^{is}(x)\bar{v}^r_{rs}] = 0$$

$\bar{G}_3$
$$\bar{v}^i_{jkl} - \frac{1}{n}[\delta^i_k \bar{v}^r_{rjl} + \delta^i_j \bar{v}^r_{rkl} - \omega_{jk}(x)\omega^{is}(x)\bar{v}^r_{rsl}] = 0$$

Setting $i = l$ and summing on i, we get:

$$\bar{v}^i_{jki} - \frac{1}{n}[\bar{v}^r_{rjk} + \bar{v}^r_{rkj} - \omega_{jk}(x)\omega^{pq}\bar{v}^r_{rpq}] = 0$$

Multiplying by $\omega^{jk}(x)$ and summing on j and k we get:

$$\omega^{jk}(x)\bar{v}^r_{rjk} - \frac{2}{n}\omega^{jk}(x)\bar{v}^r_{rjk} + \omega^{jk}(x)\bar{v}^r_{rjk} = 0$$

or

$$\omega^{jk}(x)\bar{v}^r_{rjk} = 0.$$

Substituting we have

$$\frac{n-2}{n}\bar{v}^r_{rjk} = 0 \qquad \text{and for } n \geq 3 \quad \bar{v}^r_{rjk} = 0$$

and finally

$$G_3 = \bar{G}_3 = 0.$$

As in the preceding problem G_2 is not involutive because it is of finite type and $G_3 = 0$. We shall prove similarly that G_2 is 2-acyclic. The linear system to solve is:

$$\bar{v}^i_{j\alpha;\,\beta\gamma} + \bar{v}^i_{j\beta;\,\gamma\alpha} + \bar{v}^i_{j\gamma;\,\alpha\beta} = 0$$

Substituting we obtain:

$$\delta^i_j(\bar{v}^r_{r\alpha;\,\beta\gamma} + \bar{v}^r_{r\beta;\,\gamma\alpha} + \bar{v}^r_{r\gamma;\,\alpha\beta}) + \delta^i_\alpha \bar{v}^r_{rj;\,\beta\gamma} + \delta^i_\beta \bar{v}^r_{rj;\,\gamma\alpha} + \delta^i_\gamma \bar{v}^r_{rj;\,\alpha\beta}$$
$$- (\omega_{j\alpha}(x)v^r_{rs;\,\beta\gamma} + \omega_{j\beta}(x)\bar{v}^r_{rs;\,\gamma\alpha} + \omega_{j\gamma}(x)\bar{v}^r_{rs;\,\alpha\beta})\omega^{is}(x) = 0$$

Setting $i = j$ and summing on i we get:

$$n(\bar{v}^r_{r\alpha;\,\beta\gamma} + \bar{v}^r_{r\beta;\,\gamma\alpha} + \bar{v}^r_{r\gamma;\,\alpha\beta}) = 0$$

and

$$\delta^i_\alpha \bar{v}^r_{rj;\,\beta\gamma} + \delta^i_\beta \bar{v}^r_{rj;\,\gamma\alpha} + \delta^i_\gamma \bar{v}^r_{rj;\,\alpha\beta} - (\omega_{j\alpha}(x)\bar{v}^r_{rs;\,\beta\gamma} + \omega_{j\beta}(x)\bar{v}^r_{rs;\,\gamma\alpha} + \omega_{j\gamma}(x)\bar{v}^r_{rs;\,\alpha\beta})\omega^{is}(x) = 0$$

Setting $i = \alpha$ and summing on i we obtain:

$$(n-3)\bar{v}^r_{rj;\,\beta\gamma} - \omega^{pq}(x)\omega_{j\beta}(x)\bar{v}^r_{rp;\,\gamma q} - \omega^{pq}(x)\omega_{j\gamma}(x)\bar{v}^r_{rp;\,q\beta} = 0$$

Multiplying by $\omega^{jk}(x)$ and summing on j we get:

$$(n-3)\omega^{jk}(x)\bar{v}^r_{rj;\,\beta\gamma} - \delta^k_\beta \omega^{pq}(x)\bar{v}^r_{rp;\,\gamma q} - \delta^k_\gamma \omega^{pq}(x)\bar{v}^r_{rp;\,q\beta} = 0$$

Setting $k = \beta$ and summing on k we obtain:

$$2(n-2)\omega^{pq}(x)\bar{v}^r_{rp;\,q\gamma} = 0 \quad \text{and} \quad \bar{v}^r_{rj;\,\beta\gamma} = 0 \quad \text{for} \quad n \geq 4.$$

It follows that G_2 is 2-acyclic for $n \geq 4$.

According to the criterion for formal integrability, $\mathscr{R}_2$ is formally integrable if and only if the map $\pi^3_2 : \mathscr{R}_3 \to \mathscr{R}_2$ is surjective.

Eliminating the third order jet coordinates, we obtain the relations:

$$\frac{\partial x^i}{\partial y^p}\frac{\partial y^q}{\partial x^j}\frac{\partial y^r}{\partial x^k}\frac{\partial y^s}{\partial x^l}\rho^p_{qrs}(y) = \rho^i_{jkl}(x) + A^i_{jkl}(x)$$

with

$$A^i_{jkl}(x) = \delta^i_k \sigma_{jl}(x) - \delta^i_l \sigma_{jk}(x) + \omega_{jl}(x)\omega^{is}(x)\sigma_{sk}(x)$$
$$- \omega_{jk}(x)\omega^{is}(x)\sigma_{sl}(x) + (\delta^i_k \omega_{jl}(x) - \delta^i_l \omega_{jk}(x))\omega^{rs}(x)\partial_r\sigma(x)\partial_s\alpha(x)$$

and

$$\sigma_{ij}(x) = \partial_{ij}\alpha(x) - \gamma^r_{ij}(x)\partial_r\sigma(x) - \partial_i\sigma(x)\partial_j\sigma(x)$$

Setting $i = l$ and summing on i we have:

$$\frac{\partial y^q}{\partial x^j}\frac{\partial y^r}{\partial x^k}\rho_{qr}(y) = \rho_{jk}(x) + A_{jk}(x)$$

with

$$A_{jk}(x) = (2-n)\sigma_{jk}(x) - \omega_{jk}(x)\omega^{rs}(x)\sigma_{rs}(x) + (1-n)\omega_{jk}(x)\omega^{rs}(x)\partial_r\alpha(x)\partial_s\alpha(x)$$

Introducing $\rho(x) = \omega^{jk}(x)\rho_{jk}(x)$ and multiplying by

$$e^{-2\alpha(x)}\omega^{jk}(x) = \omega^{rs}(y)\frac{\partial x^j}{\partial y^r}\frac{\partial x^k}{\partial y^s}$$

we have:

$$\rho(y) = e^{-2\alpha(x)}[\rho(x) - (2(n-1)\omega^{rs}(x)\sigma_{rs}(x) + n(n-1)\omega^{rs}(x)\partial_r\alpha(x)\partial_s\alpha(x))]$$

Introducing:

$$\tau^i_{jkl}(x) = \rho^i_{jkl}(x) - \frac{1}{(n-2)}(\delta^i_l\rho_{jk}(x) - \delta^i_k\rho_{jl}(x) + \omega_{jk}(x)\rho^i_l(x) - \omega_{jl}(x)\rho^i_k(x))$$

$$+ \frac{1}{(n-1)(n-2)}(\delta^i_l\omega_{jk}(x) - \delta^i_k\omega_{jl}(x))\rho(x)$$

with $\rho^i_l(x) = \omega^{ir}(x)\rho_{rl}(x)$ we obtain finally:

$$\frac{\partial x^i}{\partial y^p}\frac{\partial y^q}{\partial x^j}\frac{\partial y^r}{\partial x^k}\frac{\partial y^s}{\partial x^l}\tau^p_{qrs}(y) = \tau^i_{jkl}(x)$$

where τ^i_{jkl} is a section of a vector subbundle of $T \otimes T^* \otimes \Lambda^2 T^*$.

We see that $\pi^3_2 : \mathscr{R}_3 \to \mathscr{R}_2$ is surjective if and only if $\pi^3_1 : \mathscr{R}_3 \to \mathscr{R}_1$ is surjective. We shall compare the symbols $G^{(2)}_1$ of $\mathscr{R}^{(2)}_1$ and G_1 of $\mathscr{R}_1$.

$$\bar{G}^{(2)}_1\begin{cases}\bar{v}^r_s(\delta^s_j\tau^i_{rkl}(x) + \delta^s_k\tau^i_{jrl}(x) + \delta^s_l\tau^i_{jkr}(x) - \delta^i_r\tau^s_{jkl}(x)) = 0\\ \bar{G}_1\left\{\bar{v}^r_s\left(-\frac{2}{n}\,\omega_{ij}(x)\delta^s_r + \omega_{ir}(x)\delta^s_j + \omega_{jr}(x)\delta^s_i\right) = 0\right.\end{cases}$$

In order to have $\bar{G}^{(2)}_1 = \bar{G}_1$ and thus $G^{(2)}_1 = G_1$ each equation of the upper set must be a linear combination of the equations of the lower set.

Setting $r = s$ and summing on r we obtain, looking for the coefficients, $2\tau^i_{jkl}(x)$ for the upper set and 0 for the lower set.

It follows that a necessary and sufficient condition in order that $\mathscr{R}_2$ be formally integrable is $\tau^i_{jkl}(x) = 0$.

Under these conditions $\mathscr{R}_3$ is involutive.

13) Solved Problem

$n = 3, m = 1, X = \mathbb{R}^3$, E trivial bundle over X.

We consider the second order linear system:

$$R_2\begin{cases}u_{33} - x^2u_{11} = 0\\ u_{22} = 0\end{cases}$$

The reader will check that G_2 is not involutive but that G_3 is involutive. However he will discover that, though $\pi^3_2 : R_3 \to R_2$ is surjective, $\pi^4_3 : R_4 \to R_3$ is not surjective and will have to introduce $R_3^{(1)}$:

$$R_3^{(1)}\begin{cases} u_{333} - x^2 u_{113} = 0 \\ u_{233} - u_{11} = 0 \\ u_{223} = 0 \\ u_{222} = 0 \\ u_{133} - x^2 u_{111} = 0 \\ u_{122} = 0 \\ u_{112} = 0 \\ \dots\dots\dots\dots = 0 \end{cases}$$

One checks easily that $G_3^{(1)}$ is not involutive. The reader will check that $(R_3^{(1)})_{+1} = R_4^{(1)}$:

$$R_4^{(1)}\begin{cases} u_{3333} - (x^2)^2 u_{1111} = 0 \\ u_{2333} - u_{113} = 0 \\ u_{2233} = 0 \\ u_{2223} = 0 \\ u_{2222} = 0 \\ u_{1333} - x^2 u_{1113} = 0 \\ u_{1233} - u_{111} = 0 \\ u_{1223} = 0 \\ u_{1222} = 0 \\ u_{1133} - x^2 u_{1111} = 0 \\ u_{1123} = 0 \\ u_{1122} = 0 \\ u_{1112} = 0 \\ \dots\dots\dots\dots\dots = 0 \end{cases}$$

The reader will test that $G_4^{(1)}$ is involutive, that $(R_4^{(1)})_{+1} = R_5^{(1)}$ but that $\pi_4^5 : R_5^{(1)} \to R_4^{(1)}$ is not surjective. He will have to consider $R_4^{(2)}$, obtained by adding to the equations of $R_4^{(1)}$ the equation $u_{1111} = 0$. He will show that $G_4^{(2)}$ is not involutive.

Finally he will check that $(R_4^{(2)})_{+1} = R_5^{(2)}$ is an involutive system with $G_5^{(2)} = 0$.

Remark It is not evident "*a priori*" that the general analytic solution of R_2 only depends on a finite number of constants.

14) Solved Problem

$n = 3, m = 1, q = 2, X = \mathbb{R}^3, \mathscr{E}$ trivial bundle over X.

$$\mathscr{R}_2 \qquad \begin{cases} p_{33} = 0 \\ p_{23} + \frac{1}{2}(p_{13})^2 = 0 \end{cases}$$

$$G_2 \qquad \begin{cases} v_{33} = 0 \\ v_{23} + p_{13}v_{13} = 0 \end{cases} \qquad \dim G_2 = 4$$

The reader will show that G_2 is involutive and that the monomorphism

$$J_2(\mathscr{E}) \to J_1(J_1(\mathscr{E}))$$

permits one to write, with suitable local coordinates, the following equations for the image $\hat{\mathscr{R}}_1$ of $\mathscr{R}_2$ in $J_1(J_1(\mathscr{E}))$:

$$\hat{\mathscr{R}}_1 \quad \begin{cases} \hat{p}_3^3 = 0 \\ \hat{p}_3^2 + \frac{1}{2}(\hat{p}_3^1)^2 = 0 \\ \hat{p}_3^0 - \hat{y}^3 = 0, \hat{p}_2^0 - \hat{y}^2 = 0, \hat{p}_1^0 - \hat{y}^1 = 0 \\ \hat{p}_3^1 - \hat{p}_1^3 = 0, \hat{p}_3^2 - \hat{p}_2^3 = 0, \hat{p}_2^1 - \hat{p}_1^2 = 0 \end{cases}$$

setting

$$\begin{cases} \hat{y}^0 = y, \hat{y}^1 = p_1, \hat{y}^2 = p_2, \hat{y}^3 = p_3 \\ \hat{p}_i^0 = p_i, \hat{p}_j^i = p_{ij}, \ldots . \end{cases}$$

The symbol $\hat{G}_1$ of $\hat{\mathscr{R}}_1$ is defined by the equations:

$$\hat{G}_1 \quad \begin{cases} \hat{v}_3^3 \\ \hat{v}_3^2 + \hat{p}_3^1 \hat{v}_3^1 = 0 \\ \hat{v}_3^0 = 0, \hat{v}_2^0 = 0, \hat{v}_1^0 = 0 \\ \hat{v}_3^1 - \hat{v}_1^3 = 0, \hat{v}_3^2 - \hat{v}_2^3 = 0, \hat{v}_2^1 - \hat{v}_1^2 = 0 \end{cases}$$

The reader will check directly that $\dim G_2 = \dim \hat{G}_1 = 4$ and that $\hat{G}_1$ is involutive (Hint: $\alpha_2^1 = \hat{\alpha}_1^1 = 3, \alpha_2^2 = \hat{\alpha}_1^2 = 1, \alpha_2^3 = \hat{\alpha}_1^3 = 0$).

15) Problem

The same question as for the preceding solved problem for the system:

$$\begin{cases} p_{33} = 0 \\ p_{23} + \frac{1}{2}(p_{13})^2 = 0 \\ p_{22} + 2p_{12}p_{13} + (p_{13})^2 p_{11} = 0 \end{cases}$$

16) Solved Problem

$n = 2, m = 1, q = 2$, $\mathscr{E}$ trivial bundle over $X = \mathbb{R}^2$.

Study with respect to the constant λ, the second order non-linear system:

$$\mathscr{R}_2 \begin{cases} p_{22} + p_{11} + e^{2y} = 0 \\ p_{12} + \lambda p_1 \cdot p_2 \quad = 0 \end{cases}$$

Solution G_2 is defined by the equations:

$$G_2 \begin{cases} \boxed{v_{22}} + v_{11} = 0 & \boxed{1 \quad 2} \\ \boxed{v_{12}} \qquad = 0 & \boxed{1 \quad \cdot} \end{cases}$$

One checks easily that G_2 is not involutive but that $G_3 = 0$ is trivially involutive. (The reader will see that the characteristic variety is zero).

In order to obtain $\mathscr{R}_3^{(1)}$, one has to add the equation

$$p_1 \cdot p_2 e^{2y}(\lambda + 1)(\lambda - 2) = 0$$

to the equations of $\mathscr{R}_3$.

1 If $\lambda = -1$ or $\lambda = 2$, then $\mathscr{R}_3^{(1)} = \mathscr{R}_3$ because $e^{2y} \neq 0$, and $\mathscr{R}_3$ is involutive.
2 If $\lambda \neq -1, 2$ then $\mathscr{R}_3^{(1)} \neq \mathscr{R}_3$ because $e^{2y} \neq 0$. $\mathscr{R}_3^{(1)}$ is a sub-bundle of $J_3(\mathscr{E})$ if $p_1 \cdot p_2 \neq 0$.

Under these conditions, as $G_3^{(1)} = 0$ is involutive, we have to introduce $\mathscr{R}_3^{(2)}$.

The reader will check that $\mathscr{R}_3^{(2)}$ is an involutive system with two components of the same dimension, $\mathscr{R}_3'^{(2)}$ and $\mathscr{R}_3''^{(2)}$, obtained one from the other by a permutation of 1 and 2.

We have:

$$\mathscr{R}_3'^{(2)}\begin{cases} p_{222} & = 0 \\ p_{122} & = 0 \\ p_{112} & = 0 \\ p_{111} + 2p_1 \cdot e^{2y} & = 0 \\ p_{22} & = 0 \\ p_{12} & = 0 \\ p_{11} + e^{2y} & = 0 \\ p_2 & = 0 \end{cases} \quad \boxed{\begin{matrix} 1 & 2 \\ 1 & \cdot \\ 1 & \cdot \\ 1 & \cdot \\ \cdot & \cdot \\ \cdot & \cdot \\ \cdot & \cdot \\ \cdot & \cdot \end{matrix}}$$

The reader will check that $\mathscr{R}_3'^{(2)}$ and $\mathscr{R}_3''^{(2)}$ are involutive and have no common solution.

17) Problem

$n = 2$, $m = 3$, $q = 1$, $\mathscr{E}$ trivial bundle over $\mathbb{R}^2$.

We consider the system $\mathscr{R}_1 \subset J_1(\mathscr{E})$:

$$\mathscr{R}_1\begin{cases} \Phi^3 \equiv p_2^2 + p_2^3 - p_1^3 - p_1^2 = 0 \\ \Phi^2 \equiv p_2^1 - p_2^3 - p_1^3 - p_1^2 = 0 \\ \Phi^1 \equiv p_1^1 \qquad - 2p_1^3 - p_1^2 = 0 \end{cases}$$

1 Show that this system is involutive and compute the characters of its symbol G_1.

$$\left(\text{Hint: We have } \frac{\partial \Phi^1}{\partial x^2} - \frac{\partial \Phi^2}{\partial x^1} + \frac{\partial \Phi^3}{\partial x^1} \equiv 0 \quad \text{and} \quad \alpha_1^1 = 2, \alpha_1^2 = 1\right)$$

2 Prove that this analytic system has one solution $y^k = f^k(x^1, x^2)$ and only one such that:

$$f^1(0, 0) = 1,\ f^2(x^1, 0) = (x^1)^2,\ f^3(x^1, x^2) = x^1 + x^2$$

(Hint: $y^1 = (1 + x^1 + x^2)^2$, $y^2 = (x^1 + x^2)^2$, $y^3 = x^1 + x^2$)
3 Find the boards P_0, P_1 and compute $\dim C_r$, $\dim F_r$.
(Hint: $\dim F_0 = 3$, $\dim F_1 = 1$, $\dim F_2 = 0$)

18) Problem

$n = 3, m = 1, q = 2, \mathscr{E}$ trivial bundle over $\mathbb{R}^2$.

We consider the following non-linear system $\mathscr{R}_2 \subset J_2(\mathscr{E})$

$$\mathscr{R}_2 \begin{cases} p_{33} - \frac{1}{2}(p_{22})^2 = 0 \\ p_{23} - p_{22} \qquad = 0 \end{cases}$$

and the linear system $\bar{\mathscr{R}}_2 \subset \mathscr{R}_2$:

$$\bar{\mathscr{R}}_2 \begin{cases} p_{33} - \frac{1}{2} = 0 \\ p_{23} - 1 = 0 \\ p_{22} - 1 = 0 \end{cases}$$

1 Show that $\bar{\mathscr{R}}_2$ is involutive.
2 Show that the symbol G_2 of $\mathscr{R}_2$ is involutive on $\bar{\mathscr{R}}_2$ but not on $\mathscr{R}_2$.
3 Show that G_3 is involutive on $\mathscr{R}_2$. Compute the characters and find $\dim G_4, \dim G_5$.
4 Show that $\mathscr{R}_2$ is formally integrable but that G_2 is not 2-acyclic. (Take $p_{22} \neq 1$.) (Hint: The following sequence is not exact:

$$0 \to G_4 \to T^* \otimes G_3 \to \Lambda^2 T^* \otimes G_2 \to \Lambda^3 T^* \otimes T^* \otimes E \to 0)$$

5 Use the former questions in order to transform $\mathscr{R}_3$ into a first order system by means of the monomorphism $\mathscr{R}^3 \to J_1(\mathscr{R}_2)$ of affine bundles over $\mathscr{R}_2$. (Take $p_{22} \neq 1$.)

CHAPTER 4

1 Spencer families

Let $\mathscr{R}_q \subset J_q(\mathscr{E})$ be a non-linear system of order q on $\mathscr{E}$. The symbol G_q of $\mathscr{R}_q$ and its prolongations G_{q+r} are, in general, families of vector spaces over $\mathscr{R}_q$. We shall define in the next pages other such families that are very useful in order to study the formal properties of $\mathscr{R}_q$.

DEFINITION 1.1 We define the families C_r by the formulas:

$$\begin{cases} C_0 = V(\mathscr{R}_q) & r = 0 \\ C_r = \dfrac{\Lambda^r T^* \otimes C_0}{\delta(\Lambda^{r-1} T^* \otimes G_{q+1})} & 1 \leq r \leq n \\ C_r = 0 & n < r \end{cases}$$

In order to simplify the notations, we denote by δ the map $(\mathrm{id}_{\Lambda^r T^*} \otimes \varepsilon_q) \circ \delta$, we do not indicate $\mathscr{R}_q$ as a subscript under $\otimes$ and we introduce the canonical projection $\tau_r : \Lambda^r T^* \otimes C_0 \to C_r$.

For $0 < r \leq n$ we have the short exact sequences:

$$0 \longrightarrow \delta(\Lambda^{r-1} T^* \otimes G_{q+1}) \longrightarrow \Lambda^r T^* \otimes C_0 \xrightarrow{\tau_r} C_r \longrightarrow 0$$

First of all, C_0 is a vector bundle over $\mathscr{R}_q$.

Then C_1 is a vector bundle over $\mathscr{R}_q$ if and only if G_{q+1} is a vector bundle over $\mathscr{R}_q$, because for $r = 1$ we have the short exact sequence:

$$0 \longrightarrow G_{q+1} \xrightarrow{\delta} T^* \otimes C_0 \xrightarrow{\tau_1} C_1 \longrightarrow 0$$

Finally C_2 is a vector bundle over $\mathscr{R}_q$ if G_{q+1} and G_{q+2} are vector bundles over $\mathscr{R}_q$ because we have the exact sequence:

$$0 \longrightarrow G_{q+2} \xrightarrow{\delta} T^* \otimes G_{q+1} \xrightarrow{\delta} \Lambda^2 T^* \otimes C_0 \xrightarrow{\tau_2} C_2 \longrightarrow 0$$

More generally we have the following proposition:

PROPOSITION 1.2 If G_{q+1} is involutive and if G_{q+2} is a vector bundle over $\mathscr{R}_q$, then C_r is a vector bundle over $\mathscr{R}_q$, $\forall r \geq 0$.

Proof According to corollary 3.2.30, G_{q+r} is a vector bundle over $\mathscr{R}_q$, $\forall r \geq 1$. But we have the exact sequences:

$$0 \longrightarrow G_{q+r} \xrightarrow{\delta} T^* \otimes G_{q+r-1} \xrightarrow{\delta} \cdots \xrightarrow{\delta} \Lambda^r T^* \otimes C_0 \xrightarrow{\tau_r} C_r \longrightarrow 0$$

The proposition follows from an easy computation using the Euler–Poincaré formula. C.Q.F.D.

We shall now construct for $r \geq 1$, $s \geq 0$ maps:

$$\tau_r^s : S_{s+1} T^* \otimes C_{r-1} \to S_s T^* \otimes C_r$$

in order to get a sequence for $r + s = \text{cst}$, that is to say such that

$$\tau_{r+1}^s \circ \tau_r^{s+1} = 0.$$

Let us consider the following commutative diagram:

$$\begin{array}{ccc} S_{s+1} T^* \otimes \Lambda^{r-2} T^* \otimes G_{q+1} & \xrightarrow{(-1)^{r-2}\cdot\delta} & S_s T^* \otimes \Lambda^{r-1} T^* \otimes G_{q+1} \\ \big\downarrow \delta & & \big\downarrow \delta \\ S_{s+1} T^* \otimes \Lambda^{r-1} T^* \otimes G_q & \xrightarrow{(-1)^{r-1}\cdot\delta} & S_s T^* \otimes \Lambda^r T^* \otimes G_q \end{array}$$

which is explained in local coordinates as follows:

$$\begin{array}{ccc} v^k_{\mu+1_i,\nu+1_j,I}\, dx^I & \longrightarrow & v^k_{\mu+1_i,\nu+1_j,I}\, dx^I \wedge dx^j \\ \big\downarrow & & \big\downarrow \\ v^k_{\mu+1_i,\nu+1_j,I}\, dx^i \wedge dx^I & \longrightarrow & v^k_{\mu+1_i,\nu+1_j,I}\, dx^i \wedge dx^I \wedge dx^j \end{array}$$

with

$$|\mu| = q, |\nu| = s, I = (i_1, \ldots, i_{r-2})$$

and

$$A_k^{\tau\mu}(x, y, p)v^k_{\mu+1_i,\, \nu+1_j,\, I}\, dx^i \wedge dx^I \wedge dx^j = 0$$

We may also construct the maps:

$$(-1)^{r-1} \cdot \delta : S_{s+1}T^* \otimes \Lambda^{r-1}T^* \otimes C_0 \to S_sT^* \otimes \Lambda^rT^* \otimes C_0$$

The maps τ_r^s will then be defined by means of the following commutative diagram:

$$\begin{array}{ccc}
S_{s+1}T^* \otimes \Lambda^{r-2}T^* \otimes G_{q+1} & \xrightarrow{(-1)^{r-2}\cdot\delta} & S_sT^* \otimes \Lambda^{r-1}T^* \otimes G_{q+1} \\
\downarrow \delta & & \downarrow \delta \\
S_{s+1}T^* \otimes \Lambda^{r-1}T^* \otimes C_0 & \xrightarrow{(-1)^{r-1}\cdot\delta} & S_sT^* \otimes \Lambda^{r}T^* \otimes G_q \\
\downarrow \tau_{r-1} & & \downarrow \tau_r \\
S_{s+1}T^* \otimes C_{r-1} & \xrightarrow{\tau_r^s} & S_sT^* \otimes C_r \\
\downarrow & & \downarrow \\
0 & & 0
\end{array}$$

Of course we have $\tau_{r+1}^s \circ \tau_r^{s+1} = 0$ because $\delta \circ \delta = 0$, and we let the reader check himself that τ_r^s is uniquely determined from τ_r^0 by prolongation in such a way that $\tau_r^s = \rho_s(\tau_r^0)$.

PROPOSITION 1.3 We have the exact sequence:

$$0 \longrightarrow G_{q+r} \xrightarrow{\tau_0^r} S_rT^* \otimes C_0 \xrightarrow{\tau_1^{r-1}} S_{r-1}T^* \otimes C_1$$

in which τ_0^r is the map induced by the monomorphism

$$J_{q+r}(\mathscr{E}) \to J_r(J_q(\mathscr{E})).$$

Proof By definition, for $r = 1$, we have the short exact sequence:

$$0 \longrightarrow G_{q+1} \xrightarrow{\delta} T^* \otimes C_0 \xrightarrow{\tau_1} C_1 \longrightarrow 0$$

More generally we have the commutative and exact diagram:

$$\begin{array}{ccccc}
 & & & & 0 \\
 & & & & \downarrow \\
 & & 0 & \longrightarrow & S_{r-1}T^* \otimes G_{q+1} \\
 & & \downarrow & & \downarrow \delta \\
0 & \longrightarrow & S_r T^* \otimes C_0 & \xrightarrow{\delta} & S_{r-1}T^* \otimes T^* \otimes C_0 \\
 & & \| & & \downarrow \tau_1 \\
 & & S_r T^* \otimes C_0 & \xrightarrow{\tau_1^{r-1}} & S_{r-1}T^* \otimes C_1 \\
 & & \downarrow & & \downarrow \\
 & & 0 & & 0
\end{array}$$

A chase shows that ker τ_1^{r-1} is made of the elements of $S_r T^* \otimes C_0$ with components $v^k_{\mu, \nu+1_i}$ all equal to zero but those for which:

$$\begin{cases} v^k_{\mu, \nu+1_i} = v^k_{\mu+1_i, \nu} & |\mu| = q, |\nu| = r-1 \\ A_k^{\tau\mu}(x, y, p)v^k_{\mu+1_i, \nu} = 0 \end{cases}$$

As in proposition 1.9.13 one can show that these elements are identical to the elements $v^k_{\mu+\nu+1_i}$ with similar indices and belonging to G_{q+r}, as we deduce

$$A_k^{\tau\mu}(x, y, p)v^k_{\mu+\nu+1_i} = 0.$$

The map τ_0^r is described, using local coordinates by:

$$v^k_{\mu+\nu+1_i} \to v^k_{\mu, \nu+1_i}$$

and this ends the proof. C.Q.F.D.

PROPOSITION 1.4 The sequences:

$$0 \longrightarrow G_{q+r} \xrightarrow{\tau_0^r} S_r T^* \otimes C_0 \xrightarrow{\tau_1^{r-1}} S_{r-1}T^* \otimes C_1$$
$$\longrightarrow \cdots \longrightarrow T^* \otimes C_{r-1} \xrightarrow{\tau_r^0} C_r \longrightarrow 0$$

are exact if and only if G_{q+1} is involutive.

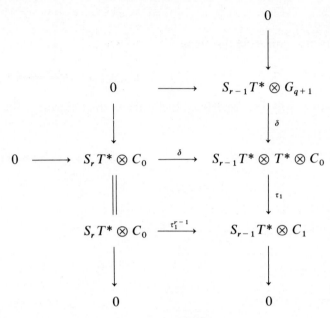

Proof This relies upon the following lemma:

LEMMA 1.5 The exterior multiplication:

$$\Lambda^r T^* \otimes T^* \to \Lambda^{r+1} T^*$$

is an epimorphism of vector bundles over X.

Proof According to proposition 3.1.4, for $r \leq n$ we have the exact sequences of vector bundles over X:

$$0 \longrightarrow S_r T^* \xrightarrow{\delta} T^* \otimes S_{r-1} T^* \longrightarrow \cdots \longrightarrow \Lambda^{r-1} T^* \otimes T^* \longrightarrow \Lambda^r T^* \longrightarrow 0$$

and this proves the lemma. C.Q.F.D.

Coming back to the proof of the proposition, we first have the commutative and exact diagram:

$$\begin{array}{ccccc}
T^* \otimes \Lambda^{r-2} T^* \otimes G_{q+1} & \longrightarrow & \Lambda^{r-1} T^* \otimes G_{q+1} & \longrightarrow & 0 \\
\downarrow \delta & & \downarrow \delta & & \\
T^* \otimes \Lambda^{r-1} T^* \otimes C_0 & \longrightarrow & \Lambda^r T^* \otimes C_0 & \longrightarrow & 0 \\
\downarrow \tau_{r-1} & & \downarrow \tau_r & & \\
T^* \otimes C_{r-1} & \xrightarrow{\tau_r^0} & C_r & \longrightarrow & 0 \\
\downarrow & & \downarrow & & \\
0 & & 0 & &
\end{array}$$

from which we deduce the surjectivity of τ_r^0. The proposition can be deduced at once from the next commutative and exact diagram on page 146:

In fact the commutativity holds because $\tau_r^s = \rho_s(\tau_r^0)$. The columns are exact because of proposition 3.1.5 and the preceding lemma. Using an induction on r, the rows are exact, unless perhaps the top one which is therefore exact at all places, unless perhaps at $T^* \otimes C_{r-1}$.

The exactness of this row can be deduced at once from the Euler–Poincaré formula because the computation of dimensions shows that $\dim \operatorname{im} \tau_{r-1}^1 = \dim \ker \tau_r^0$. As $\operatorname{im} \tau_{r-1}^1 \subset \ker \tau_r^0$ it follows that $\operatorname{im} \tau_{r-1}^1 = \ker \tau_r^0$.

C.Q.F.D.

$$
\begin{array}{ccccccccccccc}
 & & 0 & & 0 & & 0 & & & & 0 & & \\
 & & \downarrow & & \downarrow & & \downarrow & & & & \downarrow & & \\
0 & \longrightarrow & G_{q+r} & \xrightarrow{\tau_0^r} & S_r T^* \otimes C_0 & \xrightarrow{\tau_1^{r-1}} & S_{r-1} T^* \otimes C_1 & \xrightarrow{\tau_2^{r-2}} & \cdots & \xrightarrow{\tau_{r-1}^1} & T^* \otimes C_{r-1} & \xrightarrow{\tau_r^0} & C_r \longrightarrow 0 \\
 & & \downarrow{\scriptstyle \delta} & & \downarrow{\scriptstyle \delta} & & \downarrow{\scriptstyle \delta} & & & & \| & & \\
0 & \longrightarrow & T^* \otimes G_{q+r-1} & \longrightarrow & T^* \otimes S_{r-1} T^* \otimes C_0 & \longrightarrow & T^* \otimes S_{r-2} T^* \otimes C_1 & \longrightarrow & \cdots & \longrightarrow & T^* \otimes C_{r-1} & \longrightarrow & 0 \\
 & & \downarrow{\scriptstyle \delta} & & \downarrow{\scriptstyle \delta} & & \downarrow{\scriptstyle \delta} & & & & \downarrow & & \\
 & & \vdots & & \vdots & & \vdots & & & & 0 & & \\
 & & \downarrow{\scriptstyle \delta} & & \downarrow{\scriptstyle \delta} & & \downarrow{\scriptstyle \delta} & & & & & & \\
0 & \longrightarrow & \Lambda^{r-1} T^* \otimes G_{q+1} & \longrightarrow & \Lambda^{r-1} T^* \otimes T^* \otimes C_0 & \longrightarrow & \Lambda^{r-1} T^* \otimes C_1 & \longrightarrow & 0 & & & & \\
 & & \downarrow{\scriptstyle \delta} & & \downarrow{\scriptstyle \delta} & & \downarrow & & & & & & \\
0 & \longrightarrow & \Lambda^r T^* \otimes C_0 & = & \Lambda^r T^* \otimes C_0 & \longrightarrow & 0 & & & & & & \\
 & & \downarrow{\scriptstyle \tau_r} & & \downarrow & & & & & & & & \\
 & & C_r & & 0 & & & & & & & & \\
 & & \downarrow & & & & & & & & & & \\
 & & 0 & & & & & & & & & &
\end{array}
$$

The reader who would like to study concrete examples may ask why we introduce the families C_r of vector spaces over $\mathcal{R}_q$. We shall answer later on. For the moment, in the following pages, we shall introduce other kinds of families that are in fact more useful in practice than the Spencer families C_r.

For these families we shall detail the computation of their dimensions as this is a very important problem from an operational point of view, though it is a difficult one if one only deals with the families C_r.

DEFINITION 1.6 For any $q \geq 0$ and any fibered manifold $\mathcal{E}$ over X, we shall define the bundles $C_{q,r}(E)$ over $J_q(\mathcal{E})$ by the following formulas:

$$\begin{cases} C_{q,0}(E) = V(J_q(\mathcal{E})) = J_q(V(\mathcal{E})) = J_q(E) & r = 0 \\ C_{q,r}(E) = \dfrac{\Lambda^r T^* \otimes C_{q,0}(E)}{\delta(\Lambda^{r-1} T^* \otimes S_{q+1} T^* \otimes E)} & 1 \leq r \leq n \\ C_{q,r}(E) = 0 & n < r \end{cases}$$

The same construction as above can be done with the $C_{q,r}(E)$ but now $S_{q+1} T^* \otimes E$ is involutive.

REMARK 1.7 In the sequel we shall denote simply by $C_r(E)$ the restrictions of these vector bundles to $\mathcal{R}_q \subset J_q(\mathcal{E})$.

For any system $\mathcal{R}_q \subset J_q(\mathcal{E})$ of order q on $\mathcal{E}$, we have the exact sequences of vector bundles over $\mathcal{R}_q$:

$$0 \longrightarrow G_{q+r} \longrightarrow S_{q+r} T^* \otimes E \xrightarrow{\sigma_r} S_r T^* \otimes F_0$$

DEFINITION 1.8 We define the families of vector spaces over $\mathcal{R}_q$:

$$G'_r = \operatorname{im} \sigma_r \subset S_r T^* \otimes F_0$$

For $r \geq 1$ we have the short exact sequences:

$$0 \longrightarrow G_{q+r} \longrightarrow S_{q+r} T^* \otimes E \xrightarrow{\sigma_r} G'_r \longrightarrow 0$$

PROPOSITION 1.9 G'_{r+1} is the r-prolongation of G'_1 if and only if G_q is 2-acyclic. In this case we have:

$$H^s_{q+r}(G_q) = H^{s-1}_{r+1}(G'_1) \qquad \forall r \geq 0, s \geq 2$$

Moreover G'_1 is involutive if and only if G_q is involutive.

Proof According to proposition 3.2.27, we have the commutative and exact diagram of families of vector spaces over $\mathscr{R}_q$:

$$
\begin{array}{ccccccccc}
 & & 0 & & 0 & & 0 & & \\
 & & \downarrow & & \downarrow & & \downarrow & & \\
0 & \longrightarrow & G_{q+r} & \longrightarrow & S_{q+r}T^* \otimes E & \xrightarrow{\sigma_r} & G'_r & \longrightarrow & 0 \\
 & & \downarrow{\scriptstyle\delta} & & \downarrow{\scriptstyle\delta} & & \downarrow & & \\
0 & \longrightarrow & T^* \otimes G_{q+r-1} & \longrightarrow & T^* \otimes S_{q+r-1}T^* \otimes E & \xrightarrow{\sigma_{r-1}} & T^* \otimes G'_{r-1} & \longrightarrow & 0 \\
 & & \downarrow{\scriptstyle\delta} & & \downarrow{\scriptstyle\delta} & & \downarrow & & \\
 & & \vdots & & \vdots & & \vdots & & \\
 & & \downarrow{\scriptstyle\delta} & & \downarrow{\scriptstyle\delta} & & \downarrow & & \\
0 & \longrightarrow & \Lambda^{r-1}T^* \otimes G_{q+1} & \longrightarrow & \Lambda^{r-1}T^* \otimes S_{q+1}T^* \otimes E & \xrightarrow{\sigma_1} & \Lambda^{r-1}T^* \otimes G'_1 & \longrightarrow & 0 \\
 & & \downarrow{\scriptstyle\delta} & & \downarrow{\scriptstyle\delta} & & \downarrow & & \\
0 & \longrightarrow & \Lambda^r T^* \otimes G_q & \longrightarrow & \Lambda^r T^* \otimes S_q T^* \otimes E & \xrightarrow{\sigma} & \Lambda^r T^* \otimes F_0 & & \\
 & & \downarrow{\scriptstyle\delta} & & \downarrow{\scriptstyle\delta} & & & & \\
0 & \longrightarrow & \Lambda^{r+1}T^* \otimes S_{q-1}T^* \otimes E & = & \Lambda^{r+1}T^* \otimes S_{q-1}T^* \otimes E & & & &
\end{array}
$$

The rows are exact by definition and the middle column is exact because of proposition 3.1.5.

The commutativity of this diagram can be deduced from the commutativity of the diagram:

$$
\begin{array}{ccc}
\Lambda^s T^* \otimes S_{q+r+1}T^* \otimes E & \xrightarrow{\sigma_{r+1}} & \Lambda^s T^* \otimes S_{r+1}T^* \otimes F_0 \\
\downarrow{\scriptstyle\delta} & & \downarrow{\scriptstyle\delta} \\
\Lambda^{s+1}T^* \otimes S_{q+r}T^* \otimes E & \xrightarrow{\sigma_r} & \Lambda^{s+1}T^* \otimes S_r T^* \otimes F_0
\end{array}
$$

already established in proposition 3.1.5.

It remains to show that G'_1 is involutive if G_q is involutive.

From the formal point of view this is evident because of the two preceding diagrams. In fact an easy chase shows that the exactness of the left column implies the exactness of the right one.

From the operational point of view, using proposition 3.2.24 and the fact that G_q is involutive, we have:

$$(G'_1)^i = (\sigma_1(S_{q+1}T^* \otimes E))^i = \sigma_1((S_{q+1}T^* \otimes E)^i)$$

and thus the short exact sequences of families of vector spaces over $\mathscr{R}_q$:

$$0 \longrightarrow (G_{q+1})^i \longrightarrow (S_{q+1}T^* \otimes E)^i \xrightarrow{\sigma_1} (G'_1)^i \longrightarrow 0$$

This can also be shown by an induction on i, using the commutative and exact diagram:

$$
\begin{array}{ccccccccc}
 & & 0 & & 0 & & 0 & & \\
 & & \downarrow & & \downarrow & & \downarrow & & \\
0 & \longrightarrow & (G_{q+1})^i & \longrightarrow & (S_{q+1}T^* \otimes E)^i & \xrightarrow{\sigma_1} & (G'_1)^i & \longrightarrow & 0 \\
 & & \downarrow & & \downarrow & & \downarrow & & \\
0 & \longrightarrow & (G_{q+1})^{i-1} & \longrightarrow & (S_{q+1}T^* \otimes E)^{i-1} & \xrightarrow{\sigma_1} & (G'_1)^{i-1} & \longrightarrow & 0 \\
 & & {\scriptstyle \delta_i}\downarrow & & {\scriptstyle \delta_i}\downarrow & & \downarrow{\scriptstyle \delta_i} & & \\
0 & \longrightarrow & (G_q)^{i-1} & \longrightarrow & (S_q T^* \otimes E)^{i-1} & \xrightarrow{\sigma} & F_0 & & \\
 & & \downarrow & & \downarrow & & & & \\
 & & 0 & & 0 & & & &
\end{array}
$$

It is thus easy to compute the characters of G'_1, and we have

$$\alpha_1'^i = \beta_q^i + \cdots + \beta_q^n$$

The end of the proof is left to the reader as it is an easy computation using the characters. C.Q.F.D.

We shall now use the families $G'_r \subset S_r T^* \otimes F_0$ of vector spaces over $\mathscr{R}_q$.

DEFINITION 1.10 We define the families F_r of vector spaces over $\mathscr{R}_q$ by the formulas:

$$\begin{cases} F_0 = N(\iota_q) & r = 0 \\ F_r = \dfrac{\Lambda^r T^* \otimes F_0}{\delta(\Lambda^{r-1} T^* \otimes G'_1)} & 1 \le r \le n \\ F_r = 0 & n < r \end{cases}$$

When G_q is involutive we know that G'_1 is also involutive and thus we can make with the F_r the same constructions as those done with the C_r. In particular we have the exact sequences:

$$0 \to G'_r \to S_r T^* \otimes F_0 \to S_{r-1} T^* \otimes F_1 \to \cdots \to T^* \otimes F_{r-1} \to F_r \to 0$$

in which the maps

$$S_{s+1} T^* \otimes F_r \to S_s T^* \otimes F_{r+1}$$

are the s-prolongations of the surjective maps $T^* \otimes F_r \to F_{r+1}$ induced by the exterior multiplication $(-1)^r \cdot \delta$.

The use of these families F_r will be fundamental for the study of linear systems of p.d.e.

2 Formal properties

We shall now exhibit some relations between the different families of vector spaces over $\mathscr{R}_q$ that we have already considered.

First of all we have the short exact sequence:

$$0 \to V(\mathscr{R}_q) \to V(J_q(\mathscr{E}))|_{\mathscr{R}_q} \to N(\iota_q) \to 0$$

and thus the short exact sequence:

$$0 \to C_0 \to C_0(E) \to F_0 \to 0$$

Moreover, as G_q is always 1-acyclic, we have the following commutative and exact diagram of families of vector spaces pulled back over $\mathscr{R}_q$:

$$\begin{array}{ccccccccc}
 & & 0 & & 0 & & 0 & & \\
 & & \downarrow & & \downarrow & & \downarrow & & \\
0 & \longrightarrow & G_{q+1} & \longrightarrow & S_{q+1}T^* \otimes E & \longrightarrow & G'_1 & \longrightarrow & 0 \\
 & & \downarrow \delta & & \downarrow & & \downarrow & & \\
0 & \longrightarrow & T^* \otimes C_0 & \longrightarrow & T^* \otimes C_0(E) & \longrightarrow & T^* \otimes F_0 & \longrightarrow & 0 \\
 & & \downarrow \tau_1 & & \downarrow & & \downarrow & & \\
0 & \longrightarrow & C_1 & \longrightarrow & C_1(E) & \longrightarrow & F_1 & \longrightarrow & 0 \\
 & & \downarrow & & \downarrow & & \downarrow & & \\
 & & 0 & & 0 & & 0 & &
\end{array}$$

and we obtain the short exact sequence:

$$0 \rightarrow C_1 \rightarrow C_1(E) \rightarrow F_1 \rightarrow 0$$

PROPOSITION 2.1 If G_q is 2-acyclic we have the short exact sequence:

$$0 \rightarrow C_2 \rightarrow C_2(E) \rightarrow F_2 \rightarrow 0$$

Proof According to proposition 1.9, as G_q is 2-acyclic it follows that G'_2 is the 1-prolongation of G'_1 and the proposition can be deduced from a chase in the following commutative and exact diagram of families of vector spaces pulled back over $\mathscr{R}_q$:

$$\begin{array}{ccccccccc}
 & & 0 & & 0 & & 0 & & \\
 & & \downarrow & & \downarrow & & \downarrow & & \\
0 & \longrightarrow & G_{q+2} & \longrightarrow & S_{q+2}T^* \otimes E & \longrightarrow & G'_2 & \longrightarrow & 0 \\
 & & \downarrow \delta & & \downarrow \delta & & \downarrow \delta & & \\
0 & \longrightarrow & T^* \otimes G_{q+1} & \longrightarrow & T^* \otimes S_{q+1}T^* \otimes E & \longrightarrow & T^* \otimes G'_1 & \longrightarrow & 0 \\
 & & \downarrow \delta & & \downarrow \delta & & \downarrow \delta & & \\
0 & \longrightarrow & \Lambda^2 T^* \otimes C_0 & \longrightarrow & \Lambda^2 T^* \otimes C_0(E) & \longrightarrow & \Lambda^2 T^* \otimes F_0 & \longrightarrow & 0 \\
 & & \downarrow \tau_2 & & \downarrow & & \downarrow & & \\
0 & \longrightarrow & C_2 & \longrightarrow & C_2(E) & \longrightarrow & F_2 & \longrightarrow & 0 \\
 & & \downarrow & & \downarrow & & \downarrow & & \\
 & & 0 & & 0 & & 0 & &
\end{array}$$

C.Q.F.D.

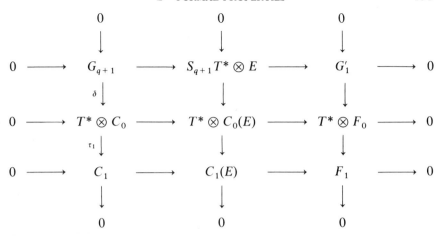

THEOREM 2.2 If G_q is s-acyclic, for $r = 0, 1, \ldots, s$ we have the short exact sequences:

$$0 \to C_r \to C_r(E) \to F_r \to 0$$

Proof According to the last proposition, we have only to consider the case $s > 2$. From proposition 1.9 we know that G'_{r+1} is the r-prolongation of G'_1 which is $(s-1)$-acyclic. The theorem can be deduced from a chase in the following commutative and exact diagram of families of vector spaces pulled back over $\mathscr{R}_q$:

$$\begin{array}{ccccccccc}
 & & 0 & & 0 & & 0 & & \\
 & & \downarrow & & \downarrow & & \downarrow & & \\
0 & \to & G_{q+r} & \to & S_{q+r}T^* \otimes E & \to & G'_r & \to & 0 \\
 & & {}^{\delta}\downarrow & & {}^{\delta}\downarrow & & \downarrow^{\delta} & & \\
 & & \vdots & & \vdots & & \vdots & & \\
 & & {}^{\delta}\downarrow & & {}^{\delta}\downarrow & & \downarrow^{\delta} & & \\
0 & \to & \Lambda^{r-1}T^* \otimes G_{q+1} & \to & \Lambda^{r-1}T^* \otimes S_{q+1}T^* \otimes E & \to & \Lambda^{r-1}T^* \otimes G'_1 & \to & 0 \\
 & & {}^{\delta}\downarrow & & {}^{\delta}\downarrow & & \downarrow^{\delta} & & \\
0 & \to & \Lambda^r T^* \otimes C_0 & \to & \Lambda^r T^* \otimes C_0(E) & \to & \Lambda^r T^* \otimes F_0 & & 0 \\
 & & {}^{\tau_r}\downarrow & & \downarrow & & \downarrow & & \\
0 & \to & C_r & \to & C_r(E) & \to & F_r & \to & 0 \\
 & & \downarrow & & \downarrow & & \downarrow & & \\
 & & 0 & & 0 & & 0 & &
\end{array}$$

C.Q.F.D.

COROLLARY 2.3 If G_{q+1} is a vector bundle over $\mathscr{R}_q$ and if G_q is involutive, we have the short exact sequences of vector bundles over $\mathscr{R}_q$:

$$0 \to C_r \to C_r(E) \to F_r \to 0$$

Proof According to proposition 3.1.16, G_{q+r} is a vector bundle over $\mathscr{R}_q$, $\forall r \geq 1$. The Euler–Poincaré formula then shows that C_r and F_r are vector bundles over $\mathscr{R}_q$ and that we have:

$$\dim C_r + \dim F_r = \dim C_r(E)$$

the last number depending only on n, m, q and not on $\mathscr{R}_q$. C.Q.F.D.

However, in practice, even when G_q is involutive it is very tedious to compute the numbers dim C_r, unless $\mathcal{R}_q$ is of finite type. In fact, in this case, from proposition 3.4.7 we know that $G_{q+r} = 0, \forall r \geq 0$ and we have

$$\dim C_r = \frac{n!}{(n-r)!r!} \dim V(\mathcal{R}_q).$$

Otherwise, the knowledge of the characters α_q^i allows one to compute dim G_{q+r}, $\forall r \geq 0$ and finally dim C_r. This tedious computation is left to the reader.

We shall now indicate a useful and quick way to compute the numbers dim F_r, at any point $(x, y, p) \in \mathcal{R}_q$ by means of dim F_0 and the characters $\alpha_q^i(x, y, p)$ of G_q when G_q is involutive. In the sequel the point $(x, y, p) \in \mathcal{R}_q$ will be fixed.

Let $P(G_q)$ be the board associated with the equations of G_q solved with respect to a δ-regular coordinate system. We complete it by dots in order to obtain the board $P(\mathcal{R}_q)$ as already explained.

The later board has n columns and dim F_0 rows; β_q^n of them have no dot; β_q^{n-1} have one dot; ..., β_q^1 have $(n-1)$ dots and the remaining ones have n dots, according to the following picture:

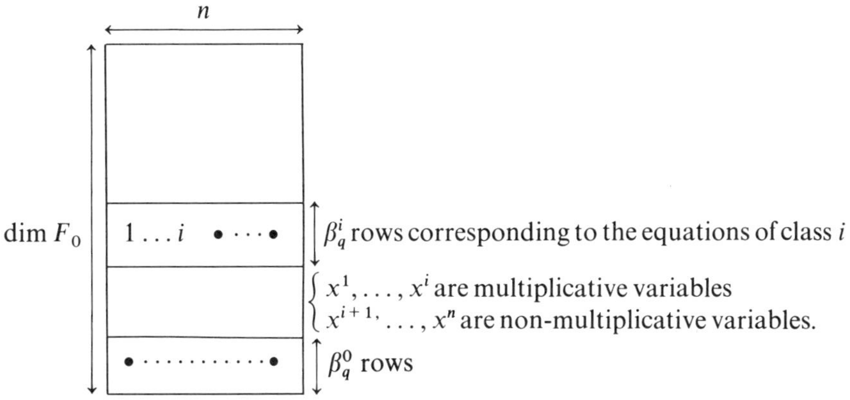

PROPOSITION 2.4 For $1 \leq r \leq n$, the numbers dim F_r can be computed easily by means of the board $P(\mathcal{R}_q)$.

Proof As G_q is involutive we have $\alpha_1'^i = \beta_q^i + \cdots + \beta_q^n$ and the character $\alpha_1'^i$ of G_1' is equal to the number of multiplicative variables appearing in the column i of $P(\mathcal{R}_q)$.

EXAMPLE 2.5 Let us compute $\alpha_1'^n$. We have the short exact sequence:

$$0 \to (G_{q+1})^{n-1} \to (S_{q+1}T^* \otimes E)^{n-1} \to (G_1')^{n-1} \to 0$$

but

$$\dim (G_{q+1})^{n-1} = \dim (G_q)^{n-1} = \alpha_q^n$$

and

$$\dim (S^{q+1}T^* \otimes E)^{n-1} = m.$$

Finally

$$\alpha_1'^n = m - \alpha_q^n = \beta_q^n.$$

Let us now use local coordinates $v_I'^\tau$ with $I = (i_1, \ldots, i_r)$ and $i_1 < \cdots < i_r$ in order to describe the fiber over $(x, y, p) \in \mathscr{R}_q$ of the vector bundle $\Lambda^r T^* \otimes F_0$.
We have the technical lemma:

LEMMA 2.6 The number $\dim F_r$ is equal to the number of multiplets $(\tau; i_1, \ldots, i_r)$ such that $x^{i_1}, \ldots, x^{i_r}$ are non-multiplicative variables for the row τ of $P(\mathscr{R}_q)$. In particular $\dim F_n = \beta_q^0$.

REMARK 2.7 This lemma is fundamental in practice as it is the only trick that allows one to compute easily the dimensions of all the vector spaces or vector bundles found in the study of any system of p.d.e. The proof is rather technical.

Proof We order the local coordinates $v_i'^\tau$ for $T^* \otimes F_0$ by means of the lexicographic order for the couples $(i, \dim F_0 - \tau)$. Any solved equation of G_1' will allow one to compute a principal component $v_i'^\tau$ by means of (x, y, p) and the parametric components of class $\leq i$ and thus of lower rank because all the dots in $P(\mathscr{R}_q)$ are in the lower right corner.
As we have the short exact sequence:

$$0 \longrightarrow G_1' \longrightarrow T^* \otimes F_0 \xrightarrow{\tau_1^0} F_1 \longrightarrow 0$$

we can index the fiber of F_1 by the couples (τ, i) such that $v_i'^\tau$ is a principal component, that is to say such that x^i is a non-multiplicative variable for the row τ of $P(\mathscr{R}_q)$.
We shall now construct $P_1 \equiv P(G_1')$ from $P_0 \equiv P(\mathscr{R}_q)$. For this we construct a board with n columns and a number of rows equal to the total number of dots in $P(\mathscr{R}_q)$, that is to say $\dim F_1$. From what has been said, the number of equations of class i of G_1' is equal to the number of dots in the column i of $P(\mathscr{R}_q)$.
Thus $x^{i+1}, \ldots, x^n$ are non-multiplicative variables for the row of $P(G_1')$ indexed by $(\tau; i)$.
We let the reader show inductively that the board $P_r = P(\ker \tau_r^0)$ has n columns and $\dim F_r$ rows and that the rows corresponding to equations

of class i_r for ker τ_r^0 can be indexed by sets $(\tau; i_1, \ldots, i_r)$ with $i_1 < \cdots < i_r$ such that $x^{i_1}, \ldots, x^{i_r}$ are non-multiplicative variables for the row τ of P_0. Moreover these rows admit $x^{i_r+1}, \ldots, x^n$ as non-multiplicative variables. This proves the lemma. C.Q.F.D.

Finally, for any element $(v'^{\tau}_{i_1,\ldots,i_r}) \in \delta(\Lambda^{r-1}T^* \otimes G'_1)$ there exists an element $(v'^{\tau}_{i_1;i_2,\ldots,i_r}) \in \Lambda^{r-1}T^* \otimes G'_1$ such that

$$\sum_{s=1}^{r} (-1)^{s-1} v'^{\tau}_{i_s;i_1,\ldots,\hat{i}_s,\ldots,i_r} = v'^{\tau}_{i_1,\ldots,i_r}$$

where we omit any letter with a hat over it.

In fact we have:

$$v'^{\tau}_{i_1,\ldots,i_r} = v'^{\tau}_{i_1;i_2,\ldots,i_r} + \sum_{s=2}^{r} (-1)^{s-1} v'^{\tau}_{i_s;i_1,\ldots,\hat{i}_s,\ldots,i_r}$$

We divide the set of components $v'^{\tau}_{i_1,\ldots,i_r}$ into two disjoint subsets: one is such that x^{i_1} is a multiplicative variable for the row τ of P_0, the other is such that x^{i_1} (and thus $x^{i_2}, \ldots, x^{i_n}$) is a non-multiplicative variable for the same row of P_0. The components of the first set are independent because $v'^{\tau}_{i_1}$ is parametric for G'_1 and the dim F_r components of the second set are thus principal components for $\delta(\Lambda^{r-1}T^* \otimes G'_1)$ because

$$\dim \delta(\Lambda^{r-1}T^* \otimes G'_1) = \dim \Lambda^r T^* \otimes F_0 - \dim F_r.$$

This proves the proposition. C.Q.F.D.

Remark 2.8 We have $\dim(\ker \tau_s^r)^{n-1} = \dim(\ker \tau_s^0)^{n-1}$ because $\tau_s^r = \rho_r(\tau_s^0)$ and the short exact sequences:

$$0 \longrightarrow \ker \tau_{r+1}^s \longrightarrow S_{s+1}T^* \otimes C_r \xrightarrow{\tau_{r+1}^s} \ker \tau_{r+2}^{s-1} \longrightarrow 0$$

thus

$$\dim C_r = \dim(\ker \tau_{r+1}^0)^{n-1} + \dim(\ker \tau_{r+2}^0)^{n-1}$$

and:

$$\begin{cases} \sum_{r=0}^{n} (-1)^r \dim C_r = \alpha_q^n \\ \sum_{r=0}^{n} (-1)^r \dim C_r(E) = m \\ \sum_{r=0}^{n} (-1)^r \dim F_r = \beta_q^n \end{cases}$$

EXERCISE 2.9 We ask the reader to check the last formula directly, using the explicit formula found above for $\dim F_r$, that is to say:

$$\dim F_r = \sum_{i=r}^{n} \frac{i!}{(i-r)!r!} \beta_q^{n-i} \quad \text{with} \quad \beta_q^0 = \dim F_0 - \sum_{i=1}^{n} \beta_q^i$$

(Hint: use the formula:

$$\sum_{r=0}^{i} (-1)^r \frac{i!}{(i-r)!r!} = 0 \quad \text{for} \quad i \geq 1 \Bigg)$$

3 Criterion theorem

We shall now indicate briefly the proof of the criterion for formal integrability, given by H. Goldschmidt (14) using the Spencer families.

Let $\mathscr{R}_q \subset J_q(\mathscr{E})$ be a system of order q on $\mathscr{E}$. By definition

$$\mathscr{R}_{q+1} = J_1(\mathscr{R}_q) \cap J_{q+1}(\mathscr{E})$$

is not in general a fibered submanifold of $J_{q+1}(\mathscr{E})$ but, when this is the case, we may identify $\mathscr{R}_{q+1}$ with its image $\hat{\mathscr{R}}_1 \subseteq J_1(\mathscr{R}_q)$ by the monomorphism

$$\rho_1(\mathrm{id}_{J_q(\mathscr{E})}) : J_{q+1}(\mathscr{E}) \to J_1(J_q(\mathscr{E})).$$

Similarly to theorem 3.3.1 we have:

PROPOSITION 3.1 The r-prolongation of the non-linear system

$$\mathscr{R}_{q+1} \subset J_1(\mathscr{R}_q)$$

of order 1 on $\mathscr{R}_q$ is $\mathscr{R}_{q+r-1}$. Moreover, if $\mathscr{S}$ is the set of solutions of $\mathscr{R}_q \subset J_q(\mathscr{E})$, then $j_q(\mathscr{S})$ is the set of solution of $\mathscr{R}_{q+1} \subset J_1(\mathscr{R}_q)$.

REMARK 3.2 Setting $\hat{\mathscr{E}} = \mathscr{R}_q$, we notice that, when $\pi_q^{q+1} : \mathscr{R}_{q+1} \to \mathscr{R}_q$ is an epimorphism, then $\hat{\pi}_0^1 : \hat{\mathscr{R}}_1 \to \hat{\mathscr{E}}$ is also an epimorphism. This method of reducing any system of order q to a system of order 1 is better than the one we have already studied. The reader may check this on some problems.

Similarly to theorem 3.3.4 we have:

PROPOSITION 3.3 The symbol $\hat{G}_1$ of $\hat{\mathscr{R}}_1$ is involutive if and only if the symbol G_q of R_q is involutive.

We shall obtain the following theorem:

THEOREM 3.4 If $\mathscr{R}_{q+1}$ is a fibered submanifold of $J_{q+1}(\mathscr{E})$, $\pi_q^{q+1}: \mathscr{R}_{q+1} \to \mathscr{R}_q$ is an epimorphism and G_q is 2-acyclic, then $\mathscr{R}_q$ is formally integrable.

Proof Under the hypothesis of the theorem, G_{q+1} is a vector bundle over $\mathscr{R}_q$ and thus, according to proposition 3.1.16, G_{q+r} is also a vector bundle over $\mathscr{R}_q$.

We have the short exact sequence:

$$0 \longrightarrow G_{q+1} \longrightarrow T^* \otimes C_0 \xrightarrow{\tau_1} C_1 \longrightarrow 0$$

But $\mathscr{R}_{q+1}$ and $J_1(\mathscr{R}_q)$ are both affine bundles over $\mathscr{R}_q$. From proposition 1.8.8, there exists an epimorphism $\Psi: J_1(\mathscr{R}_q) \to C_1$ such that we have the short exact sequence:

$$0 \longrightarrow \mathscr{R}_{q+1} \longrightarrow J_1(\mathscr{R}_q) \xrightarrow{\Psi} C_1 \longrightarrow 0$$

where $\mathscr{R}_{q+1}$ is the kernel of Ψ with respect to the zero section "0" of the vector bundle C_1 over $\mathscr{R}_q$.

It follows that $\mathscr{R}_{q+r+1} = (\mathscr{R}_{q+1})_{+r}$ is the kernel of $\rho_r(\Psi): J_{r+1}(\mathscr{R}_q) \to J_r(C_1)$ with respect to the zero section $J_r(0)$ of $J_r(C_1)$ over $J_r(\mathscr{R}_q)$, and we have $\tau_1^r = \sigma_2(\Psi)$ with $\tau_1 = \sigma(\Psi)$.

Moreover we have the following commutative and exact diagram where the names of the maps have been omitted for simplicity:

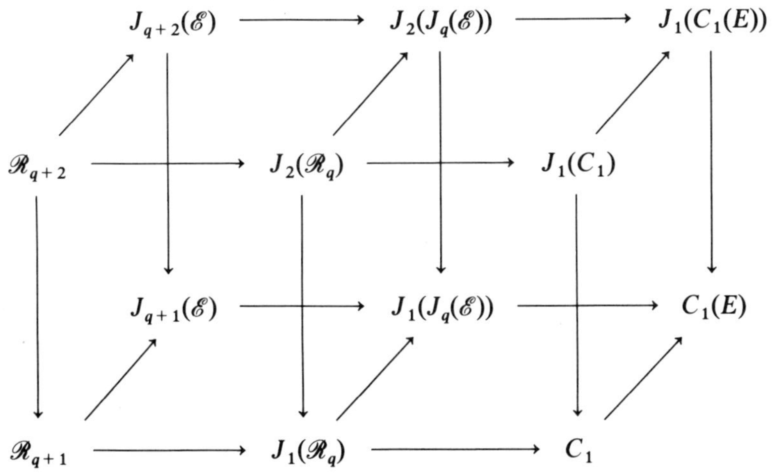

We notice that $J_{q+2}(\mathscr{E})$, $J_2(J_q(\mathscr{E}))$, $J_1(C_1(E))$, $J_2(\mathscr{R}_q)$, $J_1(C_1)$ are affine bundles respectively over $J_{q+1}(\mathscr{E})$, $J_1(J_q(\mathscr{E}))$, $C_1(E)$, $J_1(\mathscr{R}_q)$, C_1 and that we have the commutative and exact diagram of vector bundles pulled back on $\mathscr{R}_q$:

$$\begin{array}{ccccccccccc}
 & & 0 & & 0 & & 0 & & 0 & & \\
 & & \downarrow & & \downarrow & & \downarrow & & \downarrow & & \\
0 & \longrightarrow & G_{q+2} & \longrightarrow & S_2T^* \otimes C_0 & \xrightarrow{\sigma_1(\Psi)} & T^* \otimes C_1 & \longrightarrow & C_2 & \longrightarrow & 0 \\
 & & \downarrow & & \downarrow & & \downarrow & & \downarrow & & \\
0 & \longrightarrow & S_{q+2}T^* \otimes E & \longrightarrow & S_2T^* \otimes C_0(E) & \longrightarrow & T^* \otimes C_1(E) & \longrightarrow & C_2(E) & \longrightarrow & 0 \\
 & & \downarrow & & \downarrow & & \downarrow & & \downarrow & & \\
0 & \longrightarrow & G'_2 & \longrightarrow & S_2T^* \otimes F_0 & \longrightarrow & T^* \otimes F_1 & \longrightarrow & F_2 & \longrightarrow & 0 \\
 & & \downarrow & & \downarrow & & \downarrow & & \downarrow & & \\
 & & 0 & & 0 & & 0 & & 0 & &
\end{array}$$

The exactness follows from proposition 2.1 and the fact that

$$\pi_q^{q+1}(\mathscr{R}_{q+1}) = \mathscr{R}_q.$$

An easy chase allows one to get a map $\kappa(J_q(\mathscr{E})) : J_{q+1}(\mathscr{E}) \to C_2(E)$ with kernel $J_{q+2}(\mathscr{E})$. Its restriction to $\mathscr{R}_{q+1}$ is a map $\kappa(\mathscr{R}_q) : \mathscr{R}_{q+1} \to C_2$ with kernel $\mathscr{R}_{q+2}$. We have the commutative and exact diagram:

$$\begin{array}{ccccc}
\mathscr{R}_{q+2} & \longrightarrow & \mathscr{R}_{q+1} & \xrightarrow{\kappa(\mathscr{R}_q)} & C_2 \\
\downarrow & & \downarrow & & \downarrow \\
J_{q+2}(\mathscr{E}) & \longrightarrow & J_{q+1}(\mathscr{E}) & \xrightarrow{\kappa(J_q(\mathscr{E}))} & C_2(E)
\end{array}$$

As $\pi_{q+1}^{q+2} : J_{q+2}(\mathscr{E}) \to J_{q+1}(\mathscr{E})$ is an epimorphism and $C_2 \to C_2(E)$ is a monomorphism, we deduce at once that $\pi_{q+1}^{q+2} : \mathscr{R}_{q+2} \to \mathscr{R}_{q+1}$ is an epimorphism.

An easy induction on r shows that $\pi_{q+r}^{q+r+1} : \mathscr{R}_{q+r+1} \to \mathscr{R}_{q+r}$ is also an epimorphism and this proves the theorem.

We notice however that the above proof is very far from any intuitive test.

C.Q.F.D.

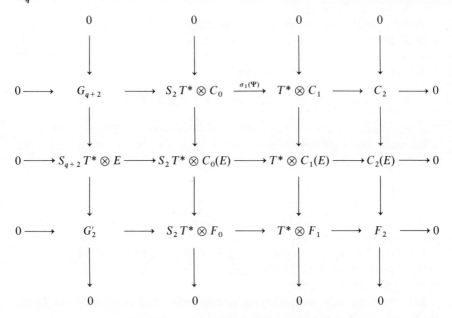

4 Analytic systems

Let $\mathscr{R}_q \subset J_q(\mathscr{E})$ be an involutive system of order q on $\mathscr{E}$. By hypothesis $\mathscr{R}_{q+r}$ is a fibered submanifold of $J_{q+r}(\mathscr{E})$, $\forall r \geq 0$ and $\pi^{q+r+1}_{q+r} : \mathscr{R}_{q+r+1} \to \mathscr{R}_{q+r}$ is an epimorphism. Moreover the symbol G_q of $\mathscr{R}_q$ is involutive.

If $\alpha^1_q, \ldots, \alpha^n_q$ are the characters of G_q, then G_{q+r} is a vector bundle over $\mathscr{R}_q$ and we have:

$$\dim G_{q+r} = \sum_{i=1}^{n} \frac{(r+i-1)!}{r!(i-1)!} \alpha^i_q$$

When one has chosen a coordinate system δ-regular at a point $(x_0, y_0, p_0) \in \mathscr{R}_q$, then it is possible to use the implicit function theorem.

In fact we can solve the equations of $\mathscr{R}_q$ and its prolongations with respect to the principal jet coordinates determined by the solved form of the equations of G_q and its prolongations G_{q+r}. Of course we have to take into account the equations of order $q-1$ of the fibered submanifold $\pi^q_{q-1}(\mathscr{R}_q) \subset J_{q-1}(\mathscr{E})$ and the short exact sequence:

$$0 \to G_q \to V(\mathscr{R}_q) \to V(\Pi^q_{q-1}(\mathscr{R}_q)) \to 0$$

For $r = 0, 1, 2, \ldots$ we can choose points of $\mathscr{R}_{q+r}$ in such a way that the point of $\mathscr{R}_{q+r+1}$ projects onto the point of $\mathscr{R}_{q+r}$, $\forall r \geq 0$.

We thus obtain a formal series:

$$y^k_0 + \sum_{|\mu|=1}^{\infty} p^k_{0\mu} \frac{(x-x_0)^\mu}{\mu!}$$

If this series is convergent and admits a sum equal to $f^k(x)$, the proposition 2.3.5 applied when $\mathscr{R}_q$ is an analytic system, then shows that the section $f : (x) \to (x, f^k(x))$ is solution of $\mathscr{R}_q$.

Remark 4.1 For an arbitrary differentiable system there is no conclusion "*a priori*". In fact, in the chapter on linear systems we shall give a celebrated example, due to H. Lewy, of an involutive linear system for which such a series may not be convergent.

From now on we shall be concerned with analytic systems. Our purpose will be to study the construction of such a Taylor development and to prove its convergence.

Proposition 4.2 The latter formal series is uniquely determined by the point $\pi^q_{q-1}(x_0, y_0, p_0) \in \pi^q_{q-1}(\mathscr{R}_q)$ and by α^i_q arbitrary formal series in i variables, for $i = 1, \ldots, n$.

Proof We use local coordinates. For $|\mu| = q$, $x^1, \ldots, x^i$ are the multiplicative variables of the p_μ^k of class i and $x^{i+1}, \ldots, x^n$ are the corresponding non-multiplicative variables.

We know that, for any multiplicative variable x^i of a principal p_μ^k, then $p_{\mu+1_i}^k$ is also a principal component of order $q + 1$ and that the principal components thus obtained are all different, according to proposition 3.2.15, and constitute the principal components of order $q + 1$. More generally $p_{\mu+\nu}^k$ with $\nu_{i+1} = 0, \ldots, \nu_n = 0$ and $|\nu| = r$ is a principal component of order $q + r$ whenever p_μ^k is a principal component of order q and class i.

The same thing is true for the parametric components. In fact, one has just to use the fact that a zero symbol is trivially involutive. Thus one obtains once and only once all the multi-index $\mu + \nu$ of order $q + r$ with $|\mu| = q$, $|\nu| = r$ from the multi-index μ of class i with $|\mu| = q$, by taking multi-index $|\nu|$ with $|\nu| = r$ and such that $\nu_{i+1} = 0, \ldots, \nu_n = 0$ for $i = 1, \ldots, n$.

Finally one obtains once and only once all the parametric components $p_{\mu+\nu}^k$ of order $q + r$ with $|\mu| = q$, $|\nu| = r$, from the parametric components p_μ^k of order q and class i, by taking ν with $|\nu| = r$ and such that

$$\nu_{i+1} = 0, \ldots, \nu_n = 0 \quad \text{for} \quad i = 1, \ldots, n.$$

As the number of principal p_μ^k of order q and class i is equal to

$$m\frac{(q + n - i - 1)!}{(q - 1)!(n - i)!} - \alpha_q^i,$$

it follows that the number of parametric p_μ^k of order q and class i is just equal to α_q^i.

If in the formal series we equate to zero the principal $p_{0\mu}^k$, there remain only the parametric $p_{0\mu}^k$ arbitrarily chosen and determined by the knowledge for $i = 1, \ldots, n$ of α_q^i formal series:

$$\sum_{\nu_{i+1}=0, \ldots, \nu_n=0} p_{0\mu+\nu}^k \frac{(x - x_0)^\nu}{\nu!}$$

This proves the proposition. C.Q.F.D.

From this proposition we deduce the important theorem:

THEOREM 4.3 (Cartan–Khäler) If $\mathcal{R}_q \subset J_q(\mathcal{E})$ is an involutive and analytic system of order q on $\mathcal{E}$, there exists one analytic solution $y^k = f^k(x)$ and only one such that:

1) $(x_0, \partial_\mu f^k(x_0))$ with $|\mu| \leq q - 1$ is a point of $\pi_{q-1}^q(\mathcal{R}_q) \subset J_{q-1}(\mathcal{E})$.
2) For $i = 1, \ldots, n$ the α_q^i parametric derivatives $\partial_\mu f^k(x)$ of class i are equal for $x^{i+1} = x_0^{i+1}, \ldots, x^n = x_0^n$ to α_q^i given analytic functions of $x^1, \ldots, x^i$.

Proof As in the preceding pages, we shall identify $\mathscr{R}_{q+1}$ with its image $\hat{\mathscr{R}}_1$ in $J_1(\mathscr{R}_q)$ by the monomorphism $\rho_1(\mathrm{id}_{J_q(\mathscr{E})}): J_{q+1}(\mathscr{E}) \to J_1(J_q(\mathscr{E}))$. Setting $\mathscr{R}_q = \hat{\mathscr{E}}$, we have $\hat{\mathscr{R}}_1 \subset J_1(\hat{\mathscr{E}})$ and we already know that there is a bijective map between the set of solutions of $\mathscr{R}_q$ and that of $\hat{\mathscr{R}}_1$.

It is easy to check that $\hat{\mathscr{R}}_1$ is also an involutive and analytic system. But now it is a first order system and moreover $\pi_0^1: \hat{\mathscr{R}}_1 \to \hat{\mathscr{E}}$ is an epimorphism.

Moreover we can identify G_{q+r} and $\hat{G}_r$ and $\hat{G}_r$ is involutive. We have the following relations:

$$\hat{\alpha}_1^i = \alpha_q^i + \cdots + \alpha_q^n = \alpha_{q+1}^i$$

and thus

$$\alpha_q^i = \hat{\alpha}_1^i - \hat{\alpha}_1^{i+1} = \hat{\beta}_1^{i+1} - \hat{\beta}_1^i.$$

Finally the proof of the above theorem is equivalent to that of the following theorem:

THEOREM 4.4 If $\mathscr{R}_1 \subset J_1(\mathscr{E})$ is a first order involutive and analytic system such that $\pi_0^1: \mathscr{R}_1 \to \mathscr{E}$ is an epimorphism, then there exists one analytic solution $y^k = f^k(x)$ and only one, such that:

1) $f^1(x), \ldots, f^{\beta_1^1}(x)$ are equal to β_1^1 given constants when $x = x_0$.
2) $f^{\beta_1^i+1}(x), \ldots, f^{\beta_1^{i+1}}(x)$ are equal, to $\beta_1^{i+1} - \beta_1^i$ given analytic functions of $x^1, \ldots, x^i$ when $x^{i+1} = x_0^{i+1}, \ldots, x^n = x_0^n$.
3) $f^{\beta_1^n+1}(x), \ldots, f^m(x)$ are $m - \beta_1^n$ given analytic functions of $x^1, \ldots, x^n$.

REMARK 4.5 In this case we have $P_0 = P(\mathscr{R}_1) = P(G_1)$ because the number of unconnected equations of $\mathscr{R}_1$ is equal to $\sum_{i=1}^n \beta_1^i$. Moreover we have

$$0 \leq \beta_1^1 \leq \cdots \leq \beta_1^n \leq m.$$

Proof We shall use the following well known particular case in which

$$\beta_1^1 = 0, \ldots, \beta_1^{n-1} = 0, \beta_1^n = m.$$

THEOREM 4.6 (Cauchy-Kowalewski) If $\mathscr{R}_1 \subset J_1(\mathscr{E})$ is a first order involutive and analytic system such that $\pi_0^1: \mathscr{R}_1 \to \mathscr{E}$ is an epimorphism and $\beta_1^1 = 0, \ldots, \beta_1^{n-1} = 0, \beta_1^n = m$, then there exists one solution $y^k = f^k(x)$ and only one such that $f^1(x), \ldots, f^m(x)$ are equal to m given analytic functions of $x^1, \ldots, x^{n-1}$ when $x^n = x_0^n$.

Proof This is elementary and can be found in many textbooks. We refer the reader to (62).

We come back to the proof of the preceding theorem and advise the reader to follow the proof while solving problem 17.

We take a coordinate system δ-regular at the point $(x_0, y_0, p_0) \in \mathcal{R}_1$ determined by the given initial conditions and we solve the equations of $\mathcal{R}_1$ in a neighbourhood of this point by means of the implicit function theorem, using the fact that the symbol G_1 of $\mathcal{R}_1$ is involutive.

We shall prove that the determination of the solution of $\mathcal{R}_1$ satisfying the given initial conditions, can be done by solving successively systems of the Cauchy–Kowalewski type.

The later systems will allow us to obtain $f^k(x^1, x_0^2, \ldots, x_0^n)$ from $f^k(x_0^1, \ldots, x_0^n)$ by using the solved equations of $\mathcal{R}_1$ of class 1, and so on, until we obtain $f^k(x^1, \ldots, x^n)$ from $f^k(x^1, \ldots, x^{n-1}, x_0^n)$ by using the solved equations of $\mathcal{R}_1$ of class n.

For simplicity we shall only detail the steps 1 and n.

Step 1 The solved equations of $\mathcal{R}_1$ of class 1 are of the type:

$$\Phi_1^\beta(x, y, p) \equiv p_1^\beta + A_1^\beta(x, y, p_1^\alpha) = 0$$

with

$$1 \le \beta \le \beta_1^1, \quad \beta_1^1 < \alpha \le m.$$

As the $f^\alpha(x^1, x_0^2, \ldots, x_0^n)$ are known because of the given initial conditions, the $f^\beta(x^1, x_0^2, \ldots, x_0^n)$ are the unique solutions of a Cauchy–Kowalewski system of the type

$$p_1^\beta + \bar{A}_1^\beta(x^1, y^\beta) = 0$$

corresponding to the given

$$f^\beta(x_0^1, \ldots, x_0^n) \quad \text{with} \quad 1 \le \beta \le \beta_1^1.$$

Step n Let us suppose that we know the solutions

$$y^k = f^k(x^1, \ldots, x^{n-1}, x_0^n)$$

of the solved equations of $\mathcal{R}_1$ of class $1, \ldots, n-1$ for $x^n = x_0^n$.

The solved equations of $\mathcal{R}_1$ of class n are of the type:

$$\Phi_n^\beta(x, y, p) \equiv p_n^\beta + A_n^\beta(x, y, p_i^\alpha) = 0$$

with

$$1 \le \beta \le \beta_1^n, \quad \beta_1^i < \alpha \le m, \quad 1 \le i \le n.$$

As the $f^\alpha(x^1, \ldots, x^n)$ are known for $\beta_1^n < \alpha \le m$ because of the given unitial conditions, the $f^\beta(x^1, \ldots, x^n)$ for $1 \le \beta \le \beta_1^n$ are the unique solutions of a Cauchy–Kowalewski system of the type:

$$p_n^\beta + \bar{A}_n^\beta(x, y^\beta, p_i^\alpha) = 0$$

with

$$1 \leq \beta \leq \beta_1^n, \quad \beta_1^i < \alpha \leq m, \quad 1 \leq i \leq n-1,$$

corresponding to the $f^\beta(x^1, \ldots, x^{n-1}, x_0^n)$ given by the induction process.

Now, because of the step n, $y^k = f^k(x)$ is a solution of the solved equations of $\mathscr{R}_1$ of class n and we have:

$$g_n^\beta(x) \equiv \Phi_n^\beta(j_1(f)(x)) \equiv 0.$$

It remains to show that $y^k = f^k(x)$ is also a solution of the other solved equations of $\mathscr{R}_1$, that is to say, that we have:

$$g_i^\beta(x) \equiv \Phi_i^\beta(j_1(f)(x)) \equiv 0 \quad \text{with} \quad 1 \leq \beta \leq \beta_1^i, 1 \leq i \leq n-1.$$

According to lemma 2.3.15, since $\pi_1^2 : \mathscr{R}_2 \to \mathscr{R}_1$ is an epimorphism, the functions $d\Phi_i^\beta/dx^n$ with $1 \leq \beta \leq \beta_1^i$ and $1 \leq i \leq n-1$ must belong to the ideal generated by the functions Φ^τ with $1 \leq \tau \leq \sum_{i=1}^n \beta_1^i$ and the functions $d\Phi_{i'}^{\beta'}/dx^j$ with $1 \leq \beta' \leq \beta_1^{i'}$, $1 \leq i' \leq n$ and $1 \leq j \leq i'$.

This fact follows easily from proposition 3.2.15 and remark 3.2.14 by eliminating the p_{ij}^k between the equations $d\Phi^\tau/dx^i$ using the board P_0.

It follows that we have identities such as:

$$\frac{d\Phi_i^\beta}{dx^n} + A_{\beta'}^{i'j}(x, y, p)\frac{d\Phi_{i'}^{\beta'}}{dx^j} + B_\tau(x, y, p)\Phi^\tau \equiv 0$$

If we set in these relations:

$$y^k = f^k(x), p_i^k = \partial_i f^k(x), p_{ij}^k = \partial_{ij} f^k(x)$$

we have

$$\frac{d\Phi^\tau}{dx^i}(j_2(f)(x)) = \partial_i \Phi^\tau(j_1(f)(x))$$

and we know that

$$\Phi_n^\beta(j_1(f)(x)) \equiv 0, \qquad \forall 1 \leq \beta \leq \beta_1^n.$$

We finally obtain, for the $g_i^\beta(x)$, a system of p.d.e. of the following type:

$$\partial_n g_i^\beta(x) + A_{\beta'}^{i'j}(x)\partial_j g_{i'}^{\beta'}(x) + B_{\beta''}^{i''}(x) g_{i''}^{\beta''}(x) = 0$$

with

$$1 \leq \beta \leq \beta_1^i, 1 \leq \beta' \leq \beta_1^{i'}, 1 \leq \beta'' \leq \beta_1^{i''}, 1 \leq i, i', i'' \leq n-1, 1 \leq j \leq i'.$$

This system, which is of the Cauchy–Kowalewski type, has one solution $g_i^\beta(x)$ and only one such that $g_i^\beta(x^1, \ldots, x^{n-1}, x_0^n) = 0$. But it has of course the solution $g_i^\beta(x) \equiv 0$.

We have $g_i^\beta(x) \equiv 0$ and thus $\Phi_i^\beta(j_1(f)(x)) \equiv 0$. C.Q.F.D.

CHAPTER 5

1 Linear systems

The reader is now familiar with the formal study of non-linear systems of p.d.e. However linear systems of p.d.e. are to be met very often and have specific properties.

We shall first briefly recall some formal results obtained in the non-linear case. Their proofs can be adapted easily from the corresponding non-linear ones. In fact, in most cases, the reader will have only to take the tensor products $\otimes$ over X instead of taking them over $\mathscr{R}_q$.

These results will then allow us to study the specific properties that are to be found in the linear case, namely the existence of some differential sequences associated with a given differential operator.

To help the reader, we have tried to keep this chapter almost self contained.

Let X be a C^∞, connected, paracompact manifold with $\dim X = n$. Let $T = T(X)$ and $T^* = T^*(X)$ be the tangent and cotangent bundles of X.

If E is a vector bundle over X, then we denote by $J_q(E)$ the bundle of q-jets of sections of E. It is a vector bundle over X and an affine bundle over $J_{q-1}(E)$ modelled on the vector bundle $S_q T^* \otimes E$ and we have the short exact sequence:

$$0 \longrightarrow S_q T^* \otimes E \xrightarrow{\varepsilon_q} J_q(E) \xrightarrow{\pi^q_{q-1}} J_{q-1}(E) \longrightarrow 0$$

In order to simplify the notations, we denote also by E the set of germs of sections of E, as the context will always make clear when differentiations are to be considered.

On an open set $U \subset X$, trivialising E, we shall adopt for E the local coordinates (x^i, u^k) and for $J_q(E)$ the local coordinates (x^i, u^k_μ) with $i = 1, \ldots, n; k = 1, \ldots, m = \dim E$

$$\mu = (\mu_1, \ldots, \mu_n); 0 \leq |\mu| = \mu_1 + \cdots + \mu_n \leq q; u^k_0 = u^k$$

We denote by $\xi : (x) \to (x, \xi^k(x))$ a local section of E over U and by $\xi_q : (x) \to (x, \xi^k_\mu(x))$ a local section of $J_q(E)$ over U.

We set $dx^I = dx^{i_1} \wedge \cdots \wedge dx^{i_r}$ with $i_1 < i_2 < \cdots < i_r$ and we adopt local coordinates $(x^i, u^k_{\mu, I})$ for the vector bundle $\Lambda^r T^* \otimes J_q(E)$.

Let E and E' be vector bundles over X. If $\Phi : J_q(E) \to E'$ is a morphism of vector bundles, we denote by $\mathscr{D} = \Phi \circ j_q$ the differential operator of order q determined by Φ.

Using local coordinates, we get:

$$\begin{array}{rccc} \Phi: & (x, u_\mu^k) & \longrightarrow & (x, A_k^{\tau\mu}(x)u_\mu^k) \\ \mathscr{D}: & \xi & \longrightarrow & \mathscr{D}\cdot\xi \\ & (x, \xi^k(x)) & \longrightarrow & (x, A_k^{\tau\mu}(x)\partial_\mu\xi^k(x)) \end{array}$$

DEFINITION 1.1 We call r-prolongation of Φ the unique morphism

$$\rho_r(\Phi): J_{q+r}(E) \to J_r(E')$$

such that

$$j_r \circ \mathscr{D} = \rho_r(\Phi) \circ j_{q+r}$$

and we denote by

$$\sigma_r(\Phi): S_{q+r}T^* \otimes E \to S_r T^* \otimes E'$$

its restriction to $S_{q+r}T^* \otimes E$.

We then introduce the families of vector spaces over X:

$$R_{q+r} = \ker \rho_r(\Phi) \quad \text{and} \quad G_{q+r} = \ker \sigma_r(\Phi).$$

We denote by Θ the set of solutions of $\mathscr{D}$, that is to say the set of local sections ξ of E such that $\mathscr{D}\cdot\xi = 0$. It follows that an element of Θ is a section ξ of E over an open set $U \subset X$, such that

$$A_k^{\tau\mu}(x)\partial_\mu\xi^k(x) = 0, \qquad \forall x \in U.$$

DEFINITION 1.2 A linear system of order q on E is a vector sub-bundle R_q of $J_q(E)$. A solution of R_q is a section ξ of E over an open set $U \subset X$ such that $j_q(\xi)$ is a section of R_q over U, that is to say

$$j_q(\xi)(x) \in R_q, \qquad \forall x \in U.$$

Contrary to what happens in the non-linear case, if we are given $R_q \subset J_q(E)$, we can always determine a vector bundle F_0 and an epimorphism $\Phi: J_q(E) \to F_0$ such that $R_q = \ker \Phi$. We have the short exact sequence:

$$0 \longrightarrow R_q \xrightarrow{\iota_q} J_q(E) \xrightarrow{\Phi} F_0 \longrightarrow 0$$

REMARK 1.3 When we have a morphism $\Phi: J_q(E) \to E'$ we shall set $F_0 = \operatorname{im}\Phi$ for reasons that will be clear later on and we still denote by $\Phi: J_q(E) \to F_0$ the induced epimorphism. F_0 in the linear case is the analogue of $N(\iota_q)$ in the non-linear case and our notations are coherent.

As in the non-linear case we have:

$$R_{q+r} = J_r(R_q) \cap J_{q+r}(E)$$

We set, in general:

$$\begin{array}{lll} R_{q-r} = J_{q-r}(E) & \text{for} & 1 \le r \le q \\ G_{q-r} = S_{q-r}T^* \otimes E & \text{for} & 1 \le r \le q \end{array}$$

We notice once more that R_{q+r} and G_{q+r} (with the exception of R_q by definition) are families of vector spaces over X and not always vector bundles for $r \geq 0$.

We have the proposition:

Proposition 1.4 If $R_q \subset J_q(E)$ is a linear system of order q on E and if R_{q+r} is a vector bundle, then $(R_{q+r})_{+s} = R_{q+r+s}$ (2.3.12).

Definition 1.5 A linear system $R_q \subset J_q(E)$ is said to be "*regular*" if R_{q+r} is a vector bundle $\forall r \geq 0$.

Definition 1.6 A regular linear system $R_q \subset J_q(E)$ is said to be "*formally integrable*" if the morphism $\pi_{q+r}^{q+r+1} : R_{q+r+1} \to R_{q+r}$ is an epimorphism, $\forall r \geq 0$.

Definition 1.7 We say that $\mathscr{D} = \Phi \cdot j_q$ is "*formally integrable*" if $R_q = \ker \Phi$ is formally integrable.

It follows that $\rho_r(\Phi)$ is a morphism of constant rank $\forall r \geq 0$.

From now on we shall suppose that R_q is regular. However, even in this case G_{q+r} is not in general a vector bundle for $r \geq 0$.

Remark 1.8 Problem 3.13 gives the simplest example of a regular system such that $\pi_q^{q+1} : R_{q+1} \to R_q$ is an epimorphism and $\pi_{q+1}^{q+2} : R_{q+2} \to R_{q+1}$ is not an epimorphism.

Now, if we are given a morphism $\Phi : J_q(E) \to E'$ we have the commutative and exact diagram:

$$
\begin{array}{ccccccc}
 & & 0 & & 0 & & 0 \\
 & & \downarrow & & \downarrow & & \downarrow \\
0 & \longrightarrow & G_{q+r+1} & \longrightarrow & S_{q+r+1}T^* \otimes E & \xrightarrow{\sigma_{r+1}(\Phi)} & S_{r+1}T^* \otimes E' \\
 & & \downarrow & & \downarrow{\scriptstyle \varepsilon_{q+r+1}} & & \downarrow{\scriptstyle \varepsilon_{r+1}} \\
0 & \longrightarrow & R_{q+r+1} & \longrightarrow & J_{q+r+1}(E) & \xrightarrow{\rho_{r+1}(\Phi)} & J_{r+1}(E') \\
 & & \downarrow & & \downarrow{\scriptstyle \pi_{q+r}^{q+r+1}} & & \downarrow{\scriptstyle \pi_r^{r+1}} \\
0 & \longrightarrow & R_{q+r} & \longrightarrow & J_{q+r}(E) & \xrightarrow{\rho_r(\Phi)} & J_r(E') \\
 & & & & \downarrow & & \downarrow \\
 & & & & 0 & & 0
\end{array}
$$

The key problem in this chapter, as can be seen from a purely "*aesthetic*" point of view, will be to complete the edge of this diagram when $\rho_r(\Phi)$ has constant rank, $\forall r \geq 0$.

For this we shall introduce, as in the non-linear case, the vector bundles over X:

$$R'_r = \operatorname{im} \rho_r(\Phi) \subset J_r(E')$$

and the family of vector spaces over X:

$$G'_r = \operatorname{im} \sigma_r(\Phi) \subset S_r T^* \otimes E'.$$

It follows easily, from the former diagram, that $\pi_r^{r+1} : R'_{r+1} \to R'_r$ is an epimorphism $\forall r \geq 0$. We have also an epimorphism $\pi_0^1 : R'_1 \to F_0$.

PROPOSITION 1.9 We have

$$R'_{r+1} \subseteq (R'_r)_{+1}, \qquad \forall r > 0.$$

Proof As $\rho_r(\Phi)$ has constant rank $\forall r > 0$, we can define the vector bundle $Q_r = \operatorname{coker} \rho_r(\Phi)$ and the epimorphism Ψ by the exact sequence:

$$0 \longrightarrow R_{q+r} \longrightarrow J_{q+r}(E) \xrightarrow{\rho_r(\Phi)} J_r(E') \xrightarrow{\Psi} Q_r \longrightarrow 0$$

The prolonged sequence is then:

$$0 \longrightarrow R_{q+r+1} \longrightarrow J_{q+r+1}(E) \xrightarrow{\rho_{r+1}(\Phi)} J_{r+1}(E') \xrightarrow{\rho_1(\Psi)} J_1(Q_r)$$

This last sequence is not in general exact and we have:

$$R'_{r+1} = \operatorname{im} \rho_{r+1}(\Phi) \subseteq \ker \rho_1(\Psi) = (R'_r)_{+1}$$

Moreover we have the commutative and exact diagram:

$$\begin{array}{ccccccc}
0 & \longrightarrow & R'_{r+1} & \longrightarrow & (R'_r)_{+1} & & \\
 & & \downarrow & & \downarrow & & \\
0 & \longrightarrow & R'_r & = & R'_r & \longrightarrow & 0 \\
 & & \downarrow & & \downarrow & & \\
 & & 0 & & 0 & &
\end{array}$$

C.Q.F.D.

As in the non-linear case, we define the morphism:

$$\delta : \Lambda^s T^* \otimes S_{r+1} T^* \to \Lambda^{s+1} T^* \otimes S_r T^*$$

using the local coordinates, by the formula:

$$\delta : (x, u_{\nu, I}\, dx^I), |\nu| = r + 1 \to (x, u_{\mu+1_i, I}\, dx^i \wedge dx^I), |\mu| = r$$

with $u_{\nu,I} = u_{\mu+1_i,I}$ whenever $\nu = \mu + 1_i$.

We set $(\delta u)_\mu = u_{\mu+1_i}\, dx^i$, where the notations are easy to understand, when $u \in S_{r+1}T^*$ and $\delta u \in T^* \otimes S_r T^*$

It follows that, $\forall \omega \in \Lambda^s T^*$ and $\forall u \in S_{r+1}T^*$ we have

$$\delta(\omega \circ u) = (-1)^s \omega \wedge \delta u$$

PROPOSITION 1.10 $\delta \circ \delta = 0$

Proof From the definition we obtain:

$$0 = u_{\mu+1_i+1_j, I}\, dx^i \wedge dx^j \wedge dx^I = u_{\mu+1_i+1_j, I}\, dx^j \wedge dx^i \wedge dx^I$$

We then have a complex, called "*δ-sequence*":

$$0 \longrightarrow S_r T^* \xrightarrow{\delta} T^* \otimes S_{r-1}T^* \xrightarrow{\delta} \cdots \xrightarrow{\delta} \Lambda^{r-1} T^* \otimes T^* \xrightarrow{\delta} \Lambda^r T^* \longrightarrow 0$$

PROPOSITION 1.11 These sequences are exact, $\forall r \geq 1$.

Proof It is similar to that of proposition 3.1.4.
If we still denote by δ the morphism:

$$\delta \otimes \mathrm{id}_E : \Lambda^s T^* \otimes S_{r+1}T^* \otimes E \to \Lambda^{s+1}T^* \otimes S_r T^* \otimes E$$

we have also the exact δ-sequences:

$$0 \longrightarrow S_r T^* \otimes E \xrightarrow{\delta} T^* \otimes S_{r-1}T^* \otimes E$$
$$\xrightarrow{\delta} \cdots \xrightarrow{\delta} \Lambda^{r-1}T^* \otimes T^* \otimes E \xrightarrow{\delta} \Lambda^r T^* \otimes E \longrightarrow 0$$

Their use will be fundamental later on.

Let $\Phi : J_q(E) \to E'$ be a morphism and consider

$$R_{q+r} = \ker \rho_r(\Phi) \quad \text{and} \quad G_{q+r} = \ker \sigma_r(\Phi).$$

Using convenient local coordinates, we can define Φ as before, then R_q and G_q by the linear equations:

$$R_q \qquad A_k^{\tau\mu}(x) u_\mu^k = 0, \qquad 0 \leq |\mu| \leq q$$

$$G_q \qquad A_k^{\tau\mu}(x) u_\mu^k = 0, \qquad |\mu| = q$$

As in the non-linear case, we can define the r-prolongation of G_q to be the kernel of the composite morphism:

$$S_{q+r}T^* \otimes E \xrightarrow{\Delta r, q} S_r T^* \otimes S_q T^* \otimes E \xrightarrow{\sigma(\Phi)} S_r T^* \otimes E'$$

The reader will check easily that:

$$\sigma_r(\Phi) = (\mathrm{id}_{S_rT^*} \otimes \sigma(\Phi)) \circ (\Delta r, q \otimes \mathrm{id}_E)$$

G_{q+r} which is also the r-prolongation of G_q is defined by the linear equations:

$$G_{q+r} \qquad A_k^{\tau\mu}(x)u_{\mu+\nu}^k = 0, \qquad |\mu| = q, |\nu| = r$$

REMARK 1.12 As already noticed we write simply $\sigma(\Phi)$ instead of $\mathrm{id}_{S_rT^*} \otimes \sigma(\Phi)$ and $\Delta r, q$ instead of $\Delta r, q \otimes \mathrm{id}_E$.

PROPOSITION 1.13 The following diagram is commutative:

$$\begin{array}{ccc} \Lambda^sT^* \otimes S_{q+r+1}T^* & \xrightarrow{\Delta r+1, q} & \Lambda^sT^* \otimes S_{r+1}T^* \otimes S_qT^* \\ \downarrow \delta & & \downarrow \delta \\ \Lambda^{s+1}T^* \otimes S_{q+r}T^* & \xrightarrow{\Delta r, q} & \Lambda^{s+1}T^* \otimes S_rT^* \otimes S_qT^* \end{array}$$

Proof The proof is just routine and is left to the reader.

Using this proposition we obtain the following commutative and exact diagram:

$$\begin{array}{ccccccc} 0 \longrightarrow & \Lambda^sT^* \otimes G_{q+r+1} & \longrightarrow & \Lambda^sT^* \otimes S_{q+r+1}T^* \otimes E & \xrightarrow{\sigma_{r+1}(\Phi)} & \Lambda^sT^* \otimes S_{r+1}T^* \otimes E' \\ & \downarrow \delta & & \downarrow \delta & & \downarrow \delta \\ 0 \longrightarrow & \Lambda^{s+1}T^* \otimes G_{q+r} & \longrightarrow & \Lambda^{s+1}T^* \otimes S_{q+r}T^* \otimes E & \xrightarrow{\sigma_r(\Phi)} & \Lambda^{s+1}T^* \otimes S_rT^* \otimes E' \end{array}$$

where the map δ on the left side is induced by the commutative right square diagram.

Using the fact that $\delta \circ \delta = 0$ we obtain a restricted complex called also "*δ-sequence*":

$$0 \longrightarrow G_{q+r} \xrightarrow{\delta} T^* \otimes G_{q+r-1}$$
$$\xrightarrow{\delta} \cdots \xrightarrow{\delta} \Lambda^rT^* \otimes G_q \xrightarrow{\delta} \delta(\Lambda^rT^* \otimes G_q) \longrightarrow 0$$

However, these sequences are not in general exact $\forall r \geq 0$ and we denote by $H_{q+r-s}^s = H_{q+r-s}^s(G_q)$ the cohomology at $\Lambda^sT^* \otimes G_{q+r-s}$ when $r \geq s$.

DEFINITION 1.14

- We say that G_q is "*s-acyclic*" if

$$H_{q+r}^0 = 0, \ldots, H_{q+r}^s = 0, \qquad \forall r \geq 0.$$

- We say that G_q is "*involutive*" if it is n-acyclic.
- We say that G_q is of "*finite type*" if $\exists r \geq 0$, $G_{q+r} = 0$.

REMARK 1.15 We say that a system $R_q \subset J_q(E)$ or the associated operator $\mathcal{D} = \Phi \circ j_q$, is of finite type if the symbol G_q of R_q is of finite type. We shall give at the end of this chapter an algebraic property of such symbols which is also specific.

REMARK 1.16 If $G_q = 0$ it is of course involutive. It is easy to see that if G_q is involutive and of finite type, then we must have $G_q = 0$. Moreover we know that G_q is always 0 and 1-acyclic, that is to say we always have the exact sequences, $\forall r \geq 0$:

$$0 \longrightarrow G_{q+r+1} \xrightarrow{\delta} T^* \otimes G_{q+r} \xrightarrow{\delta} \Lambda^2 T^* \otimes G_{q+r-1}$$

The following propositions have already been proved:

PROPOSITION 1.17 $(G_{q+r})_{+s} = G_{q+r+s}$

Proof (3.1.15)

PROPOSITION 1.18 If G_{q+1} is a vector bundle and if G_q is 2-acyclic, then G_{q+r} is a vector bundle, $\forall r \geq 1$.

Proof (3.1.16)

PROPOSITION 1.19 There exists an integer $q_0 \geq q$, depending only on $n = \dim X$, $m = \dim E$ and q such that G_{q_0} is involutive.

Proof (3.4.1)

REMARK 1.20 The last proposition is very important. However the reader will notice that if G_q is involutive, then G_{q+r} is also involutive $\forall r \geq 0$, but is not necessarily a vector bundle.

2 Formal properties

Let $R_q \subset J_q(E)$ be a system of order q on E. If the map $\pi_q^{q+1} : R_{q+1} \to R_q$ is not surjective, then R_q is certainly not formally integrable. But, even if this map is surjective, the map $\pi_{q+r}^{q+r+1} : R_{q+r+1} \to R_{q+1}$ is not necessarily surjective for $\forall r > 0$, as can be seen from problem 3.13.

The following criterion is fundamental for the applications.

CRITERION OF FORMAL INTEGRABILITY 2.1 Let $R_q \subset J_q(E)$ be a system of order q on E such that R_{q+1} is a vector bundle. If G_q is 2-acyclic and if the morphism $\pi_q^{q+1} : R_{q+1} \to R_q$ is an epimorphism then R_q is formally integrable.

DEFINITION 2.2 We say that $R_q \subset J_q(E)$ (or the associated differential operator $\mathscr{D} = \Phi \circ j_q$ uniquely determined by R_q) is "*involutive*" if its symbol G_q is involutive and if R_q is formally integrable.

REMARK 2.3 From corollary 3.2.30 it follows that G_q itself is a vector bundle over X.

REMARK 2.4 This definition is very useful because, from the theorem 3.4.1, we know that it is always possible to prolong a system a sufficient number of times for its symbol to become involutive. We have the corresponding useful criterion.

CRITERION OF INVOLUTIVENESS 2.5 Let $R_q \subset J_q(E)$ be a system of order q on E such that R_{q+1} is a vector bundle. If G_q is involutive and if the morphism $\pi_q^{q+1} : R_{q+1} \to R_q$ is an epimorphism then R_q is involutive.

If R_q is not formally integrable, however we may ask for formal solutions. If we restrict R_q to a sufficiently small open set $U \subset X$ or if R_q is sufficiently regular, we can use theorem 2.5.8 in order to get an involutive system $R_{q_s}^{(s)}$ which has the same solutions as R_q and for which it is easy to construct Taylor series for the formal solutions about a given point $x_0 \in X$.

Two cases are of particular importance:

Systems with constant coefficients

These systems can be defined by equations such as $A_k^{\tau\mu} u_\mu^k = 0$, $0 \leq |\mu| \leq q$ where the $A_k^{\tau\mu}$ are constant functions over X.

Amorphous systems

DEFINITION 2.6 A system $R_q \subset J_q(E)$ is said to be "*amorphous*" if $\forall x, x' \in X$, there exist a neighbourhood U of x, a neighbourhood U' of x' in X, an isomorphism of U onto U' and an isomorphism of $E|_U$ onto $E|_{U'}$ which is over the former one and "*preserves*" R_q.

REMARK 2.7 By preserving R_q we mean that the prolongation of the isomorphism from $E|_U$ onto $E|_{U'}$, that can be done as in the first chapter of this book, is an isomorphism from $J_q(E)|_U$ onto $J_q(E)|_{U'}$, taking $R_q|_U$ onto $R_q|_{U'}$.

In both the preceding cases, G_{q+r} and R_{q+r} are vector bundles over X, $\forall r \geq 0$. Moreover we have the two following propositions:

PROPOSITION 2.8 G_{q+r} becomes 2-acyclic for r big enough.

Proof Using proposition 4.1.9, we just need to prove that $G'_{r+1} = (G'_r)_{+1}$ for r big enough. But we have the sequence; $\forall x \in X$

$$0 \longrightarrow G'_{r+1,x} \xrightarrow{\delta} T^*_x \otimes G'_{r,x} \longrightarrow \Lambda^2 T^*_x \otimes S_{r-1} T^*_x \otimes F_{0,x}$$

It follows that

$$G'_{r+1,x} \subseteq (G'_{r,x})_{+1}, \qquad \forall x \in X.$$

From proposition 2.5.9 there exists an integer $r_0(x)$ such that

$$G'_{r+1,x} = (G'_{r,x})_{+1}, \qquad \forall r \geq r_0(x).$$

But if the latter sequence is exact at one point $x \in X$ it is clear that it is also exact at any other point $x' \in X$ and we have

$$r_0(x) = r_0(x') = r_0. \qquad \text{C.Q.F.D.}$$

PROPOSITION 2.9 R'_r is formally integrable for r big enough.

Proof Let $\bar{G}'_r$ be the symbol of R'_r, which is in general different from G'_r unless R_q is formally integrable. Using the diagram of proposition 1.9, we deduce that $\bar{G}'_{r+1} \subseteq (\bar{G}'_r)_{+1}$. As in the preceding proposition, $\exists \bar{r}_0$ such that $\bar{G}'_{r+1} = (\bar{G}'_r)_{+1}$, $\forall r \geq \bar{r}_0$. From the same diagram it then follows that $R'_{r+1} = (R'_r)_{+1}$, $\forall r \geq r_0$ and $\pi^{r+1}_r : R'_{r+1} \to R'_r$ is surjective. Finally R'_r is a vector bundle because R_{q+r} is a vector bundle $\forall r \geq 0$.

From now on we shall suppose that $R_q \subset J_q(E)$ is an involutive system of order q on E, defined by a morphism $\Phi : J_q(E) \to E'$.

As it is useful, in order to define R_q, to keep only linearly independent equations, we may suppose (and this we will do from now on) that Φ is an epimorphism, giving rise to the short exact sequence:

$$0 \longrightarrow R_q \longrightarrow J_q(E) \xrightarrow{\Phi} F_0 \longrightarrow 0$$

Because G_q is involutive, using proposition 4.1.9 we deduce that G'_{r+1} is the r-prolongation of G'_1 which is also involutive. Using the last proposition we conclude that $R'_1 \subset J_1(F_0)$ is an involutive system of order 1 on F_0.

Now let F_r be a bundle and $\mathscr{D}_r : F_{r-1} \to F_r$ be a (differential) operator $\forall r$ ranging in a finite or infinite set of consecutive integers.

DEFINITION 2.10 When $\mathscr{D}_{r+1} \circ \mathscr{D}_r = 0$, $\forall r$, we say that the sequence:

$$\cdots \longrightarrow F_{r-1} \xrightarrow{\mathscr{D}_r} F_r \xrightarrow{\mathscr{D}_{r+1}} F_{r+1} \longrightarrow \cdots$$

is of finite or infinite length, according as there is a finite or infinite number of F_r with $\dim F_r \neq 0$.

The aim of this chapter is to associate with any involutive system $R_q \subset J_q(E)$, three different finite length (differential) sequences, respectively called the "*first Spencer sequence*", the "*second Spencer sequence*" and the "*P-sequence*", and to relate them.

DEFINITION 2.11 The sequence

$$\cdots \longrightarrow F_{r-1} \xrightarrow{\mathscr{D}_r} F_r \xrightarrow{\mathscr{D}_{r+1}} F_{r+1} \longrightarrow \cdots$$

with $\mathscr{D}_r = \Psi_r \circ j_{q_r}$ is said to be "*formally exact*" at F_r if we have the exact sequences $\forall s \geq 0$:

$$J_{q_r+q_{r+1}+s}(F_{r-1}) \xrightarrow{\rho_{q_{r+1}+s}(\Psi_r)} J_{q_{r+1}+s}(F_r) \xrightarrow{\rho_s(\Psi_{r+1})} J_s(F_{r+1})$$

DEFINITION 2.12 The sequence

$$\cdots \longrightarrow F_{r-1} \xrightarrow{\mathscr{D}_r} F_r \xrightarrow{\mathscr{D}_{r+1}} F_{r+1} \longrightarrow \cdots$$

is said to be "*locally exact*" at F_r if $\operatorname{im} \mathscr{D}_r = \ker \mathscr{D}_{r+1}$, that is to say if, $\forall \eta$ section of F_r over $V \subset X$, such that $\mathscr{D}_{r+1} \cdot \eta = 0$, then $\exists \xi$ section of F_{r-1} over $U \subset V$ such that $\mathscr{D}_r \cdot \xi = \eta$ on U.

DEFINITION 2.13 The former sequence is said to be "*exact*" at F_r if it is both locally and formally exact at F_r.

DEFINITION 2.14 Thc former sequence is said to be (locally, formally) exact if it is (locally, formally) exact at F_r, $\forall r$.

3 First Spencer sequence

We recall that the Spencer operator D, already defined as an example, is a first order differential operator

$$D : J_{q+1}(E) \to T^* \otimes J_q(E)$$

defined $\forall q \geq 0$ by

$$\varepsilon_1 \circ D = j_1 \circ \pi_q^{q+1} - \rho_1(\mathrm{id}_{J_q(E)}),$$

using the commutative and exact diagram:

$$\begin{array}{ccccccccc}
0 & \longrightarrow & T^* \otimes J_{q+1}(E) & \xrightarrow{\varepsilon_1} & J_1(J_{q+1}(E)) & \xrightarrow{\pi_0^1} & J_{q+1}(E) & \longrightarrow & 0 \\
 & & \downarrow \pi_q^{q+1} & \swarrow & \downarrow J_1(\pi_q^{q+1}) & \swarrow \rho_1(\mathrm{id}_{J_q(E)}) & \downarrow \pi_q^{q+1} & & \\
0 & \longrightarrow & T^* \otimes J_q(E) & \xrightarrow{\varepsilon_1} & J_1(J_q(E)) & \xrightarrow{\pi_0^1} & J_q(E) & \longrightarrow & 0
\end{array}$$

We extend D as an operator

$$D : \Lambda^r T^* \otimes J_{q+1}(E) \to \Lambda^{r+1} T^* \otimes J_q(E),$$

setting, $\forall\omega$ section of $\Lambda^r T^*$, ξ_{q+1} section of $J_{q+1}(E)$ and $\xi_q = \pi_q^{q+1} \circ \xi_{q+1}$ section of $J_q(E)$:

$$D(\omega \otimes \xi_{q+1}) = d\omega \otimes \xi_q + (-1)^r \omega \wedge D \cdot \xi_{q+1}$$

In this formula

$$d : \Lambda^r T^* \to \Lambda^{r+1} T^*$$

is the exterior derivative

$$d : \omega = \omega_I(x)\, dx^I \to d\omega = \frac{\partial \omega_I(x)}{\partial x^i}\, dx^i \wedge dx^I$$

and we have $d \circ d = 0$.

The former definition does not depend on the given decomposition. In fact $\forall f \in C^\infty(X)$:

$$\begin{aligned} D(f\omega \otimes \xi_{q+1}) &= d(f\omega) \otimes \xi_q + (-1)^r f\omega \wedge D \cdot \xi_{q+1} \\ &= df \wedge \omega \otimes \xi_q + f\, d\omega \otimes \xi_q + (-1)^r f\omega \wedge D \cdot \xi_{q+1} \\ &= d\omega \wedge f\xi_q + (-1)^r \omega \wedge (df \otimes \xi_q + f \cdot D \cdot \xi_{q+1}) \\ &= D(\omega \otimes f\xi_{q+1}) \end{aligned}$$

because

$$j_1 \circ \pi_q^{q+1}(f \cdot \xi_{q+1}) = j_1(f) \cdot \xi_q + f \cdot j_1(\xi_q)$$

It follows that D can be defined, using local coordinates by the formula:

$$D : (x, \xi^k_{\mu, I}(x)\, dx^I), 0 \leq |\mu| \leq q + 1$$
$$(x, (\partial_i \xi^k_{\mu, I}(x) - \xi^k_{\mu + 1_i, I}(x))\, dx^i \wedge dx^I), 0 \leq |\mu| \leq q$$

We may notice that the restriction of D to $\Lambda^r T^* \otimes S_{q+1} T^* \otimes E$ is

$$-\delta : \Lambda^r T^* \otimes S_{q+1} T^* \otimes E \to \Lambda^{r+1} T^* \otimes S_q T^* \otimes E$$

because

$$\begin{aligned} \varepsilon_1 \circ D \circ \varepsilon_{q+1} &= j_1 \circ \pi_q^{q+1} \circ \varepsilon_{q+1} - \rho_1(\mathrm{id}_{J_q(E)}) \circ \varepsilon_{q+1} \\ &= -\rho_1(\mathrm{id}_{J_q(E)}) \circ \varepsilon_{q+1} = -\varepsilon_1 \circ \delta \end{aligned}$$

Moreover we have $D \circ D = 0$ because $d \circ d = 0$ and $\delta \circ \delta = 0$.

Finally, if ξ_{q+1} is a section of $J_{q+1}(E)$ such that $D \cdot \xi_{q+1} = 0$, then

$$\partial_i \xi^k_\mu(x) = \xi^k_{\mu+1_i}(x), \qquad \forall 0 \leq |\mu| \leq q$$

and

$$\xi^k_\mu(x) = \partial_\mu \xi^k(x), \qquad \forall 0 \leq |\mu| \leq q + 1.$$

It follows that we have a sequence, denoted by $S_1(E)$:

$$S_1(E) \qquad 0 \longrightarrow E \xrightarrow{j_q} J_q(E) \xrightarrow{D} T^* \otimes J_{q-1}(E) \xrightarrow{D} \cdots \longrightarrow \Lambda^n T^* \otimes J_{q-n}(E) \longrightarrow 0$$

This sequence is locally exact at $J_q(E)$ and $J_{q-r}(E) = 0$ if $r > q$.

PROPOSITION 3.1 The sequence $S_1(E)$ is not formally exact but it is locally exact.

Proof We can consider the sequence:

$$0 \longrightarrow J_{q+1}(E) \xrightarrow{\rho_1(\mathrm{id}_{J_q(E)})} J_1(J_q(E)) \xrightarrow{\Psi} T^* \otimes J_{q-1}(E) \longrightarrow 0$$

in which Ψ is the morphism such that $D = \Psi \circ j_1$.

From the last diagram defining D, or from the formulas in local coordinates, it is easy to see that Ψ is an epimorphism. Then:

$$\begin{aligned}\dim J_1(J_q(E)) &- (\dim J_{q+1}(E) + \dim T^* \otimes J_{q-1}(E)) \\ &= (n+1)\frac{(q+n)!}{q!n!}m - \frac{(q+n+1)!}{(q+1)!n!}m - n\frac{(q+n-1)!}{(q-1)!n!}m \\ &= \frac{(q+n-1)!}{(q+1)!(n-2)!}qm > 0\end{aligned}$$

It follows that the sequence is not formally exact.

REMARK 3.2 The reader will easily check that the symbol of D is involutive but that D is not formally integrable, as was shown in a particular case. This is the reason why we later introduce the second Spencer sequence in place of the first.

Now, for $q = 0$, the sequence $0 \to E = E \to 0$ is of course locally exact.

For $q = 1$, the sequence

$$0 \longrightarrow E \xrightarrow{j_1} J_1(E) \xrightarrow{D} T^* \otimes E \longrightarrow 0$$

is also locally exact.

More generally, let us consider the following commutative diagram:

$$\begin{array}{ccccccccccc}
 & & & & 0 & & 0 & & 0 & & \\
 & & & & \downarrow & & \downarrow & & \downarrow & & \\
 & & 0 & \longrightarrow & S_{q+1}T^* \otimes E & \xrightarrow{-\delta} & T^* \otimes S_q T^* \otimes E & \xrightarrow{-\delta} & \Lambda^2 T^* \otimes S_{q-1}T^* \otimes E & \xrightarrow{-\delta} & \cdots \\
 & & \downarrow & & \downarrow{\scriptstyle \varepsilon_{q+1}} & & \downarrow & & \downarrow & & \\
0 & \longrightarrow & E & \xrightarrow{j_{q+1}} & J_{q+1}(E) & \xrightarrow{D} & T^* \otimes J_q(E) & \xrightarrow{D} & \Lambda^2 T^* \otimes J_{q-1}(E) & \xrightarrow{D} & \cdots \\
 & & \| & & \downarrow{\scriptstyle \pi^{q+1}_q} & & \downarrow & & \downarrow & & \\
0 & \longrightarrow & E & \xrightarrow{j_q} & J_q(E) & \xrightarrow{D} & T^* \otimes J_{q-1}(E) & \xrightarrow{D} & \Lambda^2 T^* \otimes J_{q-2}(E) & \xrightarrow{D} & \cdots \\
 & & \downarrow & & \downarrow & & \downarrow & & \downarrow & & \\
 & & 0 & & 0 & & 0 & & 0 & &
\end{array}$$

From proposition 3.1.5 the upper row is exact. Thus, if the bottom row is locally exact, as the columns are exact, then the middle now becomes locally exact, and the proposition follows easily from an induction on q.

C.Q.F.D.

Proposition 3.3 If $\Phi : J_q(E) \to E'$ is a morphism, then we have the commutative diagram:

$$\begin{array}{ccc}
\Lambda^s T^* \otimes J_{q+r+1}(E) & \xrightarrow{\rho_{r+1}(\Phi)} & \Lambda^s T^* \otimes J_{r+1}(E') \\
\downarrow{\scriptstyle D} & & \downarrow{\scriptstyle D} \\
\Lambda^{s+1} T^* \otimes J_{q+r}(E) & \xrightarrow{\rho_r(\Phi)} & \Lambda^{s+1} T^* \otimes J_r(E')
\end{array}$$

Proof From the proposition 2.1.18 we get

$$\pi^{r+1}_r \circ \rho_{r+1}(\Phi) = \rho_r(\Phi) \circ \pi^{q+r+1}_{q+r}$$

Let ω be a section of $\Lambda^s T^*$ and ξ_{q+r+1} be a section of $J_{q+r+1}(E)$. If ξ'_{r+1} is the corresponding section of $J_{r+1}(E')$ image of ξ_{q+r+1} by $\rho_{r+1}(\Phi)$, then $\xi'_r = \pi_r^{r+1} \circ \xi'_{r+1}$ is the section of $J_r(E')$, image by $\rho_r(\Phi)$ of $\xi_{q+r} = \pi_{q+r}^{q+r+1} \circ \xi_{q+r+1}$. Then:

$$(\rho_r(\Phi) \circ D - D \circ \rho_{r+1}(\Phi))(\omega \otimes \xi_{q+r+1}) \\ = \omega \otimes (\rho_r(\Phi) \circ D - D \circ \rho_{r+1}(\Phi))(\xi_{q+r+1})$$

and we just need to prove the commutativity of the diagram when $s = 0$. In that case we have:

$$\begin{aligned} \varepsilon_1 \circ D \circ \rho_{r+1}(\Phi) &= j_1 \circ \pi_r^{r+1} \circ \rho_{r+1}(\Phi) - \rho_1(\mathrm{id}_{J_r(E')}) \circ \rho_{r+1}(\Phi) \\ &= j_1 \circ \rho_r(\Phi) \circ \pi_{q+r}^{q+r+1} - J_1(\rho_r(\Phi)) \circ \rho_1(\mathrm{id}_{J_{q+r}(E)}) \\ &= J_1(\rho_1(\Phi))[j_1 \circ \pi_{q+r}^{q+r+1} - \rho_1(\mathrm{id}_{J_{q+r}(E)})] \\ &= J_1(\rho_r(\Phi)) \circ \varepsilon_1 \circ D \\ &= \varepsilon_1 \circ \rho_r(\Phi) \circ D \end{aligned}$$

and the proposition follows from the fact that

$$\varepsilon_1 = T^* \otimes J_r(E') \to J_1(J_r(E'))$$

is a monomorphism.

Finally, for reasons that will be clear later on, we shall prove directly, using local coordinates, the existence of an operator:

$$D : \Lambda^r T^* \otimes R_{q+1} \to \Lambda^{r+1} T^* \otimes R_q$$

For a section of $\Lambda^r T^* \otimes R_{q+1}$ we have:

$$\Lambda^r T^* \otimes R_{q+1} \qquad \begin{cases} A_k^{\tau\mu}(x)\xi_{\mu+1_i, I}^k(x)\, dx^I + \partial_i A_k^{\tau\mu}(x)\xi_{\mu, I}^k(x)\, dx^I = 0 \\ A_k^{\tau\mu}(x)\xi_{\mu, I}^k(x)\, dx^I = 0 \end{cases}$$

where

$$0 \leq |\mu| \leq q \quad \text{and} \quad I = (i_1, \ldots, i_r),\ i_1 < \cdots < i_r.$$

If we take the exterior derivative of the second set of equations, we obtain:

$$A_k^{\tau\mu}(x)\partial_i \xi_{\mu, I}^k(x)\, dx^i \wedge dx^I + \partial_i A_k^{\tau\mu}(x)\xi_{\mu, I}^k(x)\, dx^i \wedge dx^I = 0$$

If we consider the exterior product by dx^i on the left of the first set of equations, substracting we get:

$$\Lambda^{r+1} T^* \otimes R_q \qquad A_k^{\tau\mu}(x)(\partial_i \xi_{\mu, I}^k(x) - \xi_{\mu+1_i, I}^k(x))\, dx^i \wedge dx^I = 0$$

C.Q.F.D.

From the preceding proposition, we easily obtain a sequence, also called a first Spencer sequence but here denoted by $S_1(\Theta)$:

$$S_1(\Theta) \qquad 0 \longrightarrow \Theta \xrightarrow{j_{q+r}} R_{q+r} \xrightarrow{D} T^* \otimes R_{q+r-1} \longrightarrow \cdots \longrightarrow \Lambda^n T^* \otimes R_{q+r-n} \longrightarrow 0$$

in which

$$\begin{aligned} R_{q+r-s} &= J_{q+r-s}(E) && \text{if} \quad r < s \leq q + r \\ &= 0 && \text{if} \quad s > q + r \end{aligned}$$

REMARK 3.4 This sequence is in general neither formally nor even locally exact, as an example, due to H. Lewy given at the end of this chapter will show.

However, the sequence is locally exact at R_{q+r} because, if a section ξ_{q+r} of R_{q+r} is such that $D \cdot \xi_{q+r} = 0$, using proposition 3.1, there must exist a section ξ of E such that $\xi_{q+r} = j_{q+r}(\xi)$ and ξ is then, by definition, a solution of R_{q+r} and finally of R_q.

An interesting property of $S_1(\Theta)$ is the following. We have the commutative diagram:

$$\begin{array}{ccccccccccccc}
 & & & & 0 & & 0 & & & & 0 & & \\
 & & & & \downarrow & & \downarrow & & & & \downarrow & & \\
 & & 0 & \longrightarrow & G_{q+r+1} & \xrightarrow{-\delta} & T^* \otimes G_{q+r} & \xrightarrow{-\delta} & \cdots & \xrightarrow{-\delta} & \Lambda^n T^* \otimes G_{q+r-n+1} & \longrightarrow & 0 \\
 & & \downarrow & & \downarrow{\scriptstyle \varepsilon_{q+r+1}} & & \downarrow & & & & \downarrow & & \\
0 & \longrightarrow & \Theta & \xrightarrow{j_{q+r+1}} & R_{q+r+1} & \xrightarrow{D} & T^* \otimes R_{q+r} & \xrightarrow{D} & \cdots & \xrightarrow{D} & \Lambda^n T^* \otimes R_{q+r-n+1} & \longrightarrow & 0 \\
 & & \| & & \downarrow{\scriptstyle \pi^{q+r+1}_{q+r}} & & \downarrow & & & & \downarrow & & \\
0 & \longrightarrow & \Theta & \xrightarrow{j_{q+r}} & R_{q+r} & \xrightarrow{D} & T^* \otimes R_{q+r-1} & \xrightarrow{D} & \cdots & \xrightarrow{D} & \Lambda^n T^* \otimes R_{q+r-n} & \longrightarrow & 0 \\
 & & \downarrow & & \downarrow & & \downarrow & & & & \downarrow & & \\
 & & 0 & & 0 & & 0 & & & & 0 & &
\end{array}$$

As G_q is involutive, the upper row is exact $\forall r \geq n - 1$ and, in this case, the lower and the middle row have the same cohomology as far as local exactness is concerned.

DEFINITION 3.5 We say that the sequence $S_1(\Theta)$ is "*stable*" when $r \geq n - 1$.

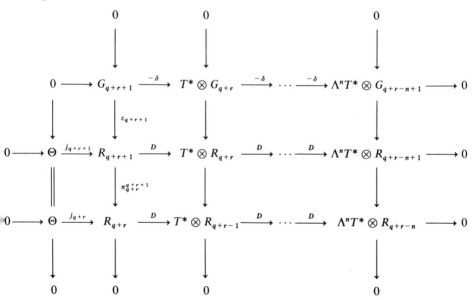

4 Second Spencer sequence

Using the fact that R_q is involutive, we may define, as in the non-linear case, the Spencer bundles:

$$\begin{cases} C_0 = R_q & r = 0 \\ C_r = \dfrac{\Lambda^r T^* \otimes C_0}{\delta(\Lambda^{r-1} T^* \otimes G_{q+1})} & 0 < r \leq n \\ C_r = 0 & r > n \end{cases}$$

REMARK 4.1 In this definition, for simplicity we do not indicate the monomorphism ε_q before δ.

PROPOSITION 4.2 The commutative and exact diagram:

$$\begin{array}{ccccccccc}
0 & \longrightarrow & \Lambda^{r-1}T^* \otimes G_{q+1} & \xrightarrow{\varepsilon_{q+1}} & \Lambda^{r-1}T^* \otimes R_{q+1} & \xrightarrow{\pi_q^{q+1}} & \Lambda^{r-1}T^* \otimes R_q & \longrightarrow & 0 \\
 & & \downarrow{\scriptstyle -\delta} & & \downarrow{\scriptstyle D} & & \downarrow{\scriptstyle D'} \quad \searrow{\scriptstyle \tau_{r-1}} & & \\
 & & & & & & \quad C_{r-1} \longrightarrow 0 & & \\
 & & & & & & \swarrow{\scriptstyle D_r} & & \\
0 & \longrightarrow & \delta(\Lambda^{r-1}T^* \otimes G_{q+1}) & \xrightarrow{\varepsilon_q} & \Lambda^r T^* \otimes R_q & \xrightarrow{\tau_r} & C_r & \longrightarrow & 0
\end{array}$$

gives rise to a sequence, denoted by $S_2(\Theta)$:

$$S_2(\Theta) \qquad 0 \longrightarrow \Theta \xrightarrow{j_q} C_0 \xrightarrow{D_1} C_1$$
$$\xrightarrow{D_2} \cdots \xrightarrow{D_n} C_n \longrightarrow 0$$

which is called the "*second Spencer sequence*".

Proof In the former diagram D' is induced by D and we have $\tau_r \circ D = D' \circ \pi_q^{q+1}$ where τ_r is the natural projection onto C_r.

If α is a section of $\delta(\Lambda^{r-2}T^* \otimes G_{q+1})$, there exists a section β of $\Lambda^{r-2}T^* \otimes G_{q+1}$ such that $\alpha = -\delta(\beta)$. As R_q is formally integrable, we can find a section γ of $\Lambda^{r-2}T^* \otimes R_{q+2}$ such that $\pi_{q+1}^{q+2}(\gamma) = \varepsilon_{q+1}(\beta)$.

It follows that we have:

$$\begin{aligned} D' \circ \varepsilon_q(\alpha) = -D' \circ \varepsilon_q \circ \delta(\beta) &= D' \circ D \circ \varepsilon_{q+1}(\beta) \\ &= D' \circ D \circ \pi_{q+1}^{q+2}(\gamma) \\ &= D' \circ \pi_q^{q+1} \circ D(\gamma) \\ &= \tau_r \circ D \circ D(\gamma) = 0 \end{aligned}$$

because $D \circ D = 0$, and there exists an operator $D_r: C_{r-1} \to C_r$. Moreover we have:

$$\begin{aligned} D_{r+1} \circ D_r \circ \tau_{r-1} \circ \pi_q^{q+2} &= D_{r+1} \circ D' \circ \pi_q^{q+1} \circ \pi_{q+1}^{q+2} \\ &= D_{r+1} \circ \tau_r \circ D \circ \pi_{q+1}^{q+2} \\ &= D' \circ D \circ \pi_{q+1}^{q+2} \\ &= D' \circ \pi_q^{q+1} \circ D \\ &= \tau_r \circ D \circ D = 0 \end{aligned}$$

As τ_{r-1} and π_q^{q+2} are epimorphisms, then we have

$$D_{r+1} \circ D_r = 0, \qquad \forall r.$$

Finally, for $r = 1$, using the relation $\pi_q^{q+1} \circ j_{q+1} = j_q$, we can show the local exactness of $S_2(\Theta)$ at C_0 by means of a diagram chasing in the following commutative diagram with exact rows:

$$\begin{array}{ccccccccc}
 & & & & 0 & & 0 & & \\
 & & & & \downarrow & & \downarrow & & \\
 & & 0 & \longrightarrow & \Theta & = & \Theta & \longrightarrow & 0 \\
 & & \downarrow & & \downarrow{\scriptstyle j_{q+1}} & & \downarrow{\scriptstyle j_q} & & \\
0 & \longrightarrow & G_{q+1} & \xrightarrow{\varepsilon_{q+1}} & R_{q+1} & \xrightarrow{\pi_q^{q+1}} & C_0 & \longrightarrow & 0 \\
 & & {\scriptstyle -\delta}\downarrow & & \downarrow{\scriptstyle D} & & \downarrow{\scriptstyle D_1} & & \\
0 & \longrightarrow & \delta(G_{q+1}) & \xrightarrow{\varepsilon_q} & T^* \otimes R_q & \xrightarrow{\tau_1} & C_1 & \longrightarrow & 0 \\
 & & \downarrow & & & & & & \\
 & & 0 & & & & & &
\end{array}$$

C.Q.F.D.

REMARK 4.3 For q fixed, we may also, as in the non-linear case, introduce the Spencer bundles:

$$\begin{cases} C_0(E) = J_q(E) & r = 0 \\ C_r(E) = \dfrac{\Lambda^r T^* \otimes C_0(E)}{\delta(\Lambda^{r-1} T^* \otimes S_{q+1} T^* \otimes E)} & 0 < r \leq n \\ C_r(E) = 0 & r > n \end{cases}$$

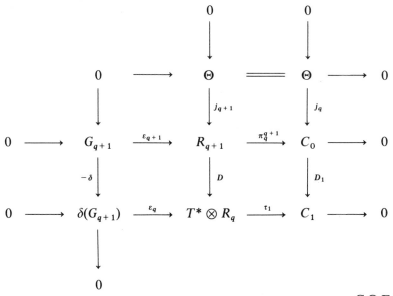

It is possible to construct as above, the second Spencer sequence $S_2(E)$:

$$S_2(E) \qquad 0 \longrightarrow E \xrightarrow{j_q} C_0(E) \xrightarrow{D_1} C_1(E) \xrightarrow{D_2} \cdots \xrightarrow{D_n} C_n(E) \longrightarrow 0$$

We shall prove later on that this sequence is exact.

5 *P*-sequence

If $\mathscr{D} = \Phi \circ j_q : E \to E'$ is an involutive operator, we may define, as in the non-linear case, the vector bundles F_r:

$$\begin{cases} F_0 = \operatorname{im} \Phi = \dfrac{J_q(E)}{R_q} & r = 0 \\[2ex] F_r = \dfrac{\Lambda^r T^* \otimes F_0}{\delta(\Lambda^{r-1} T^* \otimes G_1')} & 0 < r \leq n \\[2ex] F_r = 0 & r > n \end{cases}$$

where

$$R_q = \ker \Phi \quad \text{and} \quad G_1' = \operatorname{im} \sigma_1(\Phi).$$

We also define

$$R_1' = \operatorname{im} \rho_1(\Phi).$$

PROPOSITION 5.1 The commutative and exact diagram:

$$\begin{array}{ccccccccc}
0 & \longrightarrow & \Lambda^{r-1}T^* \otimes G_1' & \xrightarrow{\varepsilon_1} & \Lambda^{r-1}T^* \otimes R_1' & \xrightarrow{\pi_0^1} & \Lambda^{r-1}T^* \otimes F_0 & \longrightarrow & 0 \\
 & & \downarrow -\delta & & \downarrow D & & \downarrow \searrow & & \\
 & & & & & & F_{r-1} & \longrightarrow & 0 \\
 & & & & & & \downarrow \swarrow \mathscr{D}_r & & \\
0 & \longrightarrow & \delta(\Lambda^{r-1}T^* \otimes G_1') & \longrightarrow & \Lambda^r T^* \otimes F_0 & \longrightarrow & F_r & \longrightarrow & 0
\end{array}$$

gives rise to a sequence, denoted by $P(\Theta)$:

$$P(\Theta) \qquad 0 \longrightarrow \Theta \longrightarrow E \xrightarrow{\mathscr{D}} F_0 \xrightarrow{\mathscr{D}_1} F_1 \xrightarrow{\mathscr{D}_2} \cdots \xrightarrow{\mathscr{D}_n} F_n \longrightarrow 0$$

and called "*P-sequence*".

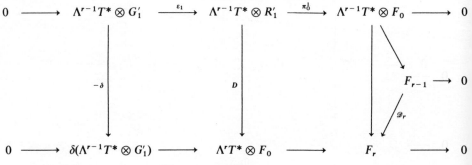

Proof Using proposition 2.9, we see that $R_1' \subset J_1(F_0)$ is formally integrable and that $\pi_0^1 : R_1' \to F_0$ is an epimorphism. It is then possible, as in the preceding proposition, to construct a sequence, denoted by $P(\Omega)$:

$$P(\Omega) \qquad 0 \longrightarrow \Omega \longrightarrow F_0 \xrightarrow{\mathscr{D}_1} F_1 \xrightarrow{\mathscr{D}_2} \cdots \xrightarrow{\mathscr{D}_n} F_n \longrightarrow 0$$

in which Ω, the set of solutions of $\mathscr{D}_1$, is also the set of solutions of R'_1.

In particular, setting $\mathscr{D}_r = \Psi_r \circ j_1$, we have the short exact sequence:

$$0 \longrightarrow R'_1 \longrightarrow J_1(F_0) \xrightarrow{\Psi_1} F_1 \longrightarrow 0$$

and thus also the exact sequence:

$$0 \longrightarrow R_{q+1} \longrightarrow J_{q+1}(E) \xrightarrow{\rho_1(\Phi)} J_1(F_0) \xrightarrow{\Psi_1} F_1 \longrightarrow 0$$

which shows that $\mathscr{D}_1 \circ \mathscr{D} = 0$ and allows us to construct the sequence $P(\Theta)$ from the sequence $P(\Omega)$.

REMARK 5.2 $P(\Omega)$ can be considered as a "*truncated*" P-sequence. In fact it is obtained from $P(\Theta)$ by "*forgetting*" $\mathscr{D}$.

As the sequences $S_2(\Theta)$, $S_2(E)$, $P(\Omega)$ each have the same finite number n of first order operators, we may look for a link between them.

THEOREM 5.3 The sequences $S_2(\Theta)$, $S_2(E)$, $P(\Omega)$ are related by the following commutative diagram in which the columns are exact and $\Phi_0, \ldots, \Phi_n$ are induced by $\Phi = \Phi_0$:

$$\begin{array}{ccccccccccccc}
 & & 0 & & 0 & & 0 & & & & 0 & & \\
 & & \downarrow & & \downarrow & & \downarrow & & & & \downarrow & & \\
0 & \longrightarrow & \Theta & \xrightarrow{j_q} & C_0 & \xrightarrow{D_1} & C_1 & \xrightarrow{D_2} & \cdots & \xrightarrow{D_n} & C_n & \longrightarrow & 0 \\
 & & \downarrow & & \downarrow & & \downarrow & & & & \downarrow & & \\
0 & \longrightarrow & E & \xrightarrow{j_q} & C_0(E) & \xrightarrow{D_1} & C_1(E) & \xrightarrow{D_2} & \cdots & \xrightarrow{D_n} & C_n(E) & \longrightarrow & 0 \\
 & & \downarrow & & \downarrow{\scriptstyle \Phi_0} & & \downarrow{\scriptstyle \Phi_1} & & & & \downarrow{\scriptstyle \Phi_n} & & \\
0 & \longrightarrow & \Omega & \longrightarrow & F_0 & \xrightarrow{\mathscr{D}_1} & F_1 & \xrightarrow{\mathscr{D}_2} & \cdots & \xrightarrow{\mathscr{D}_n} & F_n & \longrightarrow & 0 \\
 & & & & \downarrow & & \downarrow & & & & \downarrow & & \\
 & & & & 0 & & 0 & & & & 0 & &
\end{array}$$

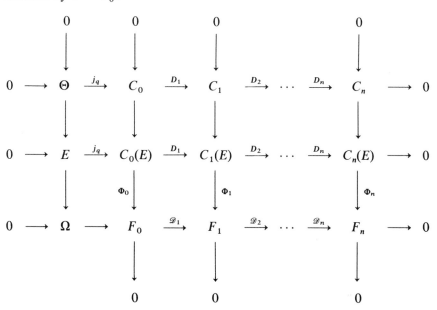

Proof The propositions 4.2 and 5.1 make it easy to construct the following commutative and exact diagram on page 185, where the edge of zeros has been omitted:

As in corollary 4.2.3 we have the short exact sequences

$$0 \longrightarrow C_r \longrightarrow C_r(E) \xrightarrow{\Phi_r} F_r \longrightarrow 0$$

The epimorphisms Φ_r are induced by the epimorphism

$$\Phi : \Lambda^r T^* \otimes J_q(E) \to \Lambda^r T^* \otimes F_0,$$

which is itself induced by the epimorphism $\Phi : J_q(E) \to F_0$ coming from the morphism $\Phi : J_q(E) \to E'$.

REMARK 5.4 In the left column, $\mathscr{D} : E \to \Omega$ is not always surjective and this is equivalent to saying that $P(\Theta)$ is not always locally exact at F_0 (see example of H. Lewy). C.Q.F.D.

We shall now give some properties of the different sequences constructed above in order to introduce them from another point of view. We shall be mainly concerned with $P(\Theta)$ which is in fact the only sequence that can be used effectively in order to study concrete problems.

THEOREM 5.5 $P(\Theta)$ is formally exact.

Proof Using proposition 4.1.4, we get the exact sequences of vector bundles, $\forall r \geq 1$:

$$0 \to G'_r \to S_r T^* \otimes F_0 \to S_{r-1} T^* \otimes F_1 \to \cdots \to T^* \otimes F_{r-1} \to F_r \to 0$$

In these sequences, the morphisms $S_{r-s+1} T^* \otimes F_{s-1} \to S_{r-s} T^* \otimes F_s$ are the $(r-s)$-prolongations of the epimorphisms $T^* \otimes F_{s-1} \to F_s$ induced by the exterior multiplication $T^* \otimes \Lambda^{s-1} T^* \to \Lambda^s T^*$.

It follows that, $\forall r \geq 0$, we have the exact sequences:

$$0 \longrightarrow G_{q+r} \longrightarrow S_{q+r} T^* \otimes E \xrightarrow{\sigma_r(\Phi)} S_r T^* \otimes F_0 \longrightarrow \cdots \longrightarrow T^* \otimes F_{r-1} \longrightarrow F_r \longrightarrow 0$$

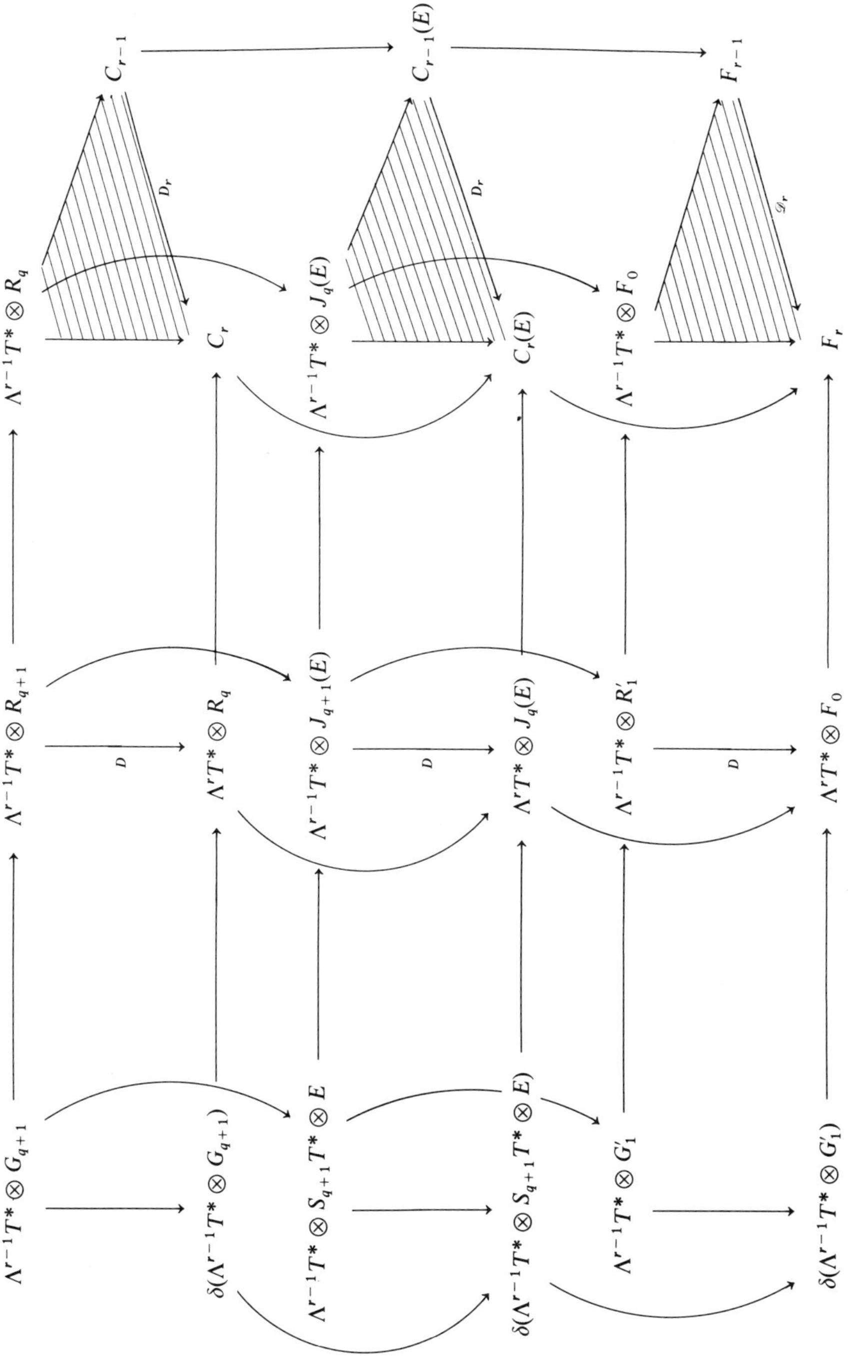

$\Lambda^{r-1}T^* \otimes G_{q+1}$
$\Lambda^{r-1}T^* \otimes R_{q+1}$
$\Lambda^{r-1}T^* \otimes R_q$
C_{r-1}
$\delta(\Lambda^{r-1}T^* \otimes G_{q+1})$
$\Lambda^r T^* \otimes R_q$
C_r
D
D_r
$\Lambda^{r-1}T^* \otimes S_{q+1}T^* \otimes E$
$\Lambda^{r-1}T^* \otimes J_{q+1}(E)$
$\Lambda^{r-1}T^* \otimes J_q(E)$
$C_{r-1}(E)$
$\delta(\Lambda^{r-1}T^* \otimes S_{q+1}T^* \otimes E)$
$\Lambda^r T^* \otimes J_q(E)$
$C_r(E)$
$\Lambda^{r-1}T^* \otimes G'_1$
$\Lambda^{r-1}T^* \otimes R'_1$
$\Lambda^{r-1}T^* \otimes F_0$
F_{r-1}
$\delta(\Lambda^{r-1}T^* \otimes G'_1)$
$\Lambda^r T^* \otimes F_0$
F_r
$\mathscr{D}_r$

But for $r = 1$ we have the following commutative and exact diagram:

$$
\begin{array}{ccccccccccc}
 & & 0 & & 0 & & 0 & & 0 & & \\
 & & \downarrow & & \downarrow & & \downarrow & & \downarrow & & \\
0 & \longrightarrow & G_{q+1} & \longrightarrow & S_{q+1}T^* \otimes E & \xrightarrow{\sigma_1(\Phi)} & T^* \otimes F_0 & \xrightarrow{\sigma(\Psi_1)} & F_1 & \longrightarrow & 0 \\
 & & \downarrow & & \downarrow{\scriptstyle \varepsilon_{q+1}} & & \downarrow{\scriptstyle \varepsilon_1} & & \| & & \\
0 & \longrightarrow & R_{q+1} & \longrightarrow & J_{q+1}(E) & \xrightarrow{\rho_1(\Phi)} & J_1(F_0) & \xrightarrow{\Psi_1} & F_1 & \longrightarrow & 0 \\
 & & \downarrow & & \downarrow{\scriptstyle \pi_q^{q+1}} & & \downarrow{\scriptstyle \pi_0^1} & & \downarrow & & \\
0 & \longrightarrow & R_q & \longrightarrow & J_q(E) & \xrightarrow{\Phi} & F_0 & \longrightarrow & 0 & & \\
 & & \downarrow & & \downarrow & & \downarrow & & & & \\
 & & 0 & & 0 & & 0 & & & &
\end{array}
$$

The theorem then follows from an induction on r, using the next commutative and exact diagram on page 187, in which the exactness of the upper row and that of the lower row leads to the exactness of the middle row.

REMARK 5.6 In the first of the two preceding diagrams we have identified $Q_1 = \operatorname{coker} \rho_1(\Phi)$ with F_1. Similarly in the second one we have identified $\operatorname{coker} \rho_1(\Psi_r)$ with F_{r+1}. In this way it becomes possible to construct $P(\Theta)$ inductively, defining the vector bundles F_r, up to an isomorphism, by the relations $F_1 = \operatorname{coker} \rho_1(\Phi)$, $F_{r+1} = \operatorname{coker} \rho_1(\Psi_r)$, where Ψ_{r+1} is the natural projection of $J_1(F_r)$ onto F_{r+1}.

COROLLARY 5.7 The sequence

$$E \xrightarrow{\mathscr{D}} F_0 \xrightarrow{\mathscr{D}_1} F_1$$

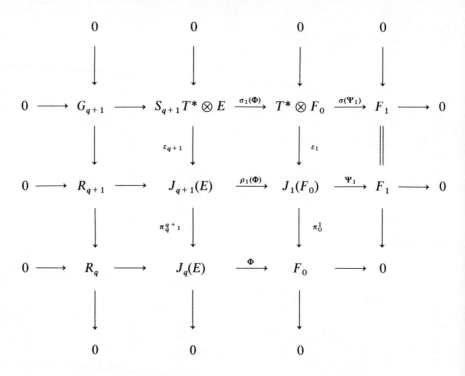

$$
\begin{array}{ccccccccccccccc}
 & & 0 & & 0 & & 0 & & & & 0 & & 0 & & \\
 & & \downarrow & & \downarrow & & \downarrow & & & & \downarrow & & \downarrow & & \\
0 & \longrightarrow & G_{q+r+1} & \longrightarrow & S_{q+r+1}T^* \otimes E & \xrightarrow{\sigma_{r+1}(\Phi)} & S_{r+1}T^* \otimes F_0 & \xrightarrow{\sigma_r(\Psi_1)} & \cdots & \longrightarrow & T^* \otimes F_r & \xrightarrow{\sigma(\Psi_{r+1})} & F_{r+1} & \longrightarrow & 0 \\
 & & \downarrow & & \downarrow & & \downarrow & & & & \downarrow & & \| & & \\
0 & \longrightarrow & R_{q+r+1} & \longrightarrow & J_{q+r+1}(E) & \xrightarrow{\rho_{r+1}(\Phi)} & J_{r+1}(F_0) & \xrightarrow{\rho_r(\Psi_1)} & \cdots & \longrightarrow & J_1(F_r) & \xrightarrow{\Psi_{r+1}} & F_{r+1} & \longrightarrow & 0 \\
 & & \downarrow & & \downarrow{\scriptstyle \pi^{q+r+1}_{q+r}} & & \downarrow & & & & \downarrow & & \downarrow & & \\
0 & \longrightarrow & R_{q+r} & \longrightarrow & J_{q+r}(E) & \xrightarrow{\rho_r(\Phi)} & J_r(F_0) & \xrightarrow{\rho_{r-1}(\Psi_1)} & \cdots & \longrightarrow & F_r & \longrightarrow & 0 & & \\
 & & \downarrow & & \downarrow & & \downarrow & & & & \downarrow & & & & \\
 & & 0 & & 0 & & 0 & & & & 0 & & & &
\end{array}
$$

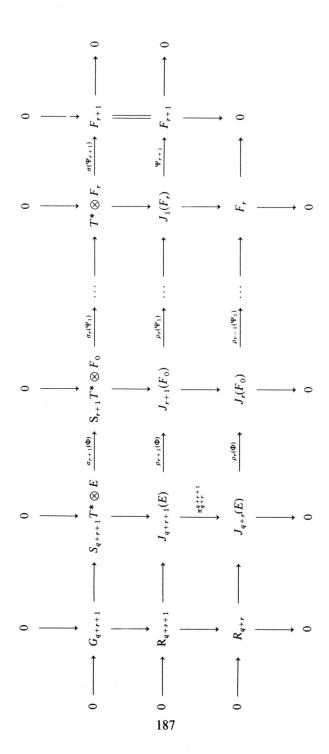

is "*minimum*", that is to say, if there exists another sequence

$$E \xrightarrow{\mathscr{D}} F_0 \xrightarrow{\mathscr{D}'_1} F'_1,$$

then there must exist an operator $\mathscr{D}' : F_1 \to F'_1$ such that $\mathscr{D}'_1 = \mathscr{D}' \circ \mathscr{D}_1$.

Proof Let $\mathscr{D}'_1 = \Psi'_1 \circ j_{q'+1}$. We have the commutative diagram:

$$\begin{array}{ccccc}
J_{q+q'+1}(E) & \xrightarrow{\rho_{q'+1}(\Phi)} & J_{q'+1}(F_0) & \xrightarrow{\rho_{q'}(\Psi_1)} & J_{q'}(F_1) \\
\| & & \| & & \downarrow \Psi' \\
J_{q+q'+1}(E) & \xrightarrow{\rho_{q'+1}(\Phi)} & J_{q'+1}(F_0) & \xrightarrow{\Psi'_1} & F'_1
\end{array}$$

As the top row is exact and $\mathscr{D}$ is regular, a diagram chase shows that there exists a morphism $\Psi : \operatorname{im} \rho_{q'}(\Psi_1) \to F'_1$ and we can extend this morphism to a morphism $\Psi' : J_{q'}(F_1) \to F'_1$ in such a way that $\Psi'_1 = \Psi' \circ \rho_{q'}(\Psi_1)$. We just have to take $\mathscr{D}' = \Psi' \circ j_{q'}$. C.Q.F.D.

REMARK 5.8 The corollary shows that a P-sequence is in some sense the minimum sequence that can be obtained from an involutive operator. This is the reason why it is of such importance.

As in the non-linear case, from proposition 4.2.4 we deduce the useful proposition:

PROPOSITION 5.9 For $0 \leq r \leq n$, the numbers $\dim F_r$ can be determined simply by means of the board $P(R_q) = P$. In particular $\dim F_r$ is equal to the number of sets $(\tau; i_1, \ldots, i_r)$ of integers such that $i_1, \ldots, i_r$ are non-multiplicative indices for the rows of index τ in P.

Similarly to the preceding theorem, we get:

THEOREM 5.10 $S_2(\Theta)$ is formally exact.

Proof The only difference is the study of the formal exactness at C_0. In fact if we identify R_{q+1} with its image in $J_1(C_0)$ by $\rho_1(\mathrm{id}_{J_q(E)})$, then the kernel of the r-prolongation of the epimorphism $J_1(C_0) \to C_1$ is equal to $J_r(R_{q+1}) \cap J_{r+1}(R_q)$ by definition, that is to say $J_{r+1}(R_q) \cap J_{q+r+1}(E) = R_{q+r+1}$ because $J_{q+r+1}(E) = J_r(J_{q+1}(E)) \cap J_{r+1}(J_q(E))$ from proposition 1.9.13. Finally the theorem follows along the same lines from an induction on r, starting for $r = 1$ from the next commutative and exact diagram where we identify C_1 and coker $\rho_1(\mathrm{id}_{J_q(E)})$:

$$\begin{array}{ccccccccc}
 & & 0 & & 0 & & 0 & & \\
 & & \downarrow & & \downarrow & & \downarrow & & \\
0 & \longrightarrow & G_{q+1} & \xrightarrow{\varepsilon_q\circ\delta} & T^*\otimes C_0 & \longrightarrow & C_1 & \longrightarrow & 0\\
 & & \varepsilon_{q+1}\downarrow & & \downarrow\varepsilon_1 & & \| & & \\
0 & \longrightarrow & R_{q+1} & \xrightarrow{\rho_1(\mathrm{id}_{J_q(E)})} & J_1(C_0) & \longrightarrow & C_1 & \longrightarrow & 0\\
 & & \pi^{q+1}_q\downarrow & & \downarrow\pi^1_0 & & \downarrow & & \\
0 & \longrightarrow & R_q & = & C_0 & \longrightarrow & 0 & & \\
 & & \downarrow & & \downarrow & & & & \\
 & & 0 & & 0 & & & &
\end{array}$$

COROLLARY 5.11 The sequence $S_2(E)$ can be considered either as a second Spencer sequence for the null operator $E \to 0$ or as a P-sequence for the operator j_q.

Proof The first point of view follows from the definition, because when $\mathscr{D}$ is a null operator, then $R_q = J_q(E)$ and $G_q = S_qT^* \otimes E$, $\forall q$. As $S_2(E)$ is formally exact it can be constructed inductively from the monomorphism:

$$\mathrm{id}_{J_q(E)} : J_q(E) \to J_q(E),$$

the r prolongation of which is the monomorphism

$$\rho_r(\mathrm{id}_{J_q(E)}) : J_{q+r}(E) \to J_r(J_q(E)).$$

C.Q.F.D.

THEOREM 5.12 The cohomology of a stable sequence $S_1(\Theta)$ at

$$\Lambda^{s+1}T^* \otimes R_{q+n+r-s-1}$$

is the same as that of the sequence $P(\Theta)$ at F_s.

Proof Using theorem 5.5 and proposition 3.1 we obtain the following commutative diagram which has exact rows except perhaps the top one and exact columns except perhaps the left one:

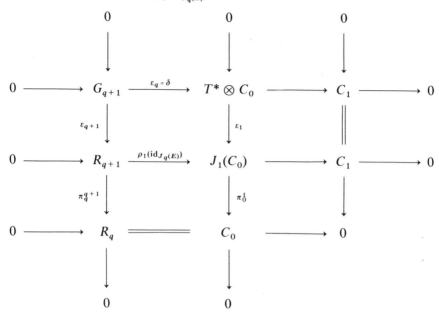

$$
\begin{array}{ccccccccccccc}
 & & 0 & & 0 & & 0 & & & & 0 & & \\
 & & \downarrow & & \downarrow & & \downarrow & & & & \downarrow & & \\
0 & \longrightarrow & \Theta & \longrightarrow & E & \xrightarrow{\mathscr{D}} & F_0 & \xrightarrow{\mathscr{D}_1} & \cdots & \xrightarrow{\mathscr{D}_n} & F_n & \longrightarrow & 0 \\
 & & \downarrow{\scriptstyle j_{q+n+r}} & & \downarrow{\scriptstyle j_{q+n+r}} & & \downarrow{\scriptstyle j_{n+r}} & & & & \downarrow{\scriptstyle j_r} & & \\
0 & \longrightarrow & R_{q+n+r} & \longrightarrow & J_{q+n+r}(E) & \longrightarrow & J_{n+r}(F_0) & \longrightarrow & \cdots & \longrightarrow & J_r(F_n) & \longrightarrow & 0 \\
 & & \downarrow{\scriptstyle D} & & \downarrow{\scriptstyle D} & & \downarrow{\scriptstyle D} & & & & \downarrow{\scriptstyle D} & & \\
0 & \longrightarrow & T^* \otimes R_{q+n+r-1} & \longrightarrow & T^* \otimes J_{q+n+r-1}(E) & \longrightarrow & T^* \otimes J_{n+r-1}(F_0) & \longrightarrow & \cdots & \longrightarrow & T^* \otimes J_{r-1}(F_n) & \longrightarrow & 0 \\
 & & \downarrow{\scriptstyle D} & & \downarrow{\scriptstyle D} & & \downarrow{\scriptstyle D} & & & & \downarrow{\scriptstyle D} & & \\
 & & \vdots & & \vdots & & \vdots & & & & \vdots & & \\
 & & \downarrow{\scriptstyle D} & & \downarrow{\scriptstyle D} & & \downarrow{\scriptstyle D} & & & & & & \\
0 & \longrightarrow & \Lambda^n T^* \otimes R_{q+r} & \longrightarrow & \Lambda^n T^* \otimes J_{q+r}(E) & \longrightarrow & \Lambda^n T^* \otimes J_r(F_0) & \longrightarrow & \cdots & & & & \\
 & & \downarrow & & \downarrow & & \downarrow & & & & & & \\
 & & 0 & & 0 & & 0 & & & & & &
\end{array}
$$

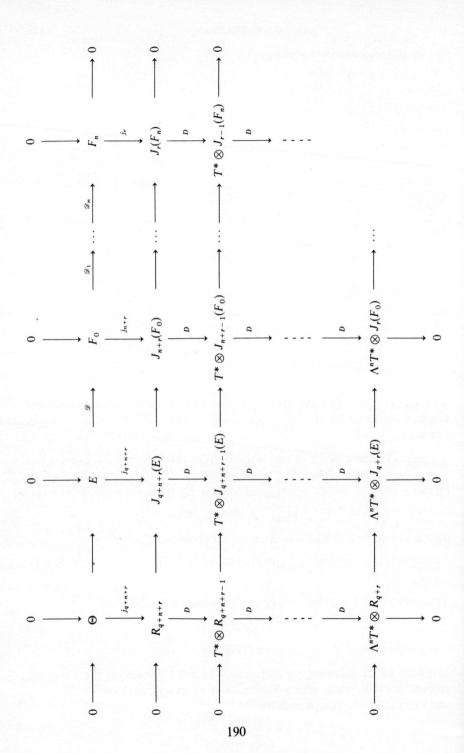

The theorem follows easily from a diagram chase. C.Q.F.D.

COROLLARY 5.13 The sequence $S_2(E)$ is exact.

Proof From the corollary 5.11 we already know that $S_2(E)$ is formally exact. But $S_1(E)$ is locally exact from the proposition 3.1. As $S_2(E)$ can be seen as a P-sequence for j_q, from the last theorem it is also locally exact.

C.Q.F.D.

From this corollary, using diagram chasing in the diagram of the theorem 5.3, we obtain:

THEOREM 5.14 The çohomology of the sequence $S_2(\Theta)$ at C_{r+1} is the same as that of the sequence $P(\Theta)$ at F_r.

REMARK 5.15 We have already computed the integers dim F_r by means of the board P associated with $\mathscr{D}$ or R_q. As $S_2(E)$ is identical to $P(0)$, we can similarly compute the numbers $C_r(E)$ by means of the board associated with j_q. This is the simplest way to compute the integers dim C_r, using the formula:

$$\dim C_r = \dim C_r(E) - \dim F_r.$$

6 Algebraic properties

We shall now consider the fibers over a given arbitrary point $x \in X$ of some vector bundles. In order to simplify the notations, we shall in general use the same letter for a vector bundle and its fiber over x. As we shall deal with local properties; we shall use local coordinates on a convenient neighbourhood U of x in X, trivialising the later bundles.

Let $\mathscr{D} = \Phi \circ j_q : E \to E'$ be an operator of order q on E.

DEFINITION 6.1 For any covector $\chi \in T^*$, with components χ_i, we define the map $\sigma_\chi(\mathscr{D}) : E \to E'$ by the composition:

$$\begin{array}{ccccc} E & \longrightarrow & S_q T^* \otimes E & \xrightarrow{\sigma(\Phi)} & E' \\ (x, u^k) & \longrightarrow & (x, \chi_\mu u^k) & \longrightarrow & (x, A_k^{\tau\mu}(x)\chi_\mu u^k) \end{array}$$

in which

$$k = 1, \ldots, \dim E; \quad \tau = 1, \ldots, \dim E'; \quad \chi_\mu = (\chi_1)^{\mu_1} \cdots (\chi_n)^{\mu_n}.$$

In particular if

$$E \xrightarrow{\mathscr{D}} E' \xrightarrow{\mathscr{D}'} E''$$

is a sequence, that is to say $\mathscr{D}' \circ \mathscr{D} = 0$, then $\sigma_\chi(\mathscr{D}') \circ \sigma_\chi(\mathscr{D}) = 0$ and

$$E \xrightarrow{\sigma_\chi(\mathscr{D})} E' \xrightarrow{\sigma_\chi(\mathscr{D}')} E''$$

is a sequence.

Our aim will be to study the matrix

$$A(x, \chi) = [A_k^\tau(x, \chi)] = [A_k^{\tau\mu}(x)\chi_\mu] : E_x \to E'_x$$

when taking different points $x \in X$ and different covectors $\chi \in T^*$.

The coefficients of this matrix are polynomials in $\chi_1, \ldots, \chi_n$, that is to say become to $\mathbb{R}[\chi] = \mathbb{R}[\chi_1, \ldots, \chi_n]$, and have coefficients in $C^\infty(X)$.

DEFINITION 6.2 A non-zero covector $\chi \in T^*$ is said to be "*characteristic*" for $\mathscr{D}$ if $\sigma_\chi(\mathscr{D})$ is not injective. $\mathscr{D}$ is said to be "*elliptic*" if it has no non-zero "*real*" characteristic covector (It may have complex ones).

REMARK 6.3 We have to understand that $\chi \in T^*$ with projection x on X is characteristic for $\mathscr{D}$ if $\sigma_\chi(\mathscr{D}) : E_x \to E'_x$ is not injective and that $\mathscr{D}$ is elliptic if one cannot find $x \in X$ and $\chi \in T^*$ with projection x on X and non-zero real components such that $\sigma_\chi(\mathscr{D}) : E_x \to E'_x$ is not injective.

EXAMPLE 6.4 $n = 2, m = 1, q = 2$, trivial bundles.

1 $\mathscr{D} \cdot \xi = \partial_{11}\xi + \partial_{22}\xi$ is elliptic because $(\chi_1)^2 + (\chi_2)^2$ has no real solution other than $\chi_1 = 0, \chi_2 = 0$.

2 $\mathscr{D} \cdot \xi = \partial_{11}\xi - \partial_{22}\xi$ is not elliptic and has non-zero real characteristic covectors satisfying $\chi_1 + \chi_2 = 0$ or $\chi_1 - \chi_2 = 0$.

Of course, in order that a non-zero covector $\chi \in T^*$ be characteristic, it must satisfy some algebraic equations that can be obtained by equating to zero some determinants of the matrix $\sigma_\chi(\mathscr{D})$.

DEFINITION 6.5 For fixed $x \in X$, the set of characteristic covectors, consider in $\mathbb{R}^n$ (or $\mathbb{C}^n$) is an algebraic variety V_x, called the "*characteristic variety*" of $\mathscr{D}$ at x. We call $V = \bigcup_{x \in X} V_x$ the characteristic variety of $\mathscr{D}$.

It follows that V is a subset of T^* and $V_x = T^*_x \cap V$.

DEFINITION 6.6 $\mathscr{D}$ is said to be "*under-determined*" ("*determined*") if $\forall x \in X, \exists \chi \in T^*_x$ such that $\sigma_\chi(\mathscr{D}) : E_x \to E'_x$ is surjective (bijective). Otherwise $\mathscr{D}$ is said to be "*over-determined*".

THEOREM 6.7 If $\mathscr{D} = \Phi \circ j_q : E \to E'$ is under determined or determined, then the morphisms $\rho_r(\Phi) : J_{q+r}(E) \to J_r(E')$ are epimorphisms $\forall r \geq 0$ and $\mathscr{D}$ is formally integrable.

Proof Let $x \in X$. As $\chi_\mu \cdot \chi_\nu = \chi_{\mu+\nu}$, $\forall \mu, \nu$, we have the following commutative diagram with $|\mu| = q$ and $|\nu| = r$.

$$\begin{array}{ccccc}
(\chi_{\mu+\nu} u^k) & S_{q+r} T_x^* \otimes E_x & \xrightarrow{\sigma_r(\Phi)} & S_r T_x^* \otimes E'_x & (\chi_\nu u'^\tau) \\
\uparrow & \uparrow & & \uparrow & \uparrow \\
(u^k) & E_x & \xrightarrow{\sigma_\chi(\mathscr{D})} & E'_x & (u'^\tau)
\end{array}$$

Take $\chi_0 \in T_x^*$ such that $\sigma_{\chi_0}(\mathscr{D})$ is surjective. By continuity, $\sigma_\chi(\mathscr{D})$ is also surjective for any covector χ sufficiently close to χ_0.

Using the diagram we see that $\operatorname{im} \sigma_r(\Phi)$ contains all the elements of $S_r T_x^* \otimes E'_x$ such as $\chi_\nu u'^\tau$, $\forall \chi$ sufficiently close to χ_0 and $\forall u'^\tau$. By linearity, it follows that $\sigma_r(\Phi)$ also contains all the derivatives of these elements $\chi_\nu u'^\tau$ with respect to the χ_i, taken at χ_0.

In particular if we effect the differentiation

$$\frac{\partial^r}{(\partial \chi_1)^{\nu_1} \cdots (\partial \chi_n)^{\nu_n}},$$

then $\sigma_r(\Phi)$ contains all the elements of $S_r T_x^* \otimes E'_x$ of components

$$(u'^\tau_\nu \neq 0, \quad u'^{\tau'}_{\nu'} = 0 \quad \text{if} \quad \tau' \neq \tau, \nu' \neq \nu)$$

and finally contains all of $S_r T_x^* \otimes E'_x$.

We conclude that $\sigma_r(\Phi)$ is an epimorphism $\forall r \geq 0$.

Using now an induction on r, starting with the case $r = 0$ which is easy to check and using the following commutative and exact diagram:

$$\begin{array}{ccccccccc}
 & & 0 & & 0 & & 0 & & \\
 & & \downarrow & & \downarrow & & \downarrow & & \\
0 & \longrightarrow & G_{q+r+1} & \longrightarrow & S_{q+r+1} T^* \otimes E & \xrightarrow{\sigma_{r+1}(\Phi)} & S_{r+1} T^* \otimes E' & \longrightarrow & 0 \\
 & & \downarrow & & \downarrow{\scriptstyle \varepsilon_{q+r+1}} & & \downarrow{\scriptstyle \varepsilon_{r+1}} & & \\
0 & \longrightarrow & R_{q+r+1} & \longrightarrow & J_{q+r+1}(E) & \xrightarrow{\rho_{r+1}(\Phi)} & J_{r+1}(E') & & \\
 & & \downarrow & & \downarrow{\scriptstyle \pi^{q+r+1}_{q+r}} & & \downarrow{\scriptstyle \pi^{r+1}_r} & & \\
0 & \longrightarrow & R_{q+r} & \longrightarrow & J_{q+r}(E) & \xrightarrow{\rho_r(\Phi)} & J_r(E') & \longrightarrow & 0 \\
 & & & & \downarrow & & \downarrow & & \\
 & & & & 0 & & 0 & &
\end{array}$$

we can conclude that $\rho_r(\Phi)$ is an epimorphism $\forall r \geq 0$. This shows that $\mathcal{D}$ is regular and that $E' = F_0$. Moreover a diagram chase also shows that $\pi^{q+r+1}_{q+r} : R_{q+r+1} \to R_{q+r}$ is an epimorphism $\forall r \geq 0$ and $\mathcal{D}$ is formally integrable. C.Q.F.D.

COROLLARY 6.8 If the operator $\mathcal{D} = \Phi \circ j_q$ is under-determined or determined, then $\mathcal{D}$ is involutive and its associated P-sequence is just:

$$0 \longrightarrow \Theta \longrightarrow E \xrightarrow{\mathcal{D}} F_0 \longrightarrow 0$$

Proof From the last theorem, $\mathcal{D}$ is formally integrable. We need only show that the symbol G_q of $R_q = \ker \Phi$ is involutive.

From the formal point of view, the involutiveness property follows immediately from the diagram of proposition 4.1.9 and the fact that $G'_r = S^r T^* \otimes E'$ is of course involutive.

What is more delicate is to find a δ-regular coordinate system on a neighbourhood U of a point $x \in X$.

If we effect the linear change of variables:

$$\bar{x}^j = a^j_i x^i, \quad \text{we get} \quad \frac{\partial}{\partial x^i} = \frac{\partial \bar{x}^j}{\partial x^i}\frac{\partial}{\partial \bar{x}^j} = a^n_i \frac{\partial}{\partial \bar{x}^n} + \cdots$$

It follows that

$$A^{\tau\mu}_k(x)\partial_\mu \xi^k = A^{\tau\mu}_k(x)(a^n_1)^{\mu_1} \cdots (a^n_n)^{\mu_n} \frac{\partial^q \xi^k}{(\partial \bar{x}^n)^q} + \cdots$$

We can introduce the covector $\chi = (x, a^n_i) \in T^*$. As it is possible to choose χ in such a way that $\sigma_\chi(\mathcal{D})$ is surjective, it becomes possible to find a coordinate system such that

$$\beta^1_q = 0, \ldots, \beta^{n-1}_q = 0, \beta^n_q = m', \quad \text{with} \quad m' = \dim E' = \dim F_0.$$

Moreover setting $r = -1$ in the above diagram we see that

$$\pi^q_{q-1} : R_q \to J_{q-1}(E)$$

is an epimorphism.

Finally

$$\dim G_{q+1} = \alpha^1_q + \cdots + n\alpha^n_q = m \frac{q + n!}{(n-1)!(q+1)!} - n \cdot m'$$

and this shows that the former coordinate system is δ-regular at x.

REMARK 6.9 Even in this case $P(\Theta)$ is not always locally exact at F_0 as will be proved using the example given by H. Lewy.

THEOREM 6.10 An operator is of finite type if and only if its characteristic variety is zero.

$$\mathscr{D} \text{ finite type} \Leftrightarrow V = \bigcup_{x \in X} V_x, \quad V_x = (0, \ldots, 0) \in T_x^*, \quad \forall x \in X$$

Proof

N.C. If $\mathscr{D} = \Phi \circ j_q : E \to E'$, for any $x \in X$ we have the diagram in the proof of theorem 6.7, linking $\sigma_r(\Phi)$ and $\sigma_\chi(\mathscr{D})$.

However, by hypothesis, for r big enough, $G_{q+r} = \ker \sigma_r(\Phi) = 0$. It follows that, for r big enough, all the maps involved in this diagram are monomorphisms if $\chi \neq 0$. Finally $\sigma_\chi(\mathscr{D})$ is always injective, unless $\chi = 0$.

S.C. $\sigma_\chi(\mathscr{D})$ is injective, $\forall \chi \neq 0$. It follows that

$$\dim E = m \leq m' = \dim E'.$$

For x fixed, among the $m \times m$ matrices that can be found in the $m \times m'$ matrix of $\sigma_\chi(\mathscr{D})$, some have a non-identically zero determinant. In fact if all the determinants are identical to zero in χ, $\sigma_\chi(\mathscr{D})$ could not be injective.

Let $[A_k^{k'\mu}(x)\chi_\mu]$ with $k' = 1, \ldots, m$ such a matrix. Among the polynomial equations defining V_x, one can find the following:

$$P(x, \chi) \equiv \det [A_k^{k'\mu}(x)\chi_\mu] = 0$$

Introducing the transposed matrix $B_{\rho'}^{\rho}(x, \chi)$ of cofactors, the elements of which are polynomials in χ, homogeneous of degree $(m - 1) \cdot q$, with coefficients in $C^\infty(X)$, we have:

$$\sum_{k'=1}^{m} B_{k'}^{\rho}(x, \chi) \cdot A_k^{k'\mu}(x)\chi_\mu = P(x, \chi) \cdot \delta_k^\rho$$

Let now $P_1(x, \chi) = 0, \ldots, P_r(x, \chi)$ be the homogeneous polynomial equations of degree $m \cdot q$ defining V, obtained by considering all the $m \times m$ submatrices of the matrix $[A_k^{\tau\mu}(x)\chi_\mu]$, $|\mu| = q$.

We need the following fundamental result from algebraic geometry.

THEOREM 6.11 (zero theorem of Hilbert) If $Q \in \mathbb{R}[\chi]$ vanishes at all the common zeros in $\mathbb{C}^n$ of the polynomials $P_1(\chi), \ldots, P_r(\chi)$ then there exist an integer $a > 0$ and polynomials $M_1(\chi), \ldots, M_r(\chi)$ such that:

$$(Q(\chi))^a \equiv M_1(\chi) \cdot P_1(\chi) + \cdots + M_r(\chi) \cdot P_r(\chi)$$

Proof We shall prove that this theorem is equivalent to the following one:

THEOREM 6.12 The polynomials $P_1(\chi), \ldots, P_r(\chi) \in \mathbb{R}[\chi]$ have no common zero in $\mathbb{C}^n$ if and only if the ideal I of $\mathbb{R}[\chi]$ that they generate contains $1 \in \mathbb{R}$.

To do this, let us introduce an auxiliary variable η and consider the polynomials:

$$P_1(\chi), \ldots, P_r(\chi), 1 - \eta \cdot Q(\chi) \in \mathbb{R}[\chi, \eta]$$

As they cannot have any common zero in $\mathbb{C}^n$, $\exists N_1, \ldots, N_r, N \in \mathbb{R}[\chi, \eta]$ such that:

$$1 = N_1 \cdot P_1 + \cdots + N_r \cdot P_r + N(1 - \eta \cdot Q)$$

Setting $\eta = 1/Q$ and pushing out the denominators, we obtain the desired result.

It remains to show that, if $P_1, \ldots, P_r$ generate a proper ideal I of $\mathbb{R}[\chi]$, they have at least one common zero in $\mathbb{C}^n$.

To do this, according to proposition 2.5.12, we notice that, $\mathbb{R}[\chi]$ being Noetherian, there exists a proper ideal J of $\mathbb{R}[\chi]$ containing I which is "*maximal*," and thus "*prime*": that is to say such that if $Q_1 \cdot Q_2 \in J$, then Q_1 or Q_2 belongs to J.

Let α_i be the J-residue of χ_i, that is to say $\alpha_i \equiv \chi_i \bmod (J)$. As J is prime,

$$\mathbb{R}[\alpha_1, \ldots, \alpha_n] = \frac{\mathbb{R}[\chi_1, \ldots, \chi_n]}{J}$$

is an integral domain and even a field, as J is maximal.

We shall show that each α_i is algebraic over $\mathbb{R}$ and that $(\alpha_1, \ldots, \alpha_n) \in \mathbb{C}^n$ is a zero of J and thus of $I \subset J$.

This is trivial if $n = 1$, because $\mathbb{R}[\alpha_1]$ cannot be a field if α_1 is transcendental over $\mathbb{R}$ because then $\mathbb{R}[\alpha_1]$ is isomorphic to $\mathbb{R}[\chi_1]$ which is the ring of polynomials in one variable.

As $\mathbb{R}[\alpha_1, \ldots, \alpha_n]$ is a field, it contains the fraction field $\mathbb{R}(\alpha_1)$ of $\mathbb{R}[\alpha_1]$ and we have:

$$\mathbb{R}[\alpha_1, \ldots, \alpha_n] = \mathbb{R}(\alpha_1)[\alpha_2, \ldots, \alpha_n].$$

By induction on n, we may suppose that $\alpha_2, \ldots, \alpha_n$ are algebraic over $\mathbb{R}(\alpha_1)$.

Let us suppose that α_1 is transcendental over $\mathbb{R}$. As $\alpha_i, 2 \leq i \leq n$ is algebraic over $\mathbb{R}(\alpha_1)$, there exists a minimal polynomial $P_i \in \mathbb{R}[\alpha_1][\chi_i]$ of degree r_i, such that $P_i(\alpha_i) = 0$. Let $A_i(\alpha_1) \neq 0$ be the coefficient of $(x_i)^{r_i}$ in P_i and

$$A(\alpha_1) = \prod_{i=2}^{n} A_i(\alpha_1) \neq 0.$$

It is easy to see that $A(\alpha_1)\alpha_i$ is "*integral*" over $\mathbb{R}[\alpha_1]$, that is to say is a zero of a polynomial of $\mathbb{R}[\alpha_1][\chi_i]$ with leading coefficient of $(\chi_i)^{r_i}$ equal to 1.

We can eliminate all the powers of the α_i greater than $r_i - 1$ from any element

$$Q = Q(\alpha_1, \ldots, \alpha_n) \in \mathbb{R}[\alpha_1, \ldots, \alpha_n].$$

Elementary linear algebra then shows that for any element Q as above, there exists an integer r, depending on Q, such that $[A(\alpha_1)]^r Q$ is integral over $\mathbb{R}[\alpha_1]$.

However, if

$$Q \in \mathbb{R}(\alpha_1) \subset \mathbb{R}[\alpha_1, \ldots, \alpha_n],$$

this is only possible if

$$[A(\alpha_1)]^r Q = B(\alpha_1) \in \mathbb{R}[\alpha_1]$$

because α_1 is transcendental over $\mathbb{R}$. Any element of $\mathbb{R}(\alpha_1)$ could be written as

$$\frac{B(\alpha_1)}{[A(\alpha_1)]^r} \quad \text{with} \quad A, B \in \mathbb{R}[\alpha_1]$$

and A fixed. From this absurdity, we conclude that α_1 is algebraic over $\mathbb{R}$ as well as $\alpha_2, \ldots, \alpha_n$ and thus $\alpha_i \in \mathbb{C}$, $\forall i = 1, \ldots, n$. C.Q.F.D.

Now, by hypothesis, $\chi = 0$ is the only common zero of

$$P_1(x, \chi), \ldots, P_r(x, \chi), \qquad \forall x \in X.$$

According to the zero theorem of Hilbert, there exists an integer $a \geq 1$ such that

$$(\chi_i)^a \equiv 0 \bmod (P_1, \ldots, P_r), \qquad \forall \iota.$$

It follows that, for r big enough, $\forall \nu$ with $|\nu| = q + r$, $\exists$ polynomials $C^\rho_\tau(x, \chi)$ such that

$$\sum_\tau C^\rho_\tau(x, \chi) A^{\tau\mu}_k(x) \chi_\mu = \chi_\nu \delta^\rho_k.$$

Thus, if we take $\partial/\partial x^i$ instead of χ_i, we get an operator

$$\sum_\tau C^\rho_\tau\left(x, \frac{\partial}{\partial x}\right)[A^{\tau\mu}_k(x)\partial_\mu \xi^k(x)] = \partial_\nu \xi^\rho(x) + \cdots$$

As ν is arbitrary with $|\nu| = q + r$, we have $G_{q+r} = 0$ for r big enough. C.Q.F.D.

We shall now associate with any operator $\mathcal{D} = \Phi \circ j_q : E \to E'$ another algebraic variety W that will allow us to specify, when $\mathcal{D}$ is involutive, some formal properties of the P-sequence for $\mathcal{D}$.

To do so we notice, as in corollary 6.8 and even when $\mathcal{D}$ is not involutive, that the rank of the matrix $\sigma_\chi(\mathcal{D})$ at any point $x \in X$ is equal to $m - \alpha^n_q(x)$, except for the covectors $\chi \in T_x^*$ that annul all the determinants of order $m - \alpha_q^n(x)$ in this matrix. We have:

$$\beta_q^n = \max_\chi \text{rank}\, \sigma_\chi(\mathscr{D}) \Leftrightarrow \alpha_q^n = \min_\chi \dim \ker \sigma_\chi(\mathscr{D})$$

DEFINITION 6.13 Such covectors are called "*systatic*". Their set at $x \in X$ is an algebraic variety W_x and we call $W = \bigcup_{x \in X} W_x$ the "*systatic variety*" of $\mathscr{D}$.

We shall indicate the relations existing between V and W:

1) If $\alpha_q^n = 0$ then $W = V \subset T^*$ because V and W are defined by the same polynomial equations.
2) If $\alpha_q^n > 0$ then $W \subset V = T^*$ because

$$\dim \ker \sigma_\chi(\mathscr{D}) \geq \alpha_q^n > 0 \qquad \forall \chi \in T^*.$$

In any case we have $W \subseteq V$ and $W \subset T^*$.

When $\mathscr{D}$ is involutive, we know that $\mathscr{D}_1, \ldots, \mathscr{D}_n$ are also involutive. However $\beta_q^n > 0$ and, according to proposition 4.1.9, $\alpha_1'^n = m - \alpha_q^n = \beta_q^n > 0$. It follows that $T^* = V(\mathscr{D}_1) = \cdots = V(\mathscr{D}_n)$ and this fact shows that the characteristic variety is not very useful for study of the P-sequence for $\mathscr{D}$.

THEOREM 6.14 When $\mathscr{D}$ is involutive, the sequence:

$$0 \longrightarrow \ker \sigma_\chi(\mathscr{D}) \longrightarrow E \xrightarrow{\sigma_\chi(\mathscr{D})} F_0 \xrightarrow{\sigma_\chi(\mathscr{D}_1)} F_1 \xrightarrow{\sigma_\chi(\mathscr{D}_2)} \cdots \xrightarrow{\sigma_\chi(\mathscr{D}_n)} F_n \longrightarrow 0$$

is exact if and only if $\chi \notin W$.

Proof

N.C. First of all $\chi \neq 0$ because if not we should have:

$$\dim \text{im}\, \sigma_\chi(\mathscr{D}) = 0 < \dim F_0 = \dim \ker \sigma_\chi(\mathscr{D}_1)$$

Thus, when the sequence is exact, we can find a coordinate system in which $\chi = (0, \ldots, 0, 1)$. From what has been said, $m - \alpha_q^n = \alpha_1'^n$ if α_q^n is the n-character of $\mathscr{D}$ and $\alpha_1'^n$ the n-character of $\mathscr{D}_1$. As this coordinate system is not δ-regular in general, we have at any point $x \in X$:

$$\begin{aligned}\dim \text{im}\, \sigma_\chi(\mathscr{D}) &= m - \dim \ker \sigma_\chi(\mathscr{D}) \leq m - \alpha_q^n \\ &= \alpha_1'^n \leq \dim \ker \sigma_\chi(\mathscr{D}_1)\end{aligned}$$

The exactness of the sequence implies its exactness at F_0 and thus $\dim \sigma_\chi(\mathscr{D}) = \alpha_q^n$ proving that $\chi \notin W$.

S.C. If $\chi \notin W$, then $\chi \neq 0$ and one can find at any point $x \in X$ a coordinate system δ-regular at this point, in which $\chi = (0, \ldots, 0, 1)$.

According to proposition 4.1.9, a system of coordinates which is δ-regular for $\mathscr{D}$ at $x \in X$ is also δ-regular for $\mathscr{D}_1, \ldots, \mathscr{D}_n$ at $x \in X$. It follows that:

$$\dim \operatorname{im} \sigma_\chi(\mathscr{D}) = m - \alpha_q^n = \alpha_q'^1 = \dim \ker \sigma_\chi(\mathscr{D}_1).$$

We can proceed similarly with $\mathscr{D}_1, \ldots, \mathscr{D}_n$ and the sequence is exact.

C.Q.F.D.

COROLLARY 6.15 When $R_q \subset J_q(E)$ is involutive, the sequence:

$$0 \longrightarrow \ker \sigma_\chi(D_1) \longrightarrow C_0 \xrightarrow{\sigma_\chi(D_1)} C_1 \xrightarrow{\sigma_\chi(D_2)} \cdots \xrightarrow{\sigma_\chi(D_n)} C_n \longrightarrow 0$$

is exact if and only if $\chi \notin W$.

Proof Using the diagram of theorem 5.3, we obtain the following commutative diagram:

$$\begin{array}{ccccccccccccc}
 & & 0 & & 0 & & 0 & & & & 0 & & \\
 & & \downarrow & & \downarrow & & \downarrow & & & & \downarrow & & \\
0 & \longrightarrow & \ker \sigma_\chi(D_1) & \longrightarrow & C_0 & \xrightarrow{\sigma_\chi(D_1)} & C_1 & \xrightarrow{\sigma_\chi(D_2)} & \cdots & \xrightarrow{\sigma_\chi(D_n)} & C_n & \longrightarrow & 0 \\
 & & \downarrow & & \downarrow & & \downarrow & & & & \downarrow & & \\
0 & \longrightarrow & E & \xrightarrow{\sigma_\chi(j_q)} & C_0(E) & \longrightarrow & C_1(E) & \longrightarrow & \cdots & \longrightarrow & C_n(E) & \longrightarrow & 0 \\
 & & \downarrow & & \downarrow{\scriptstyle \Phi_0} & & \downarrow{\scriptstyle \Phi_1} & & & & \downarrow{\scriptstyle \Phi_n} & & \\
0 & \longrightarrow & \ker \sigma_\chi(\mathscr{D}_1) & \longrightarrow & F_0 & \xrightarrow{\sigma_\chi(\mathscr{D}_1)} & F_1 & \xrightarrow{\sigma_\chi(\mathscr{D}_2)} & \cdots & \xrightarrow{\sigma_\chi(\mathscr{D}_n)} & F_n & \longrightarrow & 0 \\
 & & \downarrow & & \downarrow & & \downarrow & & & & \downarrow & & \\
 & & 0 & & 0 & & 0 & & & & 0 & &
\end{array}$$

The columns are exact except perhaps the first

As j_q is a finite type operator, $\sigma_\chi(j_q)$ is injective $\forall \chi \neq 0$. Thus

$$W(j_q) = V(j_q) = \{0\}.$$

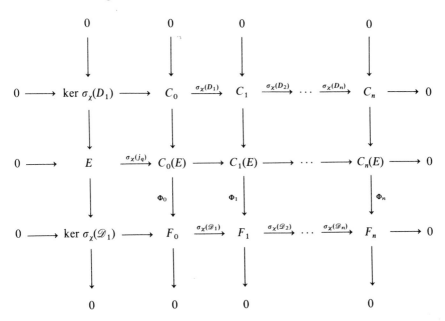

The middle row is exact $\forall\chi \neq 0$, as follows from the theorem applied to the P-sequence for j_q.

The corollary can be deduced at once from the above theorem by means of a chase. C.Q.F.D.

REMARK 6.16 As $\sigma_\chi(\mathscr{D}) = \Phi \circ \sigma_\chi(j_q)$, from the exactness of the left column we deduce:

$$\dim \ker \sigma_\chi(D_1) = m - \dim \ker \sigma_\chi(\mathscr{D}_1) = \dim \ker \sigma_\chi(\mathscr{D}).$$

In particular D_1 is elliptic whenever $\mathscr{D}$ is elliptic.

7 Examples

Many examples will be given in the second part of this book. For this reason we only give here two important ones.

d-Poincaré sequence

Let $d: \Lambda^r T^* \to \Lambda^{r+1} T^*$ be the first order differential operator defined in local coordinates by:

$$\omega = \omega_I(x)\, dx^I \to d\omega = \frac{\partial \omega_I(x)}{\partial x^i}\, dx^i \wedge dx^I$$

for any section $\omega: (x) \to (x, \omega_I(x))$ of $\Lambda^r T^*$.

It is easy to check that $d \circ d = 0$ and we obtain the d-Poincaré sequence:

$$0 \longrightarrow \Theta \longrightarrow \Lambda^0 T^* \xrightarrow{d} \Lambda^1 T^* \xrightarrow{d} \cdots \xrightarrow{d} \Lambda^n T^* \longrightarrow 0$$

We set $\Lambda^0 T^* = X \times \mathbb{R}$ as a trivial bundle over X with section belonging to $C^\infty(X)$.

PROPOSITION 7.1 The latter sequence is an exact P-sequence.

Proof By definition Θ is the set of functions locally constant on X, that is to say functions

$$f: U \subset X \to \mathbb{R} \quad \text{such that} \quad \frac{\partial f(x)}{\partial x^i} = 0 \qquad \forall x \in U.$$

It is easy to check that this linear first order system is involutive and even of finite type.

Moreover $d: \Lambda^r T^* \to \Lambda^{r+1} T^*$ is uniquely determined by a morphism $J_1(\Lambda^r T^*) \to \Lambda^{r+1} T^*$, the symbol of which is surjective $\forall r \geq 0$ as it is just the exterior multiplication $T^* \otimes \Lambda^r T^* \to \Lambda^{r+1} T^*$.

As in theorem 5.5 we conclude that the latter sequence is a P-sequence and that it is formally exact.

It remains to prove that this P-sequence is also locally exact, that is to say $\forall\alpha$ section of $\Lambda^{r+1}T^*$ over $U \subset X$ such that $d\alpha = 0$, $\exists\beta$ section of $\Lambda^r T^*$ over $U \subset X$ such that $d\beta = \alpha$ on U.

As this is a local problem, we shall restrict to the open set

$$U = \{x \in \mathbb{R}^n \,|\, |x^i| < r, r > 0\}.$$

If $f(x)$ is a C^∞ function of x defined for $0 \leq |x| < r$, $\exists g(x) = \int_0^x f(t)\,dt$ which is a C^∞ function of x, defined on the same interval and such that $dg(x)/dx = f(x)$. Moreover if f depends smoothly on certain parameters, then this remains also true for g.

Let A^i be the set of differential forms on U containing only $dx^1, \ldots, dx^i$.

If $\alpha \in A^1$, $\alpha = \alpha_1(x^1, \ldots, x^n)\,dx^1$ and if $d\alpha = 0$ then $\alpha = \alpha_1(x^1)\,dx^1$ and $\exists\beta(x^1)$ such that

$$\alpha = \frac{d\beta(x^1)}{dx^1}\,dx^1 = d\beta.$$

Now if $\alpha \in A^i$, $\alpha = dx^i \wedge \sigma + \tau$, where $\sigma, \tau \in A^{i-1}$. As $d\alpha = 0$, the coefficients of σ, τ do not contain $x^{i+1}, \ldots, x^n$ and we have for example

$$\sigma_{i_1,\ldots,i_r}(x) = \frac{\partial\gamma_{i_1,\ldots,i_r}(x)}{\partial x^i}.$$

Let

$$\gamma = \gamma_{i_1,\ldots,i_r}(x)\,dx^{i_1} \wedge \cdots \wedge dx^{i_r} \quad \text{with} \quad i_1 < \cdots < i_r.$$

Then

$$d\gamma = dx^i \wedge \sigma + \tilde{\tau} \quad \text{with} \quad \tilde{\tau} \in A^{i-1}.$$

It follows that

$$\alpha = d\gamma + (\tau - \tilde{\tau}) \quad \text{and} \quad d\alpha = d(\tau - \tilde{\tau}) = 0$$

By induction $\exists\rho$ such that $\tau - \tilde{\tau} = d\rho$ and we get

$$\alpha = d(\gamma + \rho) = d\beta.$$

C.Q.F.D.

REMARK 7.2 We have $F_r = \Lambda^{r+1}T^*$. The change of grading is due to the fact that $G_1' \neq 0$.

As $d : \Lambda^0 T^* \to \Lambda^1 T^*$ is of finite type, we have

$$\dim C_0 = \dim J_1(\Lambda^0 T^*) - \dim T^* = 1$$

and we get

$$\dim C_r = \dim \Lambda^r T^* \otimes C_0 = \frac{n!}{(n-r)!r!}.$$

According to theorem 5.14 the sequence $S_2(\Theta)$ is also exact. Moreover it is easy to see that $\sigma_\chi(d) : \Lambda^r T^* \to \Lambda^{r+1} T^*$ is just left exterior multiplication by $\chi \in T^*$. We let the reader show as an exercise that we have $\forall \chi \neq 0$ the exact sequence:

$$0 \longrightarrow \Lambda^0 T^* \xrightarrow{\sigma_\chi(d)} \Lambda^1 T^* \xrightarrow{\sigma_\chi(d)} \cdots \xrightarrow{\sigma_\chi(d)} \Lambda^n T^* \longrightarrow 0$$

Counter Example of H. Lewy

$n = 3, m = 2, q = 1, X = \mathbb{R}^3$, E, F trivial bundles over X.

We consider the operator $\mathscr{D} : E \to F$ and the inhomogeneous system $\mathscr{D} \cdot \xi = \eta$ defined in local coordinates by the equations:

$$R_1 \begin{cases} \dfrac{\partial \xi^1(x)}{\partial x^3} - \dfrac{\partial \xi^2(x)}{\partial x^1} - 2x^1 \dfrac{\partial \xi^1(x)}{\partial x^2} - 2x^3 \dfrac{\partial \xi^2(x)}{\partial x^2} = \eta^1(x) \\[2ex] \dfrac{\partial \xi^2(x)}{\partial x^3} + \dfrac{\partial \xi^1(x)}{\partial x^1} + 2x^3 \dfrac{\partial \xi^1(x)}{\partial x^2} - 2x^1 \dfrac{\partial \xi^2(x)}{\partial x^2} = \eta^2(x) \end{cases}$$

We have

$$\sigma_\chi(\mathscr{D}) = \begin{bmatrix} \chi_3 - 2x^1\chi_2 & -\chi_1 - 2x^3\chi_2 \\ \chi_1 + 2x^3\chi_2 & \chi_3 - 2x^1\chi_2 \end{bmatrix}$$

and

$$\det \sigma_\chi(\mathscr{D}) = P(x, \chi) = (\chi_3 - 2x^1\chi_2)^2 + (\chi_1 + 2x^3\chi_2)^2$$

We obtain

$$V = W = \{\chi \in T^* \mid P(x, \chi) = 0\}$$

It follows that $\sigma_{(0,0,1)}(\mathscr{D})$ is bijective $\forall x \in X$ and $\mathscr{D}$ is a determined operator.

We let the reader show that $\mathscr{D}$ is involutive. The P-sequence for $\mathscr{D}$ is simply:

$$0 \longrightarrow \Theta \longrightarrow E \xrightarrow{\mathscr{D}} F \longrightarrow 0$$

It is of course trivially formally exact.

We shall prove that it is not locally exact at F. To do this, if U is a neighbourhood of the origin in $\mathbb{R}^3$ and η a C^∞ section of F over U, let us suppose that we have found a C^∞ section ξ of E over U such that $\mathscr{D} \cdot \xi = \eta$.

We may set $\xi = \xi^1 + i\xi^2$, $\eta = \eta^1 + i\eta^2$ and introduce the new complex variable $z = x^3 + ix^1$, calling now v the variable x^2.

We get

$$\frac{\partial \xi}{\partial \bar{z}} + iz\frac{\partial \xi}{\partial v} = \frac{1}{2}\eta$$

We may also set $z = \sqrt{u}e^{i\theta}$. For u real > 0, as ξ is C^∞ on U, we have:

$$\frac{\partial \xi}{\partial \bar{z}} = z\frac{\partial \xi}{\partial u} + \frac{i}{2}\frac{z}{u}\frac{\partial \xi}{\partial \theta}$$

But

$$\int_0^{2\pi} \frac{z}{u} \cdot \frac{\partial \xi}{\partial \theta}\, d\theta = \left[\frac{z \cdot \xi}{u}\right]_0^{2\pi} - \int_0^{2\pi} \frac{\xi}{u} \cdot \frac{\partial z}{\partial \theta}\, d\theta$$

$$= -2i \int_0^{2\pi} \xi \cdot \frac{\partial z}{\partial u}\, d\theta$$

Thus

$$\int_0^{2\pi} \frac{\partial \xi}{\partial \bar{z}}\, d\theta = \frac{\partial}{\partial u}\int_0^{2\pi} z\xi(u, v; \theta)\, d\theta$$

Setting

$$\int_0^{2\pi} \sqrt{u}e^{i\theta}\xi(u, v; \theta)\, d\theta = A(u, v) + iB(u, v)$$

with $w = u + iv$ we obtain:

$$\frac{\partial}{\partial \bar{w}}(A(u, v) + iB(u, v)) = \frac{1}{2}\int_0^{2r} \eta\, d\theta$$

Taking

$$\eta^1(x) = \frac{1}{2\pi}\frac{\partial C(v)}{\partial v} \quad \text{and} \quad \eta^2(x) = 0$$

with $C(v)$ a real function which is C^∞ in v but not analytic, we have

$$\frac{\partial}{\partial \bar{w}}(A(u, v) + iB(u, v) + iC(v)) = 0$$

It follows that the function $A(u, v) + iB(u, v) + iC(v)$ is holomorphic at any point of the set $0 \le u < \varepsilon$, $-\varepsilon < v < \varepsilon$ of the complex plane, with ε conveniently chosen, except perhaps at the points $u = 0$.

Moreover this function is continuous in the same set and $A(u, v)$ cancels when $u = 0$ because at the same time $z = 0$ and $\xi^1, \xi^2 \in C^\infty(U)$.

The function $A(u, v) + iB(u, v) + iC(v)$ can be analytically prolonged on the other side of the v-axis in the complex plane (Schwarz).

Its trace $C(v)$ on the v-axis is thus analytic contrary to the hypothesis.

C.Q.F.D.

Problems

1) $n = 5, m = 1, q = 2$, E trivial bundle over $X = \mathbb{R}^5$

1 Show that the following system is involutive:

$$R_2 \begin{cases} u_{45} - u_{11} = 0 \\ u_{35} - u_{14} = 0 \\ u_{25} - u_{13} = 0 \\ u_{44} - u_{25} = 0 \\ u_{34} - u_{12} = 0 \\ u_{33} - u_{24} = 0 \end{cases}$$

2 Find the board $P(R_2)$.

3 Find the numbers dim F_r and construct the corresponding P-sequence. Find dim C_r.

(Hint: dim $F_0 = 6$, dim $F_1 = 8$, dim $F_2 = 3$, dim $F_3 = 0$)

4 Determine V and W.

2) The same questions for the systems ($n = 3$)

$$R_2 \begin{cases} u_{33} = 0 \\ u_{23} = 0 \\ u_{13} = 0 \end{cases} \qquad R_2 \begin{cases} u_{33} = 0 \\ u_{23} = 0 \\ u_{22} = 0 \end{cases}$$

3) The same questions for the system ($n = 4$)

$$R_3 \begin{cases} u_{444} - u_{113} = 0 & u_{244} - u_{133} = 0 \\ u_{344} - u_{124} = 0 & u_{234} - u_{122} = 0 \\ u_{334} - u_{123} = 0 & u_{134} - u_{112} = 0 \\ u_{333} - u_{224} = 0 & \end{cases}$$

4) Find an involutive system with the same solutions as the following systems:

$$\begin{cases} u_{33} = 0 \\ u_{22} = 0 \\ u_{112} = 0 \end{cases} \qquad \begin{cases} u_{33} = 0 \\ u_{22} = 0 \\ u_{122} = 0 \\ u_{1111} = 0 \end{cases}$$

$$\begin{cases} u_{44} = 0 \\ u_{34} = 0 \\ u_{33} = 0 \\ u_{124} = 0 \\ u_{114} = 0 \end{cases}$$

5) **1** Same question for the following system

$$R_1\begin{cases} u_4 - x^3 u_2 - u = 0 \\ u_3 - x^4 u_1 \qquad = 0 \end{cases}$$

2 Find the solution of R_1 which is a given function of x^1 for $x^2 = x_0^2$, $x^3 = x_0^3$, $x^4 = x_0^4$.

6) $n = 2$, $m = 2$, $q = 2$, trivial bundles.

1 Is the following system underdetermined, determined or overdetermined:

$$R_2\begin{cases} u_{11}^1 + u_{12}^2 - u^2 = 0 \\ u_{12}^1 + u_{22}^2 + u^1 = 0 \end{cases}$$

2 Show that the associated P-sequence is:

$$0 \longrightarrow E \xrightarrow{\mathscr{D}} F \longrightarrow 0$$

and is exact.

7) Show that any P-sequence is always exact at F_n and that $\mathscr{D}_n$ gives rise to an homogeneous system of the Cauchy–Kowalewski type.

8) Show that, if $P_1(x, \chi), \ldots, P_r(x, \chi)$ are the polynomials defining the characteristic variety of a formally integrable operator, then the polynomials:

$$\{P_\alpha, P_\beta\} = \sum_{i=1}^{n} \left(\frac{\partial P_\alpha}{\partial x^i} \frac{\partial P_\beta}{\partial \chi_i} - \frac{\partial P_\alpha}{\partial \chi_i} \frac{\partial P_\beta}{\partial x^i} \right)$$

belong to the ideal I of $\mathbb{R}_x[\chi]$ generated by $P_1, \ldots, P_r$.
(Hint: One can transform P_α and P_β into operators by setting $\partial/\partial x^i$ instead of χ_i and then consider the operator $P_\alpha \circ P_\beta - P_\beta \circ P_\alpha$.

It is then necessary to use the definition of the P_α and the fact that $\mathscr{D}$ is formally integrable.)

9) Generalise the construction of $\mathscr{D}_1$ in order to construct a formally exact sequence

$$E \xrightarrow{\mathscr{D}} E' \xrightarrow{\mathscr{D}_1} E''$$

when $\mathscr{D} = \Phi \circ j_q$ is formally integrable but when Φ is not surjective. What is the length of the sequence thus obtained?

10) Generalise the construction of 9 when Φ is not surjective and $\mathscr{D}$ is not formally integrable but satisfies one of the conditions of proposition 5.2.8.

11) One considers the following second order system: with $n = 3$, $m = 1$, trivial bundles.

$$R_2\begin{cases} u_{11} + (a(x) + \alpha(x))u_{13} + a(x)\alpha(x)u_{33} = 0 \\ u_{12} + a(x)u_{23} + b(x)u_{13} + a(x)b(x)u_{33} = 0 \\ u_{22} + (b(x) + \beta(x))u_{23} + b(x)\beta(x)u_{33} = 0 \end{cases}$$

1 Show that its symbol G_2 is involutive, $\forall a(x)$, $\alpha(x)$, $b(x)$, $\beta(x)$ and put R_2 under its solved form.

2 Determine the differentiable conditions that must be satisfied by the four above functions in order that R_2 becomes involutive.

3 Determine V and W.

4 Check the result of problem 8.

12) Prove that the system

$$R_1 : \{u_i - \partial_i a(x)u = 0\}$$

is involutive $\forall a \in C^\infty(X)$ and construct the corresponding P-sequence when $n = 1, 2, 3, 4$.

13) $n = 4$, $m = 1$, $q = 2$, trivial bundles over X.
Let us consider the linear second order system:

$$R_2\begin{cases} u_{44} + (x^2)^2 u_{22} + x^2 u_2 = 0 \\ u_{34} + x^1 x^2 u_{12} = 0 \\ u_{33} + (x^1)^2 u_{11} + x^1 u_1 = 0 \end{cases}$$

1 Prove that R_2 is formally integrable but not involutive and that R_3 is involutive if $x^1, x^2 \neq 0$.

2 Show that, if we add to the equations of R_2:

$$x^1 u_{14} - x^2 u_{23} = 0$$

we obtain an involutive system $\bar{R}_2 \subset R_2$ if $x^1, x^2 \neq 0$.

3 Show that there exist solutions of R_2 that are not solutions of $\bar{R}_2$.

(Hint: $u = x^4 \log x^1$)

4 Show that any solution of R_2, analytic in a neighbourhood of the origin in $\mathbb{R}^4$ is also a solution of $\bar{R}_2$.

Remark This problem shows that R_q may be a bundle over X when G_{q+1} and R_{q+1} are not bundles over X.

CHAPTER 6

General Comment

We will deal mainly with local and formal studies. For this reason, all the manifolds that we will consider in the sequel will be C^∞, connected and paracompact. Our aim is to consider the equations defining pseudogroups as special cases of systems of partial differential equations and to allow the reader to handle them directly, by means of local coordinates. This "*operational*" point of view is quite different from the usual one adopted in the literature on this subject.

1 Lie groups

DEFINITION 1.1 A group G is a set of elements $e, a, b, c, \ldots$, with a given map:

$$G \times G \to G : (a, b) \to a \cdot b$$

called the "*group operation*" and satisfying the following "*group axioms*":

1) The map is associative:

$$\forall a, b, c \in G : (a \cdot b) \cdot c = a \cdot (b \cdot c) = a \cdot b \cdot c$$

2) There exists an "*identity*" element e:

$$\exists e \in G, \quad \forall a \in G, \quad a \cdot e = e \cdot a = a$$

3) Each element of G has an "*inverse*":

$$\forall a \in G, \quad \exists a^{-1} \in G, \quad a \cdot a^{-1} = a^{-1} \cdot a = e$$

DEFINITION 1.2 A subgroup $H \subseteq G$ is a subset of G such that, $\forall a, b \in H$, then $a^{-1} \cdot b \in H$. A subgroup $H \subseteq G$ is called "*proper*" and noted $H \subset G$ if $H \neq G$ and $H \neq \{e\}$.

DEFINITION 1.3 A group G is called "*discrete*" if G, as a set, has only a finite number of elements.

DEFINITION 1.4 A subgroup $H \subseteq G$ is said to be "*normal*" in G, noted $H \lhd G$, if $a \cdot H = H \cdot a$, $\forall a \in G$, that is to say if, $\forall b \in H$, $\exists c \in H$ such that $a \cdot b = c \cdot a$.

Definition 1.4 A group G is called "*simple*" if it has no proper normal subgroup.

Example 1.5 The permutation group S_n of n objects is a discrete group with $n!$ elements.

A permutation is said to be odd (even) if it can be expressed as a product of an odd (even) number of transpositions.

The set of even permutations is a normal subgroup $A_n \lhd S_n$ and is simple for $n > 4$. The subgroup $\{e, (12)(34), (13)(24), (14)(23)\}$ is normal in A_4.

Definition 1.6 A group G is called a "*topological group*" if:

1) G, as a set, is a Hausdorff topological space.
2) The group operation $(a, b) \to a^{-1} \cdot b$ is continuous.

Definition 1.7 A subgroup H of a topological group G is called a "*closed subgroup*" if H is closed as a topological space.

Let X be a topological space with elements $x_0, x, y, z, \ldots$ and let G be a topological group.

Definition 1.8 If U is an open subset of X, a bijective map $f: U \to V = f(U)$ is called an "*homeomorphism*" if f and f^{-1} are respectively continuous on U and $V = f(U)$.

Definition 1.9 A topological group G is said to act on X if there is a continuous map $G \times X \to X: (a, x) \to a \cdot x$ such that:

1) $\forall x \in X,\ e \cdot x = x$.
2) $\forall a, b \in G, \forall x \in X, (a \cdot b) \cdot x = a \cdot (b \cdot x)$.

If G acts on X, G is also called a topological "*transformation group*" of X. More precisely, the action of G that we consider is called "*left action*"; a "*right action*" can be defined similarly.

Remark 1.10 It will sometimes be convenient to introduce a copy Y of X and to consider the action as a map $G \times X \to Y: (a, x) \to y = a \cdot x$.

If a topological group G acts on a topological space X, then, for every fixed $a \in G$, the map $X \to X: x \to a \cdot x$ is a homeomorphism of X onto itself since a has a continuous inverse. It follows that the action of G determines a homeomorphism of G into the group of homeomorphisms of X.

Definition 1.11 If $x_0 \in X$, the "*orbit*" of x_0 is the set

$$G(x_0) = \{x \in X \mid x = ax_0, \forall a \in G\}$$

DEFINITION 1.12 G acts "*effectively*" on X if $ax = x$, $\forall x \in X$, implies that $a = e$. G acts "*transitively*" on X if, $\forall x \in X$, $\forall y \in Y$, $\exists a \in G$ such that $y = a \cdot x$. If not G is said to act "*intransitively*".

DEFINITION 1.13 A subset $S \subseteq X$ is said to be "*invariant*" under the action of G if $ax \in S$, $\forall a \in G$ and $\forall x \in S$.

DEFINITION 1.14 Let x_0 be a fixed point of X. The set

$$G_{x_0} = \{a \in G \,|\, ax_0 = x_0\}$$

is called the "*isotropy*" group of x_0.

DEFINITION 1.15 A topological group G is called a Lie group if:

1) G is a C^∞ manifold.
2) The group operation $(a, b) \to a^{-1} \cdot b$ is C^∞.

REMARK 1.16 A Lie group is not necessarily connected. If G is a Lie group, we shall denote by G^0 the connected component of G containing the identity. G^0 is a closed normal subgroup of G. As G is homeomorphic to G^0 we shall define $\dim G = \dim G_0$. Unless mentioned we shall deal only with the component of the identity of every group we shall meet.

EXAMPLE 1.17 The group $GL(n, \mathbb{R})$ of linear transformations of $\mathbb{R}^n$ has two components according to the sign of the determinant of its elements which are $n \times n$ matrices.

Let X be a C^∞ connected paracompact manifold with $\dim X = n$.

DEFINITION 1.18 If U is an open set of X, a homeomorphism f of U onto $V = f(U)$ such that f and f^{-1} are C^∞ maps will be called a "*local diffeomorphism*". A diffeomorphism $f: X \to X$ will be called an "*automorphism*" of X and the set of automorphisms of X will be denoted $\mathrm{Aut}(X)$.

REMARK 1.19 For simplicity of notation we shall also denote by $\mathrm{Aut}(X)$ the set of local automorphisms or local diffeomorphisms of X. The context will then in general indicate clearly the open sets of definition. For the same reason we shall also write $G \times X \to X$ for a local action, instead of $G \times U \to V$ with convenient open subsets of X.

From now on we shall consider Lie groups of transformations of X and use local coordinates $(x^1, \ldots, x^n)$ for X, $(y^1, \ldots, y^n)$ for Y and $(a^1, \ldots, a^r)$ for a neighbourhood of e in G. A transformation

$$(x^i) \to y^k = f^k(a^1, \ldots, a^r; x^1, \ldots, x^n)$$

will be written simply

$$x \to y = f(a, x).$$

DEFINITION 1.20 x is the "*source*" and y the "*target*" of the transformation. In order to avoid confusion the coordinates $(a^1, \ldots, a^r)$ on G will be called "*local parameters*" or simply "*parameters*" of the transformation.

DEFINITION 1.21 If a is close to e, the resulting transformation will be called an "*infinitesimal transformation*", otherwise it will be called a "*finite transformation*".

DEFINITION 1.22 The parameters $a^1, \ldots, a^r$ are called "*effective*" if G acts effectively on X. This will always be understood in the sequel.

Let $\xi_1, \ldots, \xi_r$ with $\xi_\rho = \xi^i_\rho(x)(\partial/\partial x^i)$ be r vector fields on X, sections of $T = T(X)$.

DEFINTION 1.23 The r vector fields $\xi_1, \ldots, \xi_r$ are said to be "*dependent*" on X if there exist r constants λ^ρ not all zero and such that the identity $\sum_{\rho=1}^r \lambda^\rho \xi^i_\rho(x) = 0$ holds $\forall x \in X$. Otherwise $\xi_1, \ldots, \xi_r$ are said to be "*independent*".

DEFINITION 1.24 The r vector fields $\xi_1, \ldots, \xi_r$ are said to be "*connected*" on X if rank $[\xi^i_\rho(x)] < r, \forall x \in X$. Otherwise they are said to be "*unconnected*".

It is easy to see that the two preceding definitions do not depend on the coordinate system adopted on a neighbourhood of x in X.

EXAMPLE 1.25

1
$$\frac{\partial}{\partial x^1}, \ldots, \frac{\partial}{\partial x^n}$$

are n unconnected and thus independent vector fields on $X = \mathbb{R}^n$.

2 On the contrary

$$y\frac{\partial}{\partial z} - z\frac{\partial}{\partial y}, z\frac{\partial}{\partial x} - x\frac{\partial}{\partial z}, x\frac{\partial}{\partial y} - y\frac{\partial}{\partial x}$$

are 3 connected but independent vector fields on $X = \mathbb{R}^3$ with cartesian coordinates (x, y, z).

2 Lie fundamental theorems

The work on the above transformation groups was done mainly by the mathematicians S. Lie in 1870 and E. Cartan in 1900. We recall for our purpose the three following theorems of Lie that will be of constant use from now on.

FIRST THEOREM 2.1 If $y = f(a, x)$ are the finite transformations of a Lie group of transformations of X, the orbits $x = f(a, x_0)$ are solutions of a non-linear system of partial differential equations on $G \times X$:

$$\frac{\partial x^i}{\partial a^\sigma} = \xi^i_\rho(x)\omega^\rho_\sigma(a)$$

where the vector fields $\xi_1, \ldots, \xi_r$ are independent on X and $\det[\omega^\rho_\sigma(a)] \neq 0$.

Proof From the group axiom 1) and the action axiom 2) if $x = f(a, x_0)$ and $y = f(b, x)$ then there exists $c = b \cdot a$ such that $y = f(c, x_0)$ with $c = \varphi(b, a)$.

Let x_0 and c be fixed. Without any index, we have:

$$\begin{cases} \dfrac{\partial f(b, x)}{\partial b} db + \dfrac{\partial f(b, x)}{\partial x} dx = 0 \\[2ex] \dfrac{\partial \varphi(b, a)}{\partial b} db + \dfrac{\partial \varphi(b, a)}{\partial a} da = 0 \end{cases}$$

As $x = f(b^{-1}, y)$ we have

$$\det\left[\frac{\partial f(b, x)}{\partial x}\right] \neq 0$$

because

$$y \equiv f(b, f(b^{-1}, y)) \quad \text{and} \quad 1 = \frac{\partial f(b, x)}{\partial x}\frac{\partial f(b^{-1}, y)}{\partial y}$$

where 1 stands for the identity $n \times n$ matrix. A similar argument shows that

$$\det\left[\frac{\partial \varphi(b, a)}{\partial b}\right] \neq 0$$

and we can eliminate the db by means of Gramer's rules in order to get:

$$dx^i = \xi^i_\rho(b, x)\omega^\rho_\sigma(b, a)\, da^\sigma$$

But b can be fixed arbitrarily because c does not appear in the formulas. For example we can take $b = e$.

As above

$$\det \frac{\partial \varphi(b, a)}{\partial a} \neq 0 \quad \text{and thus} \quad \det[\omega^\rho_\sigma(e, a)] = \det[\omega^\rho_\sigma(a)] \neq 0.$$

It will be useful to introduce the $r \times r$ matrix $[\alpha^\sigma_\rho(a)]$ such that

$$\alpha^\sigma_\rho(a)\omega^\rho_\tau(a) = \delta^\sigma_\tau.$$

The proof will be completed by means of the following proposition.

PROPOSITION 2.2 The action of G on X is effective if and only if the vector fields $\xi_1, \ldots, \xi_r$ are independent on X.

Proof

N.C. Imagine that the action of G on X is effective. This means that we can choose points $x_\alpha \in X$ with $\alpha = 1, 2, \ldots$ in such a way as to obtain a unique solution $a = e$ for the equations $x_\alpha = f(a, x_\alpha)$. Then the rank of the matrix

$$\left[\frac{\partial f^i(a, x_\alpha)}{\partial a^\sigma}\right]$$

with rows indexed by σ and columns indexed by i and α, must be equal to r. As det $[\omega^\rho_\sigma(a)] \neq 0$, the rank of the similar matrix $[\xi^i_\rho(x_\alpha)]$ must also be equal to r and there cannot exist constants λ^ρ non-zero and such that $\lambda^\rho\xi^i_\rho(x_\alpha) = 0$, $\forall i, \alpha$.

S.C. Imagine that the action of G on X is not effective. Let $x_1 \in X$ and set

$$r_1 = \text{rank}\left[\frac{\partial f^i(a, x_1)}{\partial a^\sigma}\right] = \text{rank}\,[\xi^i_\rho(x_1)].$$

Choosing sufficiently many points $x_\alpha \in X$ will make the rank of the matrix $[\xi^i_\rho(x_\alpha)]$ with rows indexed by σ and columns indexed by i and α, maximum and equal to s. We have $r_1 \leq r_2 \leq \cdots \leq s < r$ by hypothesis. Then we will find non-zero constants λ^ρ such that $\lambda^\rho\xi^i_\rho(x_\alpha) = 0$, $\forall i, \alpha$. We can consider the matrix $[\xi^i_\rho(x), \xi^i_\rho(x_\alpha)]$ with n more columns. Its rank cannot be greater than s because otherwise we could find a point $x_0 \in X$ in such a way as to contradict the maximum property of s. It follows that we still have

$$\lambda^\rho\xi^i_\rho(x) = 0, \qquad \forall i, \forall x \in X.$$

C.Q.F.D.

EXAMPLE 2.3

$$G \times \mathbb{R} \to \mathbb{R} : y = (a^1 + a^2)x$$

In a similar way we have the following proposition:

PROPOSITION 2.4 The action of G on X is transitive if and only if

$$\text{rank}\,[\xi^i_\rho(x)] = n, \qquad \forall x \in X.$$

Proof

N.C. If the action of G is transitive, then $\forall x \in X$ and $\forall y \in Y$ we can solve the equation $y = f(a, x)$ in order to find the parameter a. For this we need

$$\operatorname{rank}\left[\frac{\partial f^i(a, x)}{\partial a^\sigma}\right] = n \leq r$$

because otherwise, by the implicit function theorem we could eliminate the a and find out at least one relation between x and y only contrary to the transitivity of the action. As $\det[\omega_\sigma^\rho(a)] \neq 0$ we must have $\operatorname{rank}[\xi_\rho^i(x)] = n \leq r$ and $\xi_1, \ldots, \xi_r$ must be unconnected on X.

S.C. We leave this to the reader as an exercise.

DEFINITION 2.5 The action of G on X is said to be "*simply transitive*" if $r = n$.

EXAMPLE 2.6 The following effective actions on $X = \mathbb{R}^2$ are:

- intransitive: $y^1 = x^1 + a^1, \quad y^2 = x^2$
- simply transitive: $y^1 = x^1 + a^1, \quad y^2 = x^2 + a^2$
- transitive: $y^1 = x^1 + a^1, \quad y^2 = a^3x^2 + a^2$

REMARK 2.7 We can define a left or a right effective and transitive action of G on itself by using the group law on G, that is to say the map

$$G \times G \to G : (a, b) \to c = a \cdot b$$

given in coordinates by $c = \varphi(a, b)$.

In general we will only consider left actions. In this case, the same proof as for theorem I, gives the non-linear system:

$$\frac{\partial c^\tau}{\partial a^\sigma} = \alpha_\rho^\tau(c)\omega_\sigma^\rho(a)$$

where the matrix $[\alpha_\rho^\tau(c)]$ is defined as above, because $c = a$ when $b = e$ and then $\partial c^\tau/\partial a^\sigma = \delta_\sigma^\tau$. Introducing the 1-forms $w^\rho = \omega_\sigma^\rho(a)\, da^\sigma$ with $\rho = 1, \ldots, r$ we see that $\omega_\tau^\rho(c)\, dc^\tau = \omega_\sigma^\rho(a)\, da^\sigma$ and we have the proposition:

PROPOSITION 2.8 The 1-forms $w^\rho = \omega_\sigma^\rho(a)\, da^\sigma$ are right invariant on G.

We come now to the particular case $r = 1$ and shall speak of 1-parameter groups of transformations of X.

From first theorem 2.1 we have:

$$\frac{\partial x^i}{\partial a} = \xi^i(x)\omega(a)$$

If a_0 is the value of a corresponding to the identity transformation, we can introduce a new parameter $t = \int_{a_0}^{a} \omega(b)\, db$ such that $dt = \omega(a)\, da$ in order to get a system of ordinary differential equations defining the orbits:

$$\frac{dx^i}{dt} = \xi^i(x)$$

It is known (4) that it has a unique solution equal to x_0 when $t = 0$.

For t small, close to zero, we have an infinitesimal transformation, written:

$$x^i = x_0^i + t\xi^i(x_0) + \cdots$$

For big t we shall write:

$$x = \exp(t\xi)(x_0) = f_t(x_0) = f(t, x_0)$$

The following properties are left to the reader:

$$\begin{cases} \exp(t_1\xi) \circ \exp(t_2\xi) = \exp((t_1 + t_2)\xi) \\ \exp(t\xi)^{-1} = \exp(-t\xi) \\ \exp(t\xi_1) \circ \exp(t\xi_2) \neq \exp(t(\xi_1 + \xi_2)) \end{cases}$$

REMARK 2.9 The above notation is useful to get the Taylor development with respect to t, for small t.

DEFINITION 2.10 If ξ_1 and ξ_2 are two vector fields on X sections of T, we define $[\xi_1, \xi_2]$, the "*bracket*" of ξ_1 and ξ_2 as the new section of T defined in local coordinates by:

$$([\xi_1, \xi_2])^i(x) = \xi_1^j(x)\frac{\partial \xi_2^i(x)}{\partial x^j} - \xi_2^j(x)\frac{\partial \xi_1^i(x)}{\partial x^j}$$

Now let X, Y be two manifolds with $\dim X = n$, $\dim Y = m$; $f: X \to Y$ be a map and $T(f): T(X) \to T(Y)$ the induced map.

DEFINITION 2.11 A vector field η on Y is said to be "*f-related*" to a vector field ξ on X if $T(f) \circ \xi = \eta \circ f$.

We have the commutative diagram:

$$\begin{array}{ccc} T(X) & \xrightarrow{T(f)} & T(Y) \\ \xi \uparrow\downarrow & & \uparrow\downarrow \eta \\ X & \xrightarrow{f} & Y \end{array}$$

In local coordinates, on a neighbourhood U of a point $x \in X$ and $f(U)$ of the point $y = f(x) \in Y$ we have:

$$\eta^k(f(x)) = \frac{\partial f^k(x)}{\partial x^i} \xi^i(x)$$

In particular, if $m = n$ and f is regular, that is to say rank $(T(f)) = n$, to a given vector field ξ on X corresponds a unique vector field η on Y and vice-versa. We shall write $\eta = f_*(\xi)$ or simply $\eta = f(\xi)$ where there is no confusion on the kind of geometric object. A useful case arises when $f \in \text{Aut}(X)$.

PROPOSITION 2.12 If the vector fields η_1 and η_2 on Y are respectively f-related to the vector fields ξ_1 and ξ_2 on X, then the vector field $[\eta_1, \eta_2]$ on Y is f-related to the vector field $[\xi_1, \xi_2]$ on X.

Proof Using local coordinates as above, we have:

$$\begin{aligned}
&\eta_1^l(f(x)) \frac{\partial \eta_2^k(f(x))}{\partial y^l} - \eta_2^l(f(x)) \frac{\partial \eta_1^k(f(x))}{\partial y^l} \\
&\quad = \frac{\partial f^l(x)}{\partial x^i} \xi_1^i(x) \frac{\partial \eta_2^k(f(x))}{\partial y^l} - \frac{\partial f^l(x)}{\partial x^i} \xi_2^i(x) \frac{\partial \eta_1^k(f(x))}{\partial y^l} \\
&\quad = \xi_1^i(x) \frac{\partial}{\partial x^i}\left(\frac{\partial f^k(x)}{\partial x^j} \xi_2^j(x)\right) - \xi_2^i(x) \frac{\partial}{\partial x^i}\left(\frac{\partial f^k(x)}{\partial x^j} \xi_1^j(x)\right) \\
&\quad = \frac{\partial f^k(x)}{\partial x^j}\left(\xi_1^i(x) \frac{\partial \xi_2^j(x)}{\partial x^i} - \xi_2^i(x) \frac{\partial \xi_1^j(x)}{\partial x^i}\right)
\end{aligned}$$

C.Q.F.D.

REMARK 2.13 If ξ_1, ξ_2 are vector fields on X and $f \in \text{Aut}(X)$ then we have the useful formula:

$$f([\xi_1, \xi_2]) = [f(\xi_1), f(\xi_2)]$$

PROPOSITION 2.14 If ξ_1, ξ_2, ξ_3 are vector fields on X, we have the identities:

$$[\xi_1, \xi_2] = -[\xi_2, \xi_1]$$

$$[\xi_1, [\xi_2, \xi_3]] + [\xi_2, [\xi_3, \xi_1]] + [\xi_3, [\xi_1, \xi_2]] = 0 \qquad \text{(Jacobi)}$$

Proof This straightforward computation is left to the reader.

We come back to theorem 2.1. Let us now define new parameters λ^ρ, called "*canonical parameters*", by using the following device:

Consider the system of ordinary differential equations:

$$\frac{da^\sigma}{dt} = \lambda^\rho \alpha_\rho^\sigma(a)$$

and let a_0 instead of e be the value of the a corresponding to the identity transformation of X. The later system has a unique solution $a = h(t, \lambda)$ such that $a_0 = h(0, \lambda)$ and we have $h(t, \lambda) = h(1, t\lambda)$. Introducing the vector fields $\alpha_\rho = \alpha_\rho^\sigma(a)(\partial/\partial a^\sigma)$ on G, we have $a = \exp(t\lambda^\rho \alpha_\rho)(a_0)$, that is to say

$$a^\sigma = a_0^\sigma + t\lambda^\rho \alpha_\rho^\sigma(a_0) + \cdots$$

In particular, setting $t = 1$, we have:

$$a_0 = h(0, \lambda) = h(1, 0) \quad \text{and} \quad \det \left[\frac{\partial h^\sigma(t, \lambda)}{\partial \lambda^\rho}\right]_{(1,0)} = \det\,[\alpha_\rho^\sigma(a_0)] \neq 0$$

and we can change coordinates on G, using the parameters λ instead of the parameters a. At the same time, we have

$$x = f(h(t, \lambda), x_0) \quad \text{and} \quad \frac{dx^i}{dt} = \lambda^\rho \xi_\rho^i(x).$$

Using the same notation as above we set

$$x = \exp(t\lambda^\rho \xi_\rho)(x_0)$$

DEFINITION 2.15 We say that $\xi_1, \ldots, \xi_r$ are the "*infinitesimal generators*" of the effective action of G on X.

SECOND THEOREM 2.16 If $\xi_1, \ldots, \xi_r$ are the infinitesimal generators of the effective action of a Lie group G on X, we have the relations

$$[\xi_\sigma, \xi_\tau] = c_{\sigma\tau}^\rho \xi_\rho$$

where the $c_{\sigma\tau}^\rho$ are constants, called "*structure constants*" of G.

Proof As a first remark, using proposition 6.2.2, we see that $\xi_1, \ldots, \xi_r$ are independent.

Now, as in theorem 2.1, the orbits are solutions of the non-linear first order system of p.d.e.:

$$\frac{\partial x^i}{\partial a^\sigma} = \xi_\rho^i(x)\omega_\sigma^\rho(a)$$

Let us consider on $G \times X$ the r vector fields:

$$\alpha_\rho + \xi_\rho = \alpha_\rho^\sigma(a) \frac{\partial}{\partial a^\sigma} + \xi_\rho^i(x) \frac{\partial}{\partial x^i}$$

If we set $x_0 = f^{-1}(a, x)$ in such a way that $x \equiv f(a, f^{-1}(a, x))$, then $f^{-1}(a, x)$ must be constant on the orbits and we have:

$$(\alpha_\rho + \xi_\rho) \cdot f^{-1} = 0 \qquad \forall \rho = 1, \ldots, r$$

Moreover, as $\det [\alpha_\rho^\sigma(a)] \neq 0$, the r vector fields $\alpha_\rho + \xi_\rho$ are unconnected on $G \times X$.

But

$$\operatorname{rank} \left[\frac{\partial f^{-1}(a, x)}{\partial a}, \frac{\partial f^{-1}(a, x)}{\partial x} \right] = n \qquad \forall (a, x) \in G \times X$$

as we have already seen that

$$\operatorname{rank} \left[\frac{\partial f^{-1}(a, x)}{\partial x} \right] = n.$$

It follows that there cannot be more than $(r + n) - n = r$ unconnected vector fields $\alpha_\rho + \xi_\rho$ such that $(\alpha_\rho + \xi_\rho) \cdot f^{-1} = 0$.

Now, from the definition of the bracket, we have

$$[\alpha_\sigma + \xi_\sigma, \alpha_\tau + \xi_\tau] \cdot f^{-1} = 0$$

and thus

$$[\alpha_\sigma + \xi_\sigma, \alpha_\tau + \xi_\tau] = c_{\sigma\tau}^\rho(a, x)(\alpha_\rho + \xi_\rho)$$

As the vector fields α_ρ and ξ_ρ are relative to different sets of variables, it follows that:

$$[\alpha_\sigma, \alpha_\tau] = c_{\sigma\tau}^\rho(a, x)\alpha_\rho \quad \text{and} \quad [\xi_\sigma, \xi_\tau] = c_{\sigma\tau}^\rho(a, x)\xi_\rho$$

As $\det [\alpha_\rho^\sigma(a)] \neq 0$, from the first set of relations we conclude that

$$c_{\sigma\tau}^\rho(a, x) = c_{\sigma\tau}^\rho(a)$$

and from the second set of relations we conclude that

$$\frac{\partial c_{\sigma\tau}^\rho(a)}{\partial a} \cdot \xi_\rho^i(x) = 0,$$

that is to say

$$c_{\sigma\tau}^\rho(a) = c_{\sigma\tau}^\rho = \text{cst}$$

because the vector fields $\xi_1, \ldots, \xi_r$ are independent. C.Q.F.D.

From theorem 2.16 we deduce the corollary:

COROLLARY 2.17 When the vector fields $\xi_1, \ldots, \xi_r$ are independent on X and when the vector fields $\alpha_1, \ldots, \alpha_r$ on G are such that $\det[\alpha_\rho^\sigma(a)] \neq 0$, then the non-linear first order system of p.d.e. on $G \times X$:

$$\frac{\partial x^i}{\partial a^\sigma} = \xi_\rho^i(x)\omega_\sigma^\rho(a)$$

is involutive if and only if we have the relations:

$$[\alpha_\sigma, \alpha_\tau] = c_{\sigma\tau}^\rho \alpha_\rho \quad \text{and} \quad [\xi_\sigma, \xi_\tau] = c_{\sigma\tau}^\rho \xi_\rho.$$

From proposition 2.14, we have the other important corollary:

COROLLARY 2.18 The structure constants $c_{\sigma\tau}^\rho$ satisfy the following relation:

$$c_{\sigma\tau}^\rho + c_{\tau\sigma}^\rho = 0$$

$$c_{\mu\rho}^\lambda \cdot c_{\sigma\tau}^\mu + c_{\mu\sigma}^\lambda \cdot c_{\tau\rho}^\mu + c_{\mu\tau}^\lambda \cdot c_{\rho\sigma}^\mu = 0 \qquad \text{(Jacobi)}$$

Finally we come to the celebrated:

THIRD THEOREM 2.19 If a set of $r^2(r-1)/2$ constants $c_{\sigma\tau}^\rho$ is given, satisfying the former relations, then it is possible to exhibit an analytic Lie group G having the $c_{\sigma\tau}^\rho$ as structure constants.

Proof The key idea of the proof is to exhibit the group law $G \times G \to G$ as a left action of G on itself, using the remark following proposition 2.4 and a convenient coordinate system.

Let us start with some preliminary results. We consider the finite transformation $x = f(a, x_0)$ and the system

$$\frac{\partial x^i}{\partial a^\sigma} = \xi_\rho^i(x)\omega_\sigma^\rho(a).$$

Setting $a = h(t, \lambda)$ as before, we have:

$$x = f(a, x_0) = f(h(t, \lambda), x_0) = \exp(t\lambda^\rho \xi_\rho)(x_0)$$

Then

$$\frac{\partial x^i}{\partial \lambda^\sigma} = \frac{\partial x^i}{\partial a^\tau}\frac{\partial h^\tau(t, \lambda)}{\partial \lambda^\sigma} = \xi_\rho^i(x)\omega_\tau^\rho(h(t, \lambda))\frac{\partial h^\tau(t, \lambda)}{\partial \lambda^\sigma}$$

that is to say

$$\frac{\partial x^i}{\partial \lambda^\sigma} = \xi^i_\rho(x)\chi^\rho_\sigma(t, \lambda)$$

Also we have

$$\frac{\partial x^i}{\partial t} = \lambda^\rho \xi^i_\rho(x)$$

Differentiating those equations, we get:

$$\frac{\partial^2 x^i}{\partial t \partial \lambda^\sigma} = \frac{\partial \xi^i_\tau(x)}{\partial x^j} \lambda^\rho \xi^j_\rho(x)\chi^\tau_\sigma(t, \lambda) + \xi^i_\rho(x) \frac{\partial \chi^\rho_\sigma(t, \lambda)}{\partial t}$$

$$\frac{\partial^2 x^i}{\partial t \partial \lambda^\sigma} = \xi^i_\sigma(x) + \lambda^\rho \frac{\partial \xi^i_\rho(x)}{\partial x^j} \cdot \xi^j_\tau(x) \cdot \chi^\tau_\sigma(t, \lambda)$$

Subtracting, using the fact that:

$$\xi^j_\tau(x) \frac{\partial \xi^i_\rho(x)}{\partial x^j} - \xi^j_\rho(x) \frac{\partial \xi^i_\tau(x)}{\partial x^j} = c^\sigma_{\tau\rho}\, \xi^i_\sigma(x)$$

and the fact that $\xi_1, \ldots, \xi_r$ are independent on X, we obtain:

$$\frac{\partial \chi^\rho_\sigma(t, \lambda)}{\partial t} = \delta^\rho_\sigma + c^\rho_{\nu\tau} \lambda^\tau \cdot \chi^\nu_\sigma(t, \lambda)$$

This is a system of ordinary differential equations which has a unique analytic solution with the given initial value:

$$\chi^\rho_\sigma(0, \lambda) = \omega^\rho_\tau(a_0) \frac{\partial h^\tau(0, \lambda)}{\partial \lambda^\sigma} = 0$$

Let us now define:

$$\pi^\rho_{\sigma\tau}(t, \lambda) = \frac{\partial \chi^\rho_\sigma(t, \lambda)}{\partial \lambda^\tau} - \frac{\partial \chi^\rho_\tau(t, \lambda)}{\partial \lambda^\sigma} - c^\rho_{\mu\nu}\, \chi^\mu_\sigma(t, \lambda) \cdot \chi^\nu_\tau(t, \lambda)$$

Then a straightforward computation shows that $\pi^\rho_{\sigma\tau}(t, \lambda)$ is the unique solution of the system of ordinary differential equations:

$$\frac{\partial \pi^\rho_{\sigma\tau}(t, \lambda)}{\partial t} = c^\rho_{\nu\mu} \cdot \lambda^\mu \cdot \pi^\nu_{\sigma\tau}(t, \lambda)$$

with initial value $\pi^\rho_{\sigma\tau}(0, \lambda) = 0$.

It follows that $\pi^{\rho}_{\sigma\tau}(t, \lambda) \equiv 0$.

If we introduce finally the 1-forms:

$$w^{\rho} = \chi^{\rho}_{\sigma}(1, \lambda)\, d\lambda^{\sigma} = \omega^{\rho}_{\sigma}(a)\, da^{\sigma}$$

then

$$dw^{\rho} = \frac{\partial \omega^{\rho}_{\sigma}(a)}{\partial a^{\tau}}\, da^{\tau} \wedge da^{\sigma} = \frac{1}{2}\left(\frac{\partial \omega^{\rho}_{\sigma}(a)}{\partial a^{\tau}} - \frac{\partial \omega^{\rho}_{\tau}(a)}{\partial a^{\sigma}}\right) da^{\tau} \wedge da^{\sigma}$$

and we have the equation, called "*Maurer–Cartan equation*":

$$dw^{\rho} + \tfrac{1}{2} c^{\rho}_{\sigma\tau} w^{\sigma} \wedge w^{\tau} = 0$$

Once we have determined $\chi^{\rho}_{\sigma}(1, \lambda)$, the group law is known as a solution of a non-linear first order analytic system of p.d.e., similar to the following one:

$$\frac{\partial b^{\tau}}{\partial a^{\sigma}} = \alpha^{\tau}_{\rho}(b)\omega^{\rho}_{\sigma}(a)$$

but with $\chi^{\rho}_{\sigma}(1, \lambda)$ instead of $\omega^{\rho}_{\sigma}(a)$ and the canonical parameters λ instead of a. This is possible because the system is involutive, according to the Maurer–Cartan equations, as the reader can easily check, using theorem 2.16.

C.Q.F.D.

COROLLARY 2.20 Any Lie group is differentiably equivalent to an analytic manifold with analytic group operation.

Proof The corollary follows from the existence of the map $a = h(1, \lambda)$ constructed above and from the fact that $\chi(t, \lambda)$, as already indicated, is analytic in t and λ, because it is the unique solution of a Cauchy–Kowalewski system with given Cauchy data.

REMARK 2.21 A direct proof can be given if the matrix $M(\lambda) = [c^{\rho}_{\nu\tau}\lambda^{\tau}]$ has distinct eigenvalues. In this case, the system giving $\chi^{\rho}_{\sigma}(t, \lambda)$ can be written in matrix form:

$$\frac{\partial \chi}{\partial t} = I + M(\lambda) \cdot \chi$$

and the unique solution such that $\chi(0, \lambda) = 0$ is:

$$\chi(t, \lambda) = \frac{e^{M(\lambda)t} - I}{M(\lambda)} = t + \frac{1}{2!} Mt^2 + \cdots$$

It is defined even if $\det M(\lambda) = 0$ and is analytic.

EXAMPLE 2.22 $G = \{(a^1, a^2) \in \mathbb{R}^2 \mid a^1 \neq 0\}$, $X = \mathbb{R}$

$$G \times X \to X : ((a^1, a^2), x) \to a^1 x + a^2$$

$$G \times G \to G : ((a^1, a^2), (b^1, b^2)) \to (a^1 b^1, a^1 b^2 + a^2)$$

$$\begin{cases} \dfrac{\partial x}{\partial a^1} = \dfrac{x}{a^1} - \dfrac{a^2}{a^1} \\ \\ \dfrac{\partial x}{\partial a^2} = 1 \end{cases} \Rightarrow \begin{cases} \xi_1 = x \dfrac{\partial}{\partial x} \\ \\ \xi_2 = \dfrac{\partial}{\partial x} \end{cases}$$

$$[\omega_\sigma^\rho(a)] = \begin{bmatrix} \dfrac{1}{a^1} & -\dfrac{a^2}{a^1} \\ 0 & 1 \end{bmatrix} \Rightarrow \begin{cases} \alpha_1 = a^1 \dfrac{\partial}{\partial a^1} + a^2 \dfrac{\partial}{\partial a^2} \\ \\ \alpha_2 = \dfrac{\partial}{\partial a^2} \end{cases}$$

$$[\alpha_1, \alpha_2] = -\alpha_2, \quad [\xi_1, \xi_2] = -\xi_2$$

$$w^1 = \frac{da^1}{a^1}, \quad w^2 = -\frac{a^2}{a^1} da^1 + da^2$$

Maurer–Cartan: $\begin{cases} dw^1 = 0 \\ dw^2 - w^1 \wedge w^2 = 0 \end{cases}$

$$\frac{\partial}{\partial t}\begin{bmatrix} \chi_1^1 & \chi_2^1 \\ \chi_1^2 & \chi_2^2 \end{bmatrix} = \begin{bmatrix} 1 & 0 \\ 0 & 1 \end{bmatrix} + \begin{bmatrix} 0 & 0 \\ -\lambda^2 & \lambda^1 \end{bmatrix}\begin{bmatrix} \chi_1^1 & \chi_2^1 \\ \chi_1^2 & \chi_2^2 \end{bmatrix}$$

$$\chi_\sigma^\rho(0, \lambda) = 0 \Rightarrow [\chi_\sigma^\rho(1, \lambda)] = \begin{bmatrix} 1 & 0 \\ \dfrac{\lambda^2}{\lambda^1} - \dfrac{\lambda^2}{(\lambda^1)^2}(e^{\lambda_1} - 1) & \dfrac{1}{\lambda^1}(e^{\lambda_1} - 1) \end{bmatrix}$$

$$w^1 = d\lambda^1, w^2 = \left(\frac{\lambda^2}{\lambda^1} - \frac{\lambda^2}{(\lambda^1)^2}(e^{\lambda_1} - 1)\right) d\lambda^1 + \frac{1}{\lambda^1}(e^{\lambda_1} - 1)\, d\lambda^2$$

3 Invariant foliations

Let X be a C^∞, connected, paracompact manifold with $\dim X = n$.

Let $\xi_1, \ldots, \xi_r$ or $\{\xi_\rho\}$ $\rho = 1, \ldots, r$, be r vector fields of class C^∞ on X, sections of $T = T(X)$.

In general, for each $x \in X$, we have:

$$0 \leq \operatorname{rank}\, [\xi_\rho^i(x)] = m(x) \leq \inf(n, r)$$

As $m(x)$ depends on the value of certain determinants, if x_0 is a point of X such that $m(x_0) = m$ is maximum on X, there exists an open set $U \subset X$, containing x_0 and such that $m(x) = m$, $\forall x \in U$.

DEFINITION 3.1 A point $x \in X$ is said to give "*rank h*" to $\xi_1, \ldots, \xi_r$ if rank $[\xi^i_\rho(x)] = h \leq m$. A subset (submanifold) of X is said to give the rank h to $\xi_1, \ldots, \xi_r$ if each of its points give rank h to $\xi_1, \ldots, \xi_r$.

REMARK 3.2 It has to be pointed out that the maximal subset that gives rank h to $\xi_1, \ldots, \xi_r$ is defined by equating to zero the determinants of order $> h$ in the $n \times r$ matrix $[\xi^i_\rho(x)]$, while keeping the determinants of order h different from zero. It is not always a submanifold of X.

EXAMPLE 3.3 $X = \mathbb{R}^3$ with cartesian coordinates (x, y, z)

$$\begin{cases} \xi_1 \equiv x\dfrac{\partial}{\partial x} + y\dfrac{\partial}{\partial y} \\ \xi_2 \equiv (xz - y)\dfrac{\partial}{\partial x} + (yz - x)\dfrac{\partial}{\partial y} + (1 - z^2)\dfrac{\partial}{\partial z} \\ \xi_3 = (x^2 + y^2)\dfrac{\partial}{\partial x} + 2xy\dfrac{\partial}{\partial y} - y(1 - z^2)\dfrac{\partial}{\partial z} \end{cases}$$

It is easy to check that ξ_1, ξ_2, ξ_3 are connected on X. We have

$$\xi_3 = (x + yz)\xi_1 - y\xi_2 \quad \text{and} \quad m = 2.$$

The subset of X that gives rank 1 to ξ_1, ξ_2, ξ_3 is

$$\{(x, y, z) \in \mathbb{R}^3 \,|\, (y - x)(y + x) = 0,$$
$$x(1 - z^2) = 0 \,|\, (x, y, z) \neq (0, 0, 1) \text{ or } (0, 0, -1)\}$$

It is a non-connected submanifold of $\mathbb{R}^3$ of dimension 1.

If we had considered only ξ_1 and ξ_3, then the subset of X giving rank 1 to ξ_1, ξ_3 is not a submanifold of $\mathbb{R}^3$.

EXAMPLE 3.4 $X = \mathbb{R}^3$ with cartesian coordinates (x, y, z)

$$\xi_1 \equiv x\frac{\partial}{\partial x} + z\frac{\partial}{\partial z}, \qquad \xi_2 \equiv y\frac{\partial}{\partial y} + z\frac{\partial}{\partial z}$$

$$\xi_3 \equiv y\frac{\partial}{\partial x}, \qquad \xi_4 \equiv x\frac{\partial}{\partial y}$$

Subset giving rank 2 = $\{(x, y, z) \in \mathbb{R}^3 | z = 0 | (x, y, z) \neq (0, 0, 0)\}$
Subset giving rank 1 = $\{(x, y, z) \in \mathbb{R}^3 | x = 0, y = 0 | (x, y, z) \neq (0, 0, 0)\}$
Subset giving rank 0 = $\{(x, y, z) \in \mathbb{R}^3 | (x, y, z) = (0, 0, 0)\}$

EXAMPLE 3.5 $X = \mathbb{R}^3$ with cartesian coordinates (x, y, z)

$$\begin{cases} \xi_1 \equiv (x^2 - y^2 + z^2)\dfrac{\partial}{\partial x} + 2xy\dfrac{\partial}{\partial y} + 2xz\dfrac{\partial}{\partial z} \\ \xi_2 \equiv 2xy\dfrac{\partial}{\partial x} + (y^2 - x^2 + z^2)\dfrac{\partial}{\partial y} + 2yz\dfrac{\partial}{\partial z} \\ \xi_3 \equiv 2xz\dfrac{\partial}{\partial x} + 2yz\dfrac{\partial}{\partial y} + (x^2 + y^2 + z^2)\dfrac{\partial}{\partial z} \end{cases}$$

We have

$$\det[\xi^i_\rho(x)] = (z^2 - x^2 - y^2)^3.$$

Every point of X gives rank 3 to ξ_1, ξ_2, ξ_3 if $z^2 \neq (x^2 + y^2) \neq 0$.

No point of X can give rank 2 to ξ_1, ξ_2, ξ_3.

The subset of X giving rank 1 to ξ_1, ξ_2, ξ_3 is

$$\{(x, y, z) \in \mathbb{R}^3 | z^2 - (x^2 + y^2) = 0 | (x, y, z) \neq (0, 0, 0)\}$$

It is a submanifold of $\mathbb{R}^3$ of dimension 2.

The subset of X giving rank 0 to ξ_1, ξ_2, ξ_3 is the origin $(x, y, z) = (0, 0, 0)$ of $\mathbb{R}^3$.

DEFINITION 3.6 An "*r-distribution* ∇" on X is a choice of an r-dimensional subspace ∇_x of T_x for each $x \in X$.

We shall consider only distributions on X such that, for each $x \in X$, there is a neighbourhood U of x and r vector fields $\xi_1, \ldots, \xi_r$ defined and unconnected on U, which span the distribution at each point of U. Such a distribution of vector fields will be called regular and of course we have

$$\text{rank}\,[\xi^i_\rho(x)] = r, \qquad \forall x \in U.$$

EXAMPLE 3.7 If $\xi_1, \ldots, \xi_r$ are r vector fields on X, then we have an m-distribution on any open subset of X giving the rank m to $\xi_1, \ldots, \xi_r$ with

$$m = \max_{x \in X} \text{rank}\,[\xi^i_\rho(x)].$$

DEFINITION 3.8 A vector field ξ on X is said to belong to the distribution ∇ if $\xi(x) \in \nabla_x, \forall x \in X$. It is equivalent to say that its restriction $\xi|_U$ is connected with $\xi_1, \ldots, \xi_r$ on U, when U is defined as above.

Definition 3.9 A regular distribution on X is said to be "*involutive*" if, whenever ξ_1 and ξ_2 belong to the distribution, then $[\xi_1, \xi_2]$ also belong to the distribution. In particular, on U defined as above, we must have:

$$[\xi_\sigma, \xi_\tau] = c^\rho_{\sigma\tau}(x)\xi_\rho$$

Definition 3.10 A submanifold $R \subset X$ is an "*integral manifold*" of a distribution on X if $T_x(R) = \nabla_x, \forall x \in R$.

Remark 3.11 On U we have rank $[\xi^i_\rho(x)] = r$. We may assume, without loss of generality, that the matrix $[\xi^\sigma_\rho(x)]$ for $\rho, \sigma = 1, \ldots, r$ has rank r on U and we introduce its inverse matrix $[a^\sigma_\rho(x)]$. Then we can define the distribution on U by means of the linear combinations $\bar{\xi}_\rho(x) = a^\sigma_\rho(x)\xi_\sigma(x)$. We have

$$\bar{\xi}_\rho = \frac{\partial}{\partial x^\rho} + \sum_{i>r} \bar{\xi}^i_\rho(x) \frac{\partial}{\partial x^i}$$

and $[\bar{\xi}_\sigma, \bar{\xi}_\tau]$ is connected with

$$\frac{\partial}{\partial x^{r+1}}, \ldots, \frac{\partial}{\partial x^n}.$$

When the distribution is involutive, it is easy to check that $[\bar{\xi}_\sigma, \bar{\xi}_\tau]$ is also connected with $\xi_1, \ldots, \xi_r$ and thus with $\bar{\xi}_1, \ldots, \bar{\xi}_r$. This is only possible if $[\bar{\xi}_\sigma, \bar{\xi}_\tau] = 0, \forall \sigma, \tau$.

Finally, for each U as defined above, we can always take such a basis for an involutive distribution.

Now let ξ, η be vector fields on X and define the new vector field

$$\eta_t = \exp(t\xi)(\eta).$$

Proposition 3.12 We have the useful formula:

$$\frac{d\eta_t}{dt} = [\eta_t, \xi]$$

Proof We set $y = f(t, x) = f_t(x) = \exp(t\xi)(x)$
Then, by definition we have:

$$\frac{\partial f^k(t + dt, x)}{\partial x^i} \eta^i(x) = \eta^k_{t+dt}(f(t + dt, x))$$

$$\frac{\partial f^k(t, x)}{\partial x^i} \eta^i(x) = \eta^k_t(f(t, x))$$

Substracting, dividing by dt and taking the limit as $dt \to 0$, we have:

$$\frac{\partial^2 f^k(t, x)}{\partial x^i \partial t} \eta^i(x) = \frac{d\eta_t}{dt}(y) + \frac{\partial \eta_t^k(y)}{\partial y^l} \xi^l(y)$$

but

$$\frac{\partial^2 f^k(t, x)}{\partial x^i \partial t} \eta^i(x) = \frac{\partial \xi^k(f(t, x))}{\partial x^i} \eta^i(x) = \frac{\partial \xi^k(y)}{\partial y^l} \frac{\partial f^l(t, x)}{\partial x^i} \eta^i(x)$$
$$= \frac{\partial \xi^k(y)}{\partial y^l} \eta_t^l(y).$$

and

$$\frac{d\eta_t^k(y)}{dt} = \eta_t^l(y) \frac{\partial \xi^k(y)}{\partial y^l} - \xi^l(y) \frac{\partial \eta_t^k(y)}{\partial y^l} = ([\eta_t, \xi])^k(y)$$

C.Q.F.D.

We use this proposition in the following one:

PROPOSITION 3.13 If ξ and η are vector fields on X then $[\xi, \eta] = 0$ if and only if

$$\exp(t\xi) \circ \exp(s\eta) = \exp(s\eta) \circ \exp(t\xi), \qquad \forall s, t.$$

Proof

N.C. By the preceding proposition and proposition 2.12 we have

$$\frac{d}{dt}(\exp(t\xi)(\eta)) = \exp(t\xi)([\eta, \xi]) = 0$$

because

$$\exp(t\xi)(\xi) = \xi \quad \text{and} \quad [\xi, \eta] = -[\eta, \xi] = 0.$$

Consider the one parameter group of transformations of X:

$$g_s = \exp(t\xi) \circ \exp(s\eta) \circ \exp(-t\xi)$$

Differentiation with respect to s shows that its associated vector field is just $\exp(t\xi)(\eta) = \eta$. This means that $g_s = \exp(s\eta)$ and the proof follows from the fact that $\exp(t\xi)^{-1} = \exp(-t\xi)$.

S.C. Looking for the terms up to order 2 in the Taylor development, we have:

$$g_s(x) = \exp(s\eta)(x) = x + s\eta(x) + \frac{s^2}{2}\eta(\eta(x)) + \cdots$$

$$\exp(t\xi) \circ \exp(s\eta) \circ \exp(-t\xi)(x) = x + s\eta(x) + \frac{s^2}{2}\eta(\eta(x)) + st[\eta, \xi](x) + \cdots$$

and we must have

$$[\eta, \xi](x) = 0, \qquad \forall x \in X,$$

that is to say

$$[\xi, \eta] = -[\eta, \xi] = 0. \qquad \text{C.Q.F.D.}$$

THEOREM 3.14 (Frobenius) Let ∇ be an r-dimensional involutive distribution on X. Through each $x \in X$ there passes a unique connected integral manifold of ∇.

Proof Let $x_0 \in X$ and choose a coordinate system on a sufficiently small neighbourhood U of x_0 such that

$$\xi_1, \ldots, \xi_r, \frac{\partial}{\partial x^{r+1}}, \ldots, \frac{\partial}{\partial x^n}$$

are unconnected on U and $[\xi_\sigma, \xi_\tau] = 0$ as in the remark above.

For $t_i \in \mathbb{R}$, $|t_i| < \varepsilon$ sufficiently small, the point

$$x = \exp(t_1\xi_1) \circ \cdots \circ \exp(t_r\xi_r)(x_0)$$

will be in U.

We have

$$\frac{\partial x}{\partial t_1} = \xi_1(x).$$

But, from proposition 3.13 we have

$$\exp(t_i\xi_i) \circ \exp(t_j\xi_j) = \exp(t_j\xi_j) \circ \exp(t_i\xi_i) \qquad \forall i, j = 1, \ldots, r$$

and thus also

$$\frac{\partial x}{\partial t^i} = \xi_i(x), \qquad \forall i = 1, \ldots, r.$$

Finally, as $\xi_1, \ldots, \xi_r$ are unconnected and span the r-dimensional subspace ∇_x of $T_x(X)$ at each point $x \in U$, from the implicit function theorem we have constructed in U an integral manifold through x_0 of dimension r, which is unique, connected and parametrised by $t_1, \ldots, t_r$,

Setting

$$\xi_{r+1} = \frac{\partial}{\partial x^{r+1}}, \ldots, \xi_n = \frac{\partial}{\partial x^n}$$

we can introduce the local diffeomorphism

$$x_0 \to \exp(t_1 \xi_1) \circ \cdots \circ \exp(t_n \xi_n)(x_0)$$

and adopt on U the new coordinates $(t_1, \ldots, t_n)$ instead of $(x^1, \ldots, x^n)$. The previous construction applied with $\exp(t_{r+1}\xi_{r+1}) \circ \cdots \circ \exp(t_n \xi_n)(x_0)$ instead of x_0, shows that each slice $t_{r+1} = \text{cst}, \ldots, t_n = \text{cst}$ is an integral manifold.

It remains to show that it is possible to patch together the local pieces in order to continue away from x_0, as far as possible. This is done by means of topological arguments that are out of our scope. For a detailed exposition, the reader can refer to reference 54. C.Q.F.D.

Now let ∇ be a regular distribution of vector fields on X. Then for each $x \in X$, we can find a neighbourhood $U \subset X$ and a basis $\xi_\rho = \xi^i_\rho(x)(\partial/\partial x^i)$ with $\rho = 1, \ldots, r$ for ∇ on U such that $[\xi_\sigma, \xi_\tau] = c^\rho_{\sigma\tau}(x)\xi_\rho$.

Let us consider the first order linear system of p.d.e.:

$$\xi_\rho \cdot \Phi(x) \equiv \xi^i_\rho(x)\partial_i \Phi(x) = 0$$

on the trivial vector bundle $U \times \mathbb{R}$, where $\Phi \in C^\infty(U)$ is a section over U and the summation is done for $1 \leq i \leq n$.

PROPOSITION 3.15 The distribution ∇ is involutive if and only if the above system of p.d.e. is involutive.

Proof

N.C. As ∇ is regular and involutive, we can take $\bar{\xi}_\rho$ as a basis of ∇ on U, instead of ξ_ρ, and we have

$$\bar{\xi}_\rho = \frac{\partial}{\partial x^\rho} + \sum_{i>r} \bar{\xi}^i_\rho(x) \frac{\partial}{\partial x^i} \quad \text{with} \quad [\bar{\xi}_\sigma, \bar{\xi}_\tau] = 0$$

The linear system

$$\partial_\rho \Phi(x) + \sum_{i>r} \bar{\xi}^i_\rho(x)\partial_i \Phi(x) = 0$$

is equivalent to the preceding one because rank $[\xi^i_\rho(x)] = r$ on U.

We will now show that the new system of coordinates obtained from the initial one by the permutation of indices

$$\begin{pmatrix} 1 & \cdots & r & \cdots & n \\ n & \cdots & n-r+1 & \cdots & 1 \end{pmatrix}$$

is δ-regular on U.

To show this, we need to prove that the linear system:

$$\partial_{n-r+\alpha}\Phi(x) + \sum_{i \le n-r} \xi^i_{n-r+\alpha}(x)\partial_i\Phi(x) = 0 \qquad 1 \le \alpha \le r$$

has an involutive symbol. But the equations are already in a solved form and $x^{n-r+\alpha+1}, \ldots, x^n$ are the non-multiplicative variables for the equation $n - r + \alpha$ associated to the vector field $\xi_{n-r+\alpha}$ on U.

Differentiating with respect to one of them, say $x^{n-r+\beta}$ with $\beta > \alpha$, we get:

$$\frac{\partial^2\Phi(x)}{\partial x^{n-r+\alpha}\partial x^{n-r+\beta}} + \sum_{i \le n-r} \xi^i_{n-r+\alpha}(x)\frac{\partial^2\Phi(x)}{\partial x^i\partial x^{n-r+\beta}} + \sum_{i \le n-r} \frac{\partial\xi^i_{n-r+\alpha}(x)}{\partial x^{n-r+\beta}}\frac{\partial\Phi(x)}{\partial x^i} = 0$$

But we have also, using the multiplicative variables:

$$\frac{\partial^2\Phi(x)}{\partial x^{n-r+\alpha}\partial x^{n-r+\beta}} + \sum_{j \le n-r} \xi^j_{n-r+\beta}(x)\frac{\partial^2\Phi(x)}{\partial x^j\partial x^{n-r+\alpha}} + \sum_{j \le n-r} \frac{\partial\xi^j_{n-r+\beta}(x)}{\partial x^{n-r+\alpha}}\frac{\partial\Phi(x)}{\partial x^j} = 0$$

$$\frac{\partial^2\Phi(x)}{\partial x^i\partial x^{n-1+\beta}} + \sum_{k \le n-r} \xi^k_{n-r+\beta}(x)\frac{\partial^2\Phi(x)}{\partial x^k\partial x^i} + \sum_{k \le n-r} \frac{\partial\xi^k_{n-r+\beta}(x)}{\partial x^i}\frac{\partial\Phi(x)}{\partial x^k} = 0$$

$$\frac{\partial^2\Phi(x)}{\partial x^j\partial x^{n-r+\alpha}} + \sum_{i \le n-r} \xi^i_{n-r+\alpha}(x)\frac{\partial^2\Phi(x)}{\partial x^i\partial x^j} + \sum_{i \le n-r} \frac{\partial\xi^i_{n-r+\alpha}(x)}{\partial x^j}\frac{\partial\Phi(x)}{\partial x^i} = 0$$

because the x^i (x^j) are multiplicative variables for the equations $n - r + \beta$ ($n - r + \alpha$).

The involutiveness of the symbol follows from the fact that:

$$\sum_{i,k \le n-r} \xi^i_{n-r+\alpha}(x)\xi^k_{n-r+\beta}(x)\frac{\partial^2\Phi(x)}{\partial x^i\partial x^k} - \sum_{j,k \le n-r} \xi^j_{n-r+\beta}(x)\xi^k_{n-r+\alpha}(x)\frac{\partial^2\Phi(x)}{\partial x^j\partial x^k} \equiv 0$$

The symbol and its prolongations are vector bundles on U because ∇ is regular on U.

Using now the criterion for involutiveness given in the first part of this book, we check by elimination that the conditions:

$$\begin{aligned}\sum_{i \le n-r} \frac{\partial\xi^i_{n-r+\alpha}(x)}{\partial x^{n-r+\beta}}\frac{\partial\Phi(x)}{\partial x^i} - \sum_{j \le n-r} \frac{\partial\xi^j_{n-r+\beta}(x)}{\partial x^{n-r+\alpha}}\frac{\partial\Phi(x)}{\partial x^j} \\ - \sum_{i,k \le n-r} \xi^i_{n-r+\alpha}(x)\frac{\partial\xi^k_{n-r+\beta}(x)}{\partial x^i}\frac{\partial\Phi(x)}{\partial x^k} \\ + \sum_{j,k \le n-r} \xi^j_{n-r+\beta}(x)\frac{\xi^k_{n-r+\alpha}(x)}{\partial x^j}\frac{\partial\Phi(x)}{\partial x^k} = 0\end{aligned}$$

that is to say:

$$\sum_{i \le n-r} \left(\frac{\partial \xi^i_{n-r+\alpha}(x)}{\partial x^{n-r+\beta}} - \frac{\partial \xi^i_{n-r+\beta}(x)}{\partial x^{n-r+\alpha}} + \sum_{j \le n-r} \xi^j_{n-r+\beta}(x) \frac{\partial \xi^i_{n-r+\alpha}(x)}{\partial x^j} \right.$$
$$\left. - \sum_{j \le n-r} \xi^j_{n-r+\alpha}(x) \frac{\partial \xi^i_{n-r+\beta}(x)}{\partial x^j} \right) \frac{\partial \Phi(x)}{\partial x^i} = 0$$

are satisfied because we have the commutation relations $[\xi_{n-r+\alpha}, \xi_{n-r+\beta}] = 0$ in the new coordinate system. C.Q.F.D.

S.C. It is left to the reader.

REMARK 3.16 An argument due to A. Mayer shows that the resolution of the above system of p.d.e. is equivalent to that of a system of ordinary differential equations (Caratheodory 4).

DEFINITION 3.17 When a local function $\Phi \in C^\infty(U)$ is a solution of a system like

$$\xi^i_\rho(x) \frac{\partial \Phi(x)}{\partial x^i} = 0$$

defined on U, we say that Φ is "*killed*" by the vector fields $\xi_1, \ldots, \xi_r$ on U.

DEFINITION 3.18 A non-constant local function Φ which is killed by a regular distribution ∇ is called an "*invariant*" of the distribution ∇.

REMARK 3.19 Taking $\Phi \in C^\infty(U)$ in the definition, we have to understand that Φ is killed by any basis of ∇ on U, or equivalently, by the restriction $\xi|_U$ of any vector field ξ belonging to ∇.

EXAMPLE 3.20 $X = \mathbb{R}^2$ with cartesian coordinates (x, y). Let us consider the 1-distribution

$$\xi = x \frac{\partial}{\partial x} + y \frac{\partial}{\partial y} \quad \text{on} \quad X - \{0, 0\}.$$

Then x/y is an invariant of this distribution for $y \neq 0$, and y/x is an invariant of this distribution for $x \neq 0$.

EXAMPLE 3.21 $X = \mathbb{R}^3$ with cartesian coordinates (x, y, z). Then

$$\Phi(x, y, z) \equiv z e^{xy} \in C^\infty(X)$$

is an invariant of the involutive distribution

$$\left\{ \xi_1 = \frac{\partial}{\partial x} - yz \frac{\partial}{\partial z}, \xi_2 = \frac{\partial}{\partial y} - xz \frac{\partial}{\partial z} \right\}.$$

DEFINITION 3.22 s local functions $\Phi^\tau \in C^\infty(U)$ are said to be "*connected*" on U if

$$\operatorname{rank}\left[\frac{\partial \Phi^\tau(x)}{\partial x^i}\right] < s, \qquad \forall x \in U \subset X.$$

In the contrary case they are said to be unconnected.

REMARK 3.23 If s local functions Φ^τ are connected, from the implicit function theorem we know that there is at least one local relation of the form

$$\Psi(\Phi^1(x), \ldots, \Phi^s(x)) \equiv 0$$

where Ψ is a local function defined on a convenient open subset of $\mathbb{R}^s$, and x any point of a convenient open subset of U.

PROPOSITION 3.24 An involutive r-distribution ∇ on X possesses at most $n - r$ unconnected invariants defined on an open set $U \subset X$.

Proof Imagine that we could find s unconnected invariants

$$\Phi^1, \ldots, \Phi^s \in C^\infty(U).$$

By definition there exists at least one point $x \in U$ such that

$$\operatorname{rank}\left[\frac{\partial \Phi^\tau(x)}{\partial x^i}\right] = s$$

at $x \in X$. Also, at this point x, the linear system

$$\xi^i(x)\frac{\partial \Phi^\tau(x)}{\partial x^i} = 0$$

has $n - s$ independent solutions. But, as the Φ^τ are invariants of ∇, we have

$$\xi^i_\rho(x)\frac{\partial \Phi^\tau(x)}{\partial x^i} = 0 \quad \text{with} \quad \operatorname{rank}[\xi^i_\rho(x)] = r$$

at x. Thus $r \leq n - s$ and $s \leq n - r$. C.Q.F.D.

In fact, from the Frobenius theorem, for each point $x_0 \in X$, it is possible to find a neighbourhood U of x_0 and $n - r$ invariants defined and unconnected on U. Moreover, from the preceeding proposition, it is always possible to take U in such a way that

$$\operatorname{rank}\left[\frac{\partial \Phi^\tau(x)}{\partial x^i}\right] = \operatorname{rank}\left[\frac{\partial \Phi^\tau(x_0)}{\partial x^i}\right] = n - r, \qquad \forall x \in U \subset X.$$

DEFINITION 3.25 Such a set of invariants is called a "*fundamental set*" of invariants of ∇ at $x_0 \in X$.

Let ∇ be an involutive r-distribution on X and $\Phi^1, \ldots, \Phi^{n-r}$ be a fundamental set of invariants of ∇ on $U \subset X$.

PROPOSITION 3.26 Any invariant Ψ of ∇ defined on U is connected with $\Phi^1, \ldots, \Phi^{n-r}$ and can be expressed locally as a function of $\Phi^1, \ldots, \Phi^{n-r}$.

Proof Define a map $\Phi: U \to \mathbb{R}^{n-r}$, using local coordinates (x^i) on U and (u^τ) on $\mathbb{R}^{n-r}$, with $u^\tau = \Phi^\tau(x)$, $\tau = 1, \ldots, n-r$.

Let $x_0 \in U$ and $u_0 = \Phi(x_0)$. We may suppose, without loss of generality that we have

$$\frac{\partial(\Phi^1, \ldots, \Phi^{n-r})}{\partial(x^1, \ldots, x^{n-r})} \neq 0$$

at x_0, and thus also on U, by shrinking U if necessary.

From the implicit function theorem we can express locally $x^1, \ldots, x^{n-r}$ as functions of $u^1, \ldots, u^{n-r}; x^{n-r+1}, \ldots, x^n$ and set:

$$x^i = A^i(u^1, \ldots, u^{n-r}; x^{n-r+1}, \ldots, x^n) \qquad \forall 1 \leq i \leq n-r$$

We have the identities:

$$\Phi^\tau(A^1(u; x^{n-r+1}, \ldots, x^n), \ldots, A^{n-r}(u; x^{n-r+1}, \ldots, x^n), x^{n-r+1}, \ldots, x^n) \equiv u^\tau$$

Differentiating with respect to $x^{n-r+\alpha}$ we get the other identities:

$$\sum_{i=1}^{n-r} \frac{\partial \Phi^\tau}{\partial x^i}(A^1, \ldots, A^{n-r}, x^{n-r+1}, \ldots, x^n) \frac{\partial A^i}{\partial x^{n-r+\alpha}} + \frac{\partial \Phi^\tau}{\partial x^{n-r+\alpha}}(A^1, \ldots, A^{n-r}; x^{n-r+1}, \ldots, x^n) \equiv 0.$$

As the Φ^τ and Ψ are invariants of ∇, we have

$$\xi^i_\rho(x) \frac{\partial \Phi^\tau(x)}{\partial x^i} = 0 \quad \text{and} \quad \xi^i_\rho(x) \frac{\partial \Psi(x)}{\partial x^i} = 0 \quad \text{on} \quad U.$$

In a way similar to that of proposition 3.24, using the fact that

$$\operatorname{rank}\,[\xi^i_\rho(x)] = r, \qquad \forall x \in U,$$

we must have

$$\frac{\partial \Psi(x)}{\partial x^i} \equiv M_\tau(x) \frac{\partial \Phi^\tau(x)}{\partial x^i}, \qquad \forall 1 \leq i \leq n.$$

Now we set:

$$\begin{aligned}\Psi &= \Psi(A^1(u; x^{n-r+1}, \ldots, x^n), \ldots, A^{n-r}(u; x^{n-r+1}, \ldots, x^n), x^{n-r+1}, \ldots, x^n)\\ &\equiv B(u; x^{n-r+1}, \ldots, x^n)\end{aligned}$$

Differentiating B with respect to $x^{n-r+\alpha}$ we get:

$$\begin{aligned}\frac{\partial B}{\partial x^{n-r+\alpha}} &= \sum_{i=1}^{n-r} \frac{\partial \Psi}{\partial x^i}(A, x^{n-r+1}, \ldots, x^n)\frac{\partial A^i}{\partial x^{n-r+\alpha}}\\ &\quad + \frac{\partial \Psi}{\partial x^{n-r+\alpha}}(A, x^{n-r+1}, \ldots, x^n)\\ &= \sum_{i=1}^{n-r} M_\tau(x)\frac{\partial \Phi^\tau}{\partial x^i}\frac{\partial A^i}{\partial x^{n-r+\alpha}} + \frac{\partial \Psi}{\partial x^{n-r+\alpha}}\\ &= -M_\tau(x)\frac{\partial \Phi^\tau}{\partial x^{n-r+\alpha}} + \frac{\partial \Psi}{\partial x^{n-r+\alpha}} = 0\end{aligned}$$

and finally

$$\Psi(x) \equiv B(\Phi^1(x), \ldots, \Phi^{n-r}(x)). \qquad \text{C.Q.F.D.}$$

Let now $G \times X \to X$ be an effective action of a Lie group on X. We always suppose that G coincides with the connected component of the identity e and we let $\xi_1, \ldots, \xi_r$ be the independent infinitesimal generators of the action.

These are r vector fields on X and we may suppose that $\xi_1, \ldots, \xi_m$ are unconnected. By definition this means that there exists at least one point $x_0 \in X$ such that rank $[\xi^i_\rho(x_0)] = m$ and that we can write locally:

$$\xi^i_{m+\alpha}(x) = \mathscr{E}^\rho_{m+\alpha}(x)\xi^i_\rho(x) \quad \text{with} \quad \alpha = 1, \ldots, r-m \text{ and } \rho = 1, \ldots, m.$$

In this way we can find the isotropy group G_{x_0}. In fact an infinitesimal transformation $\lambda^\rho\xi_\rho$ will leave x_0 fixed if and only if $\lambda^\rho\xi^i_\rho(x_0) = 0$ that is to say:

$$\sum_{\rho=1}^{m} (\lambda^\rho + \lambda^{m+\alpha}\mathscr{E}^\rho_{m+\alpha}(x_0))\xi^i_\rho(x_0) = 0$$

As rank $[\xi^i_\rho(x_0)] = m$ we must have $\lambda^\rho + \lambda^{m+\alpha}\mathscr{E}^\rho_{m+\alpha}(x_0) = 0$ and G_{x_0} has the $r - m$ infinitesimal generators:

$$\xi_{m+\alpha} - \sum_{\rho=1}^{m} \mathscr{E}^\rho_{m+\alpha}(x_0)\xi_\rho$$

If $x \in X$ is another point which also gives rank m to $\xi_1, \ldots, \xi_r$ and $\xi_1, \ldots, \xi_m$ then x is stable by G_{x_0} if and only if:

$$\xi^i_{m+\alpha}(x) - \sum_{\rho=1}^{m} \mathscr{E}^\rho_{m+\alpha}(x_0)\xi^i_\rho(x) \equiv \sum_{\rho=1}^{m} (\mathscr{E}^\rho_{m+\alpha}(x) - \mathscr{E}^\rho_{m+\alpha}(x_0))\xi^i_\rho(x) = 0$$

Finally G_{x_0} leaves fixed all the points giving rank m to $\xi_1, \ldots, \xi_r$ and belonging to the subset of X defined locally by the equation

$$\mathscr{E}^\rho_{m+\alpha}(x) = \mathscr{E}^\rho_{m+\alpha}(x_0).$$

DEFINITION 3.27 The local functions $\mathscr{E}^\rho_{m+\alpha}(x)$ are called the "*stationary functions*" of the transformation group which is said to be "*stationary*" at x_0 if

$$\text{rank}\left[\frac{\partial \mathscr{E}^\rho_{m+\alpha}(x_0)}{\partial x^i}\right] < n$$

and "*non-stationary*" if

$$\text{rank}\left[\frac{\partial \mathscr{E}^\rho_{m+\alpha}(x_0)}{\partial x^i}\right] = n.$$

PROPOSITION 3.28 Any stationary function can be expressed locally as the quotient of two m-rowed determinants of the $n \times r$ matrix $[\xi^i_\rho(x)]$.

Proof This useful proposition follows from elementary linear algebra. If we look at the $m \times (m + 1)$ submatrices:

$$\begin{bmatrix} \xi^1_1(x) & \cdots & \xi^1_m(x) & \xi^1_{m+\alpha}(x) \\ \vdots & & \vdots & \vdots \\ \xi^m_1(x) & \cdots & \xi^m_m(x) & \xi^m_{m+\alpha}(x) \end{bmatrix}$$

then the $(m + 1)$-rowed determinant obtained by adding one of the rows is identically zero. Expanding it, we find:

$$\xi^i_1(x) \cdot \det[\quad] + \cdots + \xi^i_m(x) \cdot \det[\quad] + \xi^i_{m+\alpha}(x) \cdot \det\begin{bmatrix} \xi^1_1(x) & \cdots & \xi^1_m(x) \\ \xi^m_1(x) & & \xi^m_m(x) \end{bmatrix} = 0$$

and the proposition follows from the fact that the last determinant is different from zero by continuity as it is supposed different from zero at x_0.

DEFINITION 3.29 We call a family of m-dimensional submanifolds of X which are described locally by equations of the form:

$$\Phi^1(x) = \Phi^1(x_0) = \text{cst}, \ldots, \Phi^{n-m}(x) = \Phi^{n-m}(x_0) = \text{cst}$$

an "*m-foliation*" on X.

DEFINITION 3.30 A foliation on X is said to be "*invariant*" by an automorphism $f \in \text{Aut}(\Psi)$ if the different manifolds of the family are just transformed among themselves by f, that is to say, if $y = f(x)$ and $y_0 = f(x_0)$, then $\Phi^\tau(y) = \Phi^\tau(y_0)$ whenever $\Phi^\tau(x) = \Phi^\tau(x_0)$.

We have the very useful proposition.

PROPOSITION 3.31 The stationary functions define locally on X a foliation which is invariant under the action of G on X.

Proof From the fundamental group property:

$$[\xi_\rho, \xi_{m+\alpha}] = c^\tau_{\rho m+\alpha}\xi_\tau \quad \text{with} \quad \rho, \tau = 1, \ldots, r \text{ and } \alpha = 1, \ldots, r - \text{m}$$

and from the identity:

$$\xi_{m+\alpha} = \mathscr{E}^\tau_{m+\alpha} \cdot \xi_\tau$$

we deduce that:

$$(\xi_\rho \cdot \mathscr{E}^\tau_{m+\alpha})\xi_\tau + \mathscr{E}^\tau_{m+\alpha}[\xi_\rho, \xi_\tau] = c^\tau_{\rho m+\alpha} \cdot \xi_\tau$$

and finally, as $\xi_1, \ldots, \xi_m$ are unconnected:

$$\xi^i_\rho(x)\frac{\partial \mathscr{E}^\tau_{m+\alpha}(x)}{\partial x^i} = c^\tau_{\rho m+\alpha} - \mathscr{E}^\sigma_{m+\alpha}(x)(c^\tau_{\rho\sigma} + c^{m+\beta}_{\rho\sigma}\mathscr{E}^\tau_{m+\beta}(x))$$

Taking then the unconnected stationary functions as new variables in such a way as to describe locally the foliation by means of slices, the above formula shows that

$$\frac{d\mathscr{E}}{dt} = A(\mathscr{E}) \quad \text{when} \quad x = \exp(t\lambda^\rho\xi_\rho)(x_0).$$

It follows that the slices are just transformed between themselves by $\exp(t\lambda^\rho\xi_\rho)$, $\forall\lambda^\rho$ with t small. C.Q.F.D.

PROPOSITION 3.32 For any $x_0 \in X$ that gives rank m to $\xi_1, \ldots, \xi_r$ there is a neighbourhood U of x_0 in X on which is defined an involutive m-distribution.

Proof Imagine that x_0 gives rank m to $\xi_1, \ldots, \xi_m$. By continuity, there is a neighbourhood U of x_0 in X such that any point $x \in U$ gives rank m to $\xi_1, \ldots, \xi_m$. Moreover, we have on U:

$$[\xi_\sigma, \xi_\tau] = (c^\rho_{\sigma\tau} + \mathscr{E}^\rho_{m+\alpha}(x)c^{m+\alpha}_{\sigma\tau})\xi_\rho$$

with $\rho, \sigma, \tau = 1, \ldots, m$. C.Q.F.D.

DEFINITION 3.33 A function $\Phi \in C^\infty(X): X \to \mathbb{R}$ is said to be "*invariant*" under the action of a Lie group G on X if:

$$\Phi(a \cdot x) = \Phi(x) \qquad \forall a \in G, \forall x \in X$$

REMARK 3.34 An equivalent definition is that Φ is constant along the orbit of any point of X.

Using the infinitesimal point of view, we have:

PROPOSITION 3.35 Φ is invariant if and only if:

$$\xi_\rho \cdot \Phi = 0 \qquad \forall \rho = 1, \ldots, r$$

Proof By theorem 2.1, along the orbit of any point $x_0 \in X$, we have:

$$\frac{\partial \Phi(x)}{\partial a^\sigma} = \frac{\partial \Phi(x)}{\partial x^i}\frac{\partial x^i}{\partial a^\sigma} = \frac{\partial \Phi(x)}{\partial x^i}\xi^i_\rho(x)\omega^\rho_\sigma(a) = 0$$

Taking $a = e$ and using the fact that $\omega^\rho_\sigma(e) = \delta^\rho_\sigma$ we get

$$\xi^i_\rho(x_0)\frac{\partial \Phi(x_0)}{\partial x^i} = 0 \qquad \forall x_0 \in X.$$

The converse is easy and we have

$$\Phi(x) = \Phi(x_0), \qquad \forall x \in G(x_0).$$

C.Q.F.D.

DEFINITION 3.36 A submanifold $R \subset X$ is said to be "*invariant*" by a transformation group G of X if $\forall a \in G$ and $x \in R$, then $a \cdot x \in R$.

If $\xi_1, \ldots, \xi_r$ are the infinitesimal generators of the transformation group, as in the proposition 3.24, we have:

PROPOSITION 3.37 A submanifold $R \subset X$ is invariant if and only if

$$\xi_\rho(x) \in T_x(R), \qquad \forall x \in R.$$

Proof

N.C. For each point $x_0 \in R$, there exists a neighbourhood U of x_0 in X such that R is defined locally on U by $n - \dim R$ equations $\Phi^\tau(x) = 0$ with

$$\text{rank}\left[\frac{\partial \Phi^\tau(x)}{\partial x^i}\right] = \text{rank}\left[\frac{\partial \Phi^\tau(x_0)}{\partial x^i}\right] = n - \dim R, \qquad \forall x \in U.$$

For small t, the point $\exp(t\lambda^\rho \xi_\rho)(x_0) \in G(x_0)$ is in U and as R is invariant we may have:

$$\Phi^\tau(\exp(t\lambda^\rho \xi_\rho)(x_0)) = 0 \qquad \forall t \text{ small}, \forall \lambda^\rho.$$

Differentiating with respect to t, for $t = 0$ we get:

$$\frac{\partial \Phi^\tau(x_0)}{\partial x^i} \lambda^\rho \xi^i_\rho(x_0) = 0, \qquad \forall \lambda^\rho$$

and thus

$$\xi^i_\rho(x_0) \frac{\partial \Phi^\tau(x_0)}{\partial x^i} = 0, \qquad \forall x_0 \in R.$$

S.C. Let $x_0 \in R \subset X$. With given λ^ρ we can construct the curve

$$x_0 \to x = f_t(x_0) = \exp(t\lambda^\rho \xi_\rho)(x_0).$$

As

$$\xi_\rho(x) \in T_x(R), \qquad \forall x \in R,$$

we can consider the restrictions $\xi_\rho|_R$ of ξ_ρ to R. These give r vector fields on R and for given λ^ρ there is a unique curve passing through x_0. Its image in X by the inclusive $R \subset X$ will be a curve $x = g_t(x_0)$ and for both curves we have:

$$\frac{dx^i}{dt} = \lambda^\rho \xi^i_\rho(x).$$

But this system of ordinary differential equations is known to have a unique solution, at least locally in a neighbourhood of x_0, such that $x = x_0$ for $t = 0$. Thus we must have $f_t \equiv g_t$ and

$$f_t(x_0) \in R, \qquad \forall x_0 \in R. \qquad \text{C.Q.F.D.}$$

Let now ξ be a vector field on X and R be a submanifold of X such that

$$\xi(x) \in T_x(R), \qquad \forall x \in R.$$

PROPOSITION 3.38 $\forall \Psi \in C^\infty(X)$, we have:

$$(\xi \cdot \Psi)|_R = \xi|_R \cdot \Psi|_R$$

Proof We shall use local coordinates on a neighbourhood U of a point $x_0 \in R$, and describe R locally by $(n - \dim R)$ equation $\Phi^\tau(x) = 0$ with

$$\operatorname{rank}\left[\frac{\partial \Phi^\tau(x)}{\partial x^i}\right] = \operatorname{rank}\left[\frac{\partial \Phi^\tau(x_0)}{\partial x^i}\right] = n - \dim R = s, \qquad \forall x \in U.$$

On U we can solve those equations with respect to $x^1, \ldots, x^s$ and set

$$x^1 = A^1(x^{s+1}, \ldots, x^n), \ldots, x^s = A^s(x^{s+1}, \ldots, x^n).$$

Taking now $X = U$ and $R = R \cap U$ in order to simplify the local notations and using $x^{s+1}, \ldots, x^n$ as local coordinates for R, we have:

$$\Psi|_R(x^{s+1}, \ldots, x^n) = \Psi(A^1(x^{s+1}, \ldots, x^n), \ldots, A^s(x^{s+1}, \ldots, x^n); x^{s+1}, \ldots, x^n)$$

$$(\xi \cdot \Psi)|_R(x^{s+1}, \ldots, x^n) = \sum_{i=1}^{n} \xi^i(A; x^{s+1}, \ldots, x^n) \frac{\partial \Psi}{\partial x^i}(A; x^{s+1}, \ldots, x^n)$$

But

$$\xi|_R = \sum_{j=s+1}^{n} \xi^j(A; x^{s+1}, \ldots, x^n) \frac{\partial}{\partial x^j}$$

and

$$\xi|_R \cdot \Psi|_R(x^{s+1}, \ldots, x^n) = \sum_{i=1}^{s} \sum_{j=s+1}^{n} \xi^j(A; x^{s+1}, \ldots, x^n) \times \frac{\partial \Psi}{\partial x^i}(A; x^{s+1}, \ldots, x^n) \frac{\partial A^i}{\partial x^j}(x^{s+1}, \ldots, x^n) + \sum_{i=s+1}^{n} \xi^i(A; x^{s+1}, \ldots, x^n) \frac{\partial \Psi}{\partial x^i}(A; x^{s+1}, \ldots, x^n)$$

Thus, substracting we deduce:

$$((\xi \cdot \Psi)|_R - \xi|_R \cdot \Psi|_R)(x^{s+1}, \ldots, x^n) = \sum_{i=1}^{s} \left(\xi^i(A; x^{s+1}, \ldots, x^n) - \sum_{j=s+1}^{n} \xi^j(A; x^{s+1}, \ldots, x^n) \frac{\partial A^i}{\partial x^j}(x^{s+1}, \ldots, x^n)\right) \frac{\partial \Psi}{\partial x^i}$$

But

$$\xi^i(A; x^{s+1}, \ldots, x^n) - \sum_{j=s+1}^{n} \xi^j(A; x^{s+1}, \ldots, x^n) \frac{\partial A^i}{\partial x^j}(x^{s+1}, \ldots, x^n) = 0$$

because $\xi(x) \in T_x(R)$ and the proposition follows. C.Q.F.D.

Proposition 3.39 If a submanifold $R \subset X$ is invariant under the action of a Lie group G on X with infinitesimal generators $\xi_1, \ldots, \xi_r$ then $\xi_1|_R, \ldots, \xi_r|_R$ are the infinitesimal generators of the "*restricted*" action of G on R.

Proof From the preceding proposition, $\forall \Psi \in C^\infty(X)$ we have

$$(\xi_\rho \cdot \Psi)|_R = \xi_\rho|_R \cdot \Psi|_R \qquad \forall \rho = 1, \ldots, r$$

It follows easily that:

$$[\xi_\sigma|_R, \xi_\tau|_R] \cdot \Psi|_R = ([\xi_\sigma, \xi_\tau] \cdot \Psi)|_R = c^\rho_{\sigma\tau}(\xi_\rho \cdot \Psi)|_R = c^\rho_{\sigma\tau}\xi_\rho|_R \cdot \Psi|_R$$

and thus

$$[\xi_\sigma|_R, \xi_\tau|_R] = c^\rho_{\sigma\tau}\xi_\rho|_R. \qquad \text{C.Q.F.D.}$$

Remark 3.40 If the original action on X is effective, then the restricted action on R is not always effective.

Example 3.41 $X = \mathbb{R}^2$ with cartesian coordinates (x, y). The line manifold

$$R = \{(x, y) \in \mathbb{R}^2 \,|\, x + y - 1 = 0\}$$

is invariant under the action of the Lie group G which has the two independent infinitesimal generators:

$$\xi_1 = (x - x^2)\frac{\partial}{\partial x} - xy\frac{\partial}{\partial y}, \qquad \xi_2 = xy\frac{\partial}{\partial x} - (y - y^2)\frac{\partial}{\partial y}$$

Then

$$\xi_1|_R = \xi_2|_R = (x - x^2)\left(\frac{\partial}{\partial x} - \frac{\partial}{\partial y}\right).$$

Definition 3.42 A distribution ∇ on X is said to be "*invariant*" by an automorphism

$$f \in \mathrm{Aut}\,(X) \quad \text{if} \quad f(\nabla_x) = \nabla_{f(x)}, \qquad \forall x \in X.$$

In particular, $\forall \xi \in \nabla$, then $f(\xi) \in \nabla$.

If U is an open set of X and $f \in \mathrm{Aut}\,(U)$, using a basis $\xi_1, \ldots, \xi_r$ of ∇ on U and defining $\eta_\rho = f(\xi_\rho)$ by the formulas:

$$\eta^j_\rho(f(x)) = \frac{\partial f^j(x)}{\partial x^i} \cdot \xi^i_\rho(x),$$

then there exists a $r \times r$ matrix $[B^\sigma_\rho(x)]$ with $\det [B^\sigma_\rho(x)] \neq 0, \forall x \in U$ such that

$$\eta^i_\rho(x) = B^\sigma_\rho(x)\xi^i_\sigma(x), \qquad \forall x \in U.$$

Remark 3.43 If Φ is an invariant of ∇, then $\Phi \circ f$ is also an invariant of ∇ because we have:

$$\xi^i_\rho(x)\,\frac{\partial\Phi(f(x))}{\partial x^j}\,\frac{\partial f^j(x)}{\partial x^i} = \eta^j_\rho(f(x))\,\frac{\partial\Phi(f(x))}{\partial x^j}$$

$$= B^\sigma_\rho(f(x))\left(\xi^i_\sigma(f(x))\,\frac{\partial\Phi(f(x))}{\partial x^i}\right) = 0$$

If now we consider a local 1-parameter group $f_t = \exp(t\eta)$ with t small, and define $\xi_t = f_t(\xi)$ for any vector field ξ on X, then we have the proposition:

Proposition 3.44 An involutive distribution ∇ on X is invariant by the 1-parameter group $f_t = \exp(t\eta)$ if and only if $[\xi, \eta] \in \nabla$, $\forall \xi \in \nabla$.

Proof

N.C. We must have $\xi_t \in \nabla$, $\forall t$ small. By continuity we must have also $(d\xi_t/dt) \in \nabla$, $\forall t$ small. Using proposition 3.12, we have $d\xi_t/dt = [\xi_t, \eta]$ and thus for $t = 0$, we must have $[\xi, \eta] \in \nabla$.

S.C. We shall prove that for t small enough $\xi_t \in \nabla$, $\forall \xi \in \nabla$ when $[\xi, \eta] \in \nabla$.

Let $x_0 \in X$ and choose local coordinates (x^i) as a neighbourhood U of x_0 in X, such that ∇ is span on U by

$$\xi_1 \equiv \frac{\partial}{\partial x^1}, \ldots, \xi_r \equiv \frac{\partial}{\partial x^r}.$$

This is possible, if U is sufficiently small, by the Frobenius theorem.

An easy computation shows that η must have locally the form:

$$\eta = \sum_{i=1}^{r} \eta^i(x^1, \ldots, x^n)\,\frac{\partial}{\partial x^i} + \sum_{i=r+1}^{n} \eta^i(x^{r+1}, \ldots, x^n)\,\frac{\partial}{\partial x^i}$$

It follows that

$$x^i = f^i(t, x_0^{r+1}, \ldots, x_0^n) \quad \text{for } i = r+1, \ldots, n.$$

As a consequence we have:

$$\xi^i_{\rho t}(x) = \frac{\partial f^i(t, x_0)}{\partial x_0^j}\,\xi^j_\rho(x_0) = \frac{\partial f^i(t, x_0)}{\partial x_0^\rho} = 0 \quad \text{if } i = r+1, \ldots, n.$$

C.Q.F.D.

Remark 3.45 From the last proposition, we see that the integral manifolds of ∇ are not invariant under the action of f_t but exchanged among themselves. In fact, locally, f_t changes any slice into another.

Now let $\xi_1, \ldots, \xi_r$ be the infinitesimal generators of the action of a Lie group G on X and let f be a local diffeomorphism of X.

DEFINITION 3.46 We say that the action of G on X is "*invariant*" by f if $\forall a \in G$ and $\forall x \in X$, there exists $b \in G$ such that $f(a \cdot x) = b \cdot f(x)$.

PROPOSITION 3.47 The action of G on X is invariant by f if and only if $f(\xi_\rho) = B_\rho^\sigma \cdot \xi_\sigma$, when B_ρ^σ is a constant $r \times r$ matrix with $\det [B_\rho^\sigma] \neq 0$.

Proof

N.C. We set

$$a \cdot x = \exp(t\lambda^\rho \xi_\rho)(x) = x + t\lambda^\rho \xi_\rho(x) + \cdots$$

From proposition 3.13, we have:

$$f(a \cdot f^{-1}(x)) = \exp(t\lambda^\rho f(\xi_\rho))(x) = b \cdot x$$

and there must exist r constants μ^σ such that $\lambda^\rho f(\xi_\rho) = \mu^\sigma \cdot \xi_\sigma$. In particular we must have $f(\xi_\rho) = B_\rho^\sigma \cdot \xi_\sigma$ and, as f is invertible, $\det [B_\rho^\sigma] \neq 0$.

S.C. The proof is similar.

As in the proposition 3.44, we can adopt the infinitesimal point of view.

PROPOSITION 3.48 The action of G on X is invariant by the 1-parameter group $f_t = \exp(t\eta)$ if and only if $[\xi_\rho, \eta] = A_\rho^\sigma \cdot \xi_\sigma$ when A_ρ^σ is a constant *$r \times r$ matrix.*

Proof

N.C. Setting $\xi_{\rho t} = f_t(\xi_\rho)$, we must have: $\xi_{\rho t} = B_\rho^\sigma(t) \cdot \xi_\sigma$ and also, for t small.

$$\frac{d\xi_{\rho t}}{dt} = \frac{dB_\rho^\sigma(t)}{dt} \cdot \xi_\sigma = [\xi_{\rho t}, \eta]$$

Taking $t = 0$, we have

$$[\xi_\rho, \eta] = \left.\frac{dB_\rho^\sigma(t)}{dt}\right|_{t=0} \xi_\sigma = A_\rho^\sigma \cdot \xi_\sigma$$

S.C. The proof is similar.

We have

$$\frac{d\xi_{\rho t}}{dt} = [\xi_{\rho t}, \eta] = \exp(t\eta)([\xi_\rho, \eta])$$
$$= \exp(t\eta)(A^\sigma_\rho \cdot \xi_\sigma)$$
$$= A^\sigma_\rho \cdot \xi_{\sigma t}$$

This is an ordinary differential equation and it is known that the solution $\xi_{\rho t} = B^\sigma_\rho(t)\xi_\rho$ such that $B^\sigma_\rho(0) = \delta^\sigma_\rho$ is unique. For small t, we have $\det[B^\sigma_\rho(t)] \neq 0$ because it is equal to 1 for $t = 0$. C.Q.F.D.

EXAMPLE 3.49 In example 3.5 above, the transformation group of $\mathbb{R}^3$ is invariant under the rotations with axis z and generator

$$x\frac{\partial}{\partial y} - y\frac{\partial}{\partial x}.$$

EXAMPLE 3.50 Any transformation group is invariant under its own finite and infinitesimal transformation, because we have

$$[\xi_\sigma, \xi_\tau] = c^\rho_{\sigma\tau}\xi_\rho \qquad \forall \rho, \sigma, \tau = 1, \ldots, r.$$

The following theorem will be very useful later on.

THEOREM 3.51 If the action of a Lie group G on X is invariant under an automorphism $f \in \operatorname{Aut}(X)$, the set of points $x \in X$ giving rank k to $\xi_1, \ldots, \xi_r$ is invariant under f.

Proof We have to show that

$$\operatorname{rank}[\xi^i_\rho(f(x))] = h \quad \text{whenever} \quad \operatorname{rank}[\xi^i_\rho(x)] = h.$$

By hypothesis we have:

$$\frac{\partial f^k(x)}{\partial x^i} \cdot \xi^i_\rho(x) = B^\sigma_\rho \cdot \xi^k_\sigma(f(x))$$

with

$$\det\left[\frac{\partial f^k(x)}{\partial x^i}\right] \neq 0 \quad \text{and} \quad \det[B^\sigma_\rho] \neq 0.$$

We shall prove that the matrices

$$[\xi^i_\rho(x)] \quad \text{and} \quad \left[\frac{\partial f^k(x)}{\partial x^i} \cdot \xi^i_\rho(x)\right]$$

have the same rank h. In fact, if we have a relation between the columns such that

$$\xi^i_{h+\alpha}(x) = \mathscr{E}^{\tau}_{h+\alpha}(x)\xi^i_{\tau}(x) \quad \text{with} \quad \tau = 1, \ldots, h \quad \text{and} \quad \alpha = 1, \ldots, r-h,$$

then we have also the corresponding relation between the columns:

$$\frac{\partial f^k(x)}{\partial x^i}\, \xi^i_{h+\alpha}(x) = \mathscr{E}^{\tau}_{h+\alpha}(x)\, \frac{\partial f^k(x)}{\partial x^i}\, \xi^i_{\tau}(x).$$

It follows that

$$\text{rank}\left[\frac{\partial f^k(x)}{\partial x^i}\, \xi^i_{\rho}(x)\right] \leq \text{rank}\ [\xi^i_{\rho}(x)].$$

As

$$f \in \text{Aut}\ (X), \quad \det\left[\frac{\partial f^k(x)}{\partial x^i}\right] \neq 0$$

and the converse is also true, that is to say

$$\text{rank}\left[\frac{\partial f^k(x)}{\partial x^i}\, \xi^i_{\rho}(x)\right] \geq \text{rank}\ [\xi^i_{\rho}(x)].$$

The same argument can be repeated for the matrices

$$[\xi^k_{\sigma}(f(x))] \quad \text{and} \quad [B^{\sigma}_{\rho} \cdot \xi^k_{\sigma}(f(x))],$$

but now using the rows. It follows that:

$$\text{rank}\ [\xi^i_{\rho}(x)] = \text{rank}\left[\frac{\partial f^k(x)}{\partial x^i}\, \xi^i_{\rho}(x)\right] = \text{rank}\ [B^{\sigma}_{\rho}\xi^k_{\sigma}(f(x))] = \text{rank}\ [\xi^k_{\sigma}(f(x))]$$

C.Q.F.D.

Using proposition 3.37 we obtain the corollary:

COROLLARY 3.52 Any submanifold of X giving rank h to $\xi_1, \ldots, \xi_r$ and invariant by the restricted action of the Lie group G on X has dimension $\geq h$.

Our final problem will be to construct explicitly and at least locally the submanifolds of X giving rank h to $\xi_1, \ldots, \xi_r$.

For this we have first to equate to zero all the determinants of order $> h$ in the matrix $[\xi^i_{\rho}(x)]$. The difficult point is that the local equations obtained by this way *do not in general* allow us to use the implicit function theorem in order to get a local submanifold of X. This fact can be clearly seen in the examples, especially in example 3.5 above.

We do not know any methods for studying the subsets of X giving rank h to $\xi_1, \dots, \xi_r$ locally, except in the special cases that we shall meet in the next chapter.

Imagine for the moment that it is possible to prove that the above subset is a local submanifold R of X. (The reader can refer to all the examples given above.)

From the last corollary R has dimension $s \geq h$ and is invariant under restricted action of G on X, because of theorem 3.51. It can be described locally by the equation:

$$\Phi^1(x^1, \dots, x^n) = 0, \dots, \Phi^{n-s}(x^1, \dots, x^n) = 0$$

Now we have by definition on R an h involutive distribution. This can easily be seen by using proposition 3.39.

Finally we can apply the Frobenius theorem to this involutive distribution and foliate R with invariant submanifolds of dimension h. The latter are clearly submanifolds because they can be constructed by using only the restriction of h unconnected vector fields among $\xi_1, \dots, \xi_r$ say $\xi_1, \dots, \xi_h$ If we use local coordinates $x^1, \dots, x^s$ on R, the above foliation can be obtained from a fundamental set of $(s - h)$ invariants of the involutive h-distribution, say $\Phi^{n-s+1}(x^1, \dots, x^s), \dots, \Phi^{n-h}(x^1, \dots, x^s)$.

To sum up, the minimal invariant submanifolds giving rank h to $\xi_1, \dots, \xi_r$ have dimension h and are described locally by the $(n - s) + (s - h) = (n - h)$ equations:

$$\Phi^1(x^1, \dots, x^n) = 0, \dots, \Phi^{n-s}(x^1, \dots, x^n) = 0,$$

$$\Phi^{n-s+1}(x^1, \dots, x^s) = c_1, \dots, \Phi^{n-h}(x^1, \dots, x^s) = c_{s-h}$$

where $c_1, \dots, c_{s-h}$ are $(s - h)$ constants.

The orbit of a given point $x_0 \in X$ giving rank h to $\xi_1, \dots, \xi_r$ is thus described locally by the $(n - h)$ equations:

$$\begin{cases} \Phi^1(x^1, \dots, x^n) = 0, \dots, \Phi^{n-s}(x^1, \dots, x^n) = 0 \\ \Phi^{n-s+1}(x^1, \dots, x^s) = \Phi^{n-s+1}(x_0^1, \dots, x_0^s) \\ \dots\dots\dots\dots\dots\dots\dots\dots\dots\dots \\ \Phi^{n-h}(x^1, \dots, x^s) = \Phi^{n-h}(x_0^1, \dots, x_0^s) \end{cases}$$

and we have of course:

$$\Phi^1(x_0^1, \dots, x_0^n) = 0, \dots, \Phi^{n-s}(x_0^1, \dots, x_0^n) = 0.$$

EXAMPLE 3.53 In example 3.5 above, we can take (x, y) as coordinates on the cone R and the restricted generators are:

$$\begin{cases} \xi_1|_R = 2x\left(x\dfrac{\partial}{\partial x} + y\dfrac{\partial}{\partial y}\right) \\ \xi_2|_R = 2y\left(x\dfrac{\partial}{\partial x} + y\dfrac{\partial}{\partial y}\right) \\ \xi_3|_R = 2\sqrt{x^2 + y^2}\left(x\dfrac{\partial}{\partial x} + y\dfrac{\partial}{\partial y}\right) \end{cases}$$

A common invariant can be y/x and the minimum invariant submanifolds giving rank 1 to ξ_1, ξ_2, ξ_3 are described locally by the equations:

$$z^2 - (x^2 + y^2) = 0 \qquad \frac{y}{x} = c = \text{cst}$$

4 Lie derivative

An r-form $\omega = \omega_I(x)\,dx^I$ is a section of the vector bundle $\Lambda^r T^* \to X$. A local r-form is a section of $\Lambda^r T^*$ over an open set $U \subset X$.

In this section we are mainly concerned with the local point of view in order to give explicit formulas using local coordinates on U.

We have already defined

$$d\omega = \frac{\partial \omega_I(x)}{\partial x^i}\,dx^i \wedge dx^I$$

which is a $(p + 1)$-form on X.

DEFINITION 4.1 Let X and Y be manifolds and $f: X \to Y$ be a map. Using local coordinates (x^i) on U and (y^k) on $V = f(U)$, we define, for any r-form ω on Y: the r-form $f(\omega)$ on X by patching together local r-forms, using the formula:

$$f(\omega) = \sum_{i_1, \ldots, i_r} \left(\sum_{k_1 < \cdots < k_r} \omega_{k_1 \ldots k_r}(f(x)) \frac{\partial y^{k_1}}{\partial x^{i_1}} \cdots \frac{\partial y^{k_r}}{\partial x^{i_r}} \right) dx^{i_1} \wedge \cdots \wedge dx^{i_r}$$

PROPOSITION 4.2 $d \circ d\omega = 0$, $f(d\omega) = df(\omega)$

Proof These straightforward computations are left to the reader.

DEFINITION 4.3 If α is an r-form and β an s-form we define the $(r+s)$-form $\alpha \wedge \beta$ by bilinearity, setting

$$(dx^{i_1} \wedge \cdots \wedge dx^{i_r}) \wedge (dx^{j_1} \wedge \cdots \wedge dx^{j_s}) = dx^{i_r} \wedge \cdots \wedge dx^{i_r} \wedge dx^{j_1} \wedge \cdots \wedge dx^{j_s}$$

PROPOSITION 4.4

$$f(\alpha \wedge \beta) = f(\alpha) \wedge f(\beta), \qquad d(\alpha \wedge \beta) = d\alpha \wedge \beta + (-1)^r \alpha \wedge d\beta$$

Proof This straightforward computation is left to the reader.

We have already proved the following theorem:

THEOREM 4.5 (Poincaré) If α is a closed $(r+1)$-form, then there exists an open set $U \subset X$ and an r-form β on U such that

$$\alpha|_U = d\beta|_U$$

We have the following theorem:

THEOREM 4.6 (Frobenius) If α is a 1-form such that $\alpha \wedge d\alpha = 0$, there exists an open set $U \subset X$ and local functions f, $g \in C^\infty(U)$ such that $\alpha|_U = f \cdot dg$.

Proof Let $U \subset X$ such that $\alpha_1(x), \ldots, \alpha_n(x)$ not all zero $\forall x \in U$. Then $\alpha = \alpha_i(x)\, dx^i$ and suppose that $\alpha_n(x) \neq 0$, $\forall x \in U$ then

$$dx^n = \frac{1}{\alpha_n(x)} \alpha - \sum_{i=1}^{n-1} \frac{\alpha_i(x)}{\alpha_n(x)} dx^i.$$

We have

$$d\alpha = \beta \wedge dx^n + \gamma \quad \text{and} \quad d\alpha = \beta \wedge \frac{\alpha}{\alpha_n(x)} + \left(\gamma + \sum_{i=1}^{n-1} \frac{\alpha_i(x)}{\alpha_n(x)} dx^i \wedge \beta\right)$$

where β and γ do not contain dx^n. As $\alpha \wedge d\alpha = 0$ we must have

$$d\alpha = \frac{\beta}{\alpha_n(x)} \wedge \alpha = \omega \wedge \alpha$$

on U.

We shall now use an induction on n.

The case $n = 1$ is trivial.

For $n = 2$ let $\alpha = \alpha_1(x^1, x^2)\, dx^1 + \alpha_2(x^1, x^2)\, dx^2$ and choose $U \subset X$ such that $\alpha_2(x) \neq 0$, $\forall x \in U$. Let $g(x^1, x^2)$ be an invariant of the distribution

$$\alpha_2(x) \frac{\partial}{\partial x^1} - \alpha_1(x) \frac{\partial}{\partial x^2}.$$

We have

$$\alpha = \frac{\alpha_1(x)}{\partial g/\partial x^1}\, dg = f \cdot dg.$$

Now, if we consider x^n as a parameter, we have:

$$(d\alpha)|_{dx^n=0} = d(\alpha|_{dx^n=0}) = \alpha|_{dx^n=0} \wedge \beta$$

By induction, there exist $f, g \in C^\infty(U)$ such that:

$$\alpha|_{dx^n=0} = f(x)\sum_{j=1}^{n-1} \frac{\partial g(x)}{\partial x^i}\, dx^i \quad \text{and} \quad \alpha = f \cdot dg + h\, dx^n.$$

where $h \in C^\infty(U)$ and $f(x)$, $h(x)$ not all zero $\forall x \in U$. Choosing $g, x^2, \ldots, x^n$ as new coordinates on U, then it follows that $d(h/f) \wedge dx^n$ must be zero when $dg + (h/f)\, dx^n$ is zero. We deduce that h/f depends only on g and x^n and we have reduced our problem to the case $n = 2$, already solved. C.Q.F.D.

If now $\xi = \xi^i(x)(\partial/\partial x^i)$ is a vector field on X and ω an r-form, we define the r-form $\omega_t = f_t(\omega)$ with $f_t = \exp(t\xi) \in \mathrm{Aut}(X)$.

DEFINITION 4.7 We call "*Lie derivative*" of ω with respect to ξ, the r-form:

$$\mathscr{L}(\xi)\omega = \left.\frac{d\omega_t}{dt}\right|_{t=0}.$$

DEFINITION 4.8 We define the $(r-1)$-form $i(\xi)\omega$ by the formula:

$$(i(\xi)\omega)_{i_1 \ldots i_{r-1}}(x) = \xi^i(x)\omega_{ii_1 \ldots i_{r-1}}(x).$$

PROPOSITION 4.9 $\mathscr{L}(\xi) = i(\xi) \circ d + d \circ i(\xi)$

Proof

$$\omega_t = \sum_{i_1,\ldots,i_r}\left(\sum_{j_1<\cdots<j_r} \omega_{j_1\ldots j_r}(f_t(x))\frac{\partial f_t^{j_1}(x)}{\partial x^{i_r}} \cdots \frac{\partial f_t^{j_r}(x)}{\partial x^{i_r}}\right) dx^{i_1} \wedge \cdots \wedge dx^{i_r}$$

But

$$\frac{d}{dt}\left(\frac{\partial f_t^j(x)}{\partial x^i}\right)\bigg|_{t=0} = \frac{\partial}{\partial x^i}\left(\frac{df_t^j(x)}{dt}\right)\bigg|_{t=0} = \frac{\partial \xi^j(f_t(x))}{\partial x^i}\bigg|_{t=0} = \frac{\partial \xi^j(x)}{\partial x^i}$$

and

$$\mathscr{L}(\xi)\omega = \sum_{i_1 < \cdots < i_r} \sum_{s=1}^{r} (-1)^{s-1} \omega_{i_1 \ldots i_r}(x) \frac{\partial \xi^{i_s}}{\partial x^i}$$
$$\times dx^i \wedge dx^{i_1} \wedge \cdots \wedge \widehat{dx^{i_s}} \wedge \cdots \wedge dx^{i_r}$$
$$+ \sum_{i_1 < \cdots < i_r} \xi^i(x) \frac{\partial \omega_{i_1 \ldots i_r}}{\partial x^i} dx^{i_1} \wedge \cdots \wedge dx^{i_r}$$

where we omit any term under a hat.

$$\mathscr{L}(\xi)\omega = \sum_{i_1 < \cdots < i_r} \sum_{s=1}^{r} (-1)^{s-1} \frac{\partial(\omega_{i_1 \ldots i_r}(x)\xi^{i_s}(x))}{\partial x^i}$$
$$\times dx^i \wedge dx^{i_1} \wedge \cdots \wedge \widehat{dx^{i_s}} \wedge \cdots \wedge dx^{i_r}$$
$$+ \sum_{i_1 < \cdots < i_r} \sum_{s=1}^{r} (-1)^s \xi^{i_s}(x) \frac{\partial \omega_{i_1 \ldots i_r}(x)}{\partial x^i}$$
$$\times dx^i \wedge dx^{i_1} \wedge \cdots \wedge \widehat{dx^{i_s}} \wedge \cdots \wedge dx^{i_r}$$
$$+ \sum_{i_1 < \cdots < i_r} \xi^i(x) \frac{\partial \omega_{i_1 \ldots i_r}}{\partial x^i} dx^{i_1} \wedge \cdots \wedge dx^{i_r}$$

But

$$d(i(\xi)\omega) = d\left(\sum_{s=1}^{r} (-1)^{s-1} \xi^{i_s}(x) \omega_{i_1 \ldots i_s \ldots i_r}(x)\, dx^{i_1} \wedge \cdots \wedge \widehat{dx^{i_s}} \wedge \cdots \wedge dx^{i_r} \right)$$
$$= \sum_{i_1 < \cdots < i_r} \sum_{s=1}^{r} (-1)^{s-1} \frac{\partial(\omega_{i_1 \ldots i_r}(x)\xi^{i_s}(x))}{\partial x^i}$$
$$\times dx^i \wedge dx^{i_1} \wedge \cdots \wedge \widehat{dx^{i_s}} \wedge \cdots \wedge dx^{i_r}$$

and

$$i(\xi)(d\omega) = i(\xi)\left(\sum_{i_1 < \ldots < i_r} \frac{\partial \omega_{i_1 \ldots i_r}(x)}{\partial x^i} dx^i \wedge dx^{i_1} \wedge \cdots \wedge dx^{i_r} \right)$$
$$= \sum_{i_1 < \cdots < i_r} \xi^i(x) \frac{\partial \omega_{i_1 \ldots i_r}(x)}{\partial x^i} dx^{i_1} \wedge \cdots \wedge dx^{i_r}$$
$$+ \sum_{i_1 < \cdots < i_r} \sum_{s=1}^{r} (-1)^s \xi^{i_s}(x) \frac{\partial \omega_{i_1 \ldots i_r}(x)}{\partial x^i}$$
$$\times dx^i \wedge dx^{i_1} \wedge \cdots \wedge \widehat{dx^{i_s}} \wedge \cdots \wedge dx^{i_r}$$

and the proposition follows. C.Q.F.D.

REMARK 4.10 From the proposition we obtain:

$$\mathscr{L}(\xi) \circ d = d \circ i(\xi) \circ d = d \circ \mathscr{L}(\xi)$$

and

$$\mathscr{L}(\xi) \cdot \Phi = i(\xi)\, d\Phi = \xi \cdot \Phi, \qquad \forall \Phi \in C^\infty(X).$$

PROPOSITION 4.11 For any r-form α and any s-form β we have

$$\mathscr{L}(\xi)(\alpha \wedge \beta) = (\mathscr{L}(\xi)\alpha) \wedge \beta + \alpha \wedge (\mathscr{L}(\xi)\beta)$$

Proof This just uses the fact that $f_t(\alpha \wedge \beta) = f_t(\alpha) \wedge f_t(\beta)$. C.Q.F.D.

Finally, one of the most important properties of the Lie derivative is expressed by the following proposition:

PROPOSITION 4.12 If ξ_1 and ξ_2 are vector fields on X, then

$$\mathscr{L}(\xi_1) \circ \mathscr{L}(\xi_2) - \mathscr{L}(\xi_2) \circ \mathscr{L}(\xi_1) = \mathscr{L}([\xi_1, \xi_2])$$

Proof From the preceding proposition we have:

$$\begin{aligned}\mathscr{L}(\xi)\omega = &\sum_{i_1 < \cdots < i_r} (\mathscr{L}(\xi)\omega_{i_1 \ldots i_r})(x)\, dx^{i_1} \wedge \cdots \wedge dx^{i_r} \\ &+ \sum_{i_1 < \cdots < i_r} \sum_{s=1}^{r} \omega_{i_1 \ldots i_r}(x)\, dx^{i_1} \wedge \cdots \wedge dx^{i_{s-1}} \\ &\wedge (\mathscr{L}(\xi)\, dx^{i_s}) \wedge dx^{i_{1+r}} \wedge \cdots \wedge dx^{i_r}\end{aligned}$$

Using remark 4.10, we get:

$$\begin{aligned}(\mathscr{L}(\xi_1) \circ \mathscr{L}(\xi_2) - \mathscr{L}(\xi_2) \circ \mathscr{L}(\xi_1))\Phi &= \xi_2 \cdot (\xi_1 \cdot \Phi) - \xi_1 \cdot (\xi_2 \cdot \Phi) \\ &= \mathscr{L}([\xi_1, \xi_2])\Phi\end{aligned}$$

and

$$\begin{aligned}&(\mathscr{L}(\xi_1) \circ \mathscr{L}(\xi_2) - \mathscr{L}(\xi_2) \circ \mathscr{L}(\xi_1))\, dx^{i_s} \\ &\quad = d((\mathscr{L}(\xi_1) \circ \mathscr{L}(\xi_2) - \mathscr{L}(\xi_2) \circ \mathscr{L}(\xi_1))x^{i_s}) \\ &\quad = d(\mathscr{L}([\xi_1, \xi_2])x^{i_s}) \\ &\quad = \mathscr{L}([\xi_1, \xi_2])\, dx^{i_s}\end{aligned}$$

and the proposition follows. C.Q.F.D.

5 Prolongation of transformations

Let X with local coordinates $(x^1, \ldots, x^n)$ and Y with local coordinates $(y^1, \ldots, y^m)$ be two C^∞, connected paracompact manifolds.

If $f: X \to Y$ is any map $x \to y = f(x)$, then we shall also call f the "*graph*" of f, that is to say the section $f: X \to X \times Y: (x) \to (x, f(x))$ of the trivial bundle $X \times Y \to X$ with projection π_X or simply π when there is no confusion.

Let $\Phi: X \times Y \to X \times Y$ be a bundle map over $\varphi \in \operatorname{Aut}(X)$, expressed in local coordinates by $(x, y) \to (\bar{x} = \varphi(x), \bar{y} = \psi(x, y))$.

If we set $y = f(x)$, we have $\bar{x} = \varphi(x)$ and we can find inverse formulas $x = \varphi^{-1}(\bar{x})$ by means of the implicit function theorem, using the fact that

$$\operatorname{rank}\left[\frac{\partial \varphi^i(x)}{\partial x^j}\right] = n, \qquad \forall x \in X.$$

Thus we have

$$\bar{y} = \psi(\varphi^{-1}(\bar{x}), f(\varphi^{-1}(\bar{x}))) = \bar{f}(\bar{x})$$

and this gives rise to the following commutative diagram:

$$\begin{array}{ccc} X \times Y & \xrightarrow{\Phi} & X \times Y \\ \downarrow\uparrow f & & \downarrow\uparrow \bar{f} \\ X & \xrightarrow{\varphi} & X \end{array}$$

We have also:

$$\bar{f}(\varphi(x)) = \psi(x, f(x)), \qquad \forall x \in X$$

THEOREM 5.1 For any section f of the trivial bundle $X \times Y \to X$ and any integer $q \geq 0$, there exist a unique section $\bar{f}$ of the trivial bundle $X \times Y \to X$ and a unique bundle map $\rho_q(\Phi): J_q(X \times Y) \to J_q(X \times Y)$ such that we have the following commutative diagram

$$\begin{array}{ccc} J_q(X \times Y) & \xrightarrow{\rho_q(\Phi)} & J_q(X \times Y) \\ \downarrow\uparrow j_q(f) & & \downarrow\uparrow j_q(\bar{f}) \\ X & \xrightarrow{\varphi} & X \end{array}$$

DEFINITION 5.2 $\rho_q(\Phi)$ is called the "*q-prolongation*" of Φ and we set $\rho_0(\Phi) = \Phi$.

Proof We shall use an induction on q, starting with the case $q = 0$ already considered.

Imagine now that we have such a map $\rho_{q-1}(\Phi)$ expressed in local coordinates by $\rho_{q-1}(\Phi) = (\varphi, \psi, \chi)$ with

$$\rho_{q-1}(\Phi)\begin{cases} \bar{p}^k_\mu = \chi^k_\mu(x, y, p) & 1 \le |\mu| \le q - 1 \\ \bar{y}^k = \psi^k(x, y) \\ \bar{x}^i = \varphi^i(x) \end{cases}$$

Then we have, for any section $f: X \to X \times Y$, a section $\bar{f}: X \to X \times Y$ such that, with $|\mu| = q - 1$:

$$\frac{\partial^{|\mu|}\bar{f}^k(\varphi(x))}{(\partial\bar{x})^\mu} = \chi^k_\mu\left(x, f(x), \frac{\partial f(x)}{\partial x}\right) = \chi^k_\mu(j_{q-1}(f)(x))$$

Differentiating totally with respect to x^i we get:

$$\frac{\partial\varphi^j(x)}{\partial x^i}\frac{\partial^{|\mu|+1}\bar{f}^k(\varphi(x))}{(\partial\bar{x})^{\mu+1_j}} = \frac{\partial\chi^k_\mu(j_{q-1}(f)(x))}{\partial x^i}$$

$$= \frac{d\chi^k_\mu}{dx^i}(j_q(f)(x))$$

Using finally the jet-coordinates we get:

$$\frac{\partial\varphi^j(x)}{\partial x^i}\bar{p}^k_{\mu+1_j} = \frac{d\chi^k_\mu}{dx^i} \quad \text{with} \quad |\mu| = q - 1$$

and this determines the map $\rho_q(\Phi)$. C.Q.F.D.

DEFINITION 5.3 Let $\Phi = (\varphi, \psi) \in \text{Aut}(X \times Y)$ with $\varphi \in \text{Aut}(X)$ and $\psi \in \text{Aut}(Y)$. If $\varphi = \text{id}_X$ we shall speak of "*target transformation*". If $\psi = \text{id}_Y$ we shall speak of "*source transformation*".

COROLLARY 5.4 For any section f of the fiber bundle $\mathscr{E} \to X$ and any integer $q \ge 0$, there exist a unique section $\bar{f}$ of $\mathscr{E}$ and a unique bundle map $\rho_q(\Phi): J_q(\mathscr{E}) \to J_q(\mathscr{E})$ such that we have the following commutative diagram:

$$\begin{array}{ccc} J_q(\mathscr{E}) & \xrightarrow{\rho_q(\Phi)} & J_q(\mathscr{E}) \\ \downarrow\uparrow j_q(f) & & \downarrow\uparrow j_q(\bar{f}) \\ X & \xrightarrow{\varphi} & X \end{array}$$

whenever $\Phi: \mathscr{E} \to \mathscr{E}$ is a bundle map over $\varphi \in \text{Aut}(X)$. Moreover $\rho_q(\Phi): J_q(\mathscr{E}) \to J_q(\mathscr{E})$ is a bundle map over $\rho_{q-1}(\Phi): J_{q-1}(\mathscr{E}) \to J_{q-1}(\mathscr{E})$.

Proof This is similar to that of the theorem as we may use a local trivialisation of $\mathscr{E}$. We leave the details to the reader. C.Q.F.D.

We have $\bar{p}^k_\mu = \chi^k_\mu(x, y, p)$ and

$$d\bar{p}^k_\mu = \frac{\partial \chi^k_\mu(x, y, p)}{\partial x^i} dx^i + \frac{\partial \chi^k_\mu(x, y, p)}{\partial y^l} dy^l + \frac{\partial \chi^k_\mu(x, y, p)}{\partial p^l_\nu} dp^l_\nu$$

But

$$\frac{\partial \varphi^j(x)}{\partial x^i} \bar{p}^k_{\mu+1_j} = \frac{\partial \chi^k_\mu(x, y, p)}{\partial x^i} + \frac{\partial \chi^k_\mu(x, y, p)}{\partial y^l} p^l_i + \frac{\partial \chi^k_\mu(x, y, p)}{\partial p^l_\nu} p^l_{\nu+1_i}$$

that is to say:

$$\bar{p}^k_{\mu+1_j} d\bar{x}^j = \frac{\partial \chi^k_\mu(x, y, p)}{\partial x^i} dx^i + \frac{\partial \chi^k_\mu(x, y, p)}{\partial y^l} p^l_i \, dx^i + \frac{\partial \chi^k_\mu(x, y, p)}{\partial p^l_\nu} p^l_{\nu+1_i} dx^i$$

Subtracting, we get:

$$d\bar{p}^k_\mu - \bar{p}^k_{\mu+1_i} d\bar{x}^i = \frac{\partial \chi^k_\mu(x, y, p)}{\partial y^l} (dy^l - p^l_i \, dx^i) + \frac{\partial \chi^k_\mu(x, y, p)}{\partial p^l_\nu} (dp^l_\nu - p^l_{\nu+1_i} dx^i)$$

Let us now consider the 1-forms $w^k_\mu = dp^k_\mu - p^k_{\mu+1_i} dx^i$ on $J_q(X \times Y)$ with $0 \leq |\mu| \leq q - 1$, setting $w^l = w^l_0 = dy^l - p^l_i \, dx^i$. We introduce $\bar{w}^k_\mu = d\bar{p}^k_\mu - \bar{p}^k_{\mu+1_i} d\bar{x}^i$ written $\bar{w}^k_\mu = \rho_q(\Phi)(w^k_\mu)$ and we have

$$\bar{w}^k_\mu = \frac{\partial \chi^k_\mu(x, y, p)}{\partial y^l} w^l + \frac{\partial \chi^k_\mu(x, y, p)}{\partial p^l_\nu} w^l_\nu$$

This leads us to the important proposition:

Proposition 5.5 If $\Phi_q : J_q(X \times Y) \to J_q(X \times Y)$ is a bundle map over $\Phi : X \times Y \to X \times Y$ then $\Phi_q = \rho_q(\Phi)$ if and only if

$$\bar{w}^k_\mu = \Phi_q(w^k_\mu) = B^{k,\nu}_{l,\mu}(x, y, p) w^l_\nu .$$

Proof

N.C. This has already been proved above.

S.C. Let f be a section of $X \times Y$ over X and $\bar{f}$ be the unique section of $X \times Y$ such that $\Phi \circ f = \bar{f} \circ \varphi$. Then we have $w^k_\mu |_{j_q(f)(X)} = 0$. It follows that we have also $\bar{w}^k_\mu |_{j_q(\bar{f})(X)} = 0$, that is to say

$$\bar{y}^k = \bar{f}^k(\bar{x}) \quad \text{and} \quad \bar{p}^k_\mu = \frac{\partial^{|\mu|} \bar{f}^k(\bar{x})}{(\partial \bar{x})^\mu}$$

if

$$y^k = f^k(x) \quad \text{and} \quad p^k_\mu = \frac{\partial^{|\mu|} f^k(x)}{(\partial x)^\mu}$$

where $(\bar{x}, \bar{y}, \bar{p}) = \Phi_q(x, y, p)$. But this is the definition of the q-prolongation of Φ and as it is unique, we have $\Phi_q = \rho_q(\Phi)$.

C.Q.F.D.

If X and Y are as above, let $\xi = \xi^i(x)(\partial/\partial x^i)$ be a vector field on X and θ be a vector field on $X \times Y$ such that the following diagram is commutative:

$$\begin{array}{ccc} T(X \times Y) & \xrightarrow{T(\pi)} & T(X) \\ \downarrow\uparrow \theta & & \downarrow\uparrow \xi \\ X \times Y & \xrightarrow{\pi} & X \end{array}$$

With local coordinates on $X \times Y$ we have:

$$\theta = \xi^i(x)\frac{\partial}{\partial x^i} + \eta^k(x, y)\frac{\partial}{\partial y^k}$$

THEOREM 5.6 There exists a unique vector field $\rho_q(\theta)$ on $J_q(X \times Y)$ such that:

$$\exp(t \cdot \rho_q(\theta)) = \rho_q(\exp(t \cdot \theta)) \qquad \forall t, q$$

Moreover the following diagram is commutative:

$$\begin{array}{ccc} T(J_q(X \times Y)) & \xrightarrow{T(\pi_0^q)} & T(X \times Y) \\ \downarrow\uparrow \rho_q(\theta) & & \downarrow\uparrow \theta \\ J_q(X \times Y) & \xrightarrow{\pi_0^q} & X \times Y \end{array}$$

Proof Let us consider $\Phi_t = \exp(t \cdot \theta) \in \operatorname{Aut}(X \times Y)$. It is a bundle map over $\varphi_t = \exp(t \cdot \xi) \in \operatorname{Aut}(X)$ and we can construct as in theorem 5.1 the unique q-prolongation $\rho_q(\Phi_t) \in \operatorname{Aut}(J_q(X \times Y))$.

Using the preceding proposition, it is easy to see that $\rho_q(\Phi_s) \circ \rho_q(\Phi_t)$ is the q-prolongation of the map $\Phi_s \circ \Phi_t = \Phi_{s+t}$. It follows that

$$\rho_q(\Phi_s) \circ \rho_q(\Phi_t) = \rho_q(\Phi_{s+t})$$

and we have a one parameter group on $J_q(X \times Y)$, which gives rise to a well defined vector field on $J_q(X \times Y)$, noted $\rho_q(\theta)$. C.Q.F.D.

However this result is not very useful and we shall prefer the following one which is more operational.

By definition we have:

$$\bar{x}^i = x^i + t\xi^i(x) + \cdots \quad \text{and} \quad \bar{y}^k = y^k + t\eta^k(x, y) + \cdots$$

If we set

$$\rho_q(\theta) = \xi^i(x)\frac{\partial}{\partial x^i} + \eta^k(x, y)\frac{\partial}{\partial y^k} + \zeta^k_\mu(x, y, p)\frac{\partial}{\partial p^k_\mu}$$

then

$$\bar{p}^k_\mu = p^k_\mu + t\zeta^k_\mu(x, y, p) + \cdots$$

But we have, by construction:

$$\left(\delta^j_i + t\frac{\partial \xi^j(x)}{\partial x^i} + \cdots\right)(p^k_{\mu+1_j} + t\zeta^k_{\mu+1_j}(x, y, p) + \cdots) = p^k_{\mu+1_i} + t\frac{d\zeta^k_\mu}{dx^i} + \cdots$$

Differentiating with respect to t, for $t = 0$ we get:

$$\zeta^k_{\mu+1_i} = \frac{d\zeta^k_\mu}{dx^i} - p^k_{\mu+1_j}\frac{\partial \xi^j(x)}{\partial x^i}$$

In this way we can define $\rho_q(\theta)$ inductively, starting with $\rho_0(\theta) = \theta$, and this process is uniquely defined. We leave to the reader the fact that it does not depend on the coordinate system. (Use propositions 4.2 and 4.9.)

Definition 5.7 We say that a vector field θ_q on $J_q(X \times Y)$ is over the vector field θ on $X \times Y$ if we have $\pi^q_0(\theta_q) = \theta$, that is to say $T(\pi^q_0) \circ \theta_q = \theta \circ \pi^q_0$.

Similarly to proposition 5.5, for infinitesimal transformation we have the proposition:

Proposition 5.8 If θ_q is a vector field on $J_q(X \times Y)$ over the vector field θ on $X \times Y$, then $\theta_q = \rho_q(\theta)$ if and only if $\mathscr{L}(\theta_q)w^k_\mu = A^{k,\nu}_{l,\mu}(x, y, p)w^l_\nu$.

Proof

N.C. This follows directly from proposition 4.9, using the definition of the Lie derivative.

S.C. Setting

$$\theta_q = \xi^i(x)\frac{\partial}{\partial x^i} + \eta^k(x, y)\frac{\partial}{\partial y^k} + \zeta^k_\mu(x, y, p)\frac{\partial}{\partial p^k_\mu}$$

we have:

$$\begin{aligned}
\mathscr{L}(\theta_q)w^k_\mu &= d\zeta^k_\mu - d(p^k_{\mu+1_j}\cdot\xi^j) - \zeta^k_{\mu+1_j}\,dx^j + \xi^j\,dp^k_{\mu+1_j}\\
&= d\zeta^k_\mu - p^k_{\mu+1_j}\frac{\partial\xi^j(x)}{\partial x^i}\,dx^i - \zeta^k_{\mu+1_j}\,dx^j\\
&= \frac{\partial\zeta^k_\mu}{\partial x^i}\,dx^i + \frac{\partial\xi^k_\mu}{\partial y^l}\,dy^l + \frac{\partial\zeta^k_\mu}{\partial p^l_\nu}\,dp^l_\nu - \left(p^k_{\mu+1_j}\frac{\partial\xi_j}{\partial x^i} + \zeta^k_{\mu+1_i}\right)dx^i\\
&= \left(\frac{d\zeta^k_\mu}{dx^i} - p^k_{\mu+1_j}\frac{\partial\xi^j}{\partial x^i} - \zeta^k_{\mu+1_i}\right)dx^i + \frac{\partial\zeta^k_\mu}{\partial y^l}\,w^l + \frac{\partial\zeta^k_\mu}{\partial p^l_\nu}\,w^l_\nu
\end{aligned}$$

It follows that we must have:

$$\frac{d\zeta^k_\mu}{dx^i} - p^k_{\mu+1_j}\frac{\partial\xi^j}{\partial x^i} - \zeta^k_{\mu+1_i} = 0$$

But this is just the constructive definition of $\rho_q(\theta)$ and we have $\theta_q = \rho_q(\theta)$ as $\theta_0 = \theta$. C.Q.F.D.

EXAMPLE 5.9 $n = 1, m = 1, q = 1$.

Finite transformation $\bar{x} = \varphi(x)$, $\bar{y} = \psi(x, y)$ then

$$\bar{p} = \frac{1}{\partial\varphi(x)/\partial x}\left(\frac{\partial\psi(x, y)}{\partial x} + p\frac{\partial\psi(x, y)}{\partial y}\right)$$

Infinitesimal transformation

$$\xi(x)\frac{\partial}{\partial x} + \eta(x, y)\frac{\partial}{\partial y}$$

then

$$\zeta(x, y, p) = \frac{\partial\eta(x, y)}{\partial x} + \frac{\partial\eta(x, y)}{\partial y}p - p\frac{\partial\xi(x)}{\partial x}.$$

If $\mathscr{E} \to X$ is a bundle, as in corollary 5.4 we have the following theorem:

THEOREM 5.10 If θ is a vector field on $\mathscr{E}$ over a vector field $\xi = \xi^i(x)(\partial/\partial x^i)$ on X, there exists a unique vector field $\rho_q(\theta)$ on $J_q(\mathscr{E})$ such that:

$$\exp(t\cdot\rho_q(\theta)) = \rho_q(\exp(t\cdot\theta)) \qquad \forall t, q$$

Moreover the following diagram is commutative:

$$\begin{array}{ccc}
T(J_q(\mathscr{E})) & \xrightarrow{T(\pi^q_0)} & T(\mathscr{E})\\
\downarrow\uparrow\, \rho_q(\theta) & & \downarrow\uparrow\, \theta\\
J_q(\mathscr{E}) & \xrightarrow{\pi^q_0} & \mathscr{E}
\end{array}$$

However, in the sequel, we will be mainly concerned with finite and infinitesimal source or target transformations.

We shall make the following remark:

REMARK 5.11 All the computations to be done in order to determine finite or infinitesimal prolongations are rational with respect to p^k_μ, $1 \leq |\mu| \leq q$. In fact the ψ^k_μ with $1 \leq |\mu| \leq q$ and the ζ^k_μ are polynomials in the p^k_μ, with C^∞ coefficients that are only functions of x and y, that is to say:

$$\psi^k_\mu, \zeta^k_\mu \in C^\infty(X \times Y) \cdot [p^l_\nu] \qquad 1 \leq |\mu|, |\nu| \leq q.$$

We now give two fundamental properties of the prolongation map ρ_q.

PROPOSITION 5.12 $\forall \Phi \in C^\infty(J_q(X \times Y))$,

$$\rho_{q+1}(\theta) \cdot \frac{d\Phi}{dx^i} = \frac{d}{dx^i}(\rho_q(\theta) \cdot \Phi) - \frac{\partial \xi^j(x)}{\partial x^i} \frac{d\Phi}{dx^j}$$

Proof In order to simplify the proof we set $p^k_0 = y^k$ and $\zeta^k_0 = \eta^k$. By definition

$$\frac{d\Phi}{dx^i} = \frac{\partial \Phi}{\partial x^i} + \frac{\partial \Phi}{\partial p^l_\nu} p^l_{\nu+1_i}$$

Then:

$$\rho_{q+1}(\theta) \cdot \frac{d\Phi}{dx^i} = \rho_q(\theta) \cdot \frac{\partial \Phi}{\partial x^i} + p^l_{\nu+1_i}\left(\rho_q(\theta) \cdot \frac{\partial \Phi}{\partial p^l_\nu}\right) + \frac{\partial \Phi}{\partial p^k_\mu} \cdot \zeta^k_{\mu+1_i}$$

$$\begin{aligned}
\rho_{q+1}(\theta) \cdot \frac{d\Phi}{dx^i} &= \xi^j \frac{\partial^2 \Phi}{\partial x^i \partial x^j} + \zeta^k_\mu \frac{\partial^2 \Phi}{\partial x^i \partial y^k_\mu} \\
&\quad + p^l_{\nu+1_i} \cdot \xi^j \frac{\partial^2 \Phi}{\partial x^j \partial p^l_\nu} + p^l_{\nu+1_i} \cdot \zeta^k_\mu \frac{\partial^2 \Phi}{\partial p^k_\mu \partial p^l_\nu} + \zeta^k_{\mu+1_i} \frac{\partial \Phi}{\partial p^k_\mu} \\
&= \frac{\partial}{\partial x^i}\left(\xi^j \frac{\partial \Phi}{\partial x^j} + \zeta^k_\mu \frac{\partial \Phi}{\partial p^k_\mu}\right) - \frac{\partial \xi^j}{\partial x^i} \frac{\partial \Phi}{\partial x^j} - \frac{\partial \zeta^k_\mu}{\partial x^i} \frac{\partial \Phi}{\partial p^k_\mu} \\
&\quad + p^l_{\nu+1_i} \frac{\partial}{\partial p^l_\nu}\left(\xi^j \frac{\partial \Phi}{\partial x^j} + \zeta^k_\mu \frac{\partial \Phi}{\partial p^k_\mu}\right) - p^l_{\nu+1_i} \frac{\partial \zeta^k_\mu}{\partial p^l_\nu} \frac{\partial \Phi}{\partial p^k_\mu} \\
&\quad + \frac{d\zeta^k_\mu}{dx^i} \frac{\partial \Phi}{\partial p^k_\mu} - p^k_{\mu+1_j} \frac{\partial \xi^j}{\partial x^i} \frac{\partial \Phi}{\partial p^k_\mu} \\
&= \frac{d}{dx^i}(\rho_q(\theta) \cdot \Phi) - \frac{\partial \xi^j}{\partial x^i} \cdot \frac{d\Phi}{dx^j}
\end{aligned}$$

C.Q.F.D.

REMARK 5.13 We have used the notation Φ because a function

$$\Phi \in C^\infty(J_q(X \times Y)) : J_q(X \times Y) \to \mathbb{R}$$

can be considered as a bundle map

$$\Phi : J_q(X \times Y) \to X \times \mathbb{R}$$

over id_X.

PROPOSITION 5.14 Whatever the vector fields θ_1 and θ_2 may be over ξ_1 and ξ_2 respectively, we have

$$\rho_q([\theta_1, \theta_2]) = [\rho_q(\theta_1), \rho_q(\theta_2)],$$

where $[\theta_1, \theta_2]$ is a vector field over $[\xi_1, \xi_2]$.

Proof Using the properties of the Lie derivative, we have, with matrix notation:

$$\mathscr{L}(\rho_q(\theta_1))w = A_1 \cdot w, \qquad \mathscr{L}(\rho_q(\theta_2))w = A_2 \cdot w$$

Then

$$\begin{aligned}\mathscr{L}([\rho_q(\theta_1), \rho_q(\theta_2)])w &= (\mathscr{L}(\rho_q(\theta_1)) \circ \mathscr{L}(\rho_q(\theta_2)) - \mathscr{L}(\rho_q(\theta_2)) \circ \mathscr{L}(\rho_q(\theta_1)))w \\ &= (\rho_q(\theta_1) \cdot A_2 - \rho_q(\theta_2) \cdot A_1 + A_2 \circ A_1 - A_1 \circ A_2)w\end{aligned}$$

From proposition 5.8, it follows that $[\rho_q(\theta_1), \rho_q(\theta_2)]$ is the q-prolongation of the infinitesimal transformation

$$\pi_0^q([\rho_q(\theta_1), \rho_q(\theta_2)]) = [\rho_0(\theta_1), \rho_0(\theta_2)] = [\theta_1, \theta_2]$$

It suffices now to use proposition 2.12 in order to see that $[\theta_1, \theta_2]$ is a vector field on $X \times Y$, over the vector field $[\xi_1, \xi_2]$ on X. C.Q.F.D.

Let X and Y be as above with $\dim X \neq \dim Y$ and consider the trivial bundle $\pi = \pi_X : X \times Y \to X$.

We shall now consider more precisely the case of an infinitesimal source transformation.

If

$$\xi = \xi^i(x) \frac{\partial}{\partial x^i}$$

is any vector field on X, we shall also call ξ the vector field

$$\xi^i(x) \frac{\partial}{\partial x^i} + 0 \frac{\partial}{\partial y}$$

induced on $X \times Y$. This is a natural way to split the exact sequence:

$$0 \longrightarrow V(X \times Y) \longrightarrow T(X \times Y) \underset{T(\pi_X)}{\overset{\curvearrowleft}{\longrightarrow}} \pi_X^{-1}(T(X)) \longrightarrow 0$$

and it does not depend on the coordinate system.

We shall now work with local coordinates: We have

$$\rho_q(\xi) = \xi^i(x)\frac{\partial}{\partial x^i} + \zeta_\mu^k(x, p)\frac{\partial}{\partial p_\mu^k}$$

with

$$\zeta_{\mu+1_i}^k = \frac{d\zeta_\mu^k}{dx^i} - p_{\mu+1_j}^k \frac{\partial \xi^j}{\partial x^i}$$

or

$$\zeta_{\mu+1_i}^k - p_{\mu+1_i+1_j}^k \xi^j = \frac{d}{dx^i}(\zeta_\mu^k - p_{\mu+1_j}^k \xi^j)$$

because

$$p_{\mu+1_i+1_j}^k = \frac{dp_{\mu+1_j}^k}{dx^i} \quad \text{and} \quad \frac{\partial \xi^j}{\partial x^i} = \frac{d\xi^j}{dx^i}.$$

Using Leibnitz's derivation formula,

$$\begin{aligned}\zeta_\mu^k(x, p) - p_{\mu+1_j}^k \xi^j(x) &= -\frac{d^{|p|}}{(dx)^\mu}(p_j^k \xi^j(x)) \\ &= -\sum_{0 \le \nu \le \mu} \frac{\mu!}{(\mu - \nu)!\,\nu!} p_{\mu-\nu+1_j}^k \partial_\nu \xi^j(x)\end{aligned}$$

and

$$\zeta_\mu^k(x, p) = -\sum_{0 < \nu \le \mu} \frac{\mu!}{(\mu - \nu)!\,\nu!} p_{\mu-\nu+1_j}^k \partial_\nu \xi^j(x)$$

where $0 < \nu \le \mu$ means

$$0 < \nu_j \le \mu_j, \qquad \forall j$$

It follows that

$$\rho_q(\xi) = \xi^i(x)\frac{\partial}{\partial x^i} + \partial_\nu \xi^j(x) A_j^\nu(q)$$

with

$$A_j^\nu(q) = -\sum_{\substack{\mu \ge \nu \\ |\mu| \le q}} \frac{\mu!}{(\mu - \nu)!\,\nu!} p_{\mu-\nu+1_j}^k \frac{\partial}{\partial p_\mu^k}$$

From the facts that:

$$\left[x^\mu \frac{\partial}{\partial x^i}, x^\nu \frac{\partial}{\partial x^j}\right] = \nu_i x^{\mu+\nu-1_i} \frac{\partial}{\partial x^j} - \mu_j x^{\mu+\nu-1_j} \frac{\partial}{\partial x^i}$$

and

$$[\rho_q(\xi_1), \rho_q(\xi_2)] = \rho_q([\xi_1, \xi_2])$$

we have:

$$[A_i^\mu(q), A_j^\nu(q)] = \begin{cases} \nu_i \dfrac{(\mu + \nu - 1_i)!}{\mu!\nu!} A_j^{\mu+\nu-1_i}(q) - \mu_j \dfrac{(\mu + \nu - 1_j)!}{\mu!\nu!} A_i^{\mu+\nu-1_j}(q) \\ \quad \text{if } \mu + \nu \leq q + 1 \\ 0 \quad \text{otherwise.} \end{cases}$$

REMARK 5.15 If $|\mu| = |\nu| = q$ then there are two cases:

$$\begin{cases} \bullet\ q > 1 \quad [A_i^\mu(q), A_j^\nu(q)] = 0 \\ \bullet\ q = 1 \quad [A_i^k(1), A_j^l(1)] = c_{i,j,r}^{k,l,h} A_h^r(1) \end{cases}$$

For any section f of $X \times Y$ and any $\varphi \in \mathrm{Aut}\,(X)$, set $\bar{f} = f \circ \varphi$. Then we define $j_q(f) \circ j_q(\varphi) = j_q(f \circ \varphi)$ and $j_q(\bar{f}) = j_q(f) \circ j_q(\varphi)$. Taking $x_0 \in X$ and φ such that $\varphi(x_0) = x_0$, then passing to the jet coordinates we have an action of a Lie group, noted $GL_q(n, \mathbb{R})$ on $J_q(X \times Y)$. In fact if

$$\varphi_1, \varphi_2 \in \mathrm{Aut}\,(X) \quad \text{and} \quad \varphi_1(x_0) = \varphi_2(x_0) = x_0$$

then the group law is given by

$$j_q(\varphi_1 \circ \varphi_2) = j_q(\varphi_1) \circ j_q(\varphi_2).$$

Taking φ_t instead of φ and $x_0 \in X$ such that $\varphi_t(x_0) = x_0$ and passing to the limit for $t \to 0$, we have the proposition:

PROPOSITION 5.16 The vector fields $A_i^\mu(q)$ on $J_q(X \times Y)$ are the infinitesimal generators of the effective action of $GL_q(\eta, \mathbb{R})$ on $J_q(X \times Y)$. When $m = n$, this action is also transitive on the open sub-bundle $I_q(X \times Y) \subset J_q(X \times Y)$ such that $\det\,[p_i^k] \neq 0$.

Proof We have only to show that the action is effective:
But

$$A_j^\nu(q) = -p_j^k \frac{\partial}{\partial p_\nu^k} - \sum_{\substack{\mu > \nu \\ |\mu| \leq q}} \frac{\mu!}{(\mu - \nu)!\nu!} p_{\mu-\nu+1_j}^k \frac{\partial}{\partial p_\mu^k}$$

If $\lambda_\nu^j A_j^\nu(q) = 0$ then it is easy to see by induction on $|\nu|$ that $\lambda_\nu^j = 0$. In fact for $|\nu| = 1$ we must have

$$\sum_{|\nu|=1} \lambda_\nu^j p_j^k \frac{\partial}{\partial p_\nu^k} = 0 \quad \text{and thus} \quad \lambda_\nu^j p_j^k = 0,$$

that is to say $\lambda_\nu^j = 0$.

Finally if $m = n$ and $\det [p_j^k] \neq 0$, then we have

$$-[p^{-1}]_k^j A_j^\nu(q) = \frac{\partial}{\partial p_\nu^k} + \cdots$$

and the proposition follows. C.Q.F.D.

In a similar way we shall consider the infinitesimal target transformation.

If

$$\eta = \eta^k(y) \frac{\partial}{\partial y^k}$$

is any vector field on Y, we shall also call η the vector field

$$0 \frac{\partial}{\partial x} + \eta^k(y) \frac{\partial}{\partial y^k}$$

induced on $X \times Y$. This is a natural way to split the exact sequence:

$$0 \longrightarrow \ker T(\pi_Y) \longrightarrow T(X \times Y) \xrightarrow{T(\pi_Y)} \pi_Y^{-1}(T(Y)) \longrightarrow 0$$

and it does not depend on the coordinate system.

We shall have

$$\rho_q(\eta) = \eta^k(y) \frac{\partial}{\partial y^k} + \zeta_\mu^k(y, p) \frac{\partial}{\partial p_\mu^k}$$

with

$$\zeta_{\mu+1_i}^k = \frac{d\zeta_\mu^k}{dx^i} \quad \text{and thus} \quad \zeta_\mu^k = \frac{d^{|\mu|}\eta^k}{(dx)^\mu},$$

but now we cannot apply the Leibnitz formula.

However, setting $\alpha = (\alpha_1, \ldots, \alpha_m)$, from the fact that:

$$\left[y^\alpha \frac{\partial}{\partial y^k}, y^\beta \frac{\partial}{\partial y^l} \right] = \beta_k y^{\alpha+\beta-1_k} \frac{\partial}{\partial y^l} - \alpha_l y^{\alpha+\beta-1_l} \frac{\partial}{\partial y^k}$$

and

$$[\rho_q(\eta_1), \rho_q(\eta_2)] = \rho_q([\eta_1, \eta_2]),$$

we have

$$\rho_q(\eta) = \eta^k(y)\frac{\partial}{\partial y^k} + \frac{\partial^{|\alpha|}\eta^k(y)}{(\partial y)^\alpha} B_k^\alpha(q)$$

with

$$[B_k^\alpha(q), B_\rho^\beta(q)] = \begin{cases} \beta_k \dfrac{(\alpha+\beta-1_k)!}{\alpha!\beta!} B_l^{\alpha+\beta-1_k}(q) - \alpha_l \dfrac{(\alpha+\beta-1_l)!}{\alpha!\beta!} B_k^{\alpha+\beta-1_l}(q) \\ \quad \text{if } 2 \le |\alpha+\beta| \le q+1 \\ 0 \quad \text{otherwise} \end{cases}$$

and the same remark as the one for the $A_k^\mu(q)$ is held there.

For any section f of $X \times Y$ and any $\psi \in \operatorname{Aut}(Y)$ set $\bar{f} = \psi \circ f$. Then we define

$$j_q(\psi) \circ j_q(f) = j_q(\psi \circ f) \quad \text{and} \quad j_q(\bar{f}) = j_q(\psi) \circ j_q(f).$$

Taking $y_0 \in Y$ and ψ such that $\psi(y_0) = y_0$, we have an action of the Lie group $GL_q(m, \mathbb{R})$ on $J_q(X \times Y)$.

As before we have the proposition:

PROPOSITION 5.17 The vector fields $B_k^\alpha(q)$ on $J_q(X \times Y)$ are the infinitesimal generators of the effective action of $GL_q(m, \mathbb{R})$ on $J_q(X \times Y)$. When $m = n$ this action is also transitive on the open sub-bundle $I_q(X \times Y) \subset J_q(X \times Y)$ such that $\det[p_i^k] \neq 0$.

Proof We have only to show that the action is effective. But:

$$B_k^\alpha(q) = B_k^{\rho_1 \ldots \rho_{|\alpha|}}(q) = p_{j_1}^{\rho_1} \ldots p_{j_{|\alpha|}}^{\rho_{|\alpha|}} \frac{\partial}{\partial p_{j_1 \ldots j_{|\alpha|}}^k}$$

$$+ \text{ terms } \frac{\partial}{\partial p_\mu^k} \quad \text{with } |\mu| > |\alpha|$$

and the proof follows similarly to that of the preceding proposition.

C.Q.F.D.

Now we have

$$\left[\xi^i(x)\frac{\partial}{\partial x^i}, \eta^k(y)\frac{\partial}{\partial y^k}\right] = 0$$

on $X \times Y$. It follows that

$$[\rho_q(\xi), \rho_q(\eta)] = \rho_q([\xi, \eta]) = 0$$

on $J_q(X \times Y)$ and

$$[A_i^\mu(q), B_k^\alpha(q)] = 0 \qquad \forall i, k, \mu, \alpha,$$

written $[A(q), B(q)] = 0$.

Let us consider the finite target transformation $\psi_t = \exp(t \cdot \eta)$. Passing to the limit for $t \to 0$ and using jet coordinates we see that the composition map for derivations

$$j_q(\psi_t \circ f) = j_q(\psi_t) \circ j_q(f)$$

induces, for any section η_q of $J_q(T(Y))$ a map:

$$\#(\eta_q) : J_q(X \times Y) \to V(J_q(X \times Y))$$

Moreover, as $V(J_q(X \times Y)) = J_q(V(X \times Y))$, we have:

$$\rho_q(\eta) = J_q(\eta) = \#(j_q(\eta))$$

Remark 5.19 We notice that $J_q(T(Y))$ is the bundle of q-jets of sections of $T(Y)$ over Y. To indicate that the sections and jets are considered with respect to Y, we shall note $J_q(T(Y)) = J_q(T)(Y)$. In particular we have $J_q(T(X)) = J_q(T)(X) = J_q(T)$ where there is no confusion. This notation is useful when Y is a copy of X.

Remark 5.19 As we use the same notation for a bundle and its set of sections, the context will always make clear the notation $\#(\eta_q)$ for a section η_q of $J_q(T(Y))$ over Y and similarly $\flat(\xi_q)$ for a section ξ_q of $J_q(T(X)) = J_q(T)$ over X, in such a way that $\rho_q(\xi) = \flat(j_q(\xi))$.

We will now indicate two important properties of the map $\#$.

Using local coordinates, we have:

$$\#(\eta_q) = \eta^k(y)\frac{\partial}{\partial y^k} + \sum_{1 \le |\mu| \le q} \zeta_\mu^k(y, p)\frac{\partial}{\partial p_\mu^k} = \eta^k(y)\frac{\partial}{\partial y^k} + \sum_{1 \le |\alpha| \le q} \eta_\alpha^k(y) B_k^\alpha(q)$$

If η_{q+1} is a section of $J_{q+1}(T(Y))$ such that $\pi_q^{q+1} \circ \eta_{q+1} = \eta_q$ then

$$\begin{aligned}\#(\eta_{q+1}) &= \eta^k(y)\frac{\partial}{\partial y^k} + \sum_{1 \le |\mu| \le q+1} \zeta_\mu^k(y, p)\frac{\partial}{\partial p_\mu^k} \\ &= \eta^k(y)\frac{\partial}{\partial y^k} + \sum_{1 \le |\alpha| \le q+1} \eta_\alpha^k(y) B_k^\alpha(q+1)\end{aligned}$$

PROPOSITION 5.20 $\forall \Phi \in C^\infty(J_q(X \times Y))$ we have the formula:

$$\#(\eta_{q+1}) \cdot \frac{d\Phi}{dx^i} = \frac{d}{dx^i}(\#(\eta_q)\cdot \Phi) - p_i^l\left[\left(\frac{\partial \eta^k(y)}{\partial y^l} - \eta_i^k(y)\right)\frac{\partial}{\partial y^k} + \left(\frac{\partial \eta_\alpha^k(y)}{\partial y^l} - \eta_{\alpha+1_l}^k(y)\right)B_k^\alpha(q)\right]\cdot \Phi$$

Proof We first notice that

$$\zeta_\nu^l(y, p) = \sum_{1 \le |\alpha| \le q} \eta_\alpha^k(y)(B_k^\alpha(q)\cdot p_\nu^l)$$

Then

$$\zeta_{\nu+1_i}^l(y, p) = \sum_{1 \le |\alpha| \le q}\left[\eta_{\alpha+1_m}^k(y)p_i^m(B_k^\alpha(q)\cdot p_\nu^l) + \eta_\alpha^k(y)\frac{d(B_k^\alpha(q)\cdot p_\nu^l)}{dx^i}\right]$$

Now

$$\frac{d}{dx^i}(\#(\eta_q)\cdot\Phi) = \frac{\partial \eta^k}{\partial y^l}p_i^l\frac{\partial\Phi}{\partial y^k} + \eta^k\frac{\partial^2\Phi}{\partial y^k\partial y^l}p_i^l + \eta^k\frac{\partial^2\Phi}{\partial y^k \partial p_\nu^l}p_{\nu+1_i}^l + \eta^k\frac{\partial^2\Phi}{\partial y^k\partial x^i}$$
$$+ \frac{\partial\eta_\alpha^k}{\partial y^l}p_i^l(B_k^\alpha(q)\cdot\Phi) + \zeta_{\mu+1_i}^k\frac{\partial\Phi}{\partial p_\mu^k} + \zeta_\mu^k\frac{\partial^2\Phi}{\partial p_\mu^k\partial x^i}$$
$$- \eta_{\alpha+1_l}^k p_i^l(B_k^\alpha(q)\cdot\Phi) + \zeta_\mu^k\left(\frac{\partial^2\Phi}{\partial y^l\partial p_\mu^k}p_i^l + \frac{\partial^2\Phi}{\partial p_\mu^k\partial p_\nu^l}p_{\nu+1_i}^l\right)$$

and

$$\#(\eta_{q+1})\cdot\frac{d\Phi}{dx^i} = \eta^k\frac{\partial^2\Phi}{\partial y^k\partial y^l}p_i^l + \eta^k\frac{\partial^2\Phi}{\partial y^k\partial p_\nu^l}p_{\nu+1_i}^l + \eta^k\frac{\partial^2\Phi}{\partial x^i\partial y^k}$$
$$+ \frac{\partial\Phi}{\partial y^l}\zeta_i^l + \frac{\partial\Phi}{\partial p_\mu^k}\zeta_{\mu+1_i}^k + \zeta_\mu^k\frac{\partial^2\Phi}{\partial x^i\partial p_\mu^k}$$
$$+ \zeta_\mu^k\frac{\partial^2\Phi}{\partial y^l\partial p_\mu^k}p_i^l + \zeta_\mu^k\frac{\partial^2\Phi}{\partial p_\mu^k\partial p_\nu^l}p_{\nu+1_i}^l$$

and the propositive follows from the fact that we have:

$$\zeta_i^l(y, p) = \eta_i^k(y)p_i^l$$

C.Q.F.D.

REMARK 5.21 We have the useful formula:

$$\frac{d\zeta_\nu^l(y, p)}{dx^i} - \zeta_{\nu+1_i}^l(y, p) = \sum_{1\le|\alpha|\le q} p_i^m\left(\frac{\partial\eta_\alpha^k(y)}{\partial y^m} - \eta_{\alpha+1_m}^k(y)\right)(B_k^\alpha(q)\cdot p_\nu^l)$$

REMARK 5.22 We notice that

$$\#(\eta_{q+1})\cdot\frac{d\Phi}{dx^i}=\frac{d}{dx^i}(\#(\eta_q)\cdot\Phi)$$

whenever $D\cdot\eta_{q+1}=0$ where D is the Spencer operator on Y. In fact, if $D\cdot\eta_{q+1}=0$ then $\eta_{q+1}=j_{q+1}(\eta)$ with $\eta=\pi_0^{q+1}\circ\eta_{q+1}$ and we just get proposition 5.12. Similarly to proposition 5.20 we have the formula:

$$\flat(\xi_{q+1})\cdot\frac{d\Phi}{dx^i}=\frac{d}{dx^i}(\flat(\xi_q)\cdot\Phi)-\xi_i^j(x)\frac{d\Phi}{dx^j}$$
$$-\left[\left(\frac{\partial\xi^j(x)}{\partial x^i}-\xi_i^j(x)\right)\frac{\partial}{\partial x^j}+\left(\frac{\partial\xi_\mu^k(x)}{\partial x^i}-\xi_{\mu+1_i}^k(x)\right)A_k^\mu(q)\right]\cdot\Phi$$

PROPOSITION 5.23 If η_q and $\bar{\eta}_q$ are two sections of $J_q(T(Y))$ over Y, then it is possible to define a bracket operation on $J_q(T(Y))$ by means of the formula:

$$[\#(\eta_q),\#(\bar{\eta}_q)]=\#([\eta_q,\bar{\eta}_q])$$

Proof Using proposition 5.17 and a straightforward computation, we have:

$$[\#(\eta_q),\#(\bar{\eta}_q)]=\sum_{\substack{0\le|\alpha|\le q\\0\le|\beta|\le q\\1\le|\alpha+\beta|\le q}}\frac{(\alpha+\beta)!}{\alpha!\,\beta!}(\eta_\alpha^l(y)\bar{\eta}_{\beta+1_l}^k(y)-\bar{\eta}_\beta^l(y)\eta_{\alpha+1_l}^k(y))B_k^{\alpha+\beta}(q)$$
$$+\eta^l(y)\left[\left(\frac{\partial\bar{\eta}^k(y)}{\partial y^l}-\bar{\eta}_l^k(y)\right)\frac{\partial}{\partial y^k}+\sum_{1\le|\alpha|\le q}\left(\frac{\partial\bar{\eta}_\alpha^k(y)}{\partial y^l}-\bar{\eta}_{\alpha+1_l}^k(y)\right)B_k^\alpha(q)\right]$$
$$-\bar{\eta}^l(y)\left[\left(\frac{\partial\eta^k(y)}{\partial y^l}-\eta_l^k(y)\right)\frac{\partial}{\partial y^k}+\sum_{1\le|\alpha|\le q}\left(\frac{\partial\eta_\alpha^k(y)}{\partial y^l}-\eta_{\alpha+1_l}^k(y)\right)B_k^\alpha(q)\right]$$

where $\eta_0^k(y)=n^k(y)$ and $B_k^0(q)=\partial/\partial y^k$. It may be noticed that the terms $\eta^l(y)\bar{\eta}_{\alpha+1_l}^k(y)$ and $\bar{\eta}^l(y)\eta_{\alpha+1_l}^k(y)$, with $|\alpha|=q$, cancel out. C.Q.F.D.

Similarly, if ξ_q and $\bar{\xi}_q$ are two sections of $J_q(T)$ over X, then it is possible to define a bracket operation on $J_q(T)$ by means of the formula:

$$[\flat(\xi_q),\flat(\bar{\xi}_q)]=\flat([\xi_q,\bar{\xi}_q])$$

because the vector fields $A_i^\rho(q)$ and $B_k^\alpha(q)$ have the same commutation relations. We obtain the formula:

$$[j_q(\xi),j_q(\bar{\xi})]=j_q([\bar{\xi},\xi])$$

CHAPTER 7

1 Finite and infinitesimal Lie equations

Let X be a C^∞, connected, paracompact manifold with dim $X = n$ and Y be a copy of X. We call $T = T(X)$ the tangent bundle of X and $T^* = T^*(X)$ the cotangent bundle of X.

We shall consider the trivial bundle $\pi : X \times Y \to X$ with projection source and its bundle of q-jets $J_q(X \times Y)$ which has a projection $\pi = \pi_X$ onto X, a projection π_Y onto Y and a projection $\pi_0^q = \pi_{X \times Y}$ onto $X \times Y$.

As already mentioned, a section f of $X \times Y$ over X can be considered, using the same notation, as the graph of a map $f : X \to Y$. This identification will be done in the sequel, and for this reason we shall also call f a "*finite transformation*" of X.

DEFINITION 1.1 As Y is a copy of X, we introduce the open sub-bundle $I_q(X \times Y) \subset J_q(X \times Y)$ of invertible transformations.

If we adopt local coordinates (x^i, y^k, p^k_μ) for $J_q(X \times Y)$ with $i, k = 1, \ldots, n$ and $1 \leq |\mu| \leq q$, then $I_q(X \times Y)$ is defined locally by the condition $\det [p^k_i] \neq 0$ and this does not depend on the coordinate system.

Let now $\mathcal{R}_q \subset I_q(X \times Y) \subset J_q(X \times Y)$ be a non-linear system of order q on $X \times Y$.

DEFINITION 1.2 We say that the solutions of $\mathcal{R}_q$ are the finite transformations of a "*Lie pseudogroup*" Γ of transformations of X if the following conditions are satisfied:

1 If f is a solution of $\mathcal{R}_q$ over $U \subset X$ and g is a solution of $\mathcal{R}_q$ over $V \subset X$ such that $f(U) \cap V \neq \varnothing$, then $g \circ f$ is a solution of $\mathcal{R}_q$ over $f^{-1}(f(U) \cap V)$. We have the picture:

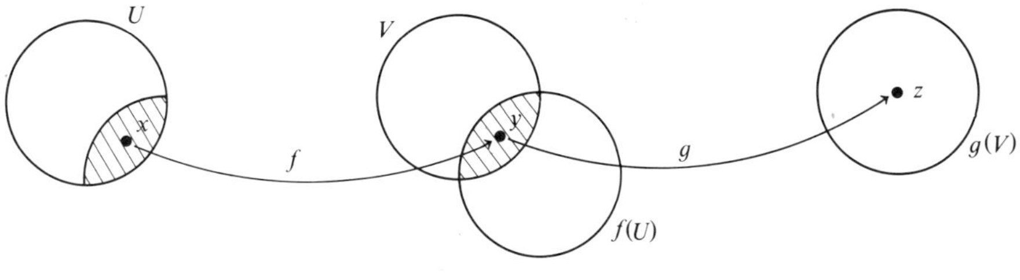

2 If f is a solution of $\mathcal{R}_q$ over $U \subset X$, then f^{-1} is a solution of $\mathcal{R}_q$ over $f(U) \subset X$.

REMARK 1.3 From **1** and **2** it follows that the identity transformation $\mathrm{id} = \mathrm{id}_X$ is a solution of $\mathcal{R}_q$.

REMARK 1.4 For simplicity we can introduce another copy Z of X and write, without any superscript:

$$y = f(x),\ z = g(y) = g(f(x)) = g \circ f(x)$$

REMARK 1.5 According to the definition, we shall say that such a system $\mathcal{R}_q$ defines Γ or that Γ is defined by $\mathcal{R}_q$, or that the equations of $\mathcal{R}_q$ are the "*finite equations*" of Γ.

Using a convenient coordinate neighbourhood $\mathcal{U}_q \subset J_q(X \times Y)$ of a point $(x_0, y_0, p_0) \in \mathcal{R}_q$ that will be in general $j_q(\mathrm{id})(x_0) = \mathrm{id}_q(x_0)$, we shall write the equations of $\mathcal{R}_q$ as in the first part of this book:

$$\mathcal{R}_q \qquad \Phi^\tau(x^i, y^k, p^k_\mu) = 0 \qquad \begin{array}{l} i, k = 1, \ldots, n \\ 1 \le |\mu| \le q \end{array}$$

or simply:

$$\mathcal{R}_q \qquad \Phi(x, y, p) = 0 \qquad 1 \le \mathrm{ord}\, p \le q$$

However if we look at the following pseudogroup of transformations of $\mathbb{R}^2$:

$$\Gamma\begin{cases} y^1 = f(x^1) \\ y^2 = g(x^2) \end{cases} \quad \text{or} \quad \begin{cases} y^1 = f(x^2) \\ y^2 = g(x^1) \end{cases}$$

where f and g are arbitrary C^∞ functions of one variable, then it is easy to see that the defining system $\mathcal{R}_1$ of Γ is the union of two disjoint submanifolds $\mathcal{R}'_1$ and $\mathcal{R}''_1$ of $I_1(X \times Y)$ with:

$$\mathcal{R}'_1: \qquad p^1_2 = 0,\ p^2_1 = 0$$

$$\mathcal{R}''_1: \qquad p^1_1 = 0,\ p^2_2 = 0$$

As a matter of fact, we shall be concerned in the sequel with the "*connected component of the identity*", that is to say the union for $x \in X$ of the connected component of the manifold $\mathcal{R}_{q,x}$ containing the point $\mathrm{id}_q(x)$.

Thus in the example just mentioned we should focus only on

$$\mathcal{R}'_1 = \bigcup_{x \in X} \mathcal{R}'_{1,x}.$$

EXAMPLE 1.6 Let us consider the action of a Lie group of transformations of X. We take the case $X = \mathbb{R}$ and denote the jet coordinates simply by $(x, y, y', y'', \ldots)$

- $y = ax \quad \Rightarrow \mathcal{R}_1 : y - xy' = 0$
- $y = x + b \quad \Rightarrow \mathcal{R}_1 : y' - 1 = 0$
- $y = ax + b \Rightarrow \mathcal{R}_2 : y'' = 0$
- $y = \dfrac{ax + b}{cx + d} \Rightarrow \mathcal{R}_3 : \dfrac{y'''}{y'} - \dfrac{3}{2}\left(\dfrac{y''}{y'}\right)^2 = 0$

  (In this case the action is not effective).

More generally, referring to the preceeding chapter, and writing the action locally as $y^k = f^k(a, x)$ then, to find out the equations of $\mathcal{R}_q$ we have to eliminate the parameters a between the equation $p^k_\mu = \partial_\mu f^k(a, x)$ and we have to take q sufficiently big in order to do so. It is easy to see that the system of p.d.e. thus obtained is of finite type.

EXAMPLE 1.7 This example uses well known facts about the Jacobian matrix of a transformation:

A pseudogroup Γ can be defined by

$$\mathcal{R}_1: \qquad \det\,[p^k_i] = 1.$$

The condition $\mathcal{R}_1 \subset I_1(X \times Y)$ is automatically satisfied.

EXAMPLE 1.8 The set of holomorphic transformations of the complex plane is also a pseudogroup. In fact its defining equations are just the classical Cauchy–Riemann equations:

$$\frac{\partial y^2}{\partial x^2} - \frac{\partial y^1}{\partial x^1} = 0, \quad \frac{\partial y^1}{\partial x^2} + \frac{\partial y^2}{\partial x^1} = 0$$

EXAMPLE 1.9 Many examples arise in physics from the existence of a given covariant tensor field on X. Let ω be that tensor field, considered as a section of a certain tensor bundle over X; then the corresponding system of p.d.e. is of the first order and can be written $f^*\omega = \omega$ with classical notations. Of course we have $(g \circ f)^* = f^* \circ g^*$ and it follows that $(g \circ f)^*\omega = \omega$ if we have also $g^*\omega = \omega$. We may notice however that such a system is not in general formally integrable, unless ω is a closed form, because, using proposition 6.42, we can deduce from it other first order equations that can be written $f^*\,d\omega = d\omega$.

EXAMPLE 1.10 Many cases are more difficult to look at and we cannot find the former situations. For example if we are given the finite transformations of Γ by means of arbitrary constants and arbitrary functions of $1, 2, \ldots, n$ variables it may be difficult to find out the defining finite equations. In what follows we take $X = \mathbb{R}^3$.

$$\Gamma: \qquad \mathbb{R}^3 \to \mathbb{R}^3 : y^1 = f(x^1), \quad y^2 = x^2 f'(x^1), \quad y^3 = x^3 + x^2 \frac{f''(x^1)}{f'(x^1)}$$

We have the defining first order system of p.d.e.

$$\mathcal{R}_1: \qquad p_3^3 - 1 = 0, \quad p_3^2 = 0, \quad p_3^1 = 0, \quad p_2^3 - \frac{y^3 - x^3}{x^2} = 0$$

$$p_2^2 - \frac{y^2}{x^2} = 0, \quad p_2^1 = 0, \quad p_1^2 - (y^3 - x^3)\frac{y^2}{x^2} = 0. \quad p_1^1 - \frac{y^2}{x^2} = 0.$$

The reader will easily check that $\mathcal{R}_1$ is involutive.

EXAMPLE 1.11 We take $X = \mathbb{R}^2$ and consider:

$$\Gamma_1: \qquad y^1 = x^1 + a, \qquad y^2 = x^2 + b(x^1)$$

$$\Gamma_2: \qquad y^1 = f(x^1), \qquad y^2 = x^2 f'(x^1)$$

The reader will easily discover the corresponding defining equations of Γ_1 and Γ_2. Though it seems that there is no link at all between Γ_1 and Γ_2, we shall see later on that in fact there is such a link.

Looking at example 1.9, we see that, as $\mathcal{R}_q$ is not, in the definition, supposed to be formally integrable, it may be difficult, even in the analytic case, to construct its solution explicitly as above, or at least its formal solutions, unless we are dealing with particular cases.

For this reason we shall sometimes prefer the formal definition that follows.

But first we need some comments.

It is possible to define on $J_q(X \times Y)$ a composition law by using the same formulas as the ones that are to be found when composing differentiations. In fact, if f and g are two sections of $X \times Y$ over X, then we just use the defining relation $j_q(g) \circ j_q(f) = j_q(g \circ f)$. This formula gives rise to a composition law (of course whenever it is defined) $(g_q, f_q) \to g_q \circ f_q$ but we shall in general consider the composition law for the fibers and not for the sections.

If we set $J_q(x, y) = (\pi_0^q)^{-1}(x, y)$ we shall write the composition law as follows:

$$J_q(y, z) \circ J_q(x, y) \to J_q(x, z)$$

Now, in order to avoid any misunderstanding about the indices, we shall sometimes write, $\partial y/\partial x$ instead of the jet coordinates p.

EXAMPLE 1.12 $X = \mathbb{R}^2$. We have 3 ways to write the Jacobian:

$$p_1^1 p_2^2 - p_2^1 p_1^2 \quad \text{or} \quad \frac{\partial(y^1, y^2)}{\partial(x^1, x^2)} \quad \text{or} \quad \frac{\partial y^1}{\partial x^1}\frac{\partial y^2}{\partial x^2} - \frac{\partial y^1}{\partial x^2}\frac{\partial y^2}{\partial x^1}$$

In the sequel, if we do want to write the composition law and to specify at the same time both the source and target, we shall keep simply $q = 1$ and write:

$$\frac{\partial z}{\partial x} = \frac{\partial z}{\partial y} \cdot \frac{\partial y}{\partial x}$$

We are now ready to give the other definition:

DEFINITION 1.13 A system $\mathcal{R}_q \subset I_q(X \times Y)$ is said to be a non-linear system of "*finite Lie equations*" if the following conditions are satisfied:

1 $\mathcal{R}_q(y, z) \circ \mathcal{R}_q(x, y) \subset \mathcal{R}_q(x, z) \qquad \forall x, y, z$

2 $\forall (x, y, p) \in \mathcal{R}_q$ then $(y, x, p^{-1}) \in \mathcal{R}_q$

REMARK 1.14 In **2** we have to understand that, if $GL_q(n, \mathbb{R})$ is defined as the Lie group with local coordinates p and composition law as above, then p^{-1} is written for the inverse of p.

REMARK 1.15 From **1** and **2** we see that

$$j_q(\mathrm{id})(x) = \mathrm{id}_q(x) \in \mathcal{R}_q \qquad \forall x \in X.$$

DEFINITION 1.16 If moreover $\mathcal{R}_q$ is such that, $\forall x_0 \in X$, $\forall y_0 \in Y$ there exists a section f of $X \times Y$, solution of $\mathcal{R}_q$ over a neighbourhood U of x_0 in X such that $y_0^k = f^k(x_0)$, then Γ is said to be "*transitive*".

REMARK 1.17 When Γ is transitive, it follows that the maps

$$\pi_0^{q+r} : \mathcal{R}_{q+r} \to X \times Y$$

are surjective $\forall r \geq 0$. This means that, using differentiations and eliminations, one cannot get from the equations of $\mathcal{R}_q$ any other equation involving only only x and y.

Using this remark we may give the following definition:

DEFINITION 1.18 If $\mathcal{R}_q$ is such that the morphism $\pi_0^q : \mathcal{R}_q \to X \times Y$ induced by the epimorphism $\pi_0^q : J_q(X \times Y) \to X \times Y$ is also an epimorphism, then $\mathcal{R}_q$ is said to be "*formally transitive*".

Of course, when $\mathscr{R}_q$ is analytic, formally integrable and formally transitive, the two definitions above are equivalent because, in this case, we can construct an analytic solution as a convergent series in a neighbourhood of x_0.

From now on we shall only consider transitive pseudogroups and formally transitive systems of Lie equations.

Coming back to the properties of $\mathscr{R}_q$ involved in the definitions 1.2 and 1.13 we see that the second always implies the first. In fact, if f is a solution of $\mathscr{R}_q$ over $U \subset X$, then by definition we have $j_q(f)(x) \in \mathscr{R}_q$, $\forall x \in U$.

However, as shown by the following instructive example, the converse is not true in general; that is to say the first property does not always imply the second.

EXAMPLE 1.19 If in example 1.10 above, we take out the equation

$$p_2^2 - \frac{y^2}{x^2} = 0,$$

we get a new system $\mathscr{R}'_1$ which is easily seen to have the same solutions as $\mathscr{R}_1$ and which is not formally integrable because $\mathscr{R}'^{(1)}_1 = \mathscr{R}_1 \subset \mathscr{R}'_1$. In fact this follows from the relations:

$$\frac{d}{dx^1}(p_2^1) - \frac{d}{dx^2}\left(p_1^1 - \frac{y^2}{x^2}\right) = \frac{1}{x^2}\left(p_2^2 - \frac{y^2}{x^2}\right)$$

The definition 1.2 applies both to $\mathscr{R}_1$ and $\mathscr{R}'_1$. However, using matrix notation, the reader will check that the definition 1.13 applies to $\mathscr{R}_1$ because:

$$\begin{bmatrix} \dfrac{z^2}{y^2} & 0 & 0 \\ (z^3 - y^3)\dfrac{z^2}{y^2} & \dfrac{z^2}{y^2} & 0 \\ A & \dfrac{z^3 - y^3}{y^2} & 1 \end{bmatrix} \circ \begin{bmatrix} \dfrac{y^2}{x^2} & 0 & 0 \\ (y^3 - x^3)\dfrac{y^2}{x^2} & \dfrac{y^2}{x^2} & 0 \\ B & \dfrac{y^3 - x^3}{x^2} & 1 \end{bmatrix}$$

$$= \begin{bmatrix} \dfrac{z^2}{x^2} & 0 & 0 \\ (z^3 - x^3)\dfrac{z^2}{x^2} & \dfrac{z^2}{x^2} & 0 \\ C & \dfrac{z^3 - x^3}{x^2} & 1 \end{bmatrix}$$

with

$$C = A\frac{y^2}{x^2} + \frac{(z^3 - y^3)(y^3 - x^3)}{x^2} + B,$$

but does not apply to $\mathscr{R}'_1$ because:

$$\begin{bmatrix} \frac{z^2}{y^2} & 0 & 0 \\ (z^3 - y^3)\frac{z^2}{y^2} & A' & 0 \\ A & \frac{z^3 - y^3}{y^2} & 1 \end{bmatrix} \circ \begin{bmatrix} \frac{y^2}{x^2} & 0 & 0 \\ (y^3 - x^3)\frac{y^2}{x^2} & B' & 0 \\ B & \frac{y^3 - x^3}{x^2} & 1 \end{bmatrix}$$

$$\neq \begin{bmatrix} \frac{z^2}{x^2} & 0 & 0 \\ (z^3 - x^3)\frac{z^2}{x^2} & C' & 0 \\ C & \frac{z^3 - x^3}{x^2} & 1 \end{bmatrix}$$

where (A, A'), (B, B') are arbitrary values given to the parametric jet-coordinates (p_2^2, p_1^3) and $(C, C)')$ the resulting values.

Finally, there is an important case in which both properties of $\mathscr{R}_q$ may hold, and are equivalent.

THEOREM 1.20 In the analytic case, if $\mathscr{R}_q$ is formally integrable, then the two properties of $\mathscr{R}_q$ involved in the definitions 1.2 and 1.13 are equivalent.

Proof It has already been proved that the second implies the first.

Conversely, any point $(x_0, y_0, p_0) \in \mathscr{R}_q$ gives rise because of the formal integrability to a formal solution of $\mathscr{R}_q$ that can be constructed by looking successively for the terms of order $q + 1, q + 2, \ldots$ in the Taylor development.

It is known from the first part of this book that, in the analytic case, such a series is convergent for any x in a sufficiently small neighbourhood U of x_0 in X and determines a solution f of $\mathscr{R}_q$ over U.

If $(y_0, z_0, p'_0) \in \mathscr{R}_q$ is another point, we can, in a similar way, construct a solution g and by shrinking U if necessary, the composition $g \circ f$ is also a solution of $\mathscr{R}_q$ over U.

The theorem now follows from the fact that

$$j_q(g \circ f)(x_0) = j_q(g)(y_0) \circ j_q(f)(x_0)$$

is just the point

$$(x_0, z_0, p'_0 \circ p_0) \in \mathcal{R}_q.$$

REMARK 1.21 In the C^∞ case, the second property is in general the only one that can be checked easily.

From now on we assume that $\mathcal{R}_q \subset J_q(X \times Y)$ is a formally transitive non-linear system of finite Lie equations.

We have now the important theorem:

THEOREM 1.22 In order that a map $\psi \in \text{Aut}(Y)$, $\psi : V \to \psi(V)$ be a finite transformation of Γ, it is necessary and sufficient that the q-prolongation of the map $(\text{id}_X, \psi) \in \text{Aut}(X \times Y)$ leaves $\mathcal{R}_q$ invariant.

Interchanging source and target we obtain.

THEOREM 1.23 In order that a map $\varphi \in \text{Aut}(X)$, $\varphi : U \to \varphi(U)$ be a finite transformation of Γ, it is necessary and sufficient that the q-prolongation of the map $(\varphi, \text{id}_Y) \in \text{Aut}(X \times Y)$ leaves $\mathcal{R}_q$ invariant.

REMARK 1.24 By "*leaves $\mathcal{R}_q$ invariant*" we have to understand "*is a diffeomorphism of $\mathcal{R}_q|_U$ onto $\mathcal{R}_q|_{\varphi(U)}$.*"

We shall prove the first theorem as it is the one that we shall use later on.

Proof Setting $\bar{y} = \psi(y)$ and $\bar{f} = \psi \circ f$, we have the following picture:

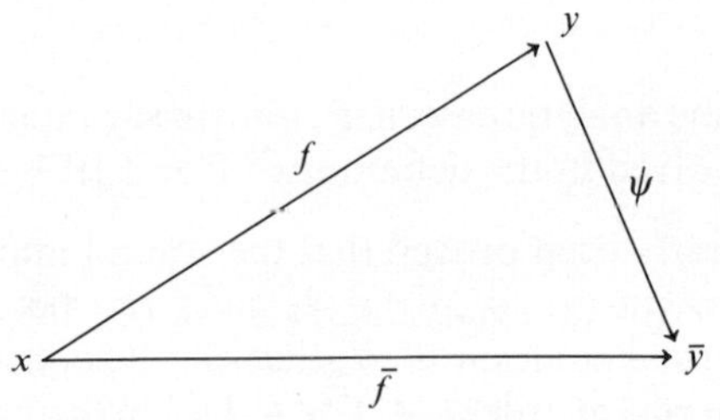

Let us consider $(x, y, p) \in \mathcal{R}_q$ and introduce $\bar{p} = (\partial\psi(y)/\partial y) \circ p$.

N.C. By definition we have $j_q(\psi)(y) \in \mathcal{R}_q$, $\forall y \in V$. From **1** in the definition 1.13, it follows that $(x, \bar{y}, \bar{p}) \in \mathcal{R}_q$.

S.C. By hypothesis, $\forall (x, y, p) \in \mathcal{R}_q$ with $y \in V$, then $(x, \bar{y}, \bar{p}) \in \mathcal{R}_q$ with $\bar{y} \in \psi(V)$. As $\partial\psi(y)/\partial y = \bar{p} \circ p^{-1}$ from **2** in the definition 1.13 and **1**, it follows that $j_q(\psi)(y) \in \mathcal{R}_q$, $\forall y \in V$ and thus ψ is a solution of $\mathcal{R}_q$ over V. C.Q.F.D.

REMARK 1.25 In particular, if:

$$\mathcal{R}_q \qquad \Phi^\tau(x, y, p) = 0$$

then the transformed equations are equivalent to $\Phi^\tau(x, \bar{y}, \bar{p}) = 0$ in the sense that they describe locally the same sub-manifold of $J_q(X \times Y)$.

EXAMPLE 1.26 If we consider the pseudogroup $\Gamma : y = ax + b$ of transformations of $\mathbb{R}$, which is defined by the equations $y'' = 0$, then, for any transformation $\bar{y} = ay + b \in \Gamma$, we have $\partial^2\bar{y}/\partial x^2 = a(\partial^2 y/\partial x^2)$ and thus $\bar{y}'' = ay''$. The system $\bar{y}'' = 0$ is equivalent to the system $y'' = 0$ because $\partial\bar{y}/\partial y = a \neq 0$.

Using the group axioms, we have seen in the preceding chapter of the book that it was possible to give a particular form to the equations defining the action of a Lie group G on X.

We may think that, using also only the group axioms, it will be possible in our case to give a particular form to the finite equations of a given pseudogroup Γ.

In the next pages we shall prove this very important fact.

In order to do so, we first need to establish the infinitesimal counterpart of the last theorems.

Using standard notations, we identify the vector field $\xi = \xi^i(x)\partial/\partial x^i$ section of $T = T(X)$ with the horizontal vector field $\xi = \xi^i(x)\partial/\partial x^i$ section of $T(X \times Y)$. In a similar manner, we identify the vector field $\eta = \eta^k(y)\partial/\partial y^k$ section of $T(Y)$ over Y, with the vertical vector field $\eta = \eta^k(y)\partial/\partial y^k$ section of $T(X \times Y)$.

Using the results of the preceding chapter, we obtain:

THEOREM 1.27 In order that ξ (or η) be an infinitesimal source (or target) transformation belonging to Γ, it is necessary and sufficient that $\rho_q(\xi)$ (or $\rho_q(\eta)$), considered as an infinitesimal transformation of $J_q(X \times Y)$, leaves $\mathcal{R}_q$ invariant.

Locally we see that $\exp(t \cdot \xi) \in \Gamma$ if and only if $(\rho_q(\xi) \cdot \Phi^\tau)(x, y, p) = 0$ whenever $(x, y, p) \in \mathcal{R}_q$ or that $\exp(t \cdot \eta) \in \Gamma$ if and only if $(\rho_q(\eta) \cdot \Phi^\tau)(x, y, p) = 0$ whenever $(x, y, p) \in \mathcal{R}_q$.

EXAMPLE 1.28 Taking example 1.26 again, we have, looking only at target transformations:

$$\eta = (Ay + B)\frac{\partial}{\partial y} \Rightarrow \rho_2(\eta) = (Ay + B)\frac{\partial}{\partial y} + A\frac{\partial}{\partial y'} \quad \text{and} \quad \rho_2(\eta) \cdot y'' = 0$$

We shall now introduce a "*linearisation process*".

We notice that, for the identity transformation, we have

$$\mathrm{id}_q(x) = j_q(\mathrm{id})(x) \qquad \forall x \in X.$$

Now, in order that an infinitesimal transformation $y^k = x^k + t\xi^k(x) + \cdots$ belongs to Γ, we must have:

$$\frac{\partial \Phi^\tau}{\partial y^k}(\mathrm{id}_q(x))\xi^k(x) + \frac{\partial \Phi^\tau}{\partial p^k_\mu}(\mathrm{id}_q(x))\partial_\mu \xi^k(x) = 0$$

These equations are easily obtained by linearisation, taking t small, then dividing by t and finally setting $t = 0$.

From another point of view, using the last theorem and the relations:

$$(\rho_q(\xi)\cdot\Phi)(\mathrm{id}_q(x)) \equiv \xi^i(x)\frac{\partial \Phi}{\partial x^i}(\mathrm{id}_q(x)) - \partial_\mu \xi^k(x)\frac{\partial \Phi}{\partial p^k_\mu}(\mathrm{id}_q(x))$$

$$(\rho_q(\eta)\cdot\Phi)(\mathrm{id}_q(y)) \equiv \eta^k(y)\frac{\partial \Phi}{\partial y^k}(\mathrm{id}_q(y)) + \partial_\mu \eta^k(y)\frac{\partial \Phi}{\partial p^k_\mu}(\mathrm{id}_q(y))$$

we obtain for an infinitesimal source transformation belonging to Γ:

$$\xi^i(x)\frac{\partial \Phi}{\partial x^i}(\mathrm{id}_q(x)) - \partial_\mu \xi^k(x)\frac{\partial \Phi}{\partial p^k_\mu}(\mathrm{id}_q(x)) = 0$$

and for an infinitesimal target transformation belonging to Γ:

$$\eta^k(y)\frac{\partial \Phi}{\partial y^k}(\mathrm{id}_q(y)) + \partial_\mu \eta^k(y)\frac{\partial \Phi}{\partial p^k_\mu}(\mathrm{id}_q(y)) = 0$$

These three linear systems of p.d.e. defining the infinitesimal transformations of Γ are in agreement. In fact, as the identity transformation $y = x$ is a solution of $\mathcal{R}_q$, we have $\Phi^\tau(\mathrm{id}_q(x)) = 0$, $\forall x \in X$.

Differentiating with respect to x, we get:

$$\frac{\partial \Phi^\tau}{\partial x^i}(\mathrm{id}_q(x)) + \frac{\partial \Phi^\tau}{\partial y^i}(\mathrm{id}_q(x)) = 0$$

Moreover, as $\mathcal{R}_q$ is a fibered submanifold of $J_q(X \times Y)$ we have:

$$\mathrm{rank}\left[\frac{\partial \Phi^\tau}{\partial y^k}(x, y, p), \frac{\partial \Phi^\tau}{\partial p^k_\mu}(x, y, p)\right] = \mathrm{rank}\left[\frac{\partial \Phi^\tau}{\partial y^k}(\mathrm{id}_q(x)), \frac{\partial \Phi^\tau}{\partial p^k_\mu}(\mathrm{id}_q(x))\right]$$

$$= \mathrm{rank}\left[\frac{\partial \Phi^\tau}{\partial p^k_\mu}(x, y, p)\right] = \mathrm{rank}\left[\frac{\partial \Phi^\tau}{\partial p^k_\mu}(\mathrm{id}_q(x))\right] = \mathrm{codim}\, \mathcal{R}_q$$

because of the formal transitivity of $\mathcal{R}_q$ and the preceding linear system for ξ defines a vector sub-bundle R_q of $J_q(T)$ that we write again:

$$R_q \qquad \xi^k(x)\frac{\partial\Phi^\tau}{\partial y^k}(\mathrm{id}_q(x)) + \xi^k_\mu(x)\frac{\partial\Phi^\tau}{\partial p^k_\mu}(\mathrm{id}_q(x)) = 0$$

DEFINITION 1.29 The linear system R_q is said to define the "*infinitesimal transformations*" of Γ. We also say that R_q is a linear system of "*infinitesimal Lie equations*". We have $R_q = \mathrm{id}_q^{-1}(V(\mathcal{R}_q))$.

REMARK 1.30 We can consider this system on $T = T(X)$ as defining the infinitesimal source transformations of Γ and denote it by $R_q = R_q(X)$. We can also consider it on $T(Y)$ as defining the infinitesimal target transformations of Γ and denote it by $R_q(Y)$.

We have the proposition:

PROPOSITION 1.31 If $\mathcal{R}_q$ is a formally transitive system of finite Lie equations, then the morphism $\pi^q_0 : R_q \to T$ induced by the epimorphism $\pi^q_0 : J_q(T) \to T$ is also an epimorphism.

Proof Let

$$\mathcal{R}_q \qquad \Phi^\tau(x, y, p) = 0$$

By hypothesis we may solve these equations with respect to some principal p, $\forall x \in X$, $\forall y \in Y$. It follows that we must have:

$$\mathrm{rank}\left[\frac{\partial\Phi^\tau}{\partial p^k_\mu}(x, y, p)\right] = \mathrm{rank}\left[\frac{\partial\Phi^\tau}{\partial p^k_\mu}(\mathrm{id}_q(x))\right] = \mathrm{codim}\,\mathcal{R}_q,\ \forall (x, y, p) \in \mathcal{R}_q$$

As $\mathrm{codim}_{J_q(X\times Y)}\,\mathcal{R}_q = \mathrm{codim}_{J_q(T)}\,R_q$, using Gramer's rules we may also solve with respect to the corresponding principal ξ^k_μ the equations for a section ξ_q of R_q:

$$R_q \qquad \frac{\partial\Phi^\tau}{\partial p^k_\mu}(\mathrm{id}_q(x))\xi^k_\mu + \xi^k\frac{\partial\Phi^\tau}{\partial y^k}(\mathrm{id}_q(x)) = 0$$

and from the rank condition, this is possible $\forall \xi = \pi^q_0 \circ \xi_q$. C.Q.F.D.

DEFINITION 1.32 We define the vector bundle R^0_q over X by the short exact sequence:

$$0 \longrightarrow R^0_q \longrightarrow R_q \xrightarrow{\pi^q_0} T \longrightarrow 0$$

and we say that the linear system R_q is "*formally transitive*".

THEOREM 1.33 If ξ_1 and ξ_2 are two sections of T solutions over a neighbourhood U of a point $x \in X$, then the bracket $[\xi_1, \xi_2]$ is also a solution of R_q over U.

DEFINITION 1.34 The differential operator $\mathscr{D}$ determined by R_q will be called a "*Lie operator*". It thus has the property that, if ξ_1 and ξ_2 are two sections of T such that $\mathscr{D} \cdot \xi_1 = 0$ and $\mathscr{D} \cdot \xi_2 = 0$, then $\mathscr{D} \cdot [\xi_1, \xi_2] = 0$. This is the most natural generalisation of the classical Lie derivative already defined.

EXAMPLE 1.35 If ω is an exterior form we may write $\mathscr{D} \cdot \xi \equiv \mathscr{L}(\xi) \cdot \omega$ as a Lie operator.

Proof From theorem 1.27 we see that both $\rho_q(\xi_1)$ and $\rho_q(\xi_2)$ are tangent to $\mathscr{R}_q$ at each point $(x, y, p) \in \mathscr{R}_q$. In particular, using proposition 6.3.39, we conclude that $[\rho_q(\xi_1), \rho_q(\xi_2)]$ is also tangent to $\mathscr{R}_q$ at each point $(x, y, p) \in \mathscr{R}_q$. But from proposition 6.5.14 we know that:

$$[\rho_q(\xi_1), \rho_q(\xi_2)] = \rho_q([\xi_1, \xi_2])$$

and thus

$$(\rho_q([\xi_1, \xi_2]) \cdot \Phi^\tau)(\mathrm{id}_q(x)) = 0 \qquad \forall x \in X$$

which concludes the proof of the theorem. C.Q.F.D.

The reader will easily prove that the infinitesimal analogue of theorem 1.22 is the following theorem, where $J_q(T(Y))$ stands for the bundle of q-jets of $T(Y)$ with respect to Y.

THEOREM 1.36 When $\mathscr{R}_q$ is a non-linear system of finite Lie equations, in order that a section η_q of $J_q(T(Y))$ over $V \subset Y$ be a section of $R_q(Y)$ over $V \subset Y$ it is necessary and sufficient that $\#(\eta_q)$ leaves $\mathscr{R}_q$ invariant, i.e. is tangent to $\mathscr{R}_q$ at any point $(x, y, p) \in \mathscr{R}_q$ with $y \in V$.

REMARK 1.37 A similar theorem can be stated with source and target interchanged, but we shall not use it in the sequel.

From the last theorem we deduce the following important one:

THEOREM 1.38 Let $\mathscr{R}_q$ be a non-linear system of finite Lie equations and consider the corresponding linearised system R_q. If η_q and $\bar{\eta}_q$ are two sections if $R_q(Y)$ over $V \subset Y$, then there exists a section $[\eta_q, \bar{\eta}_q]$ of $R_q(Y)$ over $V \subset Y$ such that $[\#(\eta_q), \# \bar{\eta}_q] = \#([\eta_q, \bar{\eta}_q])$.

Proof As we have seen that $\#(\eta_q)$ and $\#(\bar{\eta}_q)$ are tangent to $\mathscr{R}_q$ at each point $(x, y, p) \in \mathscr{R}_q$, it follows from proposition 6.3.39 that $[\#(\eta_q), \#(\bar{\eta}_q)]$ is also tangent to $\mathscr{R}_q$ at each point $(x, y, p) \in \mathscr{R}_q$.

It remains to show that it is the image by # of a section of $R_q(Y)$ over $V \subset Y$.

Because of the preceding theorem, it only remains to show that it is the image by # of a section of $J_q(T(Y))$ over $V \subset Y$.

By definition we have:

$$\begin{cases} \#(\eta_q) = \eta^k(y)\dfrac{\partial}{\partial y^k} + \eta^k_\mu(y)\cdot B^\mu_k(q) \\ \#(\bar{\eta}_q) = \bar{\eta}^l(y)\dfrac{\partial}{\partial y^\rho} + \bar{\eta}^l_\nu(y)\cdot B^\nu_\rho(q) \end{cases}$$

Then:

$$\begin{aligned}[\#(\eta_q), \#(\bar{\eta}_q)] &= \left(\eta^l(y)\frac{\partial\bar{\eta}^k(y)}{\partial y^l} - \bar{\eta}^l(y)\frac{\partial\eta^k(y)}{\partial y^l}\right)\frac{\partial}{\partial y^\rho} \\ &+ \left(\eta^l(y)\frac{\partial\bar{\eta}^k_\mu(y)}{\partial y^\rho} - \bar{\eta}^l(y)\frac{\partial\eta^k_\mu(y)}{\partial y^l}\right)B^\mu_k(q) \\ &+ \eta^k_\mu(y)\bar{\eta}^l_\nu(y)[B^\mu_k(q), B^\nu_l(q)]\end{aligned}$$

and the theorem follows directly from proposition 6.5.23. However we shall give an explicit formula for the last term. A straightforward computation shows that:

$$\begin{aligned}\eta^k_\mu(y)\bar{\eta}^l_\nu(y)[B^\mu_k(q), B^\nu_l(q)] = &\sum_{\substack{1\le|\mu|\le q \\ 0\le|\nu|\le q-1 \\ 1\le|\mu|+|\nu|\le q}} (\eta^k_\mu(y)\bar{\eta}^l_{\nu+1_k}(y) - \bar{\eta}^k_\mu(y)\eta^l_{\nu+1_k}(y)) \\ &\times \frac{\mu+\nu!}{\mu!\nu!} B^{\mu+\nu}_l(q)\end{aligned}$$

and this concludes the proof of the theorem as we have given the full expression of $[\eta_q, \bar{\eta}_q]$. C.Q.F.D.

REMARK 1.39 It has to be noticed that $[\eta_q, \bar{\eta}_q]$ involved the differentials of η_q and $\bar{\eta}_q$ with respect to y.

REMARK 1.40 If we interchange source and target, we may obtain a similar result because the commutation relations of the vector fields $A^\mu_k(q)$ and $B^\mu_k(q)$ are the same. The property of the linear system R_q, expressed by the preceding theorem will be written in brief:

$$[R_q, R_q] \subset R_q$$

We shall now try to get results from this property.

When we vary the section η_q, then $\#(\eta_q)$ gives rise to a dim R_q distribution $\#(R_q)$ of vector fields on $J_q(X \times Y)$.

THEOREM 1.41 If $\mathscr{R}_q$ is a non-linear system of finite Lie equations and R_q is the corresponding linearised system, then $\#(R_q)$ is an involutive distribution on $J_q(X \times Y)$.

Proof The proof can be deduced easily from that of the last theorem. We may even determine a basis explicitly by choosing the parametric and principal jet coordinates relative to a solved form of R_q. It follows that the bracket of two vector fields of such a basis is a linear combination of the vector fields of the basis with coefficients belonging to $C^\infty(Y)$. C.Q.F.D.

In practice we have to operate as follows:

1 Find out $\mathscr{R}_q \subset I_q(X \times Y) \subset J_q(X \times Y)$
2 Obtain $R_q \subset J_q(T)$ by linearisation.
3 Consider $R_q(Y) \subset J_q(T(Y))$ by using η and y instead of ξ and x.
4 Exhibit a solved form.
5 Construct a basis of $\#(R_q)$.

The following example will make clear this process, though we just indicate the results.

EXAMPLE 1.42

1 We take Γ and $\mathscr{R}_1$ as in example 1.10, with $X = U \subset \mathbb{R}^3$, we have for $x^2 \neq 0$:

$$\mathscr{R}_1 \begin{cases} p_3^3 - 1 = 0, \quad p_3^2 = 0, \quad p_3^1 = 0, \quad p_2^3 - \dfrac{y_3 - x_3}{x^2} = 0 \\[2ex] p_2^2 - \dfrac{y^2}{x^2} = 0, \quad p_2^1 = 0, \quad p_1^2 - (y^3 - x^3)\dfrac{y^2}{x^2} = 0, \quad p_1^1 - \dfrac{y^2}{x^2} = 0 \end{cases}$$

2 We obtain for $x^2 \neq 0$, the formally transitive equations.

$$R_1 \begin{cases} \xi_3^3 = 0, \quad \xi_3^2 = 0, \quad \xi_3^1 = 0 \\[2ex] \xi_2^3 - \dfrac{\xi^3}{x^2} = 0, \quad \xi_2^2 - \dfrac{\xi^2}{x^2} = 0, \quad \xi_2^1 = 0 \\[2ex] \xi_1^2 - \xi^3 = 0, \quad \xi_1^1 - \dfrac{\xi^2}{x^2} = 0. \end{cases}$$

3 → 4 We have just to rewrite R_1, already in solved form.

5 A basis of $\#(R_1)$ can be made of the 4 vector fields:

$$\#(R_1)\begin{cases} \dfrac{\partial}{\partial y^1}, \dfrac{\partial}{\partial y^2} + \dfrac{1}{y^2}\left(p_1^1 \dfrac{\partial}{\partial p_1^1} + p_2^1 \dfrac{\partial}{\partial p_2^1} + p_3^1 \dfrac{\partial}{\partial p_3^1} \right. \\ \qquad\qquad \left. + p_1^2 \dfrac{\partial}{\partial p_1^2} + p_2^2 \dfrac{\partial}{\partial p_2^2} + p_3^2 \dfrac{\partial}{\partial p_3^2}\right) \\ \dfrac{\partial}{\partial y^3} + \dfrac{1}{y^2}\left(p_1^2 \dfrac{\partial}{\partial p_1^3} + p_2^2 \dfrac{\partial}{\partial p_2^3} + p_3^2 \dfrac{\partial}{\partial p_3^3}\right) \\ \qquad\qquad + p_1^1 \dfrac{\partial}{\partial p_1^2} + p_2^1 \dfrac{\partial}{\partial p_2^2} + p_3^1 \dfrac{\partial}{\partial p_3^2} \\ p_1^1 \dfrac{\partial}{\partial p_1^3} + p_2^1 \dfrac{\partial}{\partial p_2^3} + p_3^1 \dfrac{\partial}{\partial p_3^3} \end{cases}$$

It is easy to check that $\#(R_1)$ is an involutive distribution. We obtain in particular:

$$\left[\frac{\partial}{\partial y^2} + \frac{1}{y^2}(\quad), p_1^1 \frac{\partial}{\partial p_1^3} + p_2^1 \frac{\partial}{\partial p_2^3} + p_3^1 \frac{\partial}{\partial p_3^3}\right]$$
$$= \frac{1}{y^2}\left(p_1^1 \frac{\partial}{\partial p_1^3} + p_2^1 \frac{\partial}{\partial p_2^3} + p_3^1 \frac{\partial}{\partial p_3^3}\right)$$

EXAMPLE 1.43 $X = \mathbb{R}^2$, $\Gamma\colon y^1 = x^1 + a$, $y^2 = x^2 + f(x^1)$

1 R_1: $\quad p_1^1 - 1 = 0,\ \ p_2^1 = 0,\ \ p_2^2 - 1 = 0$

2 R_1: $\quad \xi_1^1 = 0,\ \ \xi_2^1 = 0,\ \ \xi_2^2 = 0$

3 → 4 R_1 is already in solved form.

5 $\#(R_1)$: $\quad \dfrac{\partial}{\partial y^1},\ \dfrac{\partial}{\partial y^2},\ p_1^1 \dfrac{\partial}{\partial p_1^2} + p_2^1 \dfrac{\partial}{\partial p_2^2}$

According to the preceding chapter, as we have an involutive distribution on $J_q(X \times Y)$ tangent to $\mathcal{R}_q$ at each point $(x, y, p) \in \mathcal{R}_q$, we shall apply the Frobenius theorem in order to construct $\mathcal{R}_q$ locally, in a neighbourhood of the point $\mathrm{id}_q(x)$, with a given point $x \in X$.

We notice that any vector field of $\#(R_q)$ has a zero projection on $T(X)$ and that this fact is independent of the coordinate system.

As $\dim R_q = \dim \mathcal{R}_q - n$, we see that for fixed x, $\mathcal{R}_{q;x}$ is a maximum manifold for the involutive distribution $\#(R_q)$.

According to the definition of $\#(R_q)$ we can exhibit a basis with components depending only on y, p and not on x. From the Frobenius theorem we can choose $m = \operatorname{codim} \mathscr{R}_q$ unconnected invariants $\Phi^\tau(y, p)$ of $\#(R_q)$.

As $\mathscr{R}_{q;x}$ contains the point $\mathrm{id}_q(x)$, it is defined in a neighbourhood of this point by the equations

$$\Phi^\tau(y, p) = \Phi^\tau(\mathrm{id}_q(x)) = \omega^\tau(x).$$

It follows that we may define $\mathscr{R}_q$ locally by equations of the form:

$$\mathscr{R}_q \qquad \Phi^\tau(y, p) = \omega^\tau(x)$$

DEFINITION 1.44 Because of their definition, the functions

$$\Phi^\tau \in C^\infty(J_q(X \times Y))$$

are called "*differential invariants of order q*" and they are such that, if $\bar{y} = \psi(y)$ is any finite transformation of Γ, using the notations of theorem 1.22, we have:

$$\Phi^\tau\left(\bar{y}, \frac{\partial \bar{y}}{\partial x}\right) = \Phi^\tau\left(y, \frac{\partial y}{\partial x}\right), \qquad \forall (x, y, p) \in J_q(X \times Y)$$

The functions $\omega^\tau \in C^\infty(X)$ are called the "*values*" of the differential invariant at the q-jet of the identity. Finally, the particular form given to the equations of $\mathscr{R}_q$ is called "*Lie form*".

REMARK 1.45 The Lie form has the advantage to push out the x in the right members.

The Lie form for the finite equations of a Lie pseudogroup Γ will be fundamental in the sequel. We ask the reader to find it and to draw a picture in the case $X = \mathbb{R}$, $\Gamma : y = ax$.

2 Special and general Lie equations

From now on we shall work mainly using local coordinates. For simplicity we shall not however always specify the coordinate domains.

We also note that, up to now, we have not required that the systems $\mathscr{R}_q$ or R_q be formally integrable or involutive.

Let us consider an arbitrary section ξ of T. Then

$$\rho_q(\xi) = \xi^i(x)\frac{\partial}{\partial x^i} + \partial_\mu \xi^k(x) A_k^\mu(q)$$

From proposition 6.5.17 we have $[A_k^\mu(q), B_l^\nu(q)] = 0$ and thus it is easy to check that:

$$[A_k^\mu(q), B(q)] = 0 \quad \text{and} \quad [\rho_q(\xi), B(q)] = 0$$

for any vector field $B(q)$ of a basis of $\#(R_q)$, as it has already been constructed.

Now, for any differential invariant Φ^τ, we have:

$$[A_k^\mu(q), B(q)] \cdot \Phi^\tau = -B(q) \cdot (A_k^\mu(q) \cdot \Phi^\tau) = 0 \quad \text{because} \quad B(q) \cdot \Phi^\tau = 0.$$

As the Φ^τ has been chosen in such a way as to constitute a maximum set of unconnected invariants of $\#(R_q)$, we must have locally a functional relation such as:

$$A_k^\mu(q) \cdot \Phi^\tau \equiv L_k^{\tau\mu}(\Phi^1, \ldots, \Phi^m)$$

We may introduce, on a convenient open set of $\mathbb{R}^m$, the vector fields

$$L_k^\mu = L_k^{\tau\mu}(u) \frac{\partial}{\partial u^\tau}$$

using new variables

$$u^1, \ldots, u^m \quad \text{with} \quad u = (u^1, \ldots, u^m) \in \mathbb{R}^m.$$

PROPOSITION 2.1 The vector fields L_k^μ are the infinitesimal generators of a locally transitive Lie group of transformations in m variables, with

$$n \cdot \frac{(n+q)!}{q!n!} - n$$

parameters.

Proof According to their definition, the vector fields L_k^μ have the same commutation relations as the vector fields $A_k^\mu(q)$ and thus generate a Lie group of transformation, as this follows from the first fundamental theorem 6.2.1.

Now, if the matrix of the components of the L_k^μ on the basis $\partial/\partial u^\tau$, that is to say $[L_k^{\tau\mu}(u)]$, has maximum rank less than m, we may find a non-constant function $h(u)$ such that

$$L_k^\mu \cdot h = L_k^{\tau\mu}(u) \frac{\partial h(u)}{\partial u^\tau} = 0$$

and we should have:

$$A_k^\mu(q) \cdot h(\Phi^1, \ldots, \Phi^m) \equiv 0.$$

As we have already seen that the vector fields $A_k^\mu(q)$ are the infinitesimal generators of a transitive action on the fiber of $J_q(X \times Y)$, the function $h(\Phi^1(y, p), \ldots, \Phi^m(y, p))$ must not contain p and we should obtain an invariant function $\Psi(y)$. Among the equations of $\mathcal{R}_q$ we could find $\Psi(y) = \Psi(x)$, contradicting the formal transitivity.

It follows that the infinitesimal generators L_k^μ have rank m and generate a transitive action, at least on an open set of $\mathbb{R}^m$ where the matrix $[L_k^{\tau\mu}(u)]$ has maximum rank equal to m. C.Q.F.D.

EXAMPLE 2.2 We consider again the example of theorem 1.41. We may take:

$$\Phi^1 \equiv p_1^1, \quad \Phi^2 \equiv p_2^1, \quad \Phi^3 \equiv p_1^1 p_2^2 - p_2^1 p_1^2$$

as a maximum set of differential invariants of order 1. Then

$$\begin{cases} -L_1^1 = u^1 \dfrac{\partial}{\partial u^1} + u^3 \dfrac{\partial}{\partial u^3} & -L_2^1 = u^1 \dfrac{\partial}{\partial u^2} \\ -L_1^2 = u^2 \dfrac{\partial}{\partial u^1} & -L_2^2 = u^2 \dfrac{\partial}{\partial u^2} + u^3 \dfrac{\partial}{\partial u^3} \end{cases}$$

and the matrix $[L_k^{\tau\mu}(u)]$ of the components is, up to the sign:

$$\begin{bmatrix} u^1 & u^2 & 0 & 0 \\ 0 & 0 & u^1 & u^2 \\ u^3 & 0 & 0 & u^3 \end{bmatrix}$$

Thus the action is transitive, unless $\{u^3 = 0\}$ or $\{u^1 = 0, u^2 = 0\}$. We let the reader draw the corresponding picture in $\mathbb{R}^3$.

REMARK 2.3 The L_k^μ do not always generate an effective action. For example, we may take $X = \mathbb{R}^2$ and Γ the Lie pseudogroup defined by the non-linear system:

$$\mathcal{R}_1: \qquad p_1^1 p_2^2 - p_2^1 p_1^2 = 1$$

We have the same set of components as in the bottom row of the above matrix and $L_1^1 - L_2^2 = 0$.

If now we effect an infinitesimal source transformation

$$x \to \bar{x} = x + t\xi(x) + \cdots$$

then the Φ^τ are transformed according to:

$$\Phi^\tau \to \Phi^\tau + t\partial_\mu \xi^k(x) L_k^{\tau\mu}(\Phi^1, \ldots, \Phi^m) + \cdots$$

and we may write this transformation law, introducing the variables $u^1, \ldots, u^m$:

$$u^\tau \to \bar{u}^\tau = u^\tau + t\partial_\mu \xi^k(x) L_k^{\tau\mu}(u) + \cdots$$

EXAMPLE 2.4 If we consider the Jacobian as a differential invariant in the remark above, we have:

$$u \to \bar{u} = u + t(\partial_1 \xi^1(x) + \partial_2 \xi^2(x))u + \cdots$$

Similarly, passing to the finite point of view, if we effect a finite source transformation $x \to \bar{x} = \varphi(x)$ and if we just write first order jets for simplicity, then, using the composition rule for differentiation, that is to say

$$\frac{\partial y}{\partial \bar{x}} \cdot \frac{\partial \varphi(x)}{\partial x} = \frac{\partial y}{\partial x}, \ldots$$

and the fact that the source x does not appear explicitly in the differential invariants, we may express the new transformed differential invariants $\Phi^\tau(y, \partial y/\partial \bar{x})$ as functions of the initial ones $\Phi^\tau(y, \partial y/\partial x)$ and the derivatives $\partial\varphi(x)/\partial x$, writing without any index:

$$\Phi\left(y, \frac{\partial y}{\partial \bar{x}}\right) = G\left(\Phi\left(y, \frac{\partial y}{\partial x}\right), \frac{\partial \varphi(x)}{\partial x}\right)$$

We leave the details to the reader as an exercise.

As above, introducing the variables $u^1, \ldots, u^m$ we may say that, under a change of source the differential invariants are transformed according to the following transformation law:

$$\mathscr{F}\begin{cases} \bar{u} = G(u, \partial_\mu \varphi^k(x)) \\ \bar{x} = \varphi(x) \end{cases}$$

EXAMPLE 2.5 We consider again the last example. We have:

$$\frac{\partial(y^1, y^2)}{\partial(\bar{x}^1, \bar{x}^2)} = \frac{1}{\partial(\varphi^1(x), \varphi^2(x))/\partial(x^1, x^2)} \cdot \frac{\partial(y^1, y^2)}{\partial(x^1, x^2)}$$

and

$$\begin{cases} \bar{u} = \dfrac{u}{\partial(\varphi^1(x), \varphi^2(x))/\partial(x^1, x^2)} \\ \bar{x} = \varphi(x) \end{cases}$$

In particular, because the source variables do not appear explicitly in the differential invariants, we may consider, as finite source transformations, the transition functions of X associated to an open covering $\{U_\alpha\}$ and thus construct, according to theorem 1.1.18, a fiber bundle $\mathscr{F}$ over X, by patching together trivial bundles such as $U_\alpha \times$ open set of $\mathbb{R}^m$.

The dimension of the fiber of $\mathscr{F}$ is m. The Lie group acting on the fiber is $GL_q(n, \mathbb{R})$. The group law is just the composition law of jets of order q of invertible transformations and the infinitesimal generators of the action are the L_k^μ.

Convenient local coordinates for $\mathscr{F}$ are (x^i, u^τ). Moreover

$$\omega : (x) \to (x, \omega^\tau(x))$$

is a local section of this bundle that can be expressed in different coordinate systems, using the transition functions of $\mathscr{F}$ given above.

However, as our study is dealing mainly with local questions, in order to simplify the notations, we shall just say that ω is a section of $\mathscr{F}$, shrinking X if necessary to an open set $U \subset X$.

Of course, as we may choose different maximum sets of unconnected differential invariants, the bundle $\mathscr{F}$ is defined up to an X-isomorphism.

Moreover, as a tensor bundle is a very particular case of such a bundle $\mathscr{F}$, we shall adopt the following definition:

DEFINITION 2.6 The bundle $\mathscr{F}$ will be called a "*bundle of geometric objects*".

EXAMPLE 2.7 The transformation law of the Christoffel symbols, already given in the first part of this book provides a good example of a bundle of geometric objects which is not a tensor bundle. The reader will find many other examples in the sequel.

Summarising what we have said, we obtain:

THEOREM 2.8 If $\mathscr{R}_q$ is any non-linear, formally transitive system of finite Lie equations defining a pseudogroup Γ, then it is possible to construct a bundle $\mathscr{F}$ of geometric objects, which is defined up to an X-isomorphism, an epimorphism $\Phi: J_q(X \times Y) \to \mathscr{F}$ and a section ω of $\mathscr{F}$ over X such that we have the following exact sequence of fibered manifolds over X:

$$0 \longrightarrow \mathscr{R}_q \longrightarrow I_q(X \times Y) \underset{\omega \circ \pi}{\overset{\Phi}{\rightrightarrows}} \mathscr{F}$$

where we have denoted by Φ the restriction to $I_q(X \times Y) \subset J_q(X \times Y)$ of the epimorphism $\Phi: (x, y, p) \to (x, \Phi^\tau(y, p))$ defined by means of a convenient maximum set of unconnected differential invariants of Γ.

REMARK 2.9 The above proof, though it may be not rigorous enough for a global study, is quite sufficient for us, as we will be concerned with local problems where one just needs to know what happens under any change of coordinates.

However, though the arguments involved will appear later on, we shall now give a precise geometric definition of $\mathscr{F}$ that will be found in the sequel.

If we identify a map $f: X \to Y$ with its graph $f: X \to X \times Y$ we shall identify $J_q(X \times Y)$ with the bundle $J_q(X, Y)$ of q-jets of maps from X to Y. As before we shall denote by $I_q(X \times Y) = I_q(X, Y)$ the corresponding bundle of q-jets of invertible maps.

We shall now introduce the bundle $J_q(X, y_0)$ of q-jets of maps from X to a fixed target $y_0 \in Y$ with structure group $GL_q(n, \mathbb{R})$.

Finally, with any non-linear system $\mathscr{R}_q \subset I_q(X \times Y)$ of finite Lie equations, we associate the non-linear system $\mathscr{R}_q(X, y_0) \subset I_q(X, y_0)$ defining the q-jets of the finite transformations of Γ with fixed target $y_0 \in Y$. Taking $x_0 = y_0$ we may also introduce the Lie group

$$\mathscr{R}_q(x_0, y_0) \subset I_q(x_0, y_0) = GL_q(n, \mathbb{R}).$$

This Lie group acts on $\mathscr{R}_q(X, y_0)$ and $I_q(X, y_0)$ by the ordinary composition law of jets, the operation being produced by jets with source and target at y_0.

Though this is not evident a priori, we have:

$$\mathscr{F} = I_q(X, y_0)/\mathscr{R}_q(x_0, y_0) \qquad \text{with } x_0 = y_0$$

Defining R_q^0 as the kernel of the epimorphism $\pi_0^q : R_q \to T$, we shall even prove later on that one just needs in fact the knowledge of the Lie algebra $R_{q;x_0}^0$ of $\mathscr{R}_q(x_0, y_0)$ in order to construct $\mathscr{F}$.

At this time the projection of $\mathscr{R}_q(X, y_0)$ gives rise to a section ω of $\mathscr{F}$. Of course we may act on $I_q(X, y_0)$ and thus on $\mathscr{F}$ by source transformations.

All what we have done from now on was to find out a convenient system of local coordinates, on a neighbourhood of the point $(x_0, u_0) = \omega(x_0) \in \mathscr{F}$, in such a way that a change of these local coordinates was determined by the change of the differential invariants.

Using these local coordinates we detail the above definitions:

$$\mathscr{R}_q(X, y_0) \qquad \Phi^\tau(y_0, p) = \omega^\tau(x)$$

$$\mathscr{R}_q(x_0, y_0) \qquad \Phi^\tau(y_0, p) = \omega^\tau(x_0) \quad \text{with } x_0 = y_0$$

A useful example in the sequel will be the case

$$q = 1, \quad \mathscr{R}_q(x_0, y_0) = \mathrm{id} \subset GL(n, \mathbb{R}), \qquad R_1^0 = 0.$$

Then it is easy to see that $\mathscr{F}$ is the fibered product over X of n copies of T^*. If $X = \mathbb{R}^n$ we may take for Γ the pseudogroup of translations defined by $\partial y^k/\partial x^i = \delta_i^k$ in cartesian coordinates.

Finally, using a condensed notation, we can exhibit the following sequence:

$$\mathrm{id} \longrightarrow \Gamma \longrightarrow \mathrm{Aut}\,(X) \overset{\Phi \circ j_q}{\underset{\omega \circ \pi}{\rightrightarrows}} \mathscr{F}$$

where we denote simply by Γ the set of invertible sections of $X \times Y$ over any open set $U \subset X$ that belong to the kernel of the non-linear differential operator $\Phi \circ j_q$ with respect to the section ω of $\mathscr{F}$ over U.

In fact, if f is such a section, we have:

$$\Phi^\tau(f^k(x), \partial_\mu f^k(x)) \equiv \omega^\tau(x) \qquad \forall x \in U \subset X$$

The following converse theorem is very useful because it describes the kind of situation which is usually found in physics.

CONVERSE THEOREM 2.10 To any section ω of a given bundle $\mathscr{F}$ of geometric objects of order q, there corresponds a non-linear system $\mathscr{R}_q$ of finite Lie equations. The invertible solutions of $\mathscr{R}_q$ are the finite transformations of a Lie pseudogroup Γ.

Proof Let

$$\mathscr{F}\begin{cases}\bar{u} = G(u, \partial_\mu \varphi^k(x))\\ \bar{x} = \varphi(x)\end{cases}$$

be the transition functions of $\mathscr{F}$.

For any section ω of $\mathscr{F}$ over X we get locally the following linear system:

$$\mathscr{R}_q \qquad \omega(y) = G(\omega(x), p^k_\mu)$$

and we have just to prove that it is a non-linear system of finite Lie equations.

This follows easily from the group property of the transition functions. In fact, if:

$$\omega(y) = G\left(\omega(x), \frac{\partial y}{\partial x}\right)$$

$$\omega(z) = G\left(\omega(y), \frac{\partial z}{\partial y}\right)$$

then

$$\omega(z) = G\left(G\left(\omega(x), \frac{\partial y}{\partial x}\right), \frac{\partial z}{\partial y}\right) = G\left(\omega(x), \frac{\partial z}{\partial y} \cdot \frac{\partial y}{\partial x}\right)$$
$$= G\left(\omega(x), \frac{\partial z}{\partial x}\right)$$

where the composition of jets of invertible transformations is done according to the usual rules.

Finally:

$$\omega(x) = G^{-1}\left(\omega(y), \frac{\partial y}{\partial x}\right) = G\left(\omega(y), \frac{\partial x}{\partial y}\right)$$

with

$$\left(x, x, \frac{\partial x}{\partial y} \cdot \frac{\partial y}{\partial x}\right) = \mathrm{id}_q(x). \qquad \text{C.Q.F.D.}$$

EXAMPLE 2.11 The best known examples are the following bundles of geometric objects;

$$T, \quad T^*, \quad S_2 T^*, \quad \Lambda^2 T^*, \quad \Lambda^n T^*.$$

PROPOSITION 2.12 If we omit the index τ and introduce the inverse formulas

$$u = G^{-1}(\bar{u}, \partial_\mu \varphi^k(x))$$

then we have

$$\Phi(y, p^k_\mu) \equiv G^{-1}(\omega(y), p^k_\mu)$$

Proof By definition, using a condensed notation, we have

$$\Phi\left(y, \frac{\partial y}{\partial x}\right) \equiv G^{-1}\left(\Phi\left(y, \frac{\partial y}{\partial \bar{x}}\right), \frac{\partial \bar{x}}{\partial x}\right)$$

where

$$\frac{\partial y}{\partial \bar{x}} \cdot \frac{\partial \bar{x}}{\partial x} = \frac{\partial y}{\partial x}.$$

Taking for $\partial y/\partial \bar{x}$ the q-jet of the identity at $y = \bar{x}$, we obtain the desired result. C.Q.F.D.

REMARK 2.13 This proposition will be a key trick in the sequel. From now on, when there is no confusion, we shall write simply Φ instead of G^{-1}.

DEFINITION 2.14 We say that a pseudogroup $\bar{\Gamma}$ is "*similar*" to a pseudogroup Γ if $\bar{\Gamma} = \varphi \circ \Gamma \circ \varphi^{-1}$ with $\varphi \in \mathrm{Aut}\,(X)$.

The former properties of the equations defining $\mathscr{R}_q$ will allow us to study all the pseudogroups $\bar{\Gamma}$ similar to Γ. For this we have to consider, at the same time, a finite source transformation $\bar{x} = \varphi(x)$ and the finite target transformation $\bar{y} = \varphi(y)$.

DEFINITION 2.15 We define the new section $\bar{\omega} = \varphi(\omega)$ of $\mathscr{F}$ by:

$$\bar{\omega}(\varphi(x)) \equiv G\left(\omega(x), \frac{\partial \varphi(x)}{\partial x}\right)$$

LEMMA 2.16 $(\psi \circ \varphi)(\omega) = \psi(\varphi(\omega)), \qquad \forall \varphi, \psi \in \mathrm{Aut}\,(X).$

Proof Setting $y = \varphi(x)$ and $z = \psi(y)$ we have:

$$\begin{aligned}(\psi \circ \phi)(\omega)(z) = G\left(\omega(x), \frac{\partial z}{\partial x}\right) &= G\left(\omega(x), \frac{\partial z}{\partial y} \cdot \frac{\partial y}{\partial x}\right)\\ &= G\left(G\left(\omega(x), \frac{\partial y}{\partial x}\right), \frac{\partial z}{\partial y}\right)\\ &= G\left(\varphi(\omega)(y), \frac{\partial z}{\partial y}\right)\\ &= \psi(\varphi(\omega))(z).\end{aligned}$$

C.Q.F.D.

IMPORTANT REMARK 2.17 The preceding definition is a key to the understanding of the notion of Lie derivative that will be used in the sequel. It gives a process associating, with any $\varphi \in \mathrm{Aut}\,(X)$, an action on $\mathscr{F}$. This is in fact a deep idea and we ask the reader to think about it. He may consider the following analogy: in the plane $\mathbb{R}^2$, we can rotate a cartesian frame with a certain angle or rotate a vector, backwards with the same angle, without changing its components. This is the kind of argument to be considered hereafter, where we use the transition laws of a bundle of geometric objects in order to get an action on it.

THEOREM 2.18 If the non-linear system defining Γ is:

$$\mathscr{R}_q \qquad \Phi(\omega(y), p) = \omega(x)$$

then the non-linear system defining $\bar{\Gamma}$ is:

$$\bar{\mathscr{R}}_q = \mathrm{Ad}\,\varphi(\mathscr{R}_q) \qquad \Phi(\bar{\omega}(y), p) = \bar{\omega}(x)$$

Proof Using the preceding lemma we have:

$$\begin{aligned}\bar{\Gamma}(\bar{\omega}) = (\varphi \circ \Gamma \circ \varphi^{-1})(\bar{\omega}) &= \varphi(\Gamma(\varphi^{-1}(\bar{\omega})))\\ &= \varphi(\Gamma(\omega))\\ &= \varphi(\omega) = \bar{\omega}\end{aligned}$$

C.Q.F.D.

Let us consider now any section ξ of T. If $\varphi_t = \exp\,(t\xi)$ we can define a deformed section ω_t of $\mathscr{F}$ with

$$\omega_t = \omega + t\underset{1}{\Omega} + \cdots$$

by the identity

$$\omega_t(x) \equiv G^{-1}\left(\omega(\varphi_t(x)), \frac{\partial \varphi_t(x)}{\partial x}\right)$$

For t small, we may linearise this relation and get, looking only at the first order terms in t:

$$\underset{1}{\Omega}^\tau(x) \equiv -\partial_\mu \xi^k(x) \cdot L_k^{\tau\mu}(\omega(x)) + \xi^i(x)\frac{\partial \omega^\tau(x)}{\partial x^i}$$

Now, from the definitions, we see that ξ is an infinitesimal transformation of Γ if and only if $\exp\,(t\xi)$ is a finite transformation of Γ, that is to say if $\omega_t = \varphi_t^{-1}(\omega) = \omega$. In this way we obtain the following fundamental proposition.

PROPOSITION 2.19 We can write the equations of the linear system R_q in the following form, called again "*Lie form*":

$$R_q \qquad \underset{1}{\Omega^\tau}(x) \equiv -\xi^k_\mu(x)L^{\tau\mu}_k(\omega(x)) + \xi^i(x)\frac{\partial\omega^\tau(x)}{\partial x^i} = 0$$

REMARK 2.20 All the main results to be found in the sequel will arise from this Lie form for R_q which is easily seen to generalise that of the classical Lie derivative.

As a first property, we see that the source x appears in these equations only through the choice of the section ω. Now, because $\mathscr{F}$ is a bundle of geometric objects, for any other section $\bar{\omega}$ the new non-linear system

$$\bar{\mathscr{R}}_q \qquad \Phi(\bar{\omega}(y), p) = \bar{\omega}(x)$$

will be a non-linear system of finite Lie equations, the solutions of which are the finite transformations of a new pseudogroup $\bar{\Gamma}$ that will be in general different from Γ; and not even similar to it, unless we take for $\bar{\omega}$ the section defined in the definition 2.15, for example. This follows easily from the fact that Φ is written instead of G^{-1}.

At the same time we can linearise $\bar{\mathscr{R}}_q$ in order to get the linear system $\bar{R}_q$:

$$\bar{R}_q \qquad \underset{1}{\bar{\Omega}^\tau}(x) \equiv -\xi^k_\mu(x)L^{\tau\mu}_k(\bar{\omega}(x)) + \xi^i(x)\frac{\partial\bar{\omega}^\tau(x)}{\partial x^i} = 0$$

DEFINITION 2.21 If $\bar{\omega} = \varphi(\omega)$ we write $\bar{R}_q = \varphi(R_q)$.

DEFINITION 2.22 The pseudogroup Γ from which we started in order to construct $\mathscr{F}$ will be called "*special*". Any other pseudogroup determined by the choice of a section of $\mathscr{F}$ will be called "*general*". The same words will also be used for the equations of the corresponding non-linear and linear systems.

We now make two important remarks.

REMARK 2.23 The words "*special*" and "*general*" come from the vocabulary used in the theory of Relativity. In fact, and the reader can easily prove it as an exercise, if we start from the pseudogroup given by the transitive action on $\mathbb{R}^4$ of the inhomogeneous Lorentz group used in "*special relativity*", we just find S_2T^* as bundle $\mathscr{F}$ of geometric objects. Of course the special section to be used corresponds to the Minkowskian metric on $\mathbb{R}^4$.

"*General relativity*" introduces other metrics, sections of S_2T^*, depending on a small parameter which is the inverse of the speed of light, and these have to satisfy a non-linear system of p.d.e. called "*Einstein's equations*".

Thus the usual Taylor development of the metric with respect to the parameter involved is just the same as the one we introduced above.

REMARK 2.24 According to the preceding remark, we may introduce, instead of ω, a section ω_t close to ω for small values of the parameter t and such that $\omega_0 = \omega$. This idea will be the basic tool of "*deformation theory*" to be explained later on.

In fact there will be a pseudogroup Γ_t corresponding to ω_t and the main problem will be to compare $\Gamma = \Gamma_0$ and Γ_t, that is to find under what conditions Γ and Γ_t are identical, similar or different.

Before going further ahead we shall give a precise way to understand the linearisation process already used. Though it is a current process to be found in physics and mechanics, we shall detail its mathematical formulation.

Let us consider the following commutative diagram:

$$\begin{array}{ccc} \mathrm{id}^{-1}(V(X \times Y)) & \longrightarrow & V(X \times Y) \\ \downarrow & & \downarrow \\ X & \xrightarrow{\mathrm{id}} & X \times Y \end{array}$$

Any section of $\mathrm{id}^{-1}(V(X \times Y))$ can be written in local coordinates as:

$$\eta : (x) \to (x, \eta^k(x)).$$

As Y is a copy of X, it follows that we can identify $\mathrm{id}^{-1}(V(X \times Y))$ and T. This will be done in the sequel. We have a useful picture to keep in mind:

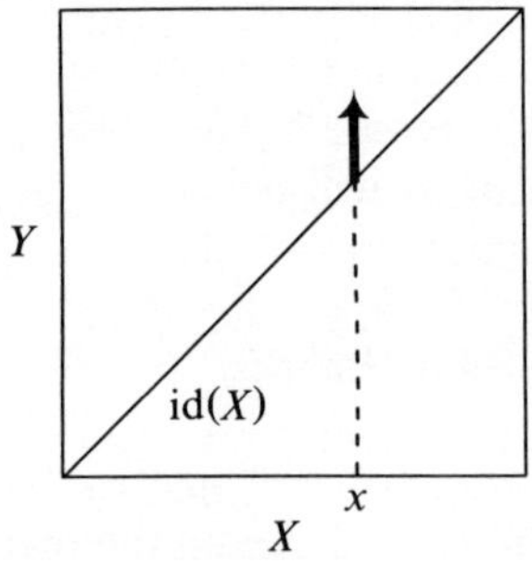

In a similar way we can consider the following commutative diagram:

$$\begin{array}{ccc} \omega^{-1}V(\mathscr{F}) & \longrightarrow & V(\mathscr{F}) \\ \downarrow & & \downarrow \\ X & \xrightarrow{\omega} & \mathscr{F} \end{array}$$

We shall define a vector bundle F_0 over X by $F_0 = \omega^{-1}(V(\mathscr{F}))$ and we have a similar picture.

Using local coordinates, we see that $\underset{1}{\Omega}$ is a section of F_0 and that the transition rules for the section of F_0 are of the following kind:

$$F_0 \begin{cases} \underset{1}{\bar{\Omega}}(\varphi(x)) = \dfrac{\partial G}{\partial u}\left(\omega(x), \dfrac{\partial\varphi(x)}{\partial x}\right)\cdot \underset{1}{\Omega}(x) \\ \bar{x} = \varphi(x) \end{cases}$$

while keeping in mind that we have to change $\omega(x)$ according to the co-ordinate domain, as $F_0 = \omega^{-1}(V(\mathscr{F}))$.

EXAMPLE 2.25 The reader will show that $F_0 = T \otimes S_2 T^*$ in the case of the bundle $\mathscr{F}$ of geometric objects with the Christoffel symbols as sections.

We may now introduce the initial part of a P-sequence:

$$0 \longrightarrow \Theta \longrightarrow T \xrightarrow{\mathscr{D}} F_0$$

with the same notations as in the first part of the book and write $\mathscr{D}\cdot\xi = \underset{1}{\Omega}$ for any section ξ of T.

Our problem will now be to construct the P-sequence when $\mathscr{D}$ is a Lie operator as above. It is thus at this point only that we have to study the prolongations of $\mathscr{R}_q$ and R_q.

PROPOSITION 2.26 If R_q is obtained from $\mathscr{R}_q$ by linearisation, then its first prolongation R_{q+1} is also obtained from the first prolongation $\mathscr{R}_{q+1}$ of $\mathscr{R}_q$ by the same process.

Proof We shall use local coordinates.

If we linearise the equations $\Phi(y^k, p^k_\mu) = \omega(x)$ of $\mathscr{R}_q$, we get for a section ξ of T solution of R_q:

$$\frac{\partial\Phi}{\partial y^k}(\mathrm{id}_q(x))\xi^k(x) + \frac{\partial\Phi}{\partial p^k_\mu}(\mathrm{id}_q(x))\partial_\mu\xi^k(x) = 0$$

Now if we differentiate totally once with respect to x^i the equations of $\mathscr{R}_q$ we get the equations of $\mathscr{R}_{q+1}$:

$$\frac{\partial\Phi}{\partial y^k}(y, p)p^k_i + \frac{\partial\Phi}{\partial p^k_\mu}(y, p)p^k_{\mu+1_i} = \frac{\partial\omega(x)}{\partial x^i}$$

If we linearise these equations, we get, for a section ξ of T solution of R_{q+1}:

$$\frac{\partial^2\Phi}{\partial y^k \partial y^l}(\mathrm{id}_q(x))\delta_i^k \xi^l(x) + \frac{\partial^2\Phi}{\partial y^k \partial p_\mu^k}(\mathrm{id}_q(x))\delta_i^k \partial_\mu \xi^k(x)$$
$$+ \frac{\partial\Phi}{\partial y^k}(\mathrm{id}_q(x))\partial_i \xi^k(x) + \frac{\partial\Phi}{\partial p_\mu^k}(\mathrm{id}_q(x))\partial_{\mu+1_i}\xi^k(x) = 0$$

that is to say:

$$\frac{\partial^2\Phi}{\partial y^i \partial y^k}(\mathrm{id}_q(x))\xi^k(x) + \frac{\partial^2\Phi}{\partial y^i \partial p_\mu^k}(\mathrm{id}_q(x))\partial_\mu \xi^k(x)$$
$$+ \frac{\partial\Phi}{\partial y^k}(\mathrm{id}_q(x))\partial_i \xi^k(x) + \frac{\partial\Phi}{\partial p_\mu^k}(\mathrm{id}_q(x))\partial_{\mu+1_i}\xi^k(x) = 0$$

The reader will check easily, using the fact that Φ contains the target y and does not contain the source x, that R_{q+1} is indeed the first prolongation of R_q.

C.Q.F.D.

REMARK 2.27 It follows that R_{q+r} is obtained from $\mathcal{R}_{q+r}$ by the same linearisation process as above. If $\mathcal{R}_q$ is regular we have:

$$R_{q+r} = \mathrm{id}_{q+r}^{-1}(V(\mathcal{R}_{q+r}))$$

THEOREM 2.28 If $\mathcal{R}_q$ is a regular non-linear system of finite Lie equations then $\mathcal{R}_{q+r}$ is also a non-linear system of finite equations, $\forall r \geq 0$.

Proof According to proposition 2.3.17, as $\mathcal{R}_q$ and $\mathcal{R}_{q+r}$ have the same set of solutions, the proof becomes trivial if we are only concerned with the definition 1.2 of a Lie pseudogroup.

Now the proof in the general case is based upon the following technical lemma for the case $r = 1$:

LEMMA 2.29 If $\mathcal{F}$ is a bundle of geometric objects over X with structure group $GL_q(n, \mathbb{R})$, then $J_1(\mathcal{F})$ is a bundle of geometric objects over X with structure group $GL_{q+1}(n, \mathbb{R})$.

Proof This can be given using a tedious but straight-forward computation of the first prolongation of the finite transition rules of $\mathcal{F}$. We leave it to the reader as an exercise.

However it is of much more interest to give the proof using the first prolongation of the infinitesimal transition rules of $\mathcal{F}$ already exhibited, that is to say:

$$\xi^i(x)\frac{\partial}{\partial x^i} + \partial_\mu \xi^k(x) L_k^{\tau\mu}(u)\frac{\partial}{\partial u^\tau}$$
$$+ \left(\partial_{\mu+1_i}\xi^k(x)L_k^{\tau\mu}(u) + \partial_\mu \xi^k(x)\frac{\partial L_k^{\tau\mu}(u)}{\partial u^\rho}u_i^\rho - u_j^\tau \frac{\partial \xi^j}{\partial x^i}\right)\frac{\partial}{\partial u_i^\tau}$$

We have just to prove that the following vector fields:

$$\begin{cases} L_k^{\tau i}(u)\dfrac{\partial}{\partial u^\tau} + \dfrac{\partial L_k^{\tau i}(u)}{\partial u^\rho}u_r^\rho \dfrac{\partial}{\partial u_r^\tau} - u_k^\tau \dfrac{\partial}{\partial u_i^\tau} \\ L_k^{\tau\mu}(u)\dfrac{\partial}{\partial u^\tau} + \dfrac{\partial L_k^{\tau\mu}(u)}{\partial u^\rho}u_r^\rho \dfrac{\partial}{\partial u_r^\tau} + \displaystyle\sum_{\alpha+1_i=\mu} L_k^{\tau\alpha}(u)\dfrac{\partial}{\partial u_i^\tau} \\ \displaystyle\sum_{\alpha+1_i=\mu} L_k^{\tau\alpha}(u)\dfrac{\partial}{\partial u_i^\tau} \end{cases}$$

which are associated to $\partial_\mu \xi^k(x)$ respectively for $|\mu| = 1$, $2 \le |\mu| \le q$ and $|\mu| = q + 1$ are the infinitesimal generators of an action of $GL_{q+1}(n, \mathbb{R})$ on the fiber of $J_1(\mathscr{F})$, with the same commutation relations as those of the vector fields $A_k^\mu(q + 1)$.

We shall just check some of those commutation relations, leaving to the reader the computation of the other ones.

$$\left[L_k^{\tau\mu}(u)\frac{\partial}{\partial u^\tau} + \frac{\partial L_k^{\tau\mu}(u)}{\partial u^\rho}u_r^\rho \frac{\partial}{\partial u_r^\tau} + \sum_{\alpha+1_i=\mu} L_k^{\tau\mu}(u)\frac{\partial}{\partial u_i^\tau}, \sum_{\beta+1_j=\nu} L_l^{\tau\beta}(u)\frac{\partial}{\partial u_j^\tau}\right]$$
$$= \sum_{\beta+1_j=\nu}([L_k^\mu, L_l^\beta])^\tau \frac{\partial}{\partial u_j^\tau} = 0$$

when $2 \le |\mu| \le q$, $|\nu| = q + 1$

$$\left[L_k^{\tau i}(u)\frac{\partial}{\partial u^\tau} + \frac{\partial L_k^{\tau i}(u)}{\partial u^\rho}u_r^\rho \frac{\partial}{\partial u_r^\tau} - u_k^\tau \frac{\partial}{\partial u_i^\tau}, \sum_{\beta+1_j=\nu} L_l^{\tau\beta}(u)\frac{\partial}{\partial u_j^\tau}\right]$$
$$= \sum_{\beta+1_j=\nu}([L_k^i, L_l^\beta])^\tau \frac{\partial}{\partial u_j^\tau} + \sum_{\beta+1_j=\nu}\delta_k^j L_l^{\tau\beta}(u)\frac{\partial}{\partial u_i^\tau}$$
$$= \sum_{\beta+1_j=\nu}\left((\beta_i + 1)L_l^{\tau\beta+1_i-1_k}(u)\frac{\partial}{\partial u_j^\tau} - \delta_l^i L_k^{\tau\beta}(u)\frac{\partial}{\partial u_j^\tau} + \delta_k^j L_l^{\tau\beta}(u)\frac{\partial}{\partial u_i^\tau}\right)$$
$$= (\nu_i + 1)\cdot \sum_{\beta+1_j=\nu} L_l^{\tau\beta+1_i-1_k}(u)\frac{\partial}{\partial u_j^\tau} - \delta_l^i \cdot \sum_{\beta+1_j=\nu} L_k^{\tau\beta}(u)\frac{\partial}{\partial u_j^\tau}$$

Because

$$(v_i + 1) = (\beta_i + \delta_i^j + 1)$$

and

$$\sum_{\beta + 1_j = v} \delta_i^j L_l^{\tau\beta + 1_i - 1_k}(u) \frac{\partial}{\partial u_j^\tau} = L_l^{\tau v - 1_k}(u) \frac{\partial}{\partial u_i^\tau} = \sum_{\beta + 1_j = v} \delta_k^j L_l^{\tau\beta}(u) \frac{\partial}{\partial u_i^\tau}$$

But

$$[A_k^i(q+1), A_l^v(q+1)] = v_k \frac{(v + 1_i - 1_k)!}{v!} A_l^{v + 1_i - 1_k}(q+1)$$

$$- \delta_l^i \frac{(v + 1_i + 1_l)!}{v!} A_k^{v + 1_i - 1_l}(q+1)$$

$$= (v_i + 1) \cdot A_l^{v + 1_i - 1_k}(q+1) - \delta_l^i \cdot A_k^v(q+1)$$

and the commutation relations are the same.

Finally we check that:

$$\left[L_k^\mu + \frac{\partial L_k^{\tau\mu}(u)}{\partial u^\rho} u_r^\rho \frac{\partial}{\partial u_r^\tau} + \sum_{\alpha + 1_i = \mu} L_k^{\tau\alpha}(u) \frac{\partial}{\partial u_i^\tau}, \right.$$

$$\left. L_l^v + \frac{\partial L_l^{\tau v}(u)}{\partial u^\rho} u_r^\rho \frac{\partial}{\partial u_r^\tau} + \sum_{\beta + 1_j = v} L_l^{\tau\beta}(u) \frac{\partial}{\partial u_j^\tau} \right]$$

$$= 0 \quad \text{if} \quad |\mu + v| \geq q + 2$$

and that:

$$\left(\sum_{\beta + 1_i = v} ([L_k^\mu, L_l^\beta])^\tau + \sum_{\alpha + 1_i = \mu} ([L_k^\alpha, L_l^v])^\tau \right) \frac{\partial}{\partial u_i^\tau} \quad \text{if } |\mu + v| \leqslant q + 1$$

$$= \sum_{\substack{\alpha + 1_i = \mu \\ \beta + 1_i = v}} \left(\beta_k \frac{(\mu + \beta - 1_k)!}{\mu! \beta!} L_l^{\tau\mu + \beta - 1_k}(u) \right.$$

$$\left. + v_k \frac{(v + \alpha - 1_k)!}{v! \alpha!} L_l^{\tau v + \alpha - 1_k}(u) \right) \frac{\partial}{\partial u_i^\tau}$$

$$- \sum_{\substack{\alpha + 1_i = \mu \\ \beta + 1_i = v}} \left(\mu_l \frac{(\mu + \beta - 1_l)!}{\mu! \beta!} L_k^{\tau\mu + \beta - 1_l}(u) \right.$$

$$\left. + \alpha_l \frac{(v + \alpha - 1_l)!}{v! \alpha!} L_k^{\tau v + \alpha - 1_l}(u) \right) \frac{\partial}{\partial u_i^\tau}$$

$$= \sum_{\gamma + 1_i = \mu + v} \left(v_k \frac{(\mu + v - 1_k)!}{\mu! v!} L_l^{\tau\gamma - 1_k}(u) \right.$$

$$\left. - \mu_l \frac{(\mu + v - 1_l)!}{\mu! v!} L_k^{\tau\gamma - 1_l}(u) \right) \frac{\partial}{\partial u_i^\tau}$$

where we have set $\gamma = \alpha + v = \mu + \beta$.

But we have:

$$[A_k^\mu(q+1), A_l^\nu(q+1)] = \nu_k \frac{(\mu+\nu-1_k)!}{\mu!\nu!} A_l^{\mu+\nu-1_k}(q+1)$$

$$- \mu_l \frac{(\mu+\nu-1_l)!}{\mu!\nu!} A_k^{\mu+\nu-1_l}(q+1)$$

and we find again the same commutation relations.

The other commutation relations are easy to check in the same way. For example we get:

$$\left[L_k^i + \frac{\partial L_k^{\tau i}(u)}{\partial u^\rho} u_r^\rho \frac{\partial}{\partial u_r^\tau} - u_k^\tau \frac{\partial}{\partial u_i^\tau}, L_l^j + \frac{\partial L_l^{\tau j}(u)}{\partial u^\tau} u_r^\tau \frac{\partial}{\partial u_r^\tau} - u_l^\tau \frac{\partial}{\partial u_j^\tau}\right]$$

$$= \delta_k^j\left(L_l^i + \frac{\partial L_l^{\tau i}(u)}{\partial u^\rho} u_r^\rho \frac{\partial}{\partial u_r^\tau} - u_l^\tau \frac{\partial}{\partial u_i^\tau}\right) - \delta_l^i\left(L_k^j + \frac{\partial L_k^{\tau j}}{\partial u^\rho} u_r^\rho \frac{\partial}{\partial u_r^\tau} - u_k^\tau \frac{\partial}{\partial u_j^\tau}\right)$$

and we know that

$$[A_k^i(q+1)A_l^j(q+1)] = \delta_k^j A_l^i(q+1) - \delta_l^i A_k^j(q+1).$$

C.Q.F.D.

Theorem 2.28 then follows easily from this lemma and the converse theorem 2.10. C.Q.F.D.

We shall now give a second proof of this lemma, using arguments that will be found again through the end of the book but in a quite different background.

It is easy to check that, to any Lie equation $R_q \subset J_q(T)$ such that $[R_q, R_q] \subset R_q$ we can associate such a bundle $\mathscr{F}$ of geometric objects, both with a section ω of it. We just need for that to integrate the involutive distribution $\#(R_q)$ on $J_q(X \times Y)$.

Conversely, to any such bundle $\mathscr{F}$ and a section ω of it, we associate a general system $R_q \subset J_q(T)$ of infinitesimal Lie equations such that $[R_q, R_q] \subset R_q$ because of its construction. Moreover the equations corresponding to $J_1(\mathscr{F})$ with section $j_1(\omega)$ are just the first prolongation R_{q+1} of R_q and we are led to prove the following equivalent lemma:

Lemma 2.29 *bis* If $[R_q, R_q] \subset R_q$ then $[R_{q+1}, R_{q+1}] \subset R_{q+1}$

Proof For two sections ξ_{q+1} and η_{q+1} of $J_{q+1}(T)$ with $\xi_q = \pi_q^{q+1} \circ \xi_{q+1}$ and $\eta_q = \pi_q^{q+1} \circ \eta_{q+1}$ we have:

$$[\xi_q, \eta_q] = \{\xi_{q+1}, \eta_{q+1}\} + i(\xi)D\eta_{q+1} - i(\eta)D\xi_{q+1}$$

where we have introduced an "*algebraic bracket*" $\{\ \}$ on $J_{q+1}(T)$ with value in $J_q(T)$ which is $C^\infty(X)$-linear, the Spencer's operator

$$D : J_{q+1}(T) \to T^* \otimes J_q(T)$$

and the interior multiplication $i(\ )$.

Using local coordinates we have:

$$(\{\xi_{q+1}, \eta_{q+1}\})^k_\nu = \sum_{\substack{0 \le |\lambda| \le q \\ 0 \le |\mu| \le q \\ \lambda + \mu = \nu}} \frac{(\lambda + \mu)!}{\lambda!\mu!} (\xi^i_\lambda \eta^k_{\mu+1_i} - \eta^i_\lambda \xi^k_{\mu+1_i})$$

Elementary properties of the binomial coefficients show that:

$$i(\zeta)D\{\xi_{q+1}, \eta_{q+1}\} = \{i(\zeta)D\xi_{q+1}, \eta_q\} + \{\xi_q, i(\zeta)D\eta_{q+1}\}$$

for any section ζ of T.

It follows that:

$$\begin{aligned} i(\zeta)D[\xi_{q+1}, \eta_{q+1}] &= \{i(\zeta)D\xi_{q+2}, \eta_{q+1}\} + \{\xi_{q+1}, i(\zeta)D\eta_{q+2}\} \\ &\quad + i(\zeta)D(i(\xi)D\eta_{q+2}) - i(\zeta)D(i(\eta)D\xi_{q+2}) \\ &= [i(\zeta)D\xi_{q+1}, \eta_q] + [\xi_q, i(\zeta)D\eta_{q+1}] \\ &\quad + i(\zeta)D(i(\xi)D\eta_{q+2}) - i(\zeta)D(i(\eta)D\xi_{q+2}) \\ &\quad - i(\xi)D(i(\zeta)D\eta_{q+2}) + i(\eta)D(i(\zeta)D\xi_{q+2}) \\ &\quad - i(i(\zeta)D\xi_1)D\eta_{q+1} + i(i(\zeta)D\eta_1)D\xi_{q+1} \end{aligned}$$

Using local coordinates for three of the six last terms we obtain:

$$\begin{aligned} &+ \zeta^i\left(\partial_i\left(\xi^r\left(\frac{\partial \eta^k_\mu}{\partial x^r} - \eta^k_{\mu+1_r}\right)\right) - \xi^r\left(\frac{\partial \eta^k_{\mu+1_i}}{\partial x^r} - \eta^k_{\mu+1_i+1_r}\right)\right) \\ &- \xi^r\left(\partial_r\left(\zeta^i\left(\frac{\partial \eta^k_\mu}{\partial x^i} - \eta^k_{\mu+1_i}\right)\right) - \zeta^i\left(\frac{\partial \eta^k_{\mu+1_r}}{\partial x^i} - \eta^k_{\mu+1_i+1_r}\right)\right) \\ &- \zeta^i\left(\frac{\partial \xi^r}{\partial x^i} - \xi^r_i\right)\left(\frac{\partial \eta^k_\mu}{\partial x^r} - \eta^k_{\mu+1_r}\right) \\ &= -\left(-\xi^i_r\zeta^r + \xi^r\frac{\partial \zeta^i}{\partial x^r}\right)\left(\frac{\partial \eta^k_\mu}{\partial x^i} - \eta^k_{\mu+1_i}\right) \end{aligned}$$

Finally, setting

$$(L(\xi_1)\zeta)^i = -\xi^i_r\zeta^r + \xi^r\frac{\partial \zeta^i}{\partial x^r},$$

we obtain the important formula:

$$\begin{aligned} i(\zeta)D[\xi_{q+1}, \eta_{q+1}] &= [i(\zeta)D\xi_{q+1}, \eta_q] + [\xi_q, i(\zeta)D\eta_{q+1}] \\ &\quad - i(L(\xi_1)\zeta)D\eta_{q+1} + i(L(\eta_1)\zeta)D\xi_{q+1} \end{aligned}$$

The lemma follows immediately from the above formula and the fact that $DR_{q+1} \subset T^* \otimes R_q$. More precisely, whenever ξ_{q+1} and η_{q+1} are sections of R_{q+1} we have:

$$\pi_q^{q+1}([\xi_{q+1}, \eta_{q+1}]) = [\xi_q, \eta_q] \subset R_q$$
$$i(\zeta)D[\xi_{q+1}, \eta_{q+1}] \subset R_q$$

An easy diagram chasing in the following commutative and exact diagram ends the proof:

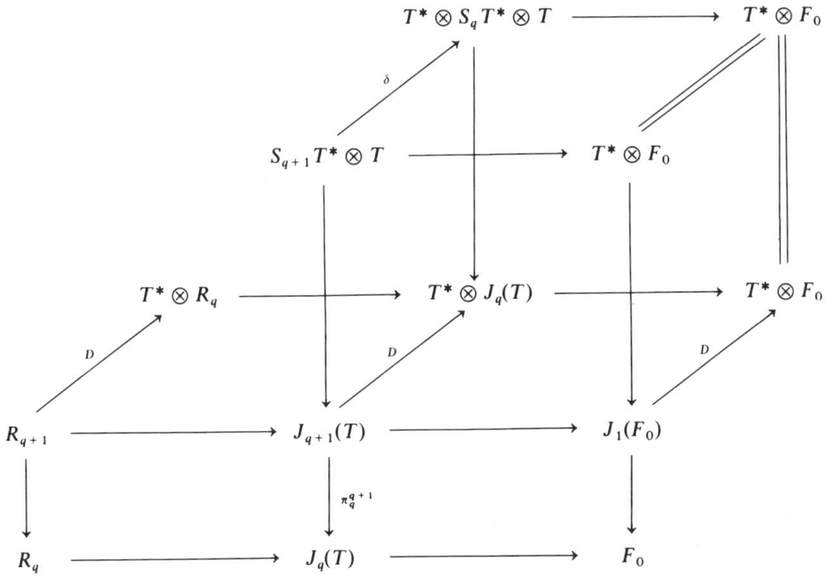

We finally note that R_q may not be formally integrable. C.Q.F.D.

Similarly, using proposition 6.3.26, proposition 6.5.20, the preceding diagram and an easy induction on r, it follows that, $\forall r \geq 0$, any differential invariant of $\#(R_{q+r})$ is functionally connected with the Φ^τ and their total derivatives up to order r.

When R_q is formally integrable, the Φ^τ are said to constitute a "*fundamental set*" of differential invariants of Γ.

We shall now look at the symbols of $\mathscr{R}_q$ and R_q.

From the first part of this book, the symbol G_q of $\mathscr{R}_q$ at any point $(x, y, p) \in \mathscr{R}_q$ is given by the following equations:

$$G_{q;\, x, y, p} \qquad \frac{\partial \Phi^\tau}{\partial p_\mu^k}(x, y, p)v_\mu^k = 0 \qquad |\mu| = q$$

From the definition of R_q it follows that the symbol of R_q at any point $x \in X$ is $G_{q;\,\mathrm{id}_q(x)}$ and the symbol of R_q is

$$\bigcup_{x \in X} G_{q;\,\mathrm{id}_q(x)},$$

also written simply, when there is no confusion:

$$G_q = \bigcup_{x \in X} G_{q;\,x}.$$

Many proofs in the sequel will rely upon the following fundamental technical lemma:

LEMMA 2.30 Whenever we have the finite transformations

$$\bar{u} = G(u, p) \quad \text{or} \quad \Phi(\bar{u}, p) = u$$

where p is considered as a coordinate for $GL_q(n, \mathbb{R})$, we have the formula:

$$-L_k^{\tau\mu}(u) = A_{k,\lambda}^{\mu,j}(p; q) \cdot (B^{-1})_{j,\nu}^{\lambda,l}(p; q) \frac{\partial u^\tau}{\partial \bar{u}^\sigma} L_l^{\sigma\nu}(\bar{u})$$

Proof We shall cut the proof into three parts:

1 Setting $y^k = x^k + t\xi^k(x) + \cdots$ and $\bar{y}^k = \bar{x}^k = \bar{x}^k + t\bar{\xi}^k(\bar{x}) + \cdots$ when $\bar{x} = \varphi(x)$ and $\bar{y} = \varphi(y)$ with $\varphi \in \mathrm{Aut}\,(X)$, we have:

$$\frac{\partial \bar{y}}{\partial x} = \frac{\partial \bar{y}}{\partial y} \cdot \frac{\partial y}{\partial x} \quad \text{and} \quad \frac{\partial \bar{y}}{\partial x} = \frac{\partial \bar{y}}{\partial \bar{x}} \cdot \frac{\partial \bar{x}}{\partial x}$$

where we have considered both source and target transformations. Looking only at the first order terms, we obtain:

$$\begin{cases} \dfrac{\partial^{|\mu|}\bar{y}^k}{\partial x^\mu} = \dfrac{\partial^{|\mu|}\bar{y}^k}{\partial y^\mu} - t\,\dfrac{\partial^{|\nu|}\xi^\rho(x)}{\partial x^\nu}\, A_{\rho,\mu}^{\nu,k}\!\left(\dfrac{\partial \varphi(y)}{\partial y}; q\right) + \cdots \\[2ex] \dfrac{\partial^{|\mu|}\bar{y}^k}{\partial x^\mu} = \dfrac{\partial^{|\mu|}\bar{x}^k}{\partial x^\mu} + t\,\dfrac{\partial^{|\nu|}\bar{\xi}^\rho(\bar{x})}{\partial \bar{x}^\nu}\, B_{\rho,\mu}^{\nu,k}\!\left(\dfrac{\partial \varphi(x)}{\partial x}; q\right) + \cdots \end{cases}$$

where we have introduced the matrices $[A_{l,\mu}^{\nu,k}(p; q)]$ and $[B_{l,\mu}^{\nu,k}(p; q)]$ of the components of the vector fields $A_l^\nu(q)$ and $B_l^\nu(q)$.

It follows that we have; setting $\partial\varphi/\partial x = p$ and keeping the development up to the first order:

$$p_{\mu+1_i}^k \cdot \xi^i - \frac{\partial^{|\nu|}\xi^l}{\partial x^\nu} A_{l,\mu}^{\nu,k}(p; q) = \frac{\partial^{|\nu|}\bar{\xi}^l}{\partial \bar{x}^\nu} B_{l,\mu}^{\nu,k}(p; q)$$

where $1 \le |\mu|, |\nu| \le q$. *This formula gives us the transition laws for the vector bundle $J_q(T)$ of geometric objects.*

2 Using the group axioms and a symbolic notation we have:

$$G\left(u, \frac{\partial \bar{y}}{\partial x}\right) = G\left(G\left(u, \frac{\partial y}{\partial x}\right), \frac{\partial \bar{y}}{\partial y}\right) = G\left(G\left(u, \frac{\partial \bar{x}}{\partial x}\right), \frac{\partial \bar{y}}{\partial \bar{x}}\right)$$

According to the first fundamental theorem of Lie, applied here to the action of $GL_q(n, \mathbb{R})$ on the fiber of $\mathscr{F}$, we have, up to the first order and writing simply $p = \partial\bar{x}/\partial x$ as coordinates for $GL_q(n, \mathbb{R})$:

$$\frac{\partial G^\tau}{\partial u^\sigma}(u, p)\frac{\partial^{|\nu|}\xi^l}{\partial x^\nu} L_l^{\sigma\nu}(u) + \frac{\partial G^\tau}{\partial p_\mu^k}(u, p)p_{\mu+1_i}^k \xi^i = \frac{\partial^{|\nu|}\bar{\xi}^l}{\partial \bar{x}^\nu} L_l^{\tau\nu}(\bar{u})$$

3 Finally, combining **1** and **2** we get:

$$\frac{\partial^{|\nu|}\xi^l}{\partial x^\nu} L_l^{\tau\nu}(u) + \frac{\partial u^\tau}{\partial \bar{u}^\sigma}\cdot\frac{\partial G^\sigma}{\partial p_\mu^k}(u, p)p_{\mu+1_i}^k \xi^i = \frac{\partial u^\tau}{\partial \bar{u}^\sigma}\frac{\partial^{|\nu|}\bar{\xi}^l}{\partial \bar{x}^\nu} L_l^{\sigma\nu}(\bar{u})$$

But we have $u \equiv \Phi(G(u, p)p)$ and

$$\frac{\partial \Phi^\tau}{\partial p_\mu^k}(\bar{u}, p) + \frac{\partial u^\tau}{\partial \bar{u}^\sigma}\cdot\frac{\partial G^\sigma}{\partial p_\mu^k}(u, p) = 0$$

Thus we obtain:

$$\frac{\partial^{|\nu|}\xi^l}{\partial x^\nu} L_l^{\tau\nu}(u) - \frac{\partial \Phi^\tau}{\partial p_\mu^k}(\bar{u}, p)p_{\mu+1_i}^k \xi^i = \frac{\partial u^\tau}{\partial \bar{u}^\sigma}\frac{\partial^{|\nu|}\bar{\xi}^l}{\partial \bar{x}^\nu} L_l^{\sigma\nu}(\bar{u})$$

in which we have to substitute:

$$\frac{\partial^{|\nu|}\bar{\xi}^l}{\partial \bar{x}^\nu} = -\frac{\partial^{|\mu|}\xi^k}{\partial x^\mu} A_{k,\lambda}^{\mu,j}(p; q)(B^{-1})_{j,\nu}^{\lambda,l}(p; q) + (B^{-1})_{k,\nu}^{\mu,l}\, p_{\mu+1_i}^k \xi^i.$$

in order to get:

$$\frac{\partial^{|\mu|}\xi^k}{\partial x^\mu}\left[L_k^{\tau\mu}(u) + A_{k,\lambda}^{\mu,j}(p; q)(B^{-1})_{j,\nu}^{\lambda,l}(p; q)\frac{\partial u^\tau}{\partial \bar{u}^\sigma} L_l^{\sigma\nu}(\bar{u})\right]$$
$$- \xi^i p_{\mu+1_i}^k\left[\frac{\partial \Phi^\tau}{\partial p_\mu^k}(\bar{u}, p) + (B^{-1})_{k,\nu}^{\mu,l}(p; q)\frac{\partial u^\tau}{\partial \bar{u}^\sigma} L_l^{\sigma\nu}(\bar{u})\right] = 0$$

As u and p are arbitrary, we must have:

$$-L_k^{\tau\mu}(u) = A_{k,\lambda}^{\mu,j}(p; q)(B^{-1})_{j,\nu}^{\lambda,l}(p; q)\frac{\partial u^\tau}{\partial \bar{u}^\sigma} L_l^{\sigma\nu}(\bar{u})$$

Now, we know that $A_k^\mu(q)$ and $B_k^\mu(q)$ are the infinitesimal generators of two transitive actions of $GL_q(n, \mathbb{R})$, with local coordinates p, when $\det[p_i^k] \neq 0$. It follows that the matrices $A = [A_{k,\nu}^{\mu,l}(p; q)]$ and $B = [B_{k,\nu}^{\mu,l}(p; q)]$ of their components both have non-zero determinants when $\det[p_i^k] \neq 0$. This we shall write simply as $\det p \neq 0$.

Finally the coefficients of $\xi^i p^k_{\mu+1_i}$ is:

$$\frac{\partial \Phi^\tau}{\partial p^k_\mu}(\bar{u}, p) - (A^{-1})^{\mu, l}_{k, \nu}(p; q) L^{\tau\mu}_l(u) = 0$$

because

$$A^{\nu, k}_{l, \mu}(p; q) \frac{\partial \Phi^\tau}{\partial p^k_\mu} \equiv A^\nu_l(q) \cdot \Phi^\tau = L^{\tau\nu}_l(u)$$

when $\Phi(\bar{u}, p) = u$. C.Q.F.D.

REMARK 2.31 The following formula is sometimes useful. From the coefficients of $\xi^i p^k_{\mu+1_i}$, we obtain:

$$\left(B^\mu_k(q) + L^{\sigma\mu}_k(\bar{u}) \frac{\partial}{\partial \bar{u}^\sigma}\right) \cdot \Phi^\tau(\bar{u}, p) \equiv 0$$

This is essentially the kind of formulas given through the proof of theorem 6.2.16.

From the definition of the differential invariants,

$$\left(\eta^k_\mu(y) B^\mu_k(q) + \eta^k(y) \frac{\partial}{\partial y^k}\right) \cdot \Phi^\tau(\omega(y), p) = 0$$

whenever

$$-\eta^k_\mu(y) L^{\tau\mu}_k(\omega(y)) + \eta^k(y) \frac{\partial \omega^\tau(y)}{\partial y^k} = 0.$$

Using the relation

$$\frac{\partial \Phi^\tau}{\partial y^k} = \frac{\partial \Phi^\tau}{\partial \bar{u}^\sigma} \cdot \frac{\partial \omega^\sigma(y)}{\partial y^k} \quad \text{with } \bar{u} = \omega(y)$$

and eliminating $\eta^k(y)$, we check of course that:

$$\eta^k_\mu(y)\left(B^\mu_k(q) + L^{\sigma\mu}_k(\bar{u}) \frac{\partial}{\partial \bar{u}^\sigma}\right) \cdot \Phi^\tau(\bar{u}, p) \equiv 0.$$

From the lemma we deduce the important:

THEOREM 2.32 The symbol of R_q is a vector bundle over X.

Proof As $\mathcal{R}_q$ is a non-linear system of formally transitive Lie equations, $\forall x, y \in U$, $\exists p$ such that:

$$\Phi^\tau(\omega(y), p) = \omega^\tau(x)$$

Let $(v^k_{i_1 \dots i_q}) \in G_{q;x}$ and $(w^l_{j_1, \dots, j_q}) \in G_{q;y}$. Then it is easy to see from the preceding lemma that:

$$\frac{\partial G^\tau}{\partial u^\sigma}(\omega(x), p)v^k_\mu L^{\sigma\mu}_k(\omega(x)) = w_\nu L^{\tau\nu}_l(\omega(y)) \quad \text{with } |\mu| = |\nu| = q$$

whenever:

$$p^l_k v^k_{i_1, \dots, i_q} = p^{j_1}_{i_1} \dots p^{j_q}_{i_q} \cdot w^l_{j_1 \dots j_q}$$

This gives an isomorphism:

$$G_{q;x} \approx G_{q;y} \qquad \forall x, y \in U$$

and the proof can be completed by patching together the open sets $U \subset X$.

C.Q.F.D.

Corollary 2.33 The symbol of R_q and its prolongations are vector bundles over X.

Proof As above we can find isomorphisms

$$G_{q+r;x} \approx G_{q+r;y} \qquad \forall x, y \in U, \forall r \geq 0.$$

and it is easy to see that these isomorphisms commute with the prolongations.

Remark 2.34 It is also easy to see that these isomorphisms commute with the δ-map. In this way we get isomorphisms:

$$H^s_{q+r}(G_{q;x}) \approx H^s_{q+r}(G_{q,y})$$

In particular, from proposition 5.2.8 we see that the prolongations G_{q+r} become in general 2-acyclic before they become involutive, for increasing values of r (see examples at the end).

Corollary 2.35 The symbol G_q of $\mathscr{R}_q$ and its prolongations G_{q+r} are vector bundles over $\mathscr{R}_q$.

Proof We have the relation:

$$A^{\mu,l}_{k,\nu}(p, q) \cdot \frac{\partial \Phi^\tau(\omega(y), p)}{\partial p^l_\nu} = L^{\tau\mu}_k(\omega(x))$$

In particular, for $|\mu| = q$, we obtain:

$$-p^l_k \cdot \frac{\partial \Phi^\tau(\omega(y), p)}{\partial p^l_\mu} = L^{\tau\mu}_k(\omega(x))$$

If now $(w_\mu^l) \in G_{q;(x,y,p)}$ and $(v_\mu^k) \in G_{q;x}$ with $|\mu| = q$, then we have an isomorphism:

$$G_{q;x} \approx G_{q;(x,y,p)}$$

given by

$$p_k^l \cdot v_\mu^k = w_\mu^\rho \quad \text{with } (x, y, p) \in \mathscr{R}_q.$$

and it is easy to see that this isomorphism commutes with the prolongations.

C.Q.F.D.

COROLLARY 2.36 We have isomorphisms:

$$R_{q;x}^0 \approx R_{q;y}^0 \qquad \forall x, y \in U$$

Proof This follows immediately from the above lemma. C.Q.F.D.

EXAMPLE 2.37 Take $X = \mathbb{R}^2$

$$\mathscr{F}\begin{cases} \bar{u}^1 \dfrac{\partial \varphi^1}{\partial x^1} + \bar{u}^2 \dfrac{\partial \varphi^2}{\partial x^1} = u^1 \\ \bar{u}^1 \dfrac{\partial \varphi^1}{\partial x^2} + \bar{u}^2 \dfrac{\partial \varphi^2}{\partial x^2} = u^2 \quad \text{when } \bar{x} = \varphi(x) \\ \bar{u}^3 \left(\dfrac{\partial \varphi^1}{\partial x^1} \dfrac{\partial \varphi^2}{\partial x^2} - \dfrac{\partial \varphi^1}{\partial x^2} \dfrac{\partial \varphi^2}{\partial x^1} \right) = u^3 \end{cases}$$

We have

$-L_2^2$	$-L_1^2$	$-L_2^1$	$-L_1^1$	
0	0	u^2	u^1	u^1
u^2	u^1	0	0	u^2
u^3	0	0	u^3	u^3

$$A_j^i(1) = -p_j^k \frac{\partial}{\partial p_i^k} \quad \text{and} \quad B_k^l(1) = p_i^l \frac{\partial}{\partial p_i^k}$$

As an exercise we invite the reader to check all the formulas given through the proof of the preceding lemma. He may study the two cases:

1) $\omega^1(x) = x^2, \quad \omega^2(x) = 0, \quad \omega^3(x) = 1$
2) $\bar{\omega}^1(x) = 1, \quad \bar{\omega}^2(x) = 0, \quad \bar{\omega}^3(x) = 1$

As $\mathscr{F}$ is a bundle of geometric objects, according to theorem 2.10, to any section ω of $\mathscr{F}$ over $U \subset X$ correspond a non-linear system $\mathscr{R}_q$ and a linear system R_q:

$$\begin{cases} \mathscr{R}_q & \Phi^\tau(\omega(y), p) = \omega^\tau(x) \\ R_q & \underset{1}{\Omega^\tau} \equiv -L_k^{\tau\mu}(\omega(x))\xi_\mu^k + \xi^i \dfrac{\partial \omega^\tau(x)}{\partial x^i} = 0 \end{cases}$$

The key idea in this chapter will be to take another section $\bar{\omega}$ of $\mathscr{F}$ over $U \subset X$ and to compare the new systems:

$$\begin{cases} \bar{\mathscr{R}}_q & \Phi^\tau(\bar{\omega}(y), p) = \bar{\omega}^\tau(x) \\ \bar{R}_q & \underset{1}{\bar{\Omega}^\tau} \equiv -L_k^{\tau\mu}(\bar{\omega}(x))\xi_\mu^k + \xi^i \dfrac{\partial \bar{\omega}^\tau(x)}{\partial x^i} = 0 \end{cases}$$

to the former ones. The comparison will also be done between the corresponding symbols G_q and $\bar{G}_q$ both with their prolongations.

This is a basis for the resolution of the famous "*equivalence problem*" stated by E. Cartan, that we will study later on.

REMARK 2.38 Broadly, the idea is "*to fix the bundle of geometric objects and to vary the section*". If ω is the special section corresponding to the special Lie pseudogroup Γ we started with in order to construct $\mathscr{F}$, then to any section $\bar{\omega}$ corresponds a Lie pseudogroup $\bar{\Gamma}$ which is in general neither identical, nor similar to Γ.

EXAMPLE 2.39 In the example 2.37 we have:

$$\Gamma: \begin{cases} y^1 = f(x^1) \\ y^2 = \dfrac{x^2}{f'(x^1)} \end{cases} \qquad \bar{\Gamma}: \begin{cases} y^1 = x^1 + a \\ y^2 = x^2 + g(x^1) \end{cases}$$

PROPOSITION 2.40 If $\mathscr{R}_q$ and R_q are formally transitive Lie equations and if $\bar{\omega}$ is any section of $\mathscr{F}$ sufficiently close to ω, then $\bar{\mathscr{R}}_q$ and $\bar{R}_q$ are formally transitive Lie equations. Moreover we have:

$$\dim \mathscr{R}_q = \dim \bar{\mathscr{R}}_q \quad \text{and} \quad \dim R_q = \dim \bar{R}_q$$

Proof According to the preceding lemma, in order to establish an isomorphism:

$$R^0_{q;x} \approx \bar{R}^0_{q;x}$$

we have just to choose p such tnat:

$$\Phi(\bar{\omega}(x), p) = \omega(x)$$

But we have already proved that the action $\bar{u} = G(u, p)$ is transitive at all points u such that rank $[L_k^{\tau\mu}(u)] = \operatorname{codim} R_q$, that is to say rank $[L_k^{\tau\mu}(u)]$ is maximum, unless u satisfies equations obtained by equating to zero some determinants of the matrix $L = [L_k^{\tau\mu}(u)]$.

Thus $\bar{u}$ must be in a sufficiently small neighbourhood of u, and in this case, as rank $[L_k^{\tau\mu}(\bar{u})]$ is equal to the number of equations of $\bar{R}_q$, by Cramer's rules we can solve them $\forall \xi^i$. It follows that:

$$\dim R_{q;x} \leq \dim \bar{R}_{q;x} \leq n + \dim \bar{R}^0_{q;x} = n + \dim R^0_{q;x} = \dim R_{q;x}$$

Finally, when p is taken as above, we have:

$$\operatorname{rank}\left[\frac{\partial \Phi^\tau(\bar{\omega}(y), p)}{\partial p^k_\mu}\right] = \operatorname{rank}\,[L_k^{\tau\mu}(\omega(y))] \quad \text{and thus} \quad \dim \mathscr{R}_q = \dim \bar{\mathscr{R}}_q.$$

C.Q.F.D.

EXAMPLE 2.41 In the last two examples, the forbidden zone is

$$\{u^1 = 0, u^2 = 0\} \bigcup \{u^3 = 0\}.$$

EXAMPLE 2.42 When $\mathscr{F} = S_2 T^*$ with local coordinates (x, u_{ij}) we must have $\det [u_{ij}] \neq 0$.

EXAMPLE 2.43 When $\mathscr{F} = \Lambda^2 T^*$ with n even, we must have a similar condition and we may speak of "2-*forms of maximum rank*".

Similarly to the preceding proposition, we have:

PROPOSITION 2.44 Under the same hypothesis as those of the preceding proposition we have isomorphisms:

$$G_{q+r} \approx \bar{G}_{q+r} \qquad \forall r \geq 0$$

$$H^s_{q+r}(G_q) \approx H^s_{q+r}(\bar{G}_q) \qquad \forall r, s \geq 0.$$

REMARK 2.45 If ω_t is a section of $\mathscr{F}$ depending on a parameter t, with $\omega_0 = \omega$, then the hypothesis above is automatically satisfied when t is sufficiently small.

REMARK 2.46 It follows in particular that $\bar{G}_q$ is 2-acyclic if G_q is 2-acyclic and that $\bar{G}_q$ is involutive if G_q is involutive, both with the same characters α_q^i.

3 Integrability conditions

We come now to the link between the two parts of this book:

If we want to suppose that R_q is formally integrable (or involutive), using the criterion of formal integrability (or involutiveness), we have just to suppose that G_q is 2-acyclic (or involutive) and that the map $\pi_q^{q+1} : R_{q+1} \to R_q$ is surjective.

In fact, according to corollary 2.35, when R_q is a linear system of formally transitive Lie equations, then G_{q+1} is a vector bundle and it follows that R_{q+1} is also a vector bundle.

From the last remark, $\bar{G}_q$ is also 2-acyclic (or involutive) and $\bar{R}_q$ is formally integrable (or involutive) if and only if $\pi_q^{q+1} : \bar{R}_{q+1} \to \bar{R}_q$ is surjective. As it is easy to see that this cannot happen for any section $\bar{\omega}$ we realise that there must exist some "*integrability conditions*" for the section in order to have the surjectivity.

In order to simplify, we shall change the notations slightly and call R_q the linear system of general infinitesimal equations, determined by an arbitrary section ω such that the determinants of proposition 2.40 do not vanish. We shall look for the integrability conditions on ω that will make the map $\pi_q^{q+1} : R_{q+1} \to R_q$ surjective. We recall that we have, for a section ξ_q of R_q and a section ξ_{q+1} of R_{q+1}, with $\pi_q^{q+1} \circ \xi_{q+1} = \xi_q$:

$$
R_{q+1}\left\{\begin{aligned}
&\underset{1}{\Omega^\tau_j}(x) \equiv -\xi^k_{\mu+1_j}(x)L^{\tau\mu}_k(\omega(x)) - \xi^k_\mu(x)\frac{\partial L^{\tau\mu}_k(\omega(x))}{\partial u^\sigma}\cdot\frac{\partial \omega^\sigma(x)}{\partial x^j}\\
&\qquad\qquad + \xi^i_j(x)\frac{\partial \omega^\tau(x)}{\partial x^i} + \xi^i(x)\frac{\partial}{\partial x^i}\left(\frac{\partial \omega^\tau(x)}{\partial x^j}\right) = 0\\
&R_q\left\{\underset{1}{\Omega^\tau}(x) \equiv -\xi^k_\mu(x)L^{\tau\mu}_k(\omega(x)) + \xi^i(x)\frac{\partial \omega^\tau(x)}{\partial x^i} = 0\right.
\end{aligned}\right.
$$

PROPOSITION 3.1 The map $\pi_q^{q+1} : R_{q+1} \to R_q$ is surjective if and only if the induced map $\pi_q^{q+1} : R^0_{q+1} \to R^0_q$ is surjective and the map $\pi_0^{q+1} : R_{q+1} \to T$ is surjective.

Proof This uses a chase in the following commutative and exact diagram.

$$
\begin{array}{ccccccccc}
 & & 0 & & 0 & & & & \\
 & & \downarrow & & \downarrow & & & & \\
0 & \longrightarrow & G_{q+1} & = & G_{q+1} & \longrightarrow & 0 & & \\
 & & \downarrow & & \downarrow & & \downarrow & & \\
0 & \longrightarrow & R^0_{q+1} & \longrightarrow & R_{q+1} & \xrightarrow{\pi^{q+1}_0} & T & \longrightarrow & 0 \\
 & & \downarrow{\scriptstyle \pi^{q+1}_q} & & \downarrow{\scriptstyle \pi^{q+1}_q} & & \| & & \\
0 & \longrightarrow & R^0_q & \longrightarrow & R_q & \xrightarrow{\pi^q_0} & T & \longrightarrow & 0 \\
 & & \downarrow & & & & \downarrow & & \\
 & & 0 & & & & 0 & &
\end{array}
$$

1 *We shall first study the surjectivity of the induced map* $\pi^{q+1}_q : R^0_{q+1} \to R^0_q$.

From the last diagram we see that the surjectivity is equivalent to the relation:

$$\dim R^0_{q+1} = \dim G_{q+1} + \dim R^0_q$$

Now we shall consider, for any section ξ of T, the infinitesimal transformation:

$$\xi^i(x)\frac{\partial}{\partial x^i} + \partial_\mu \xi^k(x) L^{\tau\mu}_k(u)\frac{\partial}{\partial u^\tau}$$

and its first prolongation:

$$\xi^i(x)\frac{\partial}{\partial x^i} + \partial_\mu \xi^k(x) L^{\tau\mu}_k(u)\frac{\partial}{\partial u^\tau}$$
$$+ \left(\partial_{\mu+1_i}\xi^k(x)L^{\tau\mu}_k(u) + \partial_\mu \xi^k(x)\frac{\partial L^{\tau\mu}_k(u)}{\partial u^\sigma}u^\sigma_i - u^\tau_j \partial_i \xi^j(x)\right)\frac{\partial}{\partial u^\tau_i}$$

The infinitesimal generators of the action of $GL_{q+1}(n, \mathbb{R})$ on the fiber of $J_1(\mathscr{F})$, which is a bundle of geometric objects according to lemma 2.29, can be written, up to the sign, in matrix form, in the following symbolic notation, where $\dim J_q(T) - \dim R_q = \dim F_0$.

$\dim S_{q+1}T^* \otimes T$	$\dim J_q(T) - n$	
$\mathrm{Inv}\,(u)$	(u^τ, u_i^τ)	} $\dim T^* \otimes F_0$
0	$L(u)$	} $\dim F_0$

We shall call the corresponding vector fields $L_{(1)k}^{\mu}$, $1 \leq |\mu| \leq q + 1$. The ones with $|\mu| = q + 1$ are given by the matrix $\mathrm{Inv}\,(u)$.

According to corollary 2.35 we have:

$$\begin{cases} \dim G_{q+1} = \dim S_{q+1}T^* \otimes T - \max_u \operatorname{rank}\,[\mathrm{Inv}\,(u)] \\ \dim R_q^0 = (\dim J_q(T) - n) - \max_u \operatorname{rank}\,[L(u)] \end{cases}$$

and we require that the rank of the preceding matrix, evaluated at $u = \omega(x)$, must be equal to

$$\max_u \operatorname{rank}\,[\mathrm{Inv}\,(u)] + \max_u \operatorname{rank}\,[L(u)]$$

As in corollary 6.3.52 we must equate certain determinants to zero to get the equations:

$\mathscr{P}_1$
$$I_*\left(u, \frac{\partial u}{\partial x}\right) = 0$$

where we write $\partial u^\tau/\partial x^i$ instead of u_i^τ in order to indicate the variables.

DEFINITION 3.2 The corresponding conditions for the section ω:

$$I_*\left(\omega(x), \frac{\partial \omega(x)}{\partial x}\right) = 0 \qquad \forall x \in U \subset X$$

are called "*integrability conditions of the first kind*".

We shall now determine these conditions. However the following study is very technical and will not be understood by the reader before he has computed many examples by himself.

According to theorem 6.3.51, as the source x is not contained explicitly in the I_*, when we effect the first prolongation of any finite transformation:

$$\bar{x} = \varphi(x), \qquad \bar{u} = G\left(u, \frac{\partial \varphi}{\partial x}\right)$$

if

$$I_*\left(u, \frac{\partial u}{\partial x}\right) = 0 \quad \text{then} \quad I_*\left(\bar{u}, \frac{\partial \bar{u}}{\partial \bar{x}}\right) = 0.$$

As a first property, we may say for this reason that the integrability conditions of the first kind do not depend on the coordinate system on $\mathscr{F}$.

A second property is given by the following lemma.

LEMMA 3.3 The integrability conditions of the first kind are quasi-linear in the u_i^τ, that is to say:

we have: $$I_*^r(u^\tau, u_i^\tau) \equiv \alpha_\tau^{ri}(u)u_i^\tau + \beta^r(u) = 0$$

or simply: $$I_*\left(u, \frac{\partial u}{\partial x}\right) \equiv \alpha(u)\frac{\partial u}{\partial x} + \beta(u) = 0$$

Proof From the commutation relations of the vector fields $A_k^\mu(q+1)$ for $|\mu| = q + 1 \geq 2$, we see that the vector fields which have Inv (u) as matrix of components, are commutative and thus generate an involutive distribution on the fiber of $J_1(\mathscr{F})$.

Let u^τ and $K^\alpha(u, u_i^\tau) \equiv \chi_\tau^{\alpha i}(u)u_i^\tau$ be a maximum set of

$$\dim F_0 + \dim T^* \otimes F_0 - \max_u \operatorname{rank} \operatorname{Inv}(u) = \dim F_0 + \dim F_1$$

unconnected invariants of this distribution in the variables (u^τ, u_i^τ).

We have:

$$\operatorname{Inv}_{k,i}^{\mu,\tau}(u)\frac{\partial}{\partial u_i^\tau} \cdot K^\alpha \equiv \operatorname{Inv}_{k,i}^{\mu,\tau}\chi_\tau^{\alpha i}(u) \equiv 0$$

with $|\mu| = q + 1$ and such a set of dim F_1 invariants K^α can be found using Cramer's rules.

According to the fact that $J_1(\mathscr{F})$ is a bundle of geometric objects with commutation relations given in lemma 2.29, we see that the vector fields of this distribution are

$$\sum_{\substack{\mu + 1_i = \text{cst} \\ |\mu| = q}} L_k^{\tau\mu}(u)\frac{\partial}{\partial u_i^\tau},$$

and that this distribution admits the vector fields $L_{(1)l}^{\ \nu} \neq \rho_1(L_l^\nu)$.

It follows that we have for $1 \leq |\mu| \leq q$:

$$L_{(1)k}^{\ \mu} \cdot K^\alpha = a_{\beta,k}^{\alpha,\mu}(u)K^\beta + b_k^\mu(u)$$

because all the components are linear or quasi-linear in u_i^τ and that $\{K^\alpha\}$ is a maximum set.

Finally we must express the fact that the matrix:

$$\underbrace{\begin{array}{|c|c|}\hline \quad 0 \quad & a(u)v + b(u) \\ \hline 0 & L(u) \\ \hline \end{array}}_{\dim S_{q+1}T^* \otimes T \quad \dim J_q(T) - n} \begin{array}{l} \} \dim F_1 \\ \} \dim F_0 \end{array}$$

has rank equal to dim F_0 = rank $[L(u)]$, setting $v = k(u, u_i^\tau)$.

Elementary linear algebra shows that the conditions thus obtained are quasi-linear in the v and finally quasi-linear in the u_i^τ. C.Q.F.D.

REMARK 3.4 When $q = 1$, we have only to consider the vector fields:

$$L_k^{\tau i}(u)\frac{\partial}{\partial u^\tau} + \left(\frac{\partial L_k^{\tau i}(u)}{\partial u^\sigma}u_j^\sigma - \delta_j^i u_k^\tau\right)\frac{\partial}{\partial u_j^\tau}$$

As their components are linear in u_i^τ, the integrability conditions of the first kind are also linear in the u_i^τ, that is to say:

$$q = 1 \Rightarrow I_*\left(u, \frac{\partial u}{\partial x}\right) \equiv \alpha(u)\frac{\partial u}{\partial x} = 0$$

This remark is important because all the well known classical examples are of this type.

The above lemma will now be used in order to compute the number of integrability conditions of the first kind.

PROPOSITION 3.5 The integrability conditions of the first kind describe, at least locally, a fibered submanifold $\mathscr{P}_1 \subset J_1(\mathscr{F})$. Moreover their number is well defined and $\leq \dim F_1$.

Proof We cannot have relations between the u only, because at this time if

$$I_*\left(u, \frac{\partial u}{\partial x}\right) \equiv \beta(u) = 0 \quad \text{and} \quad \bar{u} = G(u, p), \quad \text{then} \quad \beta(\bar{u}) = 0$$

and these conditions cannot be fulfilled by the special ω, by hypothesis.

Now, for arbitrary u, on the basis of the u_i^τ, we can select a maximum number of independent relations, say

$$\alpha_\tau^{ri}(u)u_i^\tau + \beta^r(u) = 0.$$

As they are quasi-linear combinations of the K, the number of such independent relations is surely $\leq \dim F_1$.

Finally we know that whenever

$$I_*\left(u, \frac{\partial u}{\partial x}\right) = 0$$

then as above we have

$$I_*\left(\bar{u}, \frac{\partial \bar{u}}{\partial \bar{x}}\right) = 0.$$

As $J_1(\mathscr{F})$ is an affine bundle over $\mathscr{F}$, we must have an identity:

$$I_*^r\left(\bar{u}, \frac{\partial \bar{u}}{\partial \bar{x}}\right) \equiv Q_s^r(u, p) I_*^s\left(u, \frac{\partial u}{\partial x}\right)$$

and looking at the coefficients of $\partial u/\partial x$,

$$\alpha_\sigma^{rj}(G(u, p)) \frac{\partial \bar{u}^\sigma}{\partial u^\tau}(u, p) \equiv Q_s^r(u, p)\alpha_\tau^{si}(u)p_i^j$$

with $1 \leq$ order $p \leq q$. Thus rank $[\alpha_\tau^{ri}(\bar{u})]$ = rank $[\alpha_\tau^{ri}(u)]$, this number being equal to its value when $u = \omega(x)$ and ω is special.

According to the remark above, the proof of course, is shortened when $q = 1$. C.Q.F.D.

2 *The study of the surjectivity of the map $\pi_0^{q+1} : R_{q+1} \to T$ is still more difficult*. It is based on the propositions 6.3.32 and 6.3.38. We invite the reader to effect it directly in the examples sketched hereafter.

First of all, we shall look at the distribution with matrix:

$a(u)v + b(u)$	$\left.\frac{\partial}{\partial v}\right\}\dim F_1$
$L(u)$	$\left.\frac{\partial}{\partial u}\right\}\dim F_0$

LEMMA 3.6 This distribution is involutive.

Proof From the commutation relations of the $A_k^\mu(q + 1)$ we have:

$$[L_{(1)k}^{\ \mu}, L_{(1)l}^{\ \nu}] = \begin{cases} 0 \\ \text{combination of vector fields of the distribution with} \\ \text{matrix Inv}(u) + \text{combination of } L_{(1)j}^{\ \lambda},\ 1 \leq |\lambda| \leq q \end{cases}$$

with $1 \leq |\mu|, |\nu|, |\lambda| \leq q$. Thus

$$[L_{(1)k}^{\ \mu}, L_{(1)l}^{\ \nu}] \cdot K^\alpha = \begin{cases} 0 \\ \text{combination of } L_{(1)j}^{\ \lambda} \cdot K^\alpha \text{ with } 1 \leq |\lambda| \leq q \end{cases}$$

The lemma follows from the fact that, by definition for the vector fields L, we have $L_{(1)} \cdot K = a(u)K + b(u)$. C.Q.F.D.

Then we shall restrict the above distribution to the invariant submanifold $\mathscr{P}_1$ obtained, in the coordinates (u, v) by setting:

$$\mathscr{P}_1 \qquad I_*\left(u, \frac{\partial u}{\partial x}\right) \equiv A(u)v + B(u) = 0$$

We again get a matrix, the rank of which is dim F_0, of the same kind as the one above.

On the basis of the u_i^τ we shall divide the K into two classes: the K_* relative to the linear part of the I_* and other independent K noted K_{**}.

According to proposition 6.3.39 the corresponding distribution, which now uses only the variables (u, v_{**}), is also involutive.

As its rank is equal to dim F_0, we can select, as in remark 6.3.11, dim F_0 unconnected vector fields in order to get a commutative basis of the following kind:

$$\frac{\partial}{\partial u^\tau} + (A_{\tau\beta}^\alpha(u)v_{**}^\beta + B_\tau^\alpha(u)) \frac{\partial}{\partial v_{**}^\alpha}$$

We need the following technical lemma:

LEMMA 3.7 We may find a maximum set of unconnected invariants of this distribution of the form:

$$Q_\beta^\gamma(u)v_{**}^\beta + R^\gamma(u)$$

Proof To do this, we have to solve the following system of p.d.e.:

$$\frac{\partial Q_\beta^\gamma}{\partial u^\tau} + A_{\tau\beta}^\alpha(u)Q_\alpha^\gamma = 0, \quad \frac{\partial R^\gamma}{\partial u^\tau} + B_\tau^\alpha(u)Q_\alpha^\gamma = 0$$

As the symbol of this system is trivially involutive (because it is of finite type), in order that the system becomes involutive, it is easy to check that we must have:

$$\begin{cases} \dfrac{\partial A_{\sigma\beta}^\gamma(u)}{\partial u^\tau} - \dfrac{\partial A_{\tau\beta}^\gamma(u)}{\partial u^\sigma} + A_{\tau\beta}^\alpha(u)A_{\sigma\alpha}^\gamma(u) - A_{\sigma\beta}^\alpha(u)A_{\tau\alpha}^\gamma(u) = 0 \\ \dfrac{\partial B_\sigma^\gamma(u)}{\partial u^\tau} - \dfrac{\partial B_\tau^\gamma(u)}{\partial u^\sigma} + B_\tau^\alpha(u)A_{\sigma\alpha}^\gamma(u) - B_\sigma^\alpha(u)A_{\tau\alpha}^\gamma(u) = 0 \end{cases}$$

A straightforward computation then shows that these conditions are fulfilled and can be deduced from the commutativity of the above vector fields, basis of an involutive distribution.

The integration of the latter system can be done, using the Frobenius theorem. C.Q.F.D.

REMARK 3.8 When $q = 1$, we have $\beta(u) \equiv 0$, and thus $B(u) \equiv 0$. It follows that we may take $R(u) \equiv 0$.

Substituting in these invariants the formulas for the K we find invariants, denoted by $I_{**}(u, \partial u/\partial x)$, that are quasi-linear (or linear if $q = 1$) in $\partial u/\partial x$, and such that for $1 \leq |\mu| \leq q + 1$

$$L_{(1)k}^{\;\;\mu} \cdot I_{**} = 0 \quad \text{whenever} \quad I_*\left(u, \frac{\partial u}{\partial x}\right) = 0$$

As the number of integrability conditions of the first kind is well defined and the total number of the I is equal to dim F_1 according to corollary 6.3.52, the number of invariants I_{**} is well defined.

We can now study the surjectivity of the map $\pi_0^{q+1} : R_{q+1} \to T$.

From the definition of I_* and I_{**} we have for any section ξ of T:

$$\left[\partial_\mu \xi^k(x) L_k^{\tau\mu}(u) \frac{\partial}{\partial u^\tau} + \left(\partial_{\mu+1_i} \xi^k(x) L_k^{\tau\mu}(u) + \partial_\mu \xi^k(x) \frac{\partial L_k^{\tau\mu}(u)}{\partial u^\sigma} u_i^\sigma - u_j^\tau \partial_i \xi^j(x)\right) \frac{\partial}{\partial u_i^\tau}\right] \cdot I = 0$$

when

$$I_*\left(u, \frac{\partial u}{\partial x}\right) = 0,$$

that is to say on $\mathscr{P}_1$.

It follows that:

$$\underset{1}{\Omega}^\tau(x) \cdot \frac{\partial I}{\partial u^\tau}\left(\omega(x), \frac{\partial \omega(x)}{\partial x}\right) + \underset{1}{\Omega}_i^\tau(x) \cdot \frac{\partial I}{\partial u_i^\tau}\left(\omega(x), \frac{\partial \omega(x)}{\partial x}\right) \equiv \xi^i(x) \frac{\partial}{\partial x^i}\left(I\left(\omega(x), \frac{\partial \omega(x)}{\partial x}\right)\right)$$

when

$$I_*\left(\omega(x), \frac{\partial \omega(x)}{\partial x}\right) = 0, \qquad \forall x \in U \subset X$$

Finally, when the integrability conditions of the first kind are satisfied by the section ω, that is to say when the map $\pi_q^{q+1}: R_{q+1}^0 \to R_q^0$ is surjective, then the map $\pi_0^{q+1}: R_{q+1} \to T$ is surjective if and only if

$$I_{**}\left(\omega(x), \frac{\partial\omega(x)}{\partial x}\right) = c$$

where the "c" are constants, called "*structure constants*".

DEFINITION 3.9 The conditions for the section ω:

$$I_{**}\left(\omega(x), \frac{\partial\omega(x)}{\partial x}\right) = c$$

are called "*integrability conditions of the second kind.*"

We have proved:

THEOREM 3.10 The map $\pi_q^{q+1}: R_{q+1} \to R_q$ is surjective if and only if the following dim F_1 integrability conditions for the section ω of $\mathscr{F}$ are satisfied:

$$\mathscr{R}'_1 \subset J_1(\mathscr{F}) \begin{cases} I_*\left(\omega(x), \dfrac{\partial\omega(x)}{\partial x}\right) = 0 & \text{(first kind)} \\ I_{**}\left(\omega(x), \dfrac{\partial\omega(x)}{\partial(x)}\right) = c & \text{(second kind)} \end{cases}$$

The functions $I(u^\tau, u_i^\tau)$ can be chosen quasi-linear in u_i^τ if $q > 1$ or linear if $q = 1$. Moreover the integrability conditions are invariant "*in form*" under any change of coordinates.

REMARK 3.11 When the systems of special equations are involutive, the systems of general equations are also involutive, provided the above conditions are satisfied. Now the latter identities are just equivalent to:

$$\mathscr{D}_1 \cdot \underset{1}{\Omega} = 0 \quad \text{when} \quad \mathscr{D} \cdot \xi = \underset{1}{\Omega}$$

and we have constructed the initial part of the P-sequence for an involutive Lie operator $\mathscr{D}$ by linearising the integrability conditions "*keeping the same constants c*" and setting:

$$\omega_t(x) = \omega(x) + t \cdot \underset{1}{\Omega}(x) + \cdots$$

PROPOSITION 3.12 Let $\varphi \in \mathrm{Aut}(X)$ and $\bar{\omega} = \varphi(\omega)$ defined by

$$\bar{\omega}(\varphi(x)) = G\left(\omega(x), \frac{\partial \varphi(x)}{\partial x}\right).$$

Then:

$$\begin{cases} I_*\left(\omega(x), \dfrac{\partial \omega(x)}{\partial x}\right) = 0 \\ I_{**}\left(\omega(x), \dfrac{\partial \omega(x)}{\partial x}\right) = c \end{cases} \Leftrightarrow \begin{cases} I_*\left(\bar{\omega}(x), \dfrac{\partial \bar{\omega}(x)}{\partial x}\right) = 0 \\ I_{**}\left(\bar{\omega}(x), \dfrac{\partial \bar{\omega}(x)}{\partial x}\right) = c \end{cases}$$

Proof By definition we have with $\bar{x} = \varphi(x)$:

$$I_*\left(\bar{\omega}(\bar{x}), \frac{\partial \bar{\omega}(\bar{x})}{\partial \bar{x}}\right) = 0 \quad \text{when} \quad I_*\left(\omega(x), \frac{\partial \omega(x)}{\partial x}\right) = 0$$

and at this time

$$I_{**}\left(\bar{\omega}(\bar{x}), \frac{\partial \bar{\omega}(\bar{x})}{\partial \bar{x}}\right) = I_{**}\left(\omega(x), \frac{\partial \omega(x)}{\partial x}\right) = c$$

C.Q.F.D.

REMARK 3.13 This proposition is the finite formulation of the preceding remark.

From the above proposition we deduce easily:

PROPOSITION 3.14 If R_q is formally transitive, s-acyclic and such that $\pi_q^{q+1}: R_{q+1} \to R_q$ is surjective, then the finite form $\mathscr{R}_q$ of R_q:

$$\mathscr{R}_q \qquad \Phi^\tau(\omega(y), p) = \omega^\tau(x)$$

is formally transitive, s-acyclic and such that $\pi_q^{q+1}: \mathscr{R}_{q+1} \to \mathscr{R}_q$ is surjective. Moreover we have the equivalence:

$$\begin{cases} I_*\left(\Phi, \dfrac{d\Phi}{dx}\right) = 0 \\ I_{**}\left(\Phi, \dfrac{d\Phi}{dx}\right) = c \end{cases} \Leftrightarrow \begin{cases} I_*\left(\omega(x), \dfrac{\partial \omega(x)}{\partial x}\right) = 0 \\ I_{**}\left(\omega(x), \dfrac{\partial \omega(x)}{\partial x}\right) = c \end{cases}$$

EXAMPLE 3.15 We keep the same example as the preceding ones. Up to the sign, the matrix defining R_1^0 is the following one, where we have to set $u^\tau = \omega^\tau(x)$, $u_i^\tau = \partial \omega^\tau(x)/\partial x^i$.

ξ^2_{22}	ξ^1_{22}	ξ^2_{12}	ξ^1_{12}	ξ^2_{11}	ξ^1_{11}	ξ^2_2	ξ^1_2	ξ^2_1	ξ^1_1	
0	0	u^2	u^1	0	0	u^1_2	u^1_1	u^2_2	u^1_2	u^1_2
0	0	0	0	u^2	u^1	0	0	$u^2_1 + u^1_2$	$2u^1_1$	u^1_1
u^2	u^1	0	0	0	0	$2u^2_2$	$u^1_2 + u^2_1$	0	0	u^2_2
0	0	u^2	u^1	0	0	u^2_1	u^1_1	u^2_2	u^2_1	u^2_1
u^3	0	0	u^3	0	0	$2u^3_2$	u^3_1	0	u^3_2	u^3_2
0	0	u^3	0	0	u^3	u^3_1	0	u^3_2	$2u^3_1$	u^3_1
0	0	0	0	0	0	0	0	u^2	u^1	u^1
0	0	0	0	0	0	u^2	u^1	0	0	u^2
0	0	0	0	0	0	u^3	0	0	u^3	u^3

We have for example the commutation relation between the $L_{(1)}$:

$$\left[u^2 \frac{\partial}{\partial u^2_2} + u^3 \frac{\partial}{\partial u^3_2}, u^1_2 \frac{\partial}{\partial u^1_2} + 2u^2_2 \frac{\partial}{\partial u^2_2} + u^2_1 \frac{\partial}{\partial u^2_1} + 2u^3_2 \frac{\partial}{\partial u^3_2} + u^3_1 \frac{\partial}{\partial x^3_1} + u^2 \frac{\partial}{\partial x^2} + u^3 \frac{\partial}{\partial u^3}\right] = u^2 \frac{\partial}{\partial u^2_2} + u^3 \frac{\partial}{\partial u^3_2}$$

The only invariant of the distribution with matrix Inv (u) is $K = u^1_2 - u^2_1$. The new involutive distribution to be introduced is:

ξ^2_2	ξ^1_2	ξ^2_1	ξ^1_1	
u	0	0	u	u
0	0	u^2	u^1	u^1
u^2	u^1	0	0	u^2
u^3	0	0	u^3	u^3

A commutative basis for this distribution is:

$$\frac{\partial}{\partial u^1}, \frac{\partial}{\partial u^2}, \frac{1}{u^3}\, v\frac{\partial}{\partial v} + \frac{\partial}{\partial u^3}$$

(This can be done because we must have $u^3 \neq 0$).

We find no integrability condition of the first kind and the integrability condition of the second kind is:

$$\frac{1}{\omega^3(x)}\left(\frac{\partial \omega^1(x)}{\partial x^2} - \frac{\partial \omega^2(x)}{\partial x^1}\right) = c.$$

4 Third fundamental theorem

From now on we shall assume that X has a structure of real analytic manifold compatible with its differentiable structure. In fact, as we shall be concerned with local problems, we may even take an open set $U \subset \mathbb{R}^n$ instead of X *but we require to know what happens under any change of local coordinates.*

Looking back to the local theory of Lie groups, the reader will discover that, for a Lie group G acting on a manifold X such that $\dim G = \dim X$, the bundle of geometric objects to be introduced is just the fibered product over X of n copies of T^*. He will just have to consider the special case of an additive group acting by translations $y^i = x^i + a^i$.

Now a section ω of this bundle is a set of n 1-forms $w^h = \omega^h_i(x)dx^i$.

We have, for any section ω:

$$\mathcal{R}_1 \qquad \omega^h_k(y)p^k_i = \omega^h_i(x)$$

$$R_1 \qquad \underset{1}{\Omega^h_i} \equiv \omega^h_k(x)\xi^k_i + \xi^r \frac{\partial \omega^h_i(x)}{\partial x^r} = 0$$

In order to get only vector bundles, using $\omega^h_i(x) = \delta^h_i$ as special section, we have only to ask for the inequality $\det[\omega^h_i(x)] \neq 0$. In this case we have:

$$G_{1+r} = 0, \quad R^0_{1+r} = 0 \qquad \forall r \geq 0.$$

In particular it follows that there are no integrability conditions of the first kind.

It is easy to check that the integrability conditions of the second kind are just equivalent to the Maurer–Cartan equations:

$$(I_{**}) \qquad \alpha^i_k(x)\alpha^j_l(x)\left(\frac{\partial \omega^h_j(x)}{\partial x^i} - \frac{\partial \omega^h_i(x)}{\partial x^j}\right) = c^h_{kl}$$

The symbol of this non-linear system is involutive and we have seen that this system is involutive if and only if the structure constants c^h_{kl} satisfy the Jacobi conditions.

It has been shown that the third fundamental theorem of Lie asserts that, given any set of $n^2(n-1)/2$ constants satisfying the Jacobi conditions, then it is possible to find an analytic section ω solution of the latter non-linear involutive system of order one. In fact, as this system is also analytic, we can use theorem 4.4.3 of the first part of this book in order to construct such a section.

To sum up, we may say that the third fundamental theorem of Lie is just equivalent to the analyticity and formal integrability of the non-linear system of the integrability conditions with structure constants satisfying the Jacobi relations.

Contrary to the methods developed by Cartan in his study of the Lie pseudogroups, and more recently by Spencer, Goldschmidt, Guillemin and others, we believe that the present approach, based on the work of Vessiot, is the natural generalisation of the local theory of Lie groups as described above.

In order to exhibit the analogy we shall consider the non-linear system $\mathcal{R}'_1 \subset J_1(\mathcal{F})$ of order 1 on $\mathcal{F}$:

$$\mathcal{R}'_1 \qquad \begin{cases} I_*\left(u, \dfrac{\partial u}{\partial x}\right) = 0 \\ I_{**}\left(u, \dfrac{\partial u}{\partial x}\right) = c \end{cases}$$

with fixed arbitrary structure constants.

We have

$$\operatorname{codim} \mathcal{R}'_1 = \dim F_1$$

The first analogy is in dealing with the analyticity of this system.

THEOREM 4.1 The bundle $\mathcal{F}$ can be chosen in such a way that $\mathcal{R}'_1$ becomes a non-linear analytic system.

Proof We shall use a trick in order to find out the differential invariants Φ^τ.

For this we may look at the distribution $\#(R^0_q)$ on $J_q(X \times Y)$. It is easy to check that this distribution is involutive and that $[R^0_q, R^0_q] \subset R^0_q$, this bracket being now $C^\infty(X)$-linear.

Using a solved form for R_q, which is a vector bundle over X by hypothesis, we see that the distribution $\#(R^0_q)$ may be generated by a basis of vector fields such as:

$$(1) \qquad a^k_\mu(y, p)\frac{\partial}{\partial p^k_\mu} \qquad 1 \le |\mu| \le q$$

where the components $a_\mu^k(y, p)$ are polynomials in the p with coefficients belonging to $C^\infty(Y)$.

Similarly we observe that a basis of $\#(R_q)$ may be constituted by the former vector fields and by n new ones, such as:

$$\frac{\partial}{\partial y^k} + \alpha_{k\nu}^l(y, p)\frac{\partial}{\partial p_\nu^l} \qquad 1 \leq |\nu| \leq q \tag{2}$$

because R_q is formally transitive.

As $\#(R_q)$ is an involutive distribution, we have:

(i) The bracket of two vector fields of type (1) is a linear combination with coefficients $\in C^\infty(Y)$ of vector fields of type (1).

(ii) The bracket of a vector field of type (1) with a vector field of type (2) is a linear combination with coefficients $\in C^\infty(Y)$ of vector fields of type (1) because we have just to look at the projection on $\partial/\partial y^k$.

(iii) The bracket of two vector fields of type (2) is, for the same reason, a linear combination with coefficients $\in C^\infty(Y)$ of vector fields of type (1).

Let y^k and $Q^\sigma(y, p)$ be a maximum set of unconnected invariants of the distribution $\#(R_q^0)$. According to the Frobenius integration theorem and the isomorphisms of corollary 2.36, we can fix a point $y_0 \in Y$ and choose the functions $Q^\sigma(y, p)$ such that $Q^\sigma(y_0, p)$ are analytic in p, in a neighbourhood of $\mathrm{id}_q(y_0)$.

Now, from (ii) we obtain:

$$\left[\frac{\partial}{\partial y^k} + \alpha_{k\nu}^l(y, p)\frac{\partial}{\partial p_\nu^l}, a_\mu^k(y, p)\frac{\partial}{\partial p_\mu^k}\right] \cdot Q^\sigma \equiv 0$$

and from the implicit function theorem, we have:

$$\left(\frac{\partial}{\partial y^k} + \alpha_{k\nu}^l(y, p)\frac{\partial}{\partial p_\nu^l}\right) \cdot Q^\sigma \equiv \alpha_k^\sigma(y, Q)$$

Introducing new coordinates (y, u) we get a commutative distribution:

$$\frac{\partial}{\partial y^k} + \alpha_k^\sigma(y, u)\frac{\partial}{\partial u^\sigma}$$

Using the Frobenius integration theorem again and a device of A. Mayer (4) that allows one to consider the y^k as n parameters on a curve solution of the non-linear system $\partial u^\sigma/\partial y^k = \alpha_k^\sigma(y, u)$, we can determine, uniquely, a maximum set of unconnected invariants $H^\tau(y, u)$ such that $H^\tau(y_0, u) = u^\tau$.

Our desired differential invariants are thus:

$$\Phi^\tau(y, p) \equiv H^\tau(y, Q(y, p)) \quad \text{with} \quad \Phi^\tau(y_0, p) \equiv Q^\tau(y_0, p).$$

But now, by definition we have:

$$A_k^\mu(q) \cdot \Phi^\tau \equiv L_{\mu k}^\tau(\Phi)$$

and this identity does not depend on y because of the formal transitivity of $\mathscr{R}_q$.

In particular, taking $y = y_0$ and using the implicit function theorem for analytic functions, we find that the functions $L_k^{\tau\mu}(u)$ are analytic in u, on a neighbourhood of $u_0 = \omega(y_0)$.

We may now take $x_0 = y_0 \in X$ and work over a convenient neighbourhood U of x_0 in X.

The analyticity of $\mathscr{F}$ can be deduced at once from the above result and the fact that x_0 is arbitrary.

Finally, as the vector fields L are analytic in u, the vector fields $L_{(1)}$ are quasi-linear in $\partial u/\partial x$ with coefficients analytic in u.

It follows easily that the functions $I(u, \partial u/\partial x)$ can be chosen quasi-linear in $\partial u/\partial x$ with coefficients analytic in u, because of the Frobenius integration theorem in the analytic case.

This concludes the proof of the theorem. C.Q.F.D.

From now on we shall assume that the bundle $\mathscr{F}$ is chosen in such a way. It follows that all the computations on the fibers of $J_r(\mathscr{F})$, with $r \geq 0$, will only introduce analytic functions on those fibers, defined on convenient open sets.

Let us consider again the non-linear system $\mathscr{R}'_1$ with fixed arbitrary structure constants c.

Applying again the criterion of formal integrability (or involutiveness) we do want the symbol G'_1 of $\mathscr{R}'_1$ to be 2-acyclic (or involutive) and the map $\pi_1^2 : \mathscr{R}'_2 \to \mathscr{R}'_1$ to be surjective.

From the construction of $\mathscr{R}'_1$ it follows that G'_r is a vector bundle over $\mathscr{F}$ because G_{q+r} is a vector bundle over $\mathscr{R}_q$, $\forall r \geq 0$, and that $\mathscr{R}'_1$ is an affine bundle over $\mathscr{F}$, modelled on G'_1.

REMARK 4.2 Of course we have to suppose that u is such that rank $[L(u)]$ is equal to its maximum value. This will be assumed from now on, shrinking the fiber of $\mathscr{F}$ if necessary.

As in the theory of linear systems, we shall introduce

$$R'_1 = \text{im } V_{\text{id}_q}(\Phi) \subset J_1(F_0)$$

which is a linear system of order one, depending on ω.

The symbol of R'_1 is the reciprocal image by ω over X of the symbol of $\mathscr{R}'_1$, considered as a vector bundle over $\mathscr{F}$. For simplicity we shall use the same notation G'_1 for both of them.

As by construction $\mathscr{R}'_1$ is invariant by the action of the vector fields $L_{(1)}$ on $J_1(\mathscr{F})$, the restriction of which to $\mathscr{R}'_1$ is transitive, it follows that G'_1 is 2-acyclic (or involutive) when G_q is 3-acyclic (or involutive). This will be assumed henceforth.

We now need a technical lemma.

Let $\mathscr{E}$ be a fibered manifold over X and let $\mathscr{R}_q \subset J_q(\mathscr{E})$ be a non-linear system which is invariant by the q-prolongation $\rho_q(\theta)$ of a vector field

$$\theta = \xi^i(x)\frac{\partial}{\partial x^i} + \eta^k(x, y)\frac{\partial}{\partial y^k}$$

on $\mathscr{E}$, and such that $\mathscr{R}_{q+1}$ is a fibered submanifold of $J_{q+1}(\mathscr{E})$.

LEMMA 4.3 $\mathscr{R}_{q+1}$ is invariant by $\rho_{q+1}(\theta)$.

Proof We shall use local coordinates. Let $\mathscr{R}_q$ be defined locally by the equations:

$$\mathscr{R}_q \qquad \Phi^\tau(x, y, p) = 0$$

According to proposition 5.12 we have:

$$\rho_{q+1}(\theta)\cdot\left(\frac{d\Phi^\tau}{dx^i}\right) = \frac{d}{dx^i}(\rho_q(\theta)\cdot\Phi^\tau) - \frac{\partial\xi^j(x)}{\partial x^i}\cdot\frac{d\Phi^\tau}{dx^j}$$

and we have just to prove that, if Ψ is a function on $J_q(\mathscr{E})$ which vanishes on $\mathscr{R}_q$, then $d\Psi/dx^i$ vanishes on $\mathscr{R}_{q+1}$. This is easily done using lemma 4.4.7.

In fact, locally, we have:

$$\Psi(x, y, p) \equiv A_\tau(x, y, p)\cdot\Phi^\tau(x, y, p)$$

and

$$\frac{d\Psi}{dx^i} \equiv \frac{dA_\tau}{dx^i}\cdot\Phi^\tau + A_\tau\cdot\frac{d\Phi^\tau}{dx^i} = 0$$

on $\mathscr{R}_{q+1}$. C.Q.F.D.

Now if we set

$$\theta = \xi^i(x)\frac{\partial}{\partial x^i} + \partial_\mu \xi^k(x) L_k^{\tau\mu}(u)\frac{\partial}{\partial u^\tau}$$

as a vector field on $\mathscr{F}$, then by definition $\mathscr{R}'_1$ is invariant by $\rho_1(\theta)$. It follows that $\mathscr{R}'_2$ is invariant by $\rho_2(\theta)$ and that $\mathscr{R}'^{(1)}_1$ is invariant by $\rho_1(\theta)$, for any section ξ of T. However for an arbitrary set of structure constants, we have in general $\mathscr{R}'^{(1)}_1 \neq \mathscr{R}'_1$.

It remains to study the surjectivity of the map $\pi_1^2 : \mathscr{R}'_2 \to \mathscr{R}'_1$. We can look at it using local coordinates as we have proved that the form of the equations defining $\mathscr{R}'_1$ does not depend on the trivialisation of $\mathscr{F}$.

If we have:

$$I\left(u, \frac{\partial u}{\partial x}\right) \equiv A(u)\frac{\partial u}{\partial x} + B(u)$$

we have:

$$\frac{dI}{dx} \equiv A(u)\frac{\partial^2 u}{\partial x^2} + \frac{\partial A(u)}{\partial u}\frac{\partial u}{\partial x}\frac{\partial u}{\partial x} + \frac{\partial B(u)}{\partial u}\frac{\partial u}{\partial x} = 0$$

If we eliminate the second order derivatives by linear combinations, using Gramer's rules, we get equations such as:

$$\alpha(u)\frac{\partial u}{\partial x}\frac{\partial u}{\partial x} + \beta(u)\frac{\partial u}{\partial x} = 0$$

Using the integrability conditions of the first kind, we obtain equations on $\mathscr{P}_1$, of the form:

$$\alpha(u)\frac{\partial u}{\partial x}\frac{\partial u}{\partial x} + \beta(u)\frac{\partial u}{\partial x} + \gamma(u) = 0$$

where for simplicity we keep the same notations as above for a system of local coordinates on $\mathscr{P}_1$.

PROPOSITION 4.4 The former equations can be written:

$$\alpha(u)I_{**}I_{**} + \beta(u)I_{**} + \gamma(u) = 0$$

Proof From the definition and invariance property of $\mathscr{P}_1$ we have:

$$(L_{(1)k}^{\ \nu} \cdot I_*)|_{\mathscr{P}_1} = 0 \quad \text{or} \quad L_{(1)k}^{\ \nu} \cdot I_* = M(u) \cdot I_*, \qquad 1 \leq |\nu| \leq q+1$$

in symbolic matrix notation. More generally, we have:

$$(L_{(1)k}^{\ \nu} \cdot I)|_{\mathscr{P}_1} = 0 \quad \text{or} \quad L_{(1)k}^{\ \nu} \cdot I = M(u) I_*, \qquad 1 \leq |\nu| \leq q+1$$

which introduces the invariance property of $\mathscr{R}'_1 \subset \mathscr{P}_1$. Of course, if we introduce

$$\theta = \xi^i(x)\frac{\partial}{\partial x^i} + \partial_\mu \xi^k(x) L_k^{\tau\mu}(u)\frac{\partial}{\partial u^\tau},$$

$$\begin{aligned}
\rho_2(\theta) = {} & \xi^i(x)\frac{\partial}{\partial x^i} + \partial_\mu \xi^k L_k^{\tau\mu}(u)\frac{\partial}{\partial u^\tau} \\
& + \left(\partial_{\mu+1_i}\xi^k(x) L_k^{\tau\mu}(u) + \partial_\mu \xi^k(x)\frac{\partial L_k^{\tau\mu}(u)}{\partial u^\sigma} u_i^\sigma - \partial_i \xi^j(x) u_j^\tau\right)\frac{\partial}{\partial u_i^\tau} \\
& + \left(\partial_{\mu+1_i+1_j}\xi^k(x) L_k^{\tau\mu}(u) + \partial_{\mu+1_i}\xi^k(x)\frac{\partial L_k^{\tau\mu}(u)}{\partial u^\sigma} u_j^\sigma\right. \\
& + \partial_{\mu+1_j}\xi^k(x)\frac{\partial L_k^{\tau\mu}(u)}{\partial u^\sigma} u_i^\sigma + \partial_\mu \xi^k(x)\frac{\partial^2 L_k^{\tau\mu}(u)}{\partial u^\sigma \partial u^\rho} u_i^\sigma u_j^\rho \\
& + \partial_\mu \xi^k(x)\frac{\partial L_k^{\tau\mu}(u)}{\partial u^\sigma} u_{ij}^\sigma - \partial_{ij}\xi^k(x) u_k^\tau - \partial_i \xi^k(x) u_{jk}^\tau \\
& \left. - \partial_j \xi^k(x) u_{ik}^\tau\right)\frac{\partial}{\partial u_{ij}^\tau}
\end{aligned}$$

with $1 \leq |\mu| \leq q$ and a summation done for $i \leq j$.

Looking at the case $|\nu| = q+1$, we get:

$$\begin{aligned}
L_{(2)k}^{\ \nu} = {} & \sum_{\mu+1_i=\nu} L_k^{\tau\mu}(u)\frac{\partial}{\partial u_i^\tau} + \sum_{\mu+1_i+1_j=\nu} L_k^{\tau\mu}(u)\frac{\partial}{\partial u_{ij}^\tau} \\
& + \left(\sum_{\mu+1_i=\nu}\frac{\partial L_k^{\tau\mu}(u)}{\partial u^\sigma} u_j^\sigma + \sum_{\mu+1_j=\nu}\frac{\partial L_k^{\tau\mu}(u)}{\partial u^\sigma} u_i^\sigma\right)\frac{\partial}{\partial u_{ij}^\tau}
\end{aligned}$$

If we set:

$$I^\alpha \equiv A_\tau^{\alpha i}(u) u_i^\tau + B^\alpha(u)$$

then

$$\frac{dI^\alpha}{dx^j} \equiv A_\tau^{\alpha_i}(u)u_{ij}^\tau + \frac{\partial A_\tau^{\alpha_i}(u)}{\partial u^\sigma} u_i^\tau u_j^\sigma + \frac{\partial B^\alpha(u)}{\partial u^\sigma} u_j^\sigma,$$

$$L_{(2)k}^{\nu} \cdot \frac{dI^\alpha}{dx^j} \equiv A_\tau^{\alpha_i}(u)\Bigg(\sum_{\mu+1_i+1_j=\nu} L_k^{\tau\mu}(u) + \sum_{\mu+1_i=\nu} \frac{\partial L_k^{\tau\mu}(u)}{\partial u^\sigma} \cdot u_j^\sigma$$

$$+ \sum_{\mu+1_j=\nu} \frac{\partial L_k^{\tau\mu}(u)}{\partial u^\sigma} u_i^\sigma \Bigg) + \sum_{\mu+1_i=\nu} L_k^{\tau\mu}(u) \cdot \frac{\partial A_\tau^{\alpha_i}(u)}{\partial u^\sigma} u_j^\sigma$$

$$+ \sum_{\mu+1_j=\nu} L_k^{\tau\mu}(u) \cdot \frac{\partial A_\tau^{\alpha_i}(u)}{\partial u^\tau} u_i^\sigma + \sum_{\mu+1_j=\nu} L_k^{\tau\mu}(u) \frac{\partial B^\alpha(u)}{\partial u^\tau}$$

$$\equiv \sum_{\mu+1_i=\nu} \frac{\partial (L_k^{\tau\mu}(u) \cdot A_\tau^{\alpha_i}(u))}{\partial u^\sigma} u_j^\sigma$$

$$+ \sum_{\mu+1_j=\nu} \Bigg[\sum_{\lambda+1_i=\mu} L_k^{\tau\lambda}(u) \cdot A_\tau^{\alpha_i}(u) + L_k^{\tau\mu}(u) \frac{\partial B^\alpha(u)}{\partial u^\tau}$$

$$+ \left(L_k^{\tau\mu}(x) \frac{\partial A_\sigma^{\alpha_i}(u)}{\partial u^\tau} + \frac{\partial L_k^{\tau\mu}(u)}{\partial u^\sigma} \cdot A_\tau^{\alpha_i}(u) \right) u_i^\sigma \Bigg]$$

But we have:

$$L_{(1)k}^{\nu} \cdot I^\alpha \equiv \sum_{\mu+1_i=\nu} L_k^{\tau\mu}(u) \cdot A_\tau^{\alpha_i}(u) = 0$$

since the I are quasi-linear in the K. We get:

$$L_{(2)k}^{\nu} \cdot \frac{dI^\alpha}{dx^j} \equiv \sum_{\mu+1_j=\nu} L_{(1)k}^{\mu} \cdot I^\alpha, \qquad |\nu| = q+1$$

and it follows that

$$L_{(2)k}^{\nu} \cdot \frac{dI}{dx} \equiv M(u) \cdot I_*, \quad \text{when} \quad |\nu| = q+1.$$

Now, if we consider the linear combinations such as $A_\alpha^j(u)(dI^\alpha/dx^j)$ that do not contain any second order jet-coordinate u_{ij}^τ, we have:

$$L_{(1)k}^{\nu} \cdot \left(A_\alpha^j(u) \frac{dI^\alpha}{dx^j} \right) \equiv A_\alpha^j(u) \cdot \left(L_{(2)k}^{\nu} \cdot \frac{dI^\alpha}{dx^j} \right) \equiv A(u) \cdot M(u) \cdot I_*$$

and it follows that

$$\left(L_{(1)k}^{\nu} \cdot \left(A_\alpha^j(u) \cdot \frac{dI^\alpha}{dx^j} \right) \right) \Bigg|_{\mathscr{P}_1} = 0 \qquad \forall |\nu| = q+1$$

We notice that we must have:

$$A_\alpha^j(u)A_\tau^{\alpha i}(u) + A_\alpha^i(u)A_\tau^{\alpha j}(u) = 0$$

because $u_{ij}^\tau = u_{ji}^\tau$ in the summation.

Finally,

$$L_{(1)k}^{\nu}\Big|_{\mathscr{P}_1} \cdot \left(A_\alpha^j(u)\frac{dI^\alpha}{dx^j}\right)\Big|_{\mathscr{P}_1} = \left(L_{(1)k}^{\nu} \cdot \left(A_\alpha^j(u)\frac{dI^\alpha}{dx^j}\right)\right)\Big|_{\mathscr{P}_1} = 0$$

and we know that the linear part of the I_{**} are, like the K_{**}, a maximum set of unconnected invariants in the variables u_i^τ, of the distribution $L_{(1)k}^{\nu}|_{\mathscr{P}_1}$. We then deduce the proposition by an easy linear elimination checking.

C.Q.F.D.

As a result, the map $\pi_1^2 : \mathscr{R}'_2 \to \mathscr{R}'_1$ is finally surjective if and only if the structure constants are such that:

$$J(u, c) \equiv \alpha(u) \cdot c \cdot c + \beta(u) \cdot c + \gamma(u) = 0, \qquad \forall u$$

We shall look more carefully at these relations.

First of all, as the elimination of the second order jet coordinates is done simply by linear combinations, the coefficients $A_\alpha^j(\omega(x))$ are just the ones that should be used in order to determine the symbol of $\mathscr{D}_2$. Thus, the number of unconnected conditions $J(u, c) = 0$ is $\leq \dim F_2$.

PROPOSITION 4.5 It is possible to find equivalent relations not containing u.

Proof As $\pi_0^1 : \mathscr{R}'_1 \to \mathscr{F}$ is surjective, the surjectivity of $\pi_1^2 : \mathscr{R}'_2 \to \mathscr{R}'_1$ implies the surjectivity of $\pi_0^2 : \mathscr{R}'_2 \to \mathscr{F}$, and the equations $J(u, c) = 0$ must be invariant under the action generated by the L_k^μ on the fiber of $\mathscr{F}$, for any set of structure constants.

Using the finite point of view, with $\bar{u} = G(u, p)$ and c arbitrary, this means that $J(\bar{u}, c) = 0$ whenever $J(u, c) = 0$.

As the former action is locally transitive, if we take u_0 such that $\operatorname{rank}[L(u_0)] = \dim F_0$, then, if the structure constants satisfy $J(u_0, c) = J(c) = 0$ it follows that $J(u, c) = 0$ whenever $\operatorname{rank}[L(u)] = \dim F_0$, that is to say at least on an open neighbourhood of u_0 in the fiber. As the functions $J(u, c)$ are analytic in u and of course also in c, because of the hypothesis made on $\mathscr{F}$ and therefore on the $I(u, \partial u/\partial x)$, it follows that $J(u, c) = 0$ for any u.

C.Q.F.D.

REMARK 4.6 We may choose $u_0 = \omega(x_0)$ with the same x_0 as in theorem 4.1. Moreover we can integrate the Frobenius system of lemma 3.7 using the device of A. Mayer, in order to have

$$I_{**}\left(u_0, \frac{\partial u}{\partial x}\right) = K_{**}\left(u_0, \frac{\partial u}{\partial x}\right).$$

EXAMPLE 4.7 When $q = 1$, we obtain equations such as:

$$\alpha(u) \cdot \frac{\partial u}{\partial x} \cdot \frac{\partial u}{\partial x} = 0$$

Their restrictions to $\mathscr{P}_1$ have the same form because the I_* are linear in the $\partial u/\partial x$. As the I_{**} are linear also in the K_{**}, it follows that

$$J(u, c) \equiv \alpha(u) \cdot c \cdot c = 0$$

and the conditions $J(c) = 0$ on the structure constants are homogeneous polynomials of order 2.

Finally we obtain the following theorem that completes the analogy with the theory of Lie groups:

THEOREM 4.8 In order that $\pi_1^2 ; \mathscr{R}'_2 \to \mathscr{R}'_1$ be surjective, it is necessary and sufficient that the structure constants satisfy algebraic equations or order ≤ 2, called "*Jacobi conditions*":

$\mathscr{C}$ $$J(c) = 0 \qquad \text{(Jacobi)}$$

REMARK 4.9 If ω_t with $\omega_0 = \omega$ is a "*deformation*" of ω and gives rise to an involutive system $R_q(t)$, of formally transitive infinitesimal Lie equations, then we must have, $\forall t$ small:

$$\begin{cases} I_*\left(\omega_t(x), \dfrac{\partial \omega_t(x)}{\partial x}\right) = 0 \\ I_{**}\left(\omega_t(x), \dfrac{\partial \omega_t(x)}{\partial x}\right) = c_t \end{cases}$$

and

$$J(c_t) = 0$$

Setting $c_t = c + t\underset{1}{C} + \cdots$ and linearising we get $\mathscr{D}_1 \cdot \underset{1}{\Omega} = \underset{1}{C}$ and it follows that $\mathscr{D}_2 \cdot \underset{1}{C} = 0$ using an evident notation.

However we must have:

$$\frac{\partial J(c)}{\partial c} \cdot \underset{1}{C} = 0$$

PROPOSITION 4.10 The linear equations:

$$\mathscr{D}_2 \cdot \underset{1}{C} = 0 \quad \text{and} \quad \frac{\partial J(c)}{\partial c} \cdot \underset{1}{C} = 0$$

are equivalent.

Proof According to the proof of the preceding propositions, we have symbolic identities of the following kind, on the fiber of $J_1(\mathscr{F})$:

$$A(u)\frac{dI}{dx} + [(V(u)I + W(u))I_* + J(u, I_{**})] \equiv 0$$

We now need the following remark.

REMARK 4.11 As the I are quasi-linear in u_i^τ and killed by the involutive distribution with matrix Inv (u), their transformation law under the action of $GL_q(n, \mathbb{R})$ makes it possible to introduce a new bundle of geometric objects $\mathscr{F}_1$ with coordinates (x^i, u^τ, v^α) and infinitesimal transition laws of the following kind:

$$\xi^i(x)\frac{\partial}{\partial x^i} + \partial_\mu \xi^k(x)\left(L_k^{\tau\mu}(u)\frac{\partial}{\partial u^\tau} + M_\beta^\alpha|_k^\mu(u)v_*^\beta\frac{\partial}{\partial v^\alpha}\right)$$

$\mathscr{F}_1$ is a vector bundle over $\mathscr{F}$ being the cokernel of the affine bundles $\mathscr{R}'_1$ and $J_1(\mathscr{F})$ over $\mathscr{F}$, according to the following commutative and exact diagram of affine bundles over $\mathscr{F}$:

$$\begin{array}{ccccccccc}
0 & \longrightarrow & G'_1 & \longrightarrow & T^* \otimes \mathscr{F} & \longrightarrow & \mathscr{F}_1 & \longrightarrow & 0 \\
 & & \vdots\downarrow & & \vdots\downarrow & & \vdots\downarrow & & \\
0 & \longrightarrow & \mathscr{R}'_1 & \longrightarrow & J_1(\mathscr{F}) & \longrightarrow & \mathscr{F}_1 & \longrightarrow & 0 \\
 & & \downarrow & & \downarrow & & \downarrow & & \\
 & & \mathscr{F} & = & \mathscr{F} & = & \mathscr{F} & &
\end{array}$$

Moreover we have the commutative diagram:

$$\begin{array}{ccc}
F_1 & \longrightarrow & \mathscr{F}_1 \\
\downarrow & & \downarrow \\
X & \xrightarrow{\ \omega\ } & \mathscr{F}
\end{array}$$

where F_1 is the corresponding vector bundle over X of the P-sequence for $\mathscr{D}$.

If ω is any section of $\mathscr{F}$, then we may construct a section (ω, ρ) of $\mathscr{F}_1$ by setting:

$$I\left(\omega(x), \frac{\partial\omega(x)}{\partial x}\right) = \rho(x)$$

If now ω is a section of $\mathscr{F}$ such that $\mathscr{R}_q$ and R_q are involutive, we have seen that we must have, in the given trivialisation of $\mathscr{F}$:

$$\rho_*(x) \equiv 0, \quad \rho_{**}(x) \equiv c = \text{cst}$$

We have also seen that these compatibility conditions keep the same form if we change the trivialisation.

If we deform ω, we get a deformed section ω_t of $\mathscr{F}$ and a corresponding deformed section (ω_t, ρ_t) of $\mathscr{F}_1$ with:

$$\rho_t = \rho + t\underset{1}{P} + \cdots$$

and we have

$$\mathscr{D}_2 \cdot \underset{1}{\Omega} = \underset{1}{P}.$$

Thus $\underset{1}{P}$ can be considered as a section of F_1 over X and this is the case, in particular, when $\underset{1}{P}_* = 0$, $\underset{1}{P}_{**} = \underset{1}{C}$ with the above trivialisation.

We shall see later on that it is possible to define such sections in an intrinsic way.

Finaliy it is easy to check that the condition:

$$L_{(1)}{}^{\mu}_{k} \cdot I^{\alpha} \equiv M^{\alpha}_{\beta}|^{\mu}_{k}(u) \cdot I^{\beta}_{*} \qquad 2 \leq |\mu| \leq q$$

is equivalent to the two sets of identities:

$$\begin{cases} L^{\tau\mu}_{k}(u)\dfrac{\partial A^{\alpha i}_{\sigma}(u)}{\partial u^{\tau}} + \dfrac{\partial L^{\tau\mu}_{k}(u)}{\partial u^{\sigma}} A^{\alpha i}_{\tau}(u) \equiv M^{\alpha}_{\beta}|^{\mu}_{k}(u) A^{\beta i}_{*\sigma}(u) \\[2ex] \displaystyle\sum_{\lambda + 1_i = \mu} L^{\tau\lambda}_{k}(u) A^{\alpha i}_{\tau}(u) + L^{\tau\mu}_{k}(u)\frac{\partial B^{\alpha}(u)}{\partial u^{\tau}} \equiv M^{\alpha}_{\beta}|^{\mu}_{k}(u) B^{\beta}_{*}(u) \end{cases}$$

These identities will be very useful later on.

Coming back to proposition 4.10, we may linearise the identities in order to obtain:

$$A(\omega(x))\frac{d\underset{1}{P}}{dx} + \left[B(\omega(x), c)\underset{1}{P_*} + \frac{\partial J(\omega(x), c)}{\partial u}\cdot\underset{1}{\Omega} + \frac{\partial J(\omega(x), c)}{\partial c}\cdot\underset{1}{P_{**}}\right] = 0$$

From the preceding proposition

$$\frac{\partial J(u, c)}{\partial u} = 0 \quad \text{whenever} \quad J(c) = 0.$$

We get:

$$A(\omega(x))\frac{d\underset{1}{P}}{dx} + \left[B(\omega(x), c)\underset{1}{P_*} + \frac{\partial J(\omega(x), c)}{\partial c}\cdot\underset{1}{P_{**}}\right] \equiv 0$$

As the F_1 are`determined up to an isomorphism, this linear operator $\mathscr{D}_2$ of order 1 such that $\mathscr{D}_2 \circ \mathscr{D}_1 = 0$ may be identified with the corresponding operator of the P-sequence for $\mathscr{D}$ because they have, by construction, the same symbol.

If now $\underset{1}{P_*} = 0$, $\underset{1}{P_{**}} = \underset{1}{C}$ as in the remark, we get

$$\mathscr{D}_2 \cdot \underset{1}{C} \equiv \frac{\partial J(\omega(x), c)}{\partial c}\cdot\underset{1}{C}$$

without any star-script.

The proposition then follows from the fact that:

$$J(\omega(x), c) = 0 \quad \forall x \in U \subset X \qquad \text{whenever} \quad J(c) = 0.$$

Finally, as F_2 is defined up to an isomorphism, we can choose it in order to have exactly:

$$\mathscr{D}_2 \cdot \underset{1}{C} \equiv \frac{\partial J(c)}{\partial c}\cdot\underset{1}{C}$$

This will be assumed in the sequel. C.Q.F.D.

Combining the two preceding theorems, we obtain:

THIRD FUNDAMENTAL THEOREM 4.12 If R_q is a formally transitive and involutive system of Lie equations with structure constants c, then for any point $x_0 \in X$, there exists an analytic, formally transitive and involutive system $\bar{R}_q$ of Lie equations with arbitrarily given structure constants $\bar{c} \neq c$ satisfying $J(\bar{c}) = 0$, and such that $\bar{R}^0_{q,x_0} = R^0_{q,x_0}$.

Proof As R_q is involutive and formally transitive, we can construct as in theorem 4.1 an analytic bundle $\mathscr{F}$ and find a section ω of $\mathscr{F}$, which is not in general analytic, but that satisfies the integrability conditions $\mathscr{R}'_1$ with structure constants c that are easy to compute. As R_q is involutive, then $\mathscr{R}'_1$ is an involutive non-linear analytic system, even when we take $\bar{c}$ instead of c. Let $\bar{\omega}$ be an analytic solution over a convenient open set $U \subset X$. It gives rise to the desired system $\bar{R}_q$. C.Q.F.D.

Remark 4.13 In fact, if we consider $x_0 \in X$, then we may chose $y_0 = x_0$ in the preceding theorem. As $\pi'^1_0 : \mathscr{R}'_1 \to \mathscr{F}$ is surjective, we can find $\bar{\omega}$ such that $\bar{\omega}(x_0) = \omega(x_0)$ using the Cartan–Kähler theorem and $\bar{R}^0_{q,x_0} = R^0_{q,x_0}$.

Remark 4.14 Involutiveness is not necessary. We just need the 3-acyclicity of G_q and the surjectivity of the map $\pi^{q+1}_q : R_{q+1} \to R_q$.

Example 4.15

$$\Gamma: \qquad \mathbb{R}^2 \to \mathbb{R}^2 : y^1 = \frac{ax^1 + b}{cx^1 + d}, \quad y^2 = \frac{ax^2 + b}{cx^2 + d}$$

Special finite equations in Lie form: $q = 2$.

$$\mathscr{R}_2 \begin{cases} \Phi^1 \equiv \dfrac{p^1_2}{p^1_1} = 0 & \Phi^6 \equiv \dfrac{p^1_{22}}{p^1_1} + \dfrac{2(p^1_2)^2}{p^1_1(y^2 - y^1)} = 0 \\[2ex] \Phi^2 \equiv \dfrac{p^2_1}{p^2_2} = 0 & \Phi^7 \equiv \dfrac{p^2_{22}}{p^2_2} + \dfrac{2p^2_2}{(y^1 - y^2)} = \dfrac{2}{(x^1 - x^2)} \\[2ex] \Phi^3 \equiv \dfrac{p^1_1 p^2_2}{(y^2 - y^1)^2} = \dfrac{1}{(x^2 - x^1)^2} & \Phi^8 \equiv \dfrac{p^2_{12}}{p^2_2} + \dfrac{2p^2_1}{(y^1 - y^2)} = 0 \\[2ex] \Phi^4 \equiv \dfrac{p^1_{11}}{p^1_1} + \dfrac{2p^1_1}{(y^2 - y^1)} = \dfrac{2}{(x^2 - x^1)} & \Phi^9 \equiv \dfrac{p^2_{11}}{p^2_2} + \dfrac{2(p^2_1)^2}{p^2_2(y^1 - y^2)} = 0 \\[2ex] \Phi^5 \equiv \dfrac{p^2_{12}}{p^2_1} + \dfrac{2p^1_2}{(y^2 - y^1)} = 0 & \end{cases}$$

A straightforward but tedious computation gives the general infinitesimal equations in Lie form:

$$
R_2 \left\{
\begin{aligned}
&\underset{1}{\Omega}^1 \equiv \xi^1_2 + \omega^1(x)\xi^2_2 - \omega^1(x)\xi^1_1 - (\omega^1(x))^2\xi^2_1 + \xi^i \frac{\partial \omega^1(x)}{\partial x^i} = 0 \\
&\underset{1}{\Omega}^2 \equiv \xi^2_1 + \omega^2(x)\xi^1_1 - \omega^2(x)\xi^2_2 - (\omega^2(x))^2\xi^1_2 + \xi^i \frac{\partial \omega^2(x)}{\partial x^i} = 0 \\
&\underset{1}{\Omega}^3 \equiv \omega^3(x)(\xi^1_1 + \xi^2_2 + \omega^1(x)\xi^2_1 + \omega^2(x)\xi^1_2) + \xi^i \frac{\partial \omega^3(x)}{\partial x^i} = 0 \\
&\underset{1}{\Omega}^4 \equiv \xi^1_{11} + \omega^1(x)\xi^2_{11} + \omega^4(x)\xi^1_1 + (2\omega^5(x) - \omega^1(x)\omega^4(x))\xi^2_1 \\
&\qquad + \xi^i \frac{\partial \omega^4(x)}{\partial x^i} = 0 \\
&\underset{1}{\Omega}^5 \equiv \xi^1_{12} + \omega^1(x)\xi^2_{12} + \omega^4(x)\xi^1_2 + (\omega^6(x) - \omega^1(x)\omega^5(x))\xi^2_1 \\
&\qquad + \omega^5(x)\xi^2_2 + \xi^i \frac{\partial \omega^5(x)}{\partial x^i} = 0 \\
&\underset{1}{\Omega}^6 \equiv \xi^1_{22} + \omega^1(x)\xi^2_{22} + 2\omega^6(x)\xi^2_2 + 2\omega^5(x)\xi^1_2 - \omega^6(x)\xi^1_1 \\
&\qquad - \omega^1(x)\omega^6(x)\xi^2_1 + \xi^i \frac{\partial \omega^6(u)}{\partial x^i} = 0 \\
&\underset{1}{\Omega}^7 \equiv \xi^2_{22} + \omega^2(x)\xi^1_{22} + \omega^7(x)\xi^2_2 + (2\omega^8(x) - \omega^2(x)\omega^7(x))\xi^1_2 \\
&\qquad + \xi^i \frac{\partial \omega^7(x)}{\partial x^i} = 0 \\
&\underset{1}{\Omega}^8 \equiv \xi^2_{12} + \omega^2(x)\xi^1_{12} + \omega^7(x)\xi^2_1 + (\omega^9(x) - \omega^2(x)\omega^8(x))\xi^1_2 \\
&\qquad + \omega^8(x)\xi^1_1 + \xi^i \frac{\partial \omega^8(x)}{\partial x^i} = 0 \\
&\underset{1}{\Omega}^9 \equiv \xi^2_{11} + \omega^2(x)\xi^1_{11} + 2\omega^9(x)\xi^1_1 + 2\omega^8(x)\xi^2_1 - \omega^8(x)\xi^2_2 \\
&\qquad - \omega^2(x)\omega^9(x)\xi^1_2 + \xi^i \frac{\partial \omega^9(x)}{\partial x^i} = 0
\end{aligned}
\right.
$$

A straightforward but tedious computation shows that there are 8 integrability conditions of the first kind:

$$\begin{cases} \dfrac{\partial\omega^1}{\partial x^1} - \omega^5 + \omega^1\omega^4 = 0 & \dfrac{\partial\omega^1}{\partial x^2} - \omega^6 + \omega^1\omega^5 = 0 \\[2ex] \dfrac{\partial\omega^2}{\partial x^1} - \omega^9 + \omega^2\omega^8 = 0 & \dfrac{\partial\omega^2}{\partial x^2} - \omega^8 + \omega^2\omega^7 = 0 \\[2ex] \dfrac{\partial\omega^3}{\partial x^1} - \omega^4 - \omega^8 = 0 & \dfrac{\partial\omega^3}{\partial x^2} - \omega^5 - \omega^7 = 0 \\[2ex] \dfrac{\partial(\omega^5 - \omega^1\omega^4)}{\partial x^2} - \dfrac{\partial(\omega^6 - \omega^1\omega^5)}{\partial x^1} = 0 & \\[2ex] \dfrac{\partial(\omega^9 - \omega^2\omega^8)}{\partial x^2} - \dfrac{\partial(\omega^8 - \omega^2\omega^7)}{\partial x^1} = 0 & \end{cases}$$

2 integrability conditions of the second kind:

$$\left\{ \frac{(\partial\omega^4/\partial x^2) - (\partial\omega^5/\partial x^1)}{\omega^3(1 - \omega^1\omega^2)} = c^1, \qquad \frac{(\partial\omega^8/\partial x^2) - (\partial\omega^7/\partial x^1)}{\omega^3(1 - \omega^1\omega^2)} = c^2 \right.$$

The only Jacobi condition is:

$$c^1 + c^2 = 0.$$

REMARK 4.16 The inequality conditions on ω are:

$$\omega^3(1 - \omega^1\omega^2) \neq 0$$

Specialisation:

$$\omega^1 = 0, \quad \omega^2 = 0, \quad \omega^3 = \frac{1}{(x^2 - x^1)^2}, \quad \omega^4 = \frac{2}{x^2 - x^1},$$

$$\omega^5 = 0, \quad \omega^6 = 0, \quad \omega^7 = \frac{2}{x^1 - x^2}, \quad \omega^8 = 0, \quad \omega^9 = 0$$

$$\Rightarrow c^1 = -2, \quad c^2 = 2.$$

The reader will notice that:

$$\Phi^3 \cdot (1 - \Phi^1 \cdot \Phi^2) \equiv \frac{1}{(y^2 - y^1)^2}(p_1^1 p_2^2 - p_2^1 p_1^2) = \frac{1}{(x^2 - x^1)^2}$$

and that the diagonal $x^1 = x^2$ of $\mathbb{R}^2$, invariant by Γ must be taken out of $\mathbb{R}^2$ in order to get a transitive action and to define ω^3.

5 Equivalence problem

We shall now study the famous "*equivalence problem*", stated and partially solved by Cartan. This is essentially a local problem. However, in order to simplify the expressions we shall keep to global notations.

In theorem 2.18 we have seen that, for any $\varphi \in \operatorname{Aut}(X)$, if we have a bundle $\mathscr{F}$ of geometric objects and a section ω giving rise to involutive systems $\mathscr{R}_q$ and R_q, then the section $\bar{\omega} = \varphi(\omega)$ gives rise to involutive systems

$$\bar{\mathscr{R}}_q = \operatorname{Ad} \varphi(\mathscr{R}_q) \quad \text{and} \quad \bar{R}_q = \varphi(R_q).$$

Using proposition 3.12 we see that $\bar{c} = c$, that is to say, the systems have the same structure constants. Also the pseudogroup $\bar{\Gamma}$ of solutions of $\bar{\mathscr{R}}_q$ is similar to the pseudogroup Γ of solutions of $\mathscr{R}_q$, that is to say:

$$\bar{\Gamma} = \operatorname{Ad} \varphi(\Gamma) = \varphi \circ \Gamma \circ \varphi^{-1}$$

DEFINITION 5.1 The "*finite*" equivalence problem is the converse study as it can be suggested by the result obtained in the third fundamental theorem.

PROBLEM 5.2 If $\bar{\omega}$ and ω give rise to the same structure constants $\bar{c} = c$, does there exist $\varphi \in \operatorname{Aut}(X)$ such that:

$$\bar{\Gamma} = \operatorname{Ad} \varphi(\Gamma) \quad \text{or} \quad \bar{\mathscr{R}}_q = \operatorname{Ad} \varphi(\mathscr{R}_q) \quad \text{or} \quad \bar{R}_q = \varphi(R_q)?$$

This is in fact related to the local exactness of a non-linear sequence because, as shown in proposition 2.12, we only have to solve the non-linear system:

$$\Phi^{\tau}(\bar{\omega}(y), p) = \omega(x)$$

the solutions of which are the desired φ.

As the study of such a problem is out of our scope, we shall only give one example and a counterexample:

EXAMPLE 5.3 $X = \mathbb{R}^n$ with n even.

We take $\mathscr{F} = \Lambda^2 T^*, \omega = \omega_{ij}(x)\, dx^i \wedge dx^j$.

There are only integrability conditions of the first kind:

$$d\omega = 0, \quad \frac{\partial \omega_{ij}(x)}{\partial x^k} + \frac{\partial \omega_{jk}(x)}{\partial x^i} + \frac{\partial \omega_{ki}(x)}{\partial x^j} = 0$$

and we require that $\det[\omega_{ij}(x)] \neq 0$.

Darboux showed in 1876 (53) that, if we consider the analytic section

$$\bar{\omega} = dx^1 \wedge dx^2 + \cdots + dx^{n-1} \wedge dx^n,$$

then there exists a local automorphism φ such that $\bar{\omega} = \varphi(\omega)$.

COUNTEREXAMPLE 5.4 $X = \mathbb{R}^5$

Γ = Lie pseudogroup of transformations that leave invariant:

$$\begin{cases} \lambda^1 \equiv dx^1, \quad \lambda^2 = dx^2, \quad \lambda^3 = x^2\, dx^1 - x^1\, dx^2 + dx^3 \\ \nu^1 \equiv -x^2\, dx^1 \wedge dx^2 \wedge dx^4 - x^1\, dx^1 \wedge dx^2 \wedge dx^5 \\ \qquad + dx^1 \wedge dx^3 \wedge dx^5 + dx^2 \wedge dx^3 \wedge dx^4 \\ \nu^2 = -x^1\, dx^1 \wedge dx^2 \wedge dx^4 + x^2\, dx^1 \wedge dx^2 \wedge dx^5 \\ \qquad + dx^1 \wedge dx^3 \wedge dx^4 - dx^2 \wedge dx^3 \wedge dx^5 \end{cases}$$

For the general equations, we must have five exterior forms, satisfying:

$$\lambda^1 \wedge \nu^1 + \lambda^2 \wedge \nu^2 = 0 \qquad \lambda^2 \wedge \nu^1 - \lambda^1 \wedge \nu^2 = 0$$
$$\lambda^3 \wedge \nu^1 = 0 \qquad \lambda^3 \wedge \nu^2 = 0$$

with the integrability conditions:

$$d\lambda^1 = 0, \quad d\lambda^2 = 0, \quad d\lambda^3 + 2\lambda^1 \wedge \lambda^2 = 0$$
$$d\nu^1 = 0, \quad d\nu^2 = 0$$

It is easy to check that it is possible to satisfy those conditions with:

$$\begin{cases} \bar{\lambda}^1 = \lambda^1, \quad \bar{\lambda}^2 = \lambda^2, \quad \bar{\lambda}^3 = \lambda^3 \\ \bar{\nu}^1 = \nu^1 + A(x^1, x^2, x^3)\, dx^1 \wedge dx^2 \wedge dx^3 \\ \bar{\nu}^2 = \nu^2 + B(x^1, x^2, x^3)\, dx^1 \wedge dx^2 \wedge dx^3 \end{cases}$$

The system of p.d.e. for φ^{-1} that must be solved in order to answer the finite equivalence problem is then:

$$\begin{cases} p_1^4 - p_2^5 - x^1 p_3^5 - x^2 p_3^4 = A(x^1, x^2, x^3) \\ -p_1^5 - p_2^4 - x^1 p_3^4 + x^2 p_3^5 = B(x^1, x^2, x^3) \\ p_5^5 = 1, \quad p_4^5 = 0, \quad p_4^4 = 1, \quad p_5^4 = 0 \end{cases}$$

because we may take $\bar{x}^1 = x^1, \bar{x}^2 = x^2, \bar{x}^3 = x^3$.

But this system is just, apart from some slight changes in notation, the counterexample of H. Lewy already studied in the first part of this book. We know that, for some choice of A and B, there are no C^∞ solutions and we thus cannot solve the corresponding finite equivalence problem (16a).

We finally state the "*infinitesimal*" equivalence problem:
If we take

$$\bar{\omega} = \omega_t = \omega + t\underset{1}{\Omega} + \cdots$$

then we may look for $\varphi_t = \exp(t\xi)$ linearising the integrability conditions with $\bar{c} = c_t = c$, we have to solve the equations:

$$\mathscr{D} \cdot \xi = \underset{1}{\Omega} \quad \text{with} \quad \mathscr{D}_1 \cdot \underset{1}{\Omega} = 0$$

We obtain:

THEOREM 5.5 The infinitesimal equivalence problem can be solved if and only if the P-sequence for the involutive Lie operator $\mathscr{D}$ is locally exact at F_0.

We shall give two important cases in which the infinitesimal equivalence problem can be solved.

EXAMPLE 5.6 The following definition is useful because of its physical background.

DEFINITION 5.7 A Lie pseudogroup Γ is called "*flat*" if it contains the translations. It is of course transitive.

Let now Γ be a flat Lie pseudogroup. Then $\xi^i(x) = \text{cst}$ must be a solution of its defining linear system R_q of infinitesimal equations. It follows that we must have $\partial\omega^\tau(x)/\partial x^i = 0$ and thus ω is a constant section over a convenient open set $U \subset X$. All the coefficients of $\mathscr{D}$ (and thus $\mathscr{D}_1, \ldots, \mathscr{D}_n$) are constant over U and it is known that the P-sequence for $\mathscr{D}$ is locally exact at F_r, for $r = 0, 1, \ldots, n$.

REMARK 5.8 As the differential invariants are defined at least up to a multiplication by a constant, the special section corresponding to a flat Lie pseudogroup may be expressed only by means of 0 or 1. This general fact is encountered very often in physics (Minkowskian or Euclidean section of S_2T^* in special relativity).

EXAMPLE 5.9 Lie pseudogroups of finite type.

DEFINITION 5.10 A Lie pseudogroup Γ of transformations of X is said to be of "*finite type*" if it corresponds to the action of a Lie group G on X.

PROPOSITION 5.11 If Γ is of finite type, its defining systems of p.d.e. are of finite type.

Proof As the problem is local, we shall use local coordinates. Without any loss of generality we may even take for X an open set $U \subset \mathbb{R}^n$ and assume that the action is effective.

Let

$$\xi_\rho = \xi_\rho^i(x)\frac{\partial}{\partial x^i} \quad \text{with} \quad \rho = 1, \ldots, \dim G$$

be the independent infinitesimal generators of this action, as given by the first fundamental theorem 6.2.1.

We know that any infinitesimal transformation of Γ can be written $\xi = \lambda^\rho \xi_\rho$ where the λ^ρ are constants. The linear system R_q of infinitesimal equations of Γ is thus defined by eliminating the λ^ρ between the linear equations:

$$\partial_\mu \xi^i(x) = \lambda^\rho \partial_\mu \xi_\rho^i(x)$$

with $0 \leq |\mu| \leq q$ and q big enough.

Elementary linear algebra then shows that $G_{q+1} = 0$ and that we have an isomorphism

$$R_q|_U \approx U \times \mathbb{R}^{\dim G} \quad \text{with} \quad \dim R_q = \dim G.$$

We can represent this isomorphism by a matrix $M = [M(x)]$ with rank equal to $\dim G$.

If now we define in the same way R_{q+1} by elimination, it is easy to see that $\pi_q^{q+1}: R_{q+1} \to R_q$ is surjective.

Moreover, for sections ξ_q of R_q and ξ_{q+1} of R_{q+1} with $\xi_q = \pi_q^{q+1} \circ \xi_{q+1}$, there must exist functions $\lambda^\rho(x)$ such that:

$$\xi_\mu^k(x) = \lambda^\rho(x)\partial_\mu \xi_\rho^k(x) \quad \text{with} \quad 0 \leq |\mu| \leq q + 1$$

It follows that:

$$\partial_i \xi_\mu^k(x) - \xi_{\mu+1_i}(x) = \partial_i \lambda^\rho(x)\partial_\mu \xi_\rho^k(x) \quad \text{with} \quad 0 \leq |\mu| \leq q.$$

Thus $D \cdot R_{q+1} \subset T^* \otimes R_q$ and it is easy to deduce from the latter facts that R_{q+1} is the first prolongation of R_q. C.Q.F.D.

REMARK 5.12 According to proposition 3.4.7, if G_q is involutive and of finite type, then we must have $G_q = 0$. This will be assumed throughout this example.

REMARK 5.13 Conversely, this proposition can be used in the analytic case in order to define a Lie pseudogroup of finite type. In fact, we can choose $\dim G$ independent vector fields ξ_ρ solutions of R_q by taking the parametric jet coordinates null but one at a given point $x_0 \in X$. If ξ is any solution of R_q, it follows that there exist constants λ^ρ such that $\partial_\mu \xi^k(x_0) = \lambda^\rho \partial_\mu \xi_\rho^k(x_0)$. As the vector field $\xi - \lambda^\rho \xi_\rho$ is also a solution of R_q and has null jets at x_0, we have $\xi = \lambda^\rho \xi_\rho$.

We can now use the diagram of theorem 5.5.3 in the first part of this book. We have $C^r = \Lambda^r T^* \otimes R_q$ and it is easy to compute the numbers $\dim F_r$ that depend only on n and $\dim G$.

The P-sequence will be exact if and only if the sequence $S_2(\Theta)$ is exact, where a solution of $\mathscr{D} \cdot \xi = 0$ can be written $\xi = \lambda^\rho \xi_\rho$.

If we introduce the trivial bundles $A^0 = X \times \mathbb{R}^{\dim G}$, $A^r = \Lambda^r T^* \otimes A^0$ and if we denote by $d: A^r \to A^{r+1}$ the differential operator induced in a natural way by the classical exterior derivative $d: \Lambda^r T^* \to \Lambda^{r+1} T^*$, we obtain from the Poincaré lemma an exact sequence:

$$A^0 \xrightarrow{d} A^1 \xrightarrow{d} \cdots \xrightarrow{d} A^n \longrightarrow 0$$

which is easily seen to be isomorphic to the sequence:

$$C^0 \xrightarrow{D^1} C^1 \xrightarrow{D^2} \cdots \xrightarrow{D^n} C^n \longrightarrow 0$$

because both of them are formally exact. We may write symbolically:

$$D^r \equiv D \equiv M^{-1} \circ d \circ M$$

and it follows that the above two sequences are exact. Therefore the P-sequence for a Lie operator $\mathscr{D}$ of finite type is exact. C.Q.F.D.

6 Normalizer

Let us consider, for two sections ω and $\bar{\omega}$ of $\mathscr{E}$ the corresponding linear systems R_q and $\bar{R}_q$:

$$R_q \qquad \underset{1}{\Omega^\tau} \equiv -\xi^k_\mu \cdot L^{\tau\mu}_k(\omega(x)) + \xi^i \frac{\partial \omega^\tau(x)}{\partial x^i} = 0$$

$$\bar{R}_q \qquad \underset{1}{\bar{\Omega}^\tau} \equiv -\xi^k_\mu \cdot L^{\tau\mu}_k(\bar{\omega}(x)) + \xi^i \frac{\partial \bar{\omega}^\tau(x)}{\partial x^i} = 0$$

We shall look for conditions on $\bar{\omega}$ in order to have

$$\bar{R}_q = R_q \subset J_q(T).$$

First of all we know that:

$$\forall x \in X \quad \max_u \operatorname{rank}\,[L(u)] = \operatorname{rank}\,[L(\omega(x))] = \operatorname{rank}\,[L^{\tau\mu}_k(\omega(x))] = \dim F_0$$

It follows that there exists a square submatrix $[M(u)]$ of the same maximum rank, the columns of which may correspond to the principal jet-coordinates.

The solved form of R_q will be given by the matrix $[M^{-1}(\omega(x)) \cdot L(\omega(x))]$ according to the following picture:

$$[L(u)] = \begin{array}{|c|c|} \hline M(u) & \quad \\ \hline \end{array} \qquad [M^{-1}(u) \cdot L(u)] \equiv \begin{array}{|c|c|} \hline \begin{matrix} 1 & & 0 \\ & \ddots & \\ 0 & & 1 \end{matrix} & \mathscr{E}(u) \\ \hline \end{array}$$

We shall then choose a total ordering for the jet coordinates ξ_μ^k with $1 \le |\mu| \le q$ and adopt the indices ρ, τ or σ for the principal ones, while keeping only one index α, β or γ in order to classify the vector fields

$$L_k^\mu = L_k^{\tau\mu}(u) \frac{\partial}{\partial u^\tau}.$$

Introducing the stationary functions $\mathscr{E}(u)$, we write:

$$L_{m+r} = \mathscr{E}_{m+r}^\sigma(u) L_\sigma$$

with

$$\sigma = 1, \ldots, m \quad \text{and} \quad r = 1, \ldots, \dim R_q - n.$$

It is easy to check that the stationary functions do not change if we effect linear combinations between the rows of $[L(u)]$ and, in particular, if we consider the matrix:

$$[M^{-1}(u) \cdot L(u)] = [(M^{-1})_\tau^\sigma(u) L_k^{\tau\mu}(u)]$$

According to the last picture, we have the following proposition:

PROPOSITION 6.1 $\bar{R}_q = R_q$ if and only if we have, $\forall x \in X$:

$$\begin{cases} (M^{-1})_\tau^\sigma(\bar{\omega}(x)) \cdot \dfrac{\partial \bar{\omega}^\tau(x)}{\partial x^i} = (M^{-1})_\tau^\sigma(\omega(x)) \cdot \dfrac{\partial \omega^\tau(x)}{\partial x^i} \\ \mathscr{E}_{m+r}^\sigma(\bar{\omega}(x)) = \mathscr{E}_{m+r}^\sigma(\omega(x)) \end{cases}$$

REMARK 6.2 This trick can be related to the existence of standard Grasmann coordinates (58) but for our purpose we do not need such a generality. Moreover we shall see later on that the results obtained do not depend on the choice of $[M(u)]$.

In particular, if the system R_q is given, in order to find all the sections giving the same system, we have to solve the non-linear system $\mathscr{R}_1'' \subset J_1(\mathscr{F})$:

$$\mathscr{R}_1'' \begin{cases} (M^{-1})_\tau^\sigma(u) \dfrac{\partial u^\tau}{\partial x^i} = (M^{-1})_\tau^\sigma(\omega(x)) \dfrac{\partial \omega^\tau(x)}{\partial x^i} \\ \mathscr{E}_{m+r}^\sigma(u) = \mathscr{E}_{m+r}^\sigma(\omega(x)) \end{cases}$$

that we shall write simply:

$$\mathscr{R}_1'' \qquad Q\left(u, \frac{\partial u}{\partial x}\right) = \sigma(x)$$

THEOREM 6.3 The non-linear system $\mathscr{R}_1'' \subset J_1(\mathscr{F})$ is involutive:

Proof If G_1'' is the symbol of $\mathscr{R}_1''$, we have $G_1'' = 0$ when $\det [M(u)] \neq 0$ and this is always the case because of the hypothesis made on the fiber of $\mathscr{F}$. Thus the symbol of $\mathscr{R}_1''$ is zero and trivially involutive.

It remains to prove that $\mathscr{R}_1''$ is a fibered submanifold of $J_1(\mathscr{F})$ and that the map $\pi_1^2 : \mathscr{R}_2'' \to \mathscr{R}_1''$ is surjective.

This will be done by means of the following technical lemma.

LEMMA 6.4 If $[L_\alpha, L_\beta] = c_{\alpha\beta}^\gamma \cdot L_\gamma$ we have the formula:

$$L_\alpha^\tau(u) \cdot \frac{\partial \mathscr{E}^\sigma_{m+r}}{\partial u^\tau} = (c^\sigma_{\alpha m+r} + c^{m+s}_{\alpha m+r} \cdot \mathscr{E}^\sigma_{m+s}(u))$$

$$- \mathscr{E}^\rho_{m+r}(u)(c^\sigma_{\alpha\rho} + c^{m+s}_{\alpha\rho} \cdot \mathscr{E}^\sigma_{m+s}(u))$$

Proof We have successively:

$$\begin{aligned}[L_\alpha, L_{m+r}] &= (c^\sigma_{\alpha m+r} + c^{m+s}_{\alpha m+r} \cdot \mathscr{E}^\sigma_{m+s}(u))L_\sigma \\ &= [L_\alpha, \mathscr{E}^\rho_{m+r}(u)L_\rho] \\ &= L_\alpha^\tau(u) \cdot \frac{\partial \mathscr{E}^\sigma_{m+r}(u)}{\partial u^\tau} \cdot L_\sigma + \mathscr{E}^\rho_{m+r}(u)[L_\alpha, L_\rho]\end{aligned}$$

$$[L_\alpha, L_\rho] = (c^\sigma_{\alpha\rho} + c^{m+s}_{\alpha\rho} \cdot \mathscr{E}^\sigma_{m+s}(u))L_\sigma$$

and the lemma follows because the vector fields L_σ are unconnected by hypothesis ($\det [M(u)] \neq 0$). C.Q.F.D.

Now it is easy to see, as $G_1'' = 0$, that $\mathscr{R}_1''$ is a fibered submanifold of $J_1(\mathscr{F})$ if and only if

$$\operatorname{rank}\left[\frac{\partial \mathscr{E}(u)}{\partial u}\right] = \operatorname{rank}\left[\frac{\partial \mathscr{E}^\sigma_{m+r}(u)}{\partial u^\tau}\right]$$

is constant on the fiber of $\mathscr{F}$.

Let us apply the finite transformation $\bar{u} = \exp(t\lambda^\alpha L_a) \cdot u$ to the fiber of $\mathscr{F}$. We have

$$\frac{du^\tau}{dt} = \lambda^\alpha L_\alpha^\tau(u)$$

and from the lemma:

$$\frac{d\mathscr{E}^{\sigma}_{m+r}(\bar{u}(t, u))}{dt} = \lambda^{\alpha} L^{\tau}_{\alpha}(\bar{u}) \frac{\partial \mathscr{E}^{\sigma}_{m+r}(\bar{u})}{\partial \bar{u}^{\tau}} = \text{Function of } \mathscr{E}(\bar{u}).$$

This is an ordinary differential equation and we obtain a functional relation of the following kind:

$$\mathscr{E}(\bar{u}(t, u)) \equiv H(t, \mathscr{E}(u)) \quad \text{with} \quad \mathscr{E}(\bar{u}(0, u)) = \mathscr{E}(u) = H(0, \mathscr{E}(u))$$

Finally we have:

$$\frac{\partial \bar{u}}{\partial u} \cdot \frac{\partial \mathscr{E}(\bar{u})}{\partial \bar{u}} = \frac{\partial H}{\partial \mathscr{E}} \cdot \frac{\partial \mathscr{E}(u)}{\partial u}$$

As

$$\det\left[\frac{\partial \bar{u}}{\partial u}\right] \neq 0 \quad \text{and} \quad \det\left[\frac{\partial H}{\partial \mathscr{E}}\right] \neq 0$$

we obtain:

$$\text{rank}\left[\frac{\partial \mathscr{E}(\bar{u})}{\partial \bar{u}}\right] = \text{rank}\left[\frac{\partial \mathscr{E}(u)}{\partial u}\right]$$

Thus rank $[\partial \mathscr{E}(u)/\partial u]$ is equal to its maximum value unless u is on a closed subset of the fiber of $\mathscr{F}$ invariant under the action generated by the vector fields L^{μ}_{k}. As we know that this action is transitive, it follows that rank $[\partial \mathscr{E}(u)/\partial u]$ is equal to its maximum value on the fiber of $\mathscr{F}$, because of the hypothesis made on this fiber.

We shall now check in two steps that the map $\pi^2_1 : \mathscr{R}''_2 \to \mathscr{R}''_1$ is surjective.

1 Using the preceding lemma we obtain easily:

$$\begin{aligned}\frac{\partial \mathscr{E}^{\sigma}_{m+r}(u)}{\partial x^i} &= \frac{\partial \mathscr{E}^{\sigma}_{m+r}(u)}{\partial u^{\tau}} \cdot \frac{\partial u^{\tau}}{\partial x^i} = \frac{\partial \mathscr{E}^{\sigma}_{m+r}(u)}{\partial u^{\rho}} \cdot M^{\rho}_{\sigma}(u) \cdot (M^{-1})^{\sigma}_{\tau}(u) \frac{\partial u^{\tau}}{\partial x^i} \\ &= \frac{\partial \mathscr{E}^{\sigma}_{m+r}(\omega(x))}{\partial u^{\rho}} \cdot M^{\rho}_{\sigma}(\omega(x)) \cdot (M^{-1})^{\sigma}_{\tau}(\omega(x)) \frac{\partial \omega^{\tau}(x)}{\partial x^i} \\ &= \frac{\partial \mathscr{E}^{\sigma}_{m+r}(\omega(x))}{\partial x^i}\end{aligned}$$

whenever

$$\left(x, u, \frac{\partial u}{\partial x}\right) \in \mathscr{R}''_1.$$

2 We have:

$$\frac{\partial u^\tau}{\partial x^i} = M^\tau_\sigma(u) \cdot (M^{-1})^\sigma_\rho(\omega(x)) \frac{\partial \omega^\rho(x)}{\partial x^i}$$

$$\frac{\partial^2 u^\tau}{\partial x^i \partial x^j} = \frac{\partial M^\tau_\sigma(u)}{\partial u^r} \cdot (M^{-1})^\sigma_\rho(\omega(x)) \frac{\partial \omega^\rho(x)}{\partial x^i} \frac{\partial u^r}{\partial x^j}$$

$$+ M^\tau_\sigma(u) \cdot \frac{\partial (M^{-1})^\sigma_\rho(\omega(x))}{\partial u^r} \frac{\partial \omega^r(x)}{\partial x^j} \frac{\partial \omega^\rho(x)}{\partial x^i}$$

$$+ M^\tau_\sigma(u) \cdot (M^{-1})^\sigma_\rho(\omega(x)) \cdot \frac{\partial^2 \omega^\rho(x)}{\partial x^i \partial x^j}$$

But

$$\frac{\partial M^\tau_\sigma(u)}{\partial u^r} \cdot (M^{-1})^\sigma_\rho(\omega(x)) \frac{\partial \omega^\rho(x)}{\partial x^i} \frac{\partial u^r}{\partial x^j}$$

$$= \frac{\partial M^\tau_\sigma(u)}{\partial u^r} \cdot M^r_s(u) \cdot (M^{-1})^s_t(\omega(x))(M^{-1})^\sigma_\rho(\omega(x)) \frac{\partial \omega^\rho(x)}{\partial x^i} \cdot \frac{\partial \omega^t(x)}{\partial x^j}$$

and we have to check, on $\mathscr{R}'_1$, that:

$$0 = \frac{\partial \omega^\rho(x)}{\partial x^i} \cdot \frac{\partial \omega^t(x)}{\partial x^j} \left[(M^{-1})^s_t(\omega(x)) \cdot (M^{-1})^\sigma_\rho(\omega(x)) \left(M^r_s(u) \frac{\partial M^\tau_\sigma(u)}{\partial u^r} \right. \right.$$

$$\left. \left. - M^r_\sigma(u) \frac{\partial M^\tau_s(u)}{\partial u^r} \right) + M^\tau_\sigma(u) \left(\frac{\partial (M^{-1})^\sigma_\rho(\omega(x))}{\partial u^t} - \frac{\partial (M^{-1})^\sigma_t(\omega(x))}{\partial u^\rho} \right) \right]$$

or equivalently that:

$$0 = M^\tau_\rho(u) \left[(c^\rho_{s\sigma} + c^{m+r}_{s\sigma} \cdot \mathscr{E}^\rho_{m+r}(u)) \right.$$

$$\left. + M^t_s(\omega(x)) M^r_\sigma(\omega(x)) \left(\frac{\partial (M^{-1})^\rho_r(\omega(x))}{\partial u^t} - \frac{\partial (M^{-1})^\rho_t(\omega(x))}{\partial u^r} \right) \right]$$

But

$$M^t_s(\omega(x)) \cdot M^r_\sigma(\omega(x)) \left(\frac{\partial (M^{-1})^\rho_r(\omega(x))}{\partial u^t} - \frac{\partial (M^{-1})^\rho_t(\omega(x))}{\partial u^r} \right)$$

$$= -(M^{-1})^\rho_r(\omega(x)) \cdot M^t_s(\omega(x)) \cdot \frac{\partial M^r_\sigma(\omega(x))}{\partial u^t}$$

$$+ (M^{-1})^\rho_t(\omega(x)) \cdot M^r_\sigma(\omega(x)) \frac{\partial M^t_s(\omega(x))}{\partial u^r}$$

$$= -(c^\rho_{s\sigma} + c^{m+r}_{s\sigma} \cdot \mathscr{E}^\rho_{m+r}(\omega(x)))$$

and finally:

$$0 = M^\tau_\rho(u)[c^{m+r}_{s\sigma}(\mathscr{E}^\rho_{m+r}(u) - \mathscr{E}^\rho_{m+r}(\omega(x)))] \quad \text{whenever} \quad \left(x, u, \frac{\partial u}{\partial x}\right) \in \mathscr{R}''_1$$

C.Q.F.D.

We shall now show that $\mathscr{R}''_1$ is an "*automorphic system*" (see problem 5), that is to say that all its solutions can be deduced from only one of them, by the action of a transformation group on the fiber of $\mathscr{F}$.

PROPOSITION 6.5 The Lie pseudogroup of transformations that leave invariant the 1-forms $(M^{-1})^\sigma_\tau(u)\, du^\tau$ and the 0-forms (or functions) $\mathscr{E}^\sigma_{m+r}(u)$ is a Lie group of transformations of the fiber of $\mathscr{F}$.

Proof Let $W = W^\tau(u)(\partial/\partial u^\tau)$ be an infinitesimal transformation. We must have

$$\mathscr{L}(W)((M^{-1})^\sigma_\tau(u)\, du^\tau) = 0 \quad \text{and} \quad \mathscr{L}(W)(\mathscr{E}^\sigma_{m+r}(u)) = 0$$

that is to say:

$$(M^{-1})^\sigma_\tau(u)\frac{\partial W^\tau(u)}{\partial u^\rho} + W^\tau(u)\frac{\partial (M^{-1})^\sigma_\rho(u)}{\partial u^\tau} = 0 \quad \text{and} \quad W^\tau(u)\frac{\partial \mathscr{E}^\sigma_{m+r}(u)}{\partial u^\tau} = 0$$

But this is equivalent to:

$$W^\tau(u)\frac{\partial M^\rho_\sigma(u)}{\partial u^\tau} - M^\tau_\sigma(u)\frac{\partial W^\rho(u)}{\partial u^\tau} = 0 \quad \text{and} \quad W^\tau(u)\frac{\partial \mathscr{E}^\sigma_{m+r}(u)}{\partial u^\tau} = 0$$

that is to say:

$$[W, L_\sigma] = 0 \quad \text{and} \quad W^\tau(u)\cdot\frac{\partial \mathscr{E}^\sigma_{m+r}(u)}{\partial u^\tau} = 0$$

It follows easily that $[W, L_{m+r}] = 0$.

Conversely, if we look for vector fields W such that $[W, L_\alpha] = 0$, we must have:

$$[W, L_\sigma] = 0 \quad \text{and} \quad 0 = [W, L_{m+r}] = W^\tau(u)\cdot\frac{\partial \mathscr{E}^\sigma_{m+r}(u)}{\partial u^\tau} L_\sigma$$

and this is equivalent to

$$[W, L_\sigma] = 0 \quad \text{and} \quad W^\tau(u)\frac{\partial \mathscr{E}^\sigma_{m+r}(u)}{\partial u^\tau} = 0$$

because the vector fields L_σ are unconnected.

It follows that there cannot be more than $m -$ rank $[\partial\mathscr{E}(u)/\partial u]$ unconnected such vector fields W_ν solutions.

Finally, if W_1 and W_2 are two such solutions, we have:

$$[[W_1, W_2], L_\alpha] + [[L_\alpha, W_1], W_2] + [[W_2, L_\alpha], W_1] = 0$$

by means of the Jacobi identity and thus $[W_1, W_2]$ is also a solution. It follows that

$$[W_1, W_2] = \rho^\nu_{12}(u) W_\nu .$$

Then

$$0 = [[W_1, W_2], L_\alpha] = L^\tau_\alpha(u) \cdot \frac{\partial \rho^\nu_{12}(u)}{\partial u^\tau} \cdot W_\nu$$

As the W_ν are unconnected we must have $L^\tau_\alpha(u)(\partial \rho^\nu_{12}(u)/\partial u^\tau) = 0$ and as the L_σ are unconnected, we must have $\rho^\nu_{12} = \text{cst}$. This proves that the W_ν are the infinitesimal generators of a Lie group of transformations of the fiber which is called the "*reciprocal*" of the Lie group of transformations generated by the L^μ_k. In the sequel we shall simply write $[W, L] = 0$.

C.Q.F.D.

DEFINITION 6.6 The corresponding finite transformations of the fiber of $\mathscr{F}$ will be called "*label transformations*" and noted $\bar{u} = g(a, u)$ where the number of parameters a is $\leq \dim F_0 = m$.

We can consider the W as vector fields on $\mathscr{F}$. As $[W, L] = 0$ we have

$$\left[W, \xi^i(x) \frac{\partial}{\partial x^i} + \partial_\mu \xi^k(x) L^\mu_k \right] = 0$$

and we see that the function $Q(u, \partial u/\partial x)$ on $J_1(\mathscr{F})$ are a maximum set of unconnected invariants, in the variables $(u, \partial u/\partial x)$ of the vector fields $\rho_1(W)$.

We can show, as for the Φ^τ, that the Q are transformed among themselves under any transition transformation of $\mathscr{F}$ and they give rise, in the same way, to a bundle $\tilde{\mathscr{F}}$ of geometric objects with section σ. We may consider the Q as defining a bundle map $Q : J_1(\mathscr{F}) \to \tilde{\mathscr{F}}$.

According to the converse theorem 2.10, we obtain a Lie pseudogroup $\tilde{\Gamma}$ defined by a non-linear system of finite equations:

$$\tilde{\mathscr{R}}_{q+1} \qquad \Psi(\sigma(y), p) \equiv Q\left(\Phi, \frac{d\Phi}{dx}\right) = \sigma(x)$$

with $1 \leq \operatorname{ord} p \leq q + 1$ and a commutative diagram:

$$\begin{array}{ccccccc}
\text{id} & \longrightarrow & \Gamma & \longrightarrow & \text{Aut}(X) & \underset{\omega \circ \pi}{\overset{\Phi \circ j_q}{\rightrightarrows}} & \mathscr{F} \\
 & & \downarrow & & \| & & \downarrow Q \circ j_1 \\
\text{id} & \longrightarrow & \tilde{\Gamma} & \longrightarrow & \text{Aut}(X) & \underset{\sigma \circ \pi}{\overset{\Psi \circ j_{q+1}}{\rightrightarrows}} & \tilde{\mathscr{F}}
\end{array}$$

Our problem will now be to study $\tilde{\mathscr{R}}_{q+1}$ and the corresponding linearised system $\tilde{R}_{q+1}$.

First of all, we can linearise $\mathscr{R}''_1$, setting:

$$\bar{\omega} = \omega_t = \omega + t\underset{1}{\Omega} + \cdots$$

and

$$\sigma_t = \sigma + t\underset{1}{\Sigma} + \cdots$$

As $\mathscr{R}''_1$ is an involutive system, we obtain a linear involutive operator ∇_0 and the relation $\nabla_0 \cdot \underset{1}{\Omega} = \underset{1}{\Sigma}$. We may define $\tilde{\mathscr{D}} = \nabla_0 \circ \mathscr{D}$ and obtain the exact commutative diagram:

$$\begin{array}{ccccccc}
0 & \longrightarrow & \Theta & \longrightarrow & T & \xrightarrow{\mathscr{D}} & F_0 \\
 & & \downarrow & & \| & & \downarrow \nabla_0 \\
0 & \longrightarrow & \tilde{\Theta} & \longrightarrow & T & \xrightarrow{\tilde{\mathscr{D}}} & \tilde{F}_0
\end{array}$$

where the bundle $\tilde{F}_0$ is constructed from $\tilde{\mathscr{F}}$ and σ as F_0 was constructed from $\mathscr{F}$ and ω. We have

$$\tilde{\mathscr{D}} = V_{\mathrm{id}_{q+1}}(\Psi) \circ j_{q+1}.$$

Now, from the definition of the finite equations of $\tilde{\Gamma}$, it follows that, for any finite transformation $\tilde{f}$ of $\tilde{\Gamma}$, there must exist parameters a such that $\tilde{f}(\omega) = g(a, \omega)$, because $\tilde{f}(\omega)$ and ω are two solutions of $\mathscr{R}''_1$.

Remark 6.7 We notice that for any finite transformation f of Γ we have $f(\omega) = \omega$ and thus $\Gamma \subset \tilde{\Gamma}$.

It follows that $\tilde{f}(R_q) \subset R_q$ and $\tilde{f}(\Theta) \subset \Theta$. As $\tilde{f}$ is invertible, we have $\tilde{f}(R_q) = R_q$ and $\tilde{f}(\Theta) = \Theta$.

Remark 6.8 If $\tilde{f}: U \to V$ is a local diffeomorphism of Aut (X), we recall that we have to understand:

$$\tilde{f}(R_q|_U) = R_q|_V \quad \text{and} \quad \tilde{f}(\Theta|_U) = \Theta|_V$$

but, for simplicity, we do not indicate the restriction explicitly.

Conversely, the set of finite transformations $\tilde{f}$ such that $\tilde{f}(R_q) = R_q$ is clearly a Lie pseudogroup, because we can easily construct its defining finite equations, which constitute in fact a set of differential polynomials (see example hereafter).

DEFINITION 6.9 A Lie pseudogroup that can be defined by non-linear equations belonging to $C^\infty(X \times Y)[p]$ is called an "*algebraic Lie pseudogroup*".

EXAMPLES 6.10 All the classical examples.

COUNTEREXAMPLE 6.11 We leave the reader show that the following pseudogroup of transformation of $\mathbb{R}^3$ is not algebraic:

$$X = \mathbb{R}^3 \qquad \Gamma: y^1 = x^1 + a, \quad y^2 = x^2 + f(x^1),$$
$$y^3 = x^3 \exp\left(\frac{\partial f(x^1)}{\partial x^1}\right) + g(x^1, x^2)$$

Moreover there must exist parameters a such that $\tilde{f}(\omega) = g(a, \omega)$ and this pseudogroup is clearly $\tilde{\Gamma}$ because $\tilde{f}(\sigma) = \sigma$.

DEFINITION 6.12 $\tilde{\Gamma}$ is called the "*normalizer*" of Γ in Aut (X). We shall also write $\tilde{\Gamma} = N(\Gamma)$ and $\tilde{\Theta} = N(\Theta)$.

REMARK 6.13 Roughly speaking, it is the "*biggest*" Lie pseudogroup in which Γ is normal. In fact, if $\tilde{\Gamma}^{-1} \circ \Gamma \circ \tilde{\Gamma} = \Gamma$, then $\Gamma(\tilde{\Gamma}(\omega)) = \tilde{\Gamma}(\omega)$ and it follows that $\tilde{\Gamma}(\omega) = g(a, \omega)$.

However $\tilde{\Gamma}$ may have many components and, as for Γ, we shall restrict our study to the component of the identity.

We may also obtain $\tilde{\mathcal{R}}_{q+1}$ directly from knowledge of R_q by elementary computations involving the formulas of lemma 2.30 and then obtain $\tilde{R}_{q+1}$ by linearisation.

EXAMPLE 6.14 $\Gamma: y = ax, \quad \mathcal{R}_1: \dfrac{y'}{y} = \dfrac{1}{x}, \quad R_1: \dfrac{\xi'}{x} - \dfrac{\xi}{(x)^2} = 0$

Let us define $\eta(y)$ by

$$\eta(\tilde{f}(x)) = \frac{\partial \tilde{f}(x)}{\partial x} \cdot \xi(x)$$

We must have

$$\frac{1}{y}\frac{\partial \eta(y)}{\partial y} - \frac{\eta(y)}{(y)^2} = 0 \quad \text{whenever} \quad \frac{1}{x}\frac{\partial \xi(x)}{\partial x} - \frac{\xi(x)}{(x)^2} = 0.$$

As

$$\frac{\partial \eta(\tilde{f}(x))}{\partial y} \cdot \frac{\partial \tilde{f}(x)}{\partial x} = \frac{\partial^2 \tilde{f}(x)}{\partial x^2} \cdot \xi(x) + \frac{\partial \tilde{f}(x)}{\partial x} \cdot \frac{\partial \xi(x)}{\partial x}$$

we obtain

$$\tilde{\mathscr{R}}_2: \quad -\frac{y''}{y'} + \frac{y'}{y} = \frac{1}{x}, \qquad \tilde{R}_2: \quad -\xi'' + \frac{\xi'}{x} - \frac{\xi}{(x)^2} = 0$$

and finally $\tilde{\Gamma}: y = ax^b$ with a, b arbitrary constants.

EXAMPLE 6.15 When $\Gamma: \mathbb{R}^2 \to \mathbb{R}^2$ is defined by $\mathscr{R}_1: p_2^1 = 0$, $p_1^2 = 0$ there are two components for $\tilde{\mathscr{R}}_2$ and it is possible to pass from one to the other by the finite transformation $y^1 = x^2$, $y^2 = x^1$.

We can obtain in this way a chain of linear systems of infinitesimal Lie equations:

$$R_q, \quad \tilde{R}_{q+1}, \quad \tilde{\tilde{R}}_{q+2}, \quad \ldots$$

increasing the order of the system by one each time. In general this chain is not stationary.

EXAMPLE 6.16 All the "*classical*" examples to be found in the literature stop after one step only. We have the following picture for the well known "*primitive*" ones of E. Cartan, where the arrows denote inclusion:

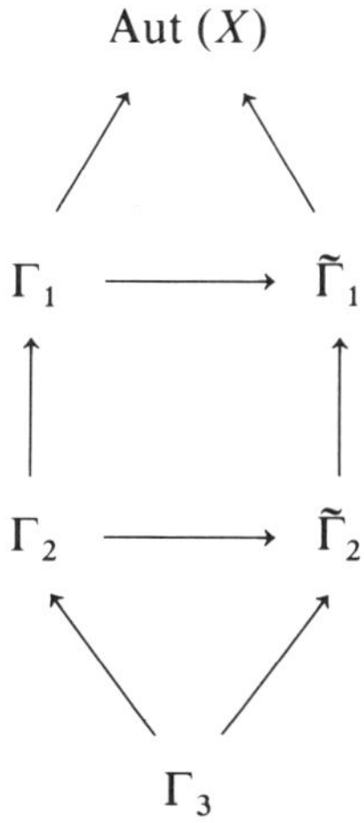

$\Gamma_1(\omega) = \omega, \quad \omega = dx^1 \wedge \cdots \wedge dx^n, \quad \tilde{\Gamma}_1(\omega) = a\omega, \quad a \text{ cst}, \quad \tilde{\tilde{\Gamma}}_1 = \tilde{\Gamma}_1$

$\Gamma_2(\omega) = \omega, \quad \omega = dx^1 \wedge dx^2 + \cdots + dx^{n-1} \wedge dx^n, \quad \tilde{\Gamma}_2(\omega) = a\omega,$
$a \text{ cst}, \quad \tilde{\tilde{\Gamma}}_2 = \tilde{\Gamma}_2$

Γ_3 Lie pseudogroup of contact transformations, $\quad \tilde{\Gamma}_3 = \Gamma_3$

For Γ_3 see problem 1.

EXAMPLE 6.17 An example that stops after two steps is provided by the following action on $X = \mathbb{R}^3$ with 3 infinitesimal generators:

$$\frac{\partial}{\partial x^1}, \quad \frac{\partial}{\partial x^2} + x^1 \frac{\partial}{\partial x^3}, \quad \frac{\partial}{\partial x^3}$$

A straightforward but tedious computation shows that $\tilde{\Gamma}$ is given by the action on $\mathbb{R}^3$ with the 9 infinitesimal generators:

$$\frac{\partial}{\partial x^1}, \quad \frac{\partial}{\partial x^2}, \quad \frac{\partial}{\partial x^3}, \quad x^1 \frac{\partial}{\partial x^3}, \quad x^2 \frac{\partial}{\partial x^3}$$

$$x^1 \frac{\partial}{\partial x^1} + x^3 \frac{\partial}{\partial x^3} + x^3 \frac{\partial}{\partial x^3}, \quad x^2 \frac{\partial}{\partial x^1} + \frac{(x^2)^2}{2} \frac{\partial}{\partial x^3}, \quad x^1 \frac{\partial}{\partial x^2} + \frac{(x^1)^2}{2} \frac{\partial}{\partial x^3}$$

Another straightforward but even more tedious computation finally shows that $\tilde{\tilde{\Gamma}}$ is given by the action on $\mathbb{R}^3$ generated by 10 infinitesimal generators constituted by the 9 former ones and by

$$(x^1 x^2 - 2x^3) \frac{\partial}{\partial x^3}.$$

REMARK 6.18 If $\Gamma_1 \subset \Gamma_2$ we do not in general have $\tilde{\Gamma}_1 \subset \tilde{\Gamma}_2$. However, if $\Gamma_1 \lhd \Gamma_2$ then of course $\Gamma_1 \lhd \Gamma_2 \subset \tilde{\Gamma}_1 \subset \mathrm{Aut}\,(X)$.

$$\left(\Gamma_1 : y = ax, \quad \tilde{\Gamma}_1 : y = ax^b, \quad \Gamma_2 = \tilde{\Gamma}_2 : y = \frac{ax + b}{cx + d}\right)$$

EXAMPLE 6.19 We have already defined $\tilde{\Theta} = N(\Theta)$. Changing the notation slightly we define $\Theta(0) = \Theta$ and inductively $\Theta(l) \subseteq \Theta(l + 1) = N(\Theta(l))$.

Let $\Theta(l)$ be defined on $\mathbb{R}^3$ as follows:

$$\Theta(l) \supset \xi = \begin{cases} a \\ f(x^1) \\ (P_l(x^1) + f'(x^1))x^3 + u(x^1)\, e^{x^2} \end{cases}$$

a is an arbitrary constant, $f(x^1)$ and $u(x^1)$ are arbitrary functions of x^1 with $f'(x^1) = \partial f(x^1)/\partial x^1$ and $P_l(x^1)$ is an arbitrary polynomial in x^1 of degree l.

We shall show that

$$N(\Theta(l)) = \Theta(l + 1) \neq \Theta(l) \qquad \forall l \geq 0.$$

In order to do this it is first of all easy to check that $\xi \in N(\Theta(l))$ must have the following form:

$$\xi = \begin{cases} \alpha x^1 + b \\ \beta x^2 + g(x^1) \\ \xi^3(x^1, x^2, x^3) \end{cases}$$

in which α, β, b are arbitrary constants and $g(x^1)$ is an arbitrary function of x^1. One may compute $\tilde{\xi}^3(x^1, x^2, x^3)$ by the condition $[\xi, \tilde{\xi}] \in \Theta(l)$, that is to say:

$$a\frac{\partial\tilde{\xi}^3}{\partial x^1} + f(x^1)\frac{\partial\tilde{\xi}^3}{\partial x^2} + \xi^3\frac{\partial\tilde{\xi}^3}{\partial x^3} - \tilde{\xi}^3\frac{\partial\xi^3}{\partial x^3} - \beta x^2 u(x^1)e^{x^2}$$
$$= (R_l(x^1) + ag''(x^1) + (\beta - \alpha)f'(x^1))x^3 + w(x^1)e^{x^2}$$

where $w(x^1)$ is a convenient function of x^1 and $R_l(x^1)$ a convenient polynomial in x^1 of degree l.

We must have:

$$a\frac{\partial^2\tilde{\xi}^3}{\partial x^1\partial x^3} + f(x^1)\frac{\partial^2\tilde{\xi}^3}{\partial x^2\partial x^3} + \xi^3\frac{\partial^2\tilde{\xi}^3}{\partial x^3\partial x^3} = R_l(x^1) + ag''(x^1) + (\beta - \alpha)f'(x^1)$$

Setting $a = 0$ and $f(x^1) \equiv 0$ we deduce $\tilde{\xi}^3 = A(x^1, x^2)x^3 + B(x^1, x^2)$ with

$$a\frac{\partial A}{\partial x^1} + f(x^1)\frac{\partial A}{\partial x^2} = R_l(x^1) + ag''(x^1) + (\beta - \alpha)f'(x^1)$$

We have thus

$$a\frac{\partial^2 A}{\partial x^1\partial x^2} + f(x^1)\frac{\partial^2 A}{\partial x^2\partial x^2} = 0$$

and it follows that

$$\frac{\partial A}{\partial x^2} = \text{cst.}$$

Setting

$$a = 0 \quad \text{and} \quad f(x^1) = e^{\pm x^1}$$

we obtain

$$\frac{\partial A}{\partial x^2} = 0 \quad \text{and} \quad \alpha = \beta.$$

Thus

$$\tilde{\xi}^3 = (Q_{l+1}(x^1) + g'(x^1))x^3 + B(x^1, x^2).$$

Finally, with convenient functions:

$$a\frac{\partial B}{\partial x^1} + f(x^1)\frac{\partial B}{\partial x^2} - \beta x^2 u(x^1)e^{x^2} - B(x^1, x^2)(P_l(x^1) + f'(x^1)) = \overline{w}(x^1)e^{x^2}$$

Setting

$$a = 0, \quad f(x^1) \equiv 0, \quad P_l(x^1) \equiv 1, \quad u(x^1) \equiv 0$$

we get:

$$\xi^3 = (Q_{l+1}(x^1) + g'(x^1))x^3 + v(x^1)e^{x^2}$$

and $\beta u(x^1) = 0$ which gives $\alpha = \beta = 0$. C.Q.F.D.

In order to end this chapter we shall exhibit some relationship between R_q and $\tilde{R}_{q+1}$.

PROPOSITION 6.20 $\mathscr{R}'_1 \cap \mathscr{R}''_1$ is a non-linear involutive system of order 1 on $\mathscr{F}$.

Proof As $\bar{R}_q = R_q$ is of course formally integrable, when $\bar{\omega} = g(a, \omega)$ we must have:

$$\begin{cases} I_*\left(\omega(x), \dfrac{\partial\omega(x)}{\partial x}\right) = 0 \\ I_{**}\left(\omega(x), \dfrac{\partial\omega(x)}{\partial x}\right) = c \end{cases} \Rightarrow \begin{cases} I_*\left(\bar{\omega}(x), \dfrac{\partial\bar{\omega}(x)}{\partial x}\right) = 0 \\ I_{**}\left(\bar{\omega}(x), \dfrac{\partial\bar{\omega}(x)}{\partial x}\right) = \bar{c} \end{cases}$$

but $\bar{c}$ may be different from c. In order to compare c and $\bar{c}$ we need the following technical lemma:

LEMMA 6.21 The functions $I_*(u, \partial u(\partial x)$ can be chosen such that $\rho_1(W) \cdot I_* = 0$.

Proof Keeping only one index for the ξ^k_μ and L^μ_k, we obtain, with rank $[M(\omega(x))] \neq 0$:

$$R_q \qquad (M^{-1})^\sigma_\tau(\omega(x))\underset{1}{\Omega^\tau} \equiv \xi^\sigma + \xi^{m+r} \cdot \mathscr{E}^\sigma_{m+r}(\omega(x)) + \xi^i(M^{-1})^\sigma_\tau(\omega(x))\frac{\partial\omega^\tau(x)}{\partial x^i} = 0$$

Using the fact that:

$$\frac{\partial\mathscr{E}^\sigma_{m+r}(\omega(x))}{\partial x^i} = \frac{\partial\mathscr{E}^\sigma_{m+r}(\omega(x))}{\partial u^\tau} \cdot \frac{\partial\omega^\tau(x)}{\partial x^i}$$
$$= \left(M^\tau_\rho(\omega(x))\frac{\partial\mathscr{E}^\sigma_{m+r}(\omega(x))}{\partial u^\tau}\right)\left((M^{-1})^\rho_s(\omega(x))\frac{\partial\omega^s(x)}{\partial x^i}\right)$$

and lemma 6.4, we see that the study of the surjectivity of the map $\pi^{q+1}_q : R^0_{q+1} \to R^0_q$ may be carried out by dealing only with the $Q(u, \partial u/\partial x)$.

It follows that we have functional relations:

$$I_*\left(u, \frac{\partial u}{\partial x}\right) \equiv H\left(Q\left(u, \frac{\partial u}{\partial x}\right)\right) = 0$$

This proves the lemma because we know that $\rho_1(W) \cdot Q = 0$ for any infinitesimal label transformation W. C.Q.F.D.

Now, as ω is a solution of $\mathscr{R}'_1$, we have $I_*|_{\mathscr{R}''_1} = 0$, that is to say

$$I_*\left(u, \frac{\partial u}{\partial x}\right) = 0 \qquad \forall\left(x, u, \frac{\partial u}{\partial x}\right) \in \mathscr{R}''_1$$

because

$$I_*\left(u, \frac{\partial u}{\partial x}\right) \equiv H\left(Q\left(u, \frac{\partial u}{\partial x}\right)\right) = H\left(Q\left(\omega(x), \frac{\partial \omega(x)}{\partial x}\right)\right) \equiv I_*\left(\omega(x), \frac{\partial \omega(x)}{\partial x}\right) = 0$$

and we notice that:

$$\mathscr{R}'_1 \subset \mathscr{P}_1, \qquad \mathscr{R}''_1 \subset \mathscr{P}_1.$$

Using propositions 6.5.14 and 6.5 we have:

$$[\rho_1(\theta), \rho_1(W)] = \rho_1([\theta, W]) = 0$$

for any vector field

$$\theta = \xi^i(x)\frac{\partial}{\partial x^i} + \partial_\mu \xi^k(x) L_k^{\tau\mu}(u)\frac{\partial}{\partial u^\tau}$$

on $\mathscr{F}$. It follows that $[L_{(1)}, \rho_1(W)] = 0$.

As $L_{(1)}$ and $\rho_1(W)$ are tangent to $\mathscr{P}_1$, we shall now work on $\mathscr{P}_1$. For simplicity we shall not write out the restrictions. We have:

$$0 = [L_{(1)}, \rho_1(W)] \cdot I_{**} = L_{(1)} \cdot (\rho_1(W) \cdot I_{**})$$

and there must exist functional relations $\rho_1(W) \cdot I_{**} = E(I_{**})$ because the I_{**} constitute a maximum set of unconnected invariants of $L_{(1)}|_{\mathscr{P}_1}$.

We have proved the proposition:

PROPOSITION 6.22 Any finite label transformation induces a finite transformation of the structure constants

$$\bar{u} = g(a, u) \Rightarrow \bar{c} = h(a, c)$$

REMARK 6.23 This action is not effective in general.

REMARK 6.24 Using theorem 4.8 when $\bar{\omega} = g(a, \omega)$, it follows that $J(\bar{c}) = 0$ when $J(c) = 0$.

We come back to the proof of proposition 6.20.

By definition we have $\tilde{\Gamma}(\omega) = g(a, \omega)$. This means that the parameters a corresponding to any finite transformation of $\tilde{\Gamma}$ are such that $c = h(a, c)$.

We have thus to look for the infinitesimal label transformations that leave the structure constants unchanged. A set of such independent infinitesimal generators such as $w = A^r \cdot W_r$ will be determined by the linear conditions $\alpha_r(c) \cdot A^r = 0$ obtained from the linearised equations $\underset{1}{C} = \alpha_r(c) \cdot A^r$.

From what has been said, $\rho_1(w)$ leaves both $\mathscr{R}'_1$ and $\mathscr{R}''_1$ invariant and thus also $\mathscr{R}'_1 \cap \mathscr{R}''_1$. Moreover we have:

$$\dim V(\mathscr{R}'_1 \cap \mathscr{R}''_1) = \dim V(\mathscr{R}'_1) \cap V(\mathscr{R}''_1) = \text{number of independent } w.$$

that is to say, the restriction of the action of $\rho_1(w)$ to any fiber $\mathscr{R}'_{1:x} \cap \mathscr{R}''_{1:x}$ is simply transitive.

It follows that $\mathscr{R}'_1 \cap \mathscr{R}''_1$ can be constructed by integrating the involutive distribution $\rho_1(w)$, each fiber being the only leaf passing through the point

$$\left(x, \omega(x), \frac{\partial \omega(x)}{\partial x}\right) \in \mathscr{R}'_1 \cap \mathscr{R}''_1 \subset J_1(\mathscr{E}).$$

Moreover there is no problem about the rank of this distribution. In fact, if $\bar{u} = G(u, p)$ as $[L, W] = 0$ and $[L, w] = 0$, we have:

$$W^\tau(\bar{u}) = \frac{\partial \bar{u}^\tau}{\partial u^\sigma} W^\sigma(u) \quad \text{and} \quad w^\tau(\bar{u}) = \frac{\partial \bar{u}^\tau}{\partial u^\sigma} \cdot w^\sigma(u)$$

that is to say

$$\operatorname{rank}\,[W(\bar{u})] = \operatorname{rank}\,[W(u)], \qquad \operatorname{rank}\,[w(\bar{u})] = \operatorname{rank}\,[w(u)]$$

Thus rank $[W(u)]$ and rank $[w(u)]$ are equal to their respective maximum values whenever rank $[L(u)] = \dim F_0$.

A maximum set of unconnected invariants of $\rho_1(w)$ may be obtained by adding some new zero-order invariants to the invariants $Q(u, \partial u/\partial x)$ of $\rho_1(W)$. The total differentials of these new invariants are also differential invariants because

$$\rho_2(w) \circ \frac{d}{dx} = \frac{d}{dx} \circ \rho_1(w).$$

As $\mathscr{R}''_1$ is involutive, according to lemma 4.3 it follows that $\mathscr{R}'_1 \cap \mathscr{R}''_1$ is also involutive with a zero symbol.

We finally sketch the situation diagrammatically:

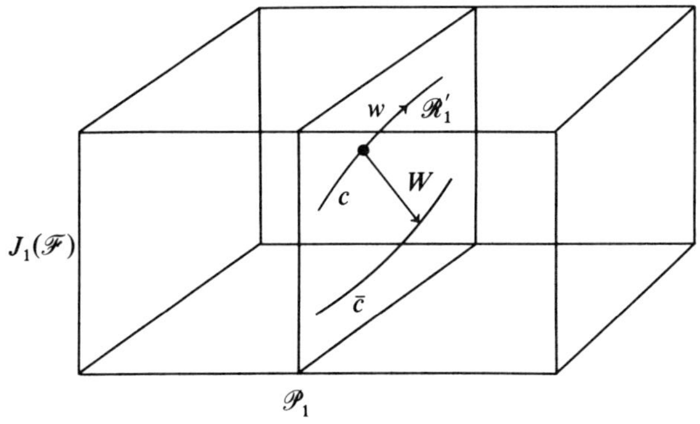

C.Q.F.D.

REMARK 6.25 As ω is a solution of $\mathscr{R}'_1$ and $\mathscr{R}''_1$ it is also a solution of $\mathscr{R}'_1 \cap \mathscr{R}''_1$ and the linearised system:

$$R'_1 \cap R''_1: \qquad \mathscr{D}_1 \cdot \underset{1}{\Omega} = 0, \qquad \nabla_0 \cdot \underset{1}{\Omega} = 0$$

is also involutive.

As a by-product we obtain the following theorem:

THEOREM 6.26 If R_q is formally integrable and G_q is 2-acyclic, then $\tilde{R}_{q+1}$ is formally integrable.

Moreover

$$\tilde{G}_{q+r} = G_{q+r} \qquad \forall r \geq 1.$$

Proof We have first to prove that $\pi^{q+2}_{q+1} : \tilde{R}_{q+2} \to \tilde{R}_{q+1}$ is an epimorphism and that $\tilde{G}_{q+1}$ is 2-acyclic. This can be deduced from a chase in the following three-dimensional commutative and exact diagram. However we shall detail this chase because it is of an unusual type. For simplicity we shall indicate the steps of the chase but not the maps involved.

Let $a \in \tilde{R}_{q+1}$ that can be considered as an element of $J_{q+1}(T)$ with zero image in $\tilde{F}$, that is to say with image in $J_1(F_0)$ an element b of R''_1. But the image of b, considered as an element of $J_1(F_0)$, is zero in F_1 because the vertical sequences are exact. It follows that $b \in R'_1 \cap R''_1$. From the above proposition, $\exists c \in (R'_1 \cap R''_1)_{+1} = R'_2 \cap R''_2$ that can be considered as an element of $J_2(F_0)$ with projection b on $J_1(F_0)$.

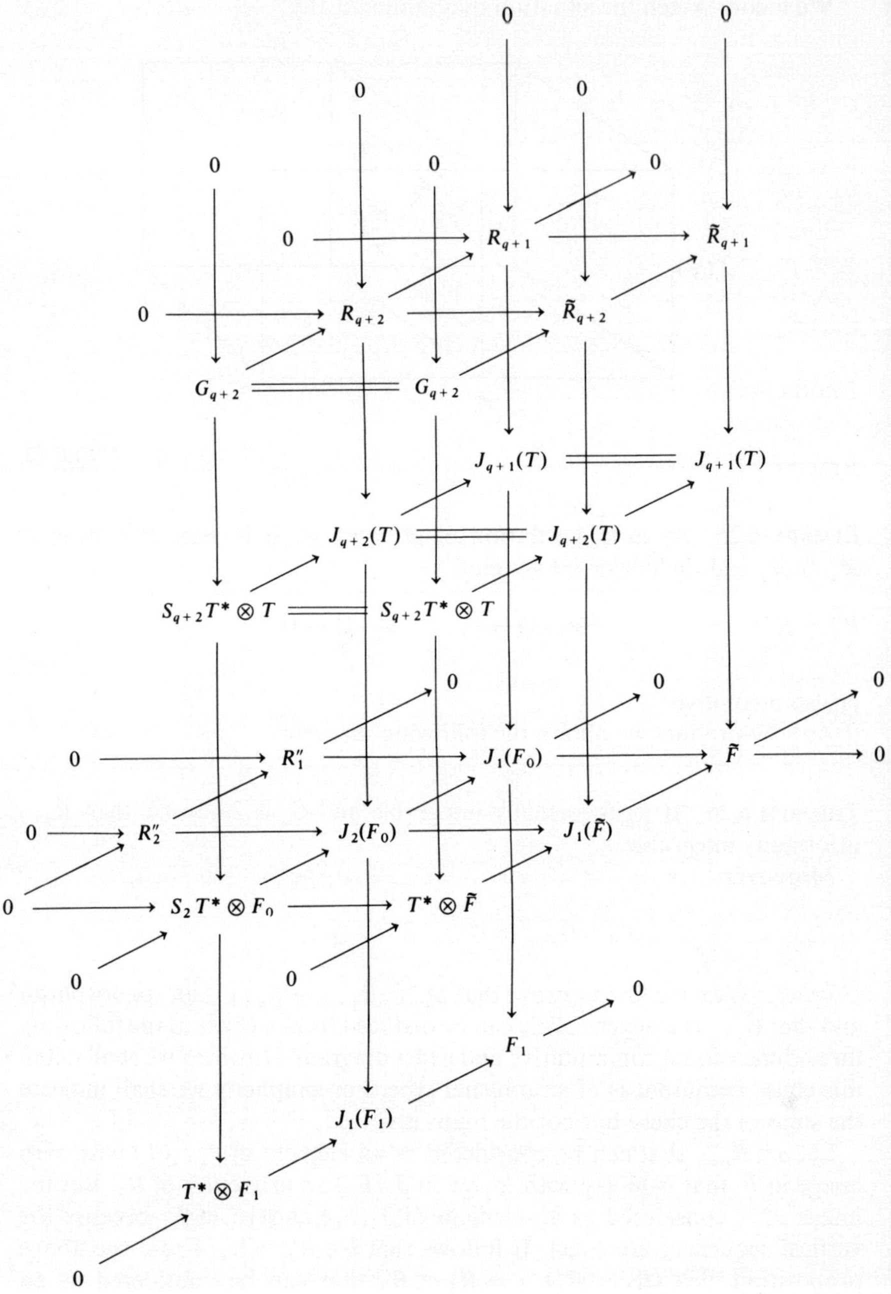
0
R_{q+1}
$\tilde{R}_{q+1}$
R_{q+2}
$\tilde{R}_{q+2}$
G_{q+2}
G_{q+2}
$J_{q+1}(T)$
$J_{q+1}(T)$
$J_{q+2}(T)$
$J_{q+2}(T)$
$S_{q+2}T^* \otimes T$
$S_{q+2}T^* \otimes T$
R_1''
$J_1(F_0)$
$\tilde{F}$
R_2''
$J_2(F_0)$
$J_1(\tilde{F})$
$S_2 T^* \otimes F_0$
$T^* \otimes \tilde{F}$
F_1
$J_1(F_1)$
$T^* \otimes F_1$

Take $d \in J_{q+2}(T)$ with projection a on $J_{q+1}(T)$. Let $\rho \in J_2(F_0)$ be the image of d. Then the projection of ρ on $J_1(F_0)$ is also b. Thus there exists an element $f = \rho - c \in S_2T^* \otimes F_0$ with zero image in $T^* \otimes F_1$ because $c \in R'_2 \cap R''_2 \subset R'_2$.

From theorems 5.5.5 and 5.2.1, as G_q is 2-acyclic and $\pi_q^{q+1} : R_{q+1} \to R_q$ is surjective, then the vertical symbol sequence on the left is exact and $\exists g \in S_{q+2}T^* \otimes T$ with image f in $S_2T^* \otimes F_0$.

Finally the image of $d - g$ in $J_2(F_0)$ is $\rho - f = c$ and $d - g \in \tilde{R}_{q+2}$ with projection $a \in \tilde{R}_{q+1}$ on $J_{q+1}(T)$.

As the symbol of ∇_0 is zero, an easy chase shows that $\tilde{G}_{q+1} = G_{q+1}$ and thus $\tilde{G}_{q+r} = G_{q+r}, \forall r \geq 1$. C.Q.F.D.

Another proof of this theorem will be given at the end of the book.

REMARK 6.27 We have

$$\dim \tilde{R}_{q+r} - \dim R_{q+r} = \dim \tilde{R}_{q+1} - \dim R_{q+1} = \dim R'_1 \cap R''_1$$

DEFINITION 6.28 We shall say that Γ is of "*finite codimension*" in $\tilde{\Gamma}$.

As the number of independent w depends on c, for a given bundle $\mathcal{F}$ of geometric objects, the codimension of Γ in $\tilde{\Gamma}$ is finite but depends on the choice of the section ω of $\mathcal{F}$ through the corresponding structure constants only.

COROLLARY 6.29 Every transitive Lie pseudogroup Γ_1 which is normal in a Lie pseudogroup Γ_2 is of finite codimension in it.

(We recall that the normality is defined by means of the infinitesimal equations, as we are dealing only with the connected component of the identity and may be expressed by $[\Theta_1, \Theta_2] \subset \Theta_1$.)

Proof We have $\Gamma_1 \lhd \Gamma_2 \subseteq \tilde{\Gamma}^1 \subset \mathrm{Aut}\,(X)$.

REMARK 6.30 This corollary can be proved directly. In fact, using the above definition of normality, we have:

$$\xi^k \eta^\rho_{\mu+1_k} \in G_q, \qquad \forall \xi^k \in T, \forall \eta^\rho_{\mu+1_k} \in \tilde{G}_{q+1}$$

with $|\mu| = q$. It follows that the map

$$\delta : \tilde{G}_{q+1} \to T^* \otimes S_qT^* \otimes T$$

factors through $T^* \otimes G_q$ in the commutative diagram with exact upper row:

$$\begin{array}{ccccccc}
0 & \longrightarrow & G_{q+1} & \xrightarrow{\delta} & T^* \otimes G_q & \xrightarrow{\delta} & \Lambda^2 T^* \otimes S_{q-1} T^* \otimes T \\
 & & \downarrow & \nearrow & \downarrow & & \| \\
0 & \longrightarrow & \tilde{G}_{q+1} & \xrightarrow{\delta} & T^* \otimes S_q T^* \otimes T & \xrightarrow{\delta} & \Lambda^2 T^* \otimes S_{q-1} T^* \otimes T
\end{array}$$

An easy chase shows that $\tilde{G}_{q+1} = G_{q+1}$.

REMARK 6.31 The corollary may not be true for intransitive Lie pseudogroups. Let $X = \mathbb{R}^2$ and consider:

$$\Gamma : y^1 = x^1 + a, \quad y^2 = x^2 + g(x^1) \qquad \Gamma_n : y^1 = x^1, \quad y^2 = x^2 + P_n(x^1)$$

where P_n is a polynomial in x^1 of order n.

We have

$$\tilde{\Gamma}_n = \tilde{\Gamma} : y^1 = \lambda x^1 + \mu, \quad y^2 = \nu x^2 + h(x^1)$$

Setting

$$\Gamma_\infty : y^1 = x^1, \quad y^2 = x^2 + g(x^1)$$

we obtain the chain:

$$\tilde{\Gamma} \rhd \Gamma \rhd \Gamma_\infty \rhd \cdots \rhd \Gamma_1 \rhd \Gamma_0 \rhd \mathrm{id}$$

7 Deformation theory of structures

Let $\mathcal{F}$ be a bundle of geometric objects over a manifold X.

In the sequel we shall assume that X is an analytic manifold and that $\mathcal{F}$ has been chosen as in theorem 4.1 and is, thus, an analytic bundle over X. All the manifolds to be considered will therefore also be analytic.

DEFINITION 7.1 By a "*structure*" over X, we understand a section ω of $\mathcal{F}$ over X.

In fact, as our problem will be a local one, we can shrink X to an open subset $U \subset X$.

DEFINITION 7.2 By a deformation of the given structure ω, we understand a section ω_t of $\mathcal{F}$, depending analytically on t and such that $\omega_0 = \omega$.

As usual we set

$$\omega_t = \omega + t \underset{1}{\Omega} + \cdots = \omega + \sum_{\nu=1}^{\infty} \frac{t^\nu}{\nu!} \underset{\nu}{\Omega}.$$

DEFINITION 7.3 We shall say that two structures are "*equivalent*" if the corresponding Lie pseudogroups are similar.

According to the preceding pages, ω and $\bar{\omega}$ are equivalent if and only if we can pass from ω to $\bar{\omega}$ by a label transformation, followed by the effect of an automorphism of X:

$$\omega \sim \bar{\omega} \Leftrightarrow \exists a, \varphi \in \operatorname{Aut}(X): \bar{\omega} = g(a, \varphi(\omega))$$

We have $g(a, \varphi(\omega)) = \varphi(g(a, \omega))$ because $[L, W] = 0$.

Now we shall assume that the systems $\mathcal{R}_q$ and R_q of Lie equations determined by ω are involutive and we shall restrict our attention to the other sections also giving rise to such involutive systems. We know that we have just, in fact, to take away a closed set from the fiber of $\mathcal{F}$.

For any such deformation ω_t of ω we must have:

$$\begin{cases} I_*\left(\omega(x), \dfrac{\partial\omega(x)}{\partial x}\right) = 0 \\[2ex] I_{**}\left(\omega(x), \dfrac{\partial\omega(x)}{\partial x}\right) = c \end{cases} \qquad \text{with } J(c) = 0$$

$$\begin{cases} I_*\left(\omega_t(x), \dfrac{\partial\omega_t(x)}{\partial x}\right) = 0 \\[2ex] I_{**}\left(\omega_t(x), \dfrac{\partial\omega_t(x)}{\partial x}\right) = c_t \end{cases} \qquad \text{with } J(c_t) = 0$$

However it may happen that $c_t \neq c$ and we know that $\bar{c} = h(a, c)$ when $\bar{\omega} \sim \omega$.

REMARK 7.4 It is sometimes better to consider a deformation of structure as a deformation of the corresponding Lie equations. In fact, when the systems $\mathcal{R}_q$ and R_q are given, we may construct $\mathcal{F}$ and its section ω. Then, to another section $\bar{\omega}$ or a deformation ω_t there will correspond the systems $\bar{\mathcal{R}}_q$ and $\bar{R}_q$ or the deformed systems $\mathcal{R}_q(t)$ and $R_q(t)$, in a natural way.

DEFINITION 7.5 Two sets of structure constants c and $\bar{c}$ are said to be "*equivalent*" if there exist parameters a such that $\bar{c} = h(a, c)$.

PROBLEM 7.6 Is it possible to find a deformation ω_t of ω satisfying the above integrability conditions and such that ω_t and ω are not equivalent, even for small t?

DEFINITION 7.7 A deformation ω_t of ω is said to be "*trivial*" whenever $\omega_t \sim \omega$.

DEFINITION 7.8 A structure is said to be "*rigid*" if it admits no non-trivial deformation.

If we are given a deformation ω_t of ω we shall divide the study of the above problem into two parts:

1 Is it possible to "*hide*" the resulting deformation c_t of c by means of a label transformation.

If this is possible, we get a new deformation $\bar{\omega}_t$ of ω satisfying the same integrability conditions.

2 Is it possible to find $\varphi_t \in \text{Aut}(X)$ such that $\bar{\omega}_t = \varphi_t(\omega)$ and $\varphi_0 = \text{id}$?

The problem **2** is in fact the equivalence problem.

An answer, though a partial one, has already been given, even in the non-analytic case: it is related to the local exactness at $\mathscr{F}$ of a non-linear sequence. This is the kind of problem studied by Spencer and others (45).

The corresponding infinitesimal problem is related to the local exactness at F_0 of a P-sequence.

To sum up we have divided the problem of deformation of structure into two other ones: the problem **1** and the equivalence problem **2**.

We shall focus now on the problem **1** which is a purely algebraic one because we have to look for deformation c_t of c satisfying the Jacobi condition $J(c_t) = 0$.

The converse problem is easily solved as it is just the third fundamental theorem. In fact, if c_t is a given deformation of c satisfying $J(c_t) = 0$, by solving the integrability conditions, we can find a deformation ω_t of ω with structure constants c_t.

We have thus changed the problem of deforming a given structure into that of deforming its structure constants, taking into account the Jacobi conditions.

We shall consider a set of structure constants as a point c on an algebraic variety $\mathscr{C}$ defined in some power of $\mathbb{R}$ by the polynomial equation $J(c) = 0$. The algebraic variety $\mathscr{C}$ may be reducible.

A deformation of c satisfying the Jacobi condition will be an analytic curve c_t on $\mathscr{C}$ passing through c for $t = 0$.

According to proposition 6.22, we have a Lie group of transformations of $\mathscr{C}$ given by the action $\bar{c} = h(a, c)$ which is not effective in general.

Our problem is to find whether there exist parameters $a(t)$ such that $c_t = h(a_t, c)$ or not.

DEFINITION 7.9 A deformation c_t of c is said to be "*trivial*" if c_t is in the orbit of c, $\forall t$.

DEFINITION 7.10 A point c is said to be "*rigid*" if any deformation of c is trivial.

REMARK 7.11 When c is rigid, any structure with structure constants c is rigid, up to the equivalence problem.

This finite problem is however difficult to study. For this reason, we shall study what happens in the tangent space $T_c(\mathscr{C})$.

By definition, for any infinitesimal label transformation with canonical infinitesimal parameters A^r, we have $\underset{1}{C} = \alpha_r(c) \cdot A^r$.

DEFINITION 7.12 We call B the vector space at c generated by the $\underset{1}{C}$ for arbitrary A.

DEFINITION 7.13 We call Z the vector space at c defined by the linear equation

$$\frac{\partial J(c)}{\partial c} \cdot \underset{1}{C} = 0.$$

As we know that $J(\bar{c}) = 0$ whenever $\bar{c} = h(a, c)$ and $J(c) = 0$, we obtain

$$B \subset T_c(\mathscr{C}) \subset Z.$$

EXAMPLE 7.14 See problem 6:

We take $\mathscr{C} \subset \mathbb{R}^2$ defined by $c^1 \cdot c^2 = 0$ and the action

$$\bar{c}^1 = \frac{a^1}{a^2} \cdot c^1, \qquad \bar{c}^2 = \frac{1}{a^1} \cdot c^2.$$

At the identity transformation we have $a^1 = a^2 = 1$. Setting $a^r = 1 + tA^r + \cdots$ it follows that:

$$\underset{1}{C}^1 = (A^1 - A^2)c^1, \qquad \underset{1}{C}^2 = -A^1 \cdot c^2$$

and for

$$c = (0, 0) : B = 0, \quad T_{(0,0)}(\mathscr{C}) = c^1 \text{ axis} + c^2 \text{ axis}, \quad Z = \mathbb{R}^2.$$

REMARK 7.15 It may however be difficult to look for $T_c(\mathscr{C})$ because, according to the Hilbert theorem 5.6.11, the ideal of $\mathbb{R}[c]$ generated by the polynomials $J(c)$ may contain powers of some polynomials not belonging to that ideal, that is to say, may not be equal to its radical.

EXAMPLE 7.16 If $\mathscr{C} \subset \mathbb{R}^2$ is defined by $(c^2 - c^1)^2 = 0$ then

$$T_c(\mathscr{C}) = \mathbb{R} \subset Z = \mathbb{R}^2, \qquad \forall c \in \mathscr{C}.$$

We obtain easily the rigidity theorem:

THEOREM 7.17 A sufficient condition of rigidity is $B = Z$.

A more subtle problem is the following.

DEFINITION 7.18 An element $\underset{1}{C}$ of Z is said to be "*integrable*" if there exists a deformation c_t of c such that

$$J(c_t) = 0 \quad \text{and} \quad \left.\frac{dc_t}{dt}\right|_{t=0} = \underset{1}{C}.$$

It is said to be "*formally integrable*" if there exists a formal power series

$$c_t = c + t\underset{1}{C} + \sum_{\nu=2}^{\infty} \frac{t^\nu}{\nu!} \underset{\nu}{C}$$

solution of $J(c_t) = 0$.

EXAMPLE 7.19 Take $c = (0, 0)$ in the above example. Looking up to the second order in t we have:

$$\left(c^1 + t\underset{1}{C}^1 + \frac{t^2}{2}\underset{2}{C}^1 + \cdots\right)\left(c^2 + t\underset{1}{C}^2 + \frac{t^2}{2}\underset{2}{C}^2 + \cdots\right)$$

$$= t\left(c^1\underset{1}{C}^2 + c^2\underset{1}{C}^1\right) + \frac{t^2}{2}\left(c^1\underset{2}{C}^2 + 2\underset{1}{C}^1\underset{1}{C}^2 + c^2\underset{2}{C}^1\right) + \cdots$$

and it follows that we must have $\underset{1}{C}^1 \cdot \underset{1}{C}^2 = 0$. The element $(1, 1) \in Z$ is not integrable.

We shall focus now on the existence of such a formal deformation without looking at the convergence of the corresponding series.

To do this, we shall use a trick. Introducing two parameters s and t, then c_{s+t} becomes a deformation of c_t and we have:

$$c_{s+t} = \sum_{\nu=0}^{\infty} \frac{(s+t)^\nu}{\nu!} \cdot \underset{\nu}{C} = \sum_{\nu=0}^{\infty} \sum_{\substack{\lambda+\mu=\nu \\ \lambda \geq 0 \\ \mu \geq 0}} \frac{s^\lambda}{\lambda!} \cdot \frac{t^\mu}{\mu!} \underset{\nu}{C}$$

$$= c_t + s\left(\sum_{\nu=0}^{\infty} \frac{t^\nu}{\nu} \underset{\nu+1}{C}\right) + \cdots$$

where, for simplicity, we set $\underset{0}{C} = c$.

Now we have

$$\mathscr{D}_2 \cdot \underset{1}{C} \equiv \frac{\partial J(c)}{\partial c} \cdot \underset{1}{C} = 0$$

Using the third fundamental theorem, we can find a deformation ω_t of ω and, for any value of t sufficiently small, we can construct a P-sequence, depending now on t. In particular $\mathcal{D}_{r+1}(t) \circ \mathcal{D}_r(t) = 0$ and we notice that for t sufficiently small $\dim F_r(t) = \dim F_r$.

It follows that we have, $\forall t$ small:

$$\mathcal{D}_2(t) \cdot \left(\sum_{\nu=0}^{\infty} \frac{t^\nu}{\nu!} \underset{\nu+1}{C} \right) \equiv \frac{\partial J(c_t)}{\partial c} \cdot \left(\sum_{\nu=0}^{\infty} \frac{t^\nu}{\nu!} \underset{\nu+1}{C} \right) = 0$$

In particular we obtain:

$$\nu = 0 \qquad \mathcal{D}_2 \cdot \underset{1}{C} = 0$$

$$\nu = 1 \qquad \mathcal{D}_2 \cdot \underset{2}{C} + \frac{\partial^2 J(c)}{\partial c \partial c} \cdot \underset{1}{C} \cdot \underset{1}{C} = 0 \qquad \text{(Hessian of } J)$$

Using the Leibnitz rule for derivation, we get the integrability condition of order ν:

$$\sum_{\substack{\lambda+\mu=\nu \\ \lambda,\mu \geq 0}} \frac{\nu!}{\lambda!\mu!} \left(\frac{\partial^\lambda \mathcal{D}_2(t)}{\partial t^\lambda} \right) \bigg|_{t=0} \cdot \underset{\mu+1}{C} = 0$$

that we transform into the following ones:

$$\mathcal{D}_2 \cdot \underset{\nu+1}{C} = - \sum_{\substack{\lambda+\mu=\nu \\ \lambda>0 \\ \mu \geq 0}} \frac{\nu!}{\lambda!\mu!} \left(\frac{\partial^\lambda \mathcal{D}_2(t)}{\partial t^\lambda} \right) \bigg|_{t=0} \cdot \underset{\mu+1}{C}$$

where the right member involves only c and $\underset{1}{C}, \ldots, \underset{\nu}{C}$.

As we have seen in the last example, the resolution of such inductively related linear systems is not always possible and may sometimes bring "*obstructions*" to the deformation at a certain order.

Let us now consider the operator $\mathcal{D}_3$ of the P-sequence constructed with ω. We have

$$\mathcal{D}_3 \circ \mathcal{D}_2 = 0 \quad \text{and} \quad \mathcal{D}_3 \circ \mathcal{D}_2 \cdot \underset{\nu+1}{C} = 0, \qquad \forall \underset{\nu+1}{C}.$$

Deriving now with respect to t the identity $\mathcal{D}_3(t) \circ \mathcal{D}_2(t) = 0$, using the Leibnitz formula we get.

$$\sum_{\substack{\lambda+\mu=\nu \\ \lambda,\mu \geq 0}} \frac{\nu!}{\lambda!\mu!} \left(\frac{\partial^\lambda \mathcal{D}_3(t)}{\partial t^\lambda} \right) \bigg|_{t=0} \circ \left(\frac{\partial^\mu \mathcal{D}_2(t)}{\partial t^\mu} \right) \bigg|_{t=0} = 0$$

Finally we obtain:

$$\mathscr{D}_3 \circ - \left(\sum_{\substack{\lambda+\mu=\nu \\ \lambda>0 \\ \mu\geq 0}} \frac{\nu!}{\lambda!\mu!} \left(\frac{\partial^\lambda \mathscr{D}_2(t)}{\partial t^\lambda}\right)\Bigg|_{t=0} \cdot \underset{\mu+1}{C} \right)$$

$$\equiv - \sum_{\substack{\lambda+\mu=\nu \\ \lambda>0 \\ \mu\geq 0}} \frac{\nu!}{\lambda!\mu!} \mathscr{D}_3 \circ \left(\frac{\partial^\lambda \mathscr{D}_2(t)}{\partial t^\lambda}\right)\Bigg|_{t=0} \cdot \underset{\mu+1}{C}$$

$$\equiv \sum_{\substack{\lambda+\mu=\nu \\ \lambda>0 \\ \mu\geq 0}} \frac{\nu!}{\lambda!\mu!} \left[\sum_{\substack{\alpha+\beta=\lambda \\ \alpha>0 \\ \beta\geq 0}} \frac{\lambda!}{\alpha!\beta!} \left(\frac{\partial^\alpha \mathscr{D}_3(t)}{\partial t^\alpha}\right)\Bigg|_{t=0} \circ \left(\frac{\partial^\beta \mathscr{D}_2(t)}{\partial t^\beta}\right)\Bigg|_{t=0} \cdot \underset{\mu+1}{C} \right]$$

$$\equiv \sum_{\alpha=1}^{\nu} \frac{\nu!}{\alpha!(\nu-\alpha)!} \left(\frac{\partial^\alpha \mathscr{D}_3(t)}{\partial t^\alpha}\right)\Bigg|_{t=0} \circ \left[\sum_{\substack{\beta+\mu=\nu-\alpha \\ \mu\geq 0 \\ \beta\geq 0}} \frac{(\nu-\alpha)!}{\beta!\mu!} \left(\frac{\partial^\beta \mathscr{D}_2(t)}{\partial t^\beta}\right)\Bigg|_{t=0} \cdot \underset{\mu+1}{C} \right]$$

$$\equiv 0$$

because of the integrability conditions up to order $\nu - 1$.

We can work over a fixed trivialising coordinate domain $U \subset X$ because the form of the Jacobi conditions do not depend on the coordinate system.

When we let $\mathscr{D}_3$ act on the constant sections of F_2 over U, we get a linear map from the fiber of F_2 to the fiber of F_3 and we have:

THEOREM 7.20 Each element of Z is formally integrable if any constant section of F_2 killed by $\mathscr{D}_3$ is equal to $\mathscr{D}_2 \cdot \underset{1}{C}$ for some constants $\underset{1}{C}$.

Later we shall give an intrinsic formulation of this theorem.

EXAMPLE 7.21 Let $X = \mathbb{R}^n$ and consider the Lie pseudogroup Γ of transformations of X given by the action of a Lie group G on X with $\dim G = \dim X$.

According to the fundamental theorems of Lie, in that case the bundle $\mathscr{F}$ of geometric objects is the fibered product of T^* by itself, n-times.

The corresponding general finite equations are:

$$\mathscr{R}_1 \qquad \omega_k^\rho(y) \frac{\partial y^k}{\partial x^i} = \omega_i^\rho(x)$$

The general infinitesimal equations are:

$$R_1 \qquad \underset{1}{\Omega_i^\rho} \equiv \omega_k^\rho(x)\xi_i^k + \xi^r \frac{\partial \omega_i^\rho(x)}{\partial x^r} = 0$$

If we set $\alpha_k^i(x)\omega_j^k(x) = \delta_j^i$ we obtain the solved form:

$$R_1 \qquad \frac{\partial \xi^k}{\partial x^i} + \alpha_\rho^k(x)\frac{\partial \omega_i^\rho(x)}{\partial x^r}\cdot \xi^r = 0$$

We have already exhibited the integrability conditions and the Jacobi conditions.

As $G_1 = 0$, Γ is a pseudogroup of finite type and the P-sequence is exact with

$$\dim F_{r-1} = \frac{n\cdot n!}{(n-r)!r!}$$

The equivalence problem **2** can always be solved in this case by means of the Frobenius integration theorem so that we have only to study the deformation of the structure constants.

In order to determine the label transformations, we have to solve the equations:

$$\bar{\alpha}_\rho^k(x)\frac{\partial \bar{\omega}_i^\rho(x)}{\partial x^r} = \alpha_\rho^k(x)\frac{\partial \omega_i^\rho(x)}{\partial x^r}$$

or equivalently the Pfaffian system:

$$\bar{\alpha}_\rho^k d(\bar{\omega}_i^\rho) = \alpha_\rho^k d(\omega_i^\rho)$$

We get

$$d(\bar{\omega}_i^\rho) + \bar{\omega}_k^\rho \omega_i^m d(\alpha_m^k) = 0$$

$$\alpha_m^i d(\bar{\omega}_i^\rho) + \bar{\omega}_i^\rho d(\alpha_m^i) = 0$$

$$d(\alpha_m^i \bar{\omega}_i^\rho) = 0$$

The finite label transformations are thus:

$$\bar{\omega}_i^k(x) = a_\rho^k \cdot \omega_i^\rho(x), \quad a_\rho^k = \text{cst}$$

The corresponding action on the structure constants is:

$$\bar{c}_{\rho m}^k = (a^{-1})_\rho^j (a^{-1})_m^h a_i^k c_{jh}^i$$

where $a = [a_\rho^k]$, and it is easy to check that $J(\bar{c})$ whenever $J(c) = 0$.

The deformation theory that we have sketched above is just the same as the classical deformation theory of Lie algebras (11, 31).

In particular, a tedious but straightforward computation, which is left to the reader, allows one to demonstrate the rigidity and integrability theorems, in this case.

This example clearly shows that the third fundamental theorem and the deformation theory of structures establish a first step towards a complete analogy between the theory of Lie groups and that of Lie pseudogroups.

8 Deformation cohomology

We come now to the ultimate idea that will complete the desired analogy.

We first need a definition:

DEFINITION 8.1 (15.2) We say that a vector bundle F over X is "*associated*" with $J_q(T)$ if there exists a first order differential operator $L(\xi_q): F \to F$ such that:

1) $L(\xi_q + \bar{\xi}_q) = L(\xi_q) + L(\bar{\xi}_q)$
2) $L(f \cdot \xi_q) = f \cdot L(\xi_q)$
3) $[L(\xi_q), L(\bar{\xi}_q)] = L([\xi_q, \bar{\xi}_q])$
4) $L(\xi_q)(f \cdot \eta) = f \cdot L(\xi_q) \cdot \eta + (\xi \cdot f) \cdot \eta$

for any section ξ_q, $\bar{\xi}_q$ of $J_q(T)$, any section ξ of T such that $\xi = \pi_0^q \circ \xi_q$, any section η of F and any $f \in C^\infty(X)$.

DEFINTION 8.2 We call $\mathscr{L}(\xi) = L(j_q(\xi))$ a "*Lie derivative*".

We have the following properties:

1) $\mathscr{L}(\xi + \bar{\xi}) = \mathscr{L}(\xi) + \mathscr{L}(\bar{\xi})$
3) $[\mathscr{L}(\xi), \mathscr{L}(\bar{\xi})] = \mathscr{L}(\xi) \circ \mathscr{L}(\bar{\xi}) - \mathscr{L}(\bar{\xi}) \circ \mathscr{L}(\xi) = \mathscr{L}([\xi, \bar{\xi}])$
4) $\mathscr{L}(\xi) \cdot (f \cdot \eta) = f\mathscr{L}(\xi) \cdot \eta + (\xi \cdot f) \cdot \eta$

EXAMPLE 8.3 $\Lambda^r T^*$ is associated with $J_1(T)$.

REMARK 8.4 Let $\mathscr{R}_q \subset J_q(X \times Y)$ be a non-linear involutive system of finite Lie equations and R_q the corresponding linearised system of infinitesimal Lie equations. If a vector bundle over X is associated with $J_q(T)$, then of course it is associated with $R_q \subset J_q(T)$ as we know that $[R_q, R_q] \subset R_q$.

THEOREM 8.5 The vector bundles F_r of the P-sequence for $\mathscr{D}$ are associated with R_q.

Proof We have already proved that $\mathscr{R}'_1 \subset J_1(\mathscr{F})$ is a non-linear involutive system of order one on $\mathscr{F}$ and also an affine bundle over $\mathscr{F}$ because its defining equations can be chosen quasi-linear in the u_i^τ.

According to the construction of the Spencer families in the first part of this book, the corresponding bundles $\mathscr{F}_0 = V(\mathscr{F}), \mathscr{F}_1, \ldots, \mathscr{F}_n$ constructed inductively for $\mathscr{R}'_1$ are vector bundles over $\mathscr{F}$.

This is evident for $\mathscr{F}_0$ and it has been proved for $\mathscr{F}_1$ as we have exhibited the infinitesimal transition laws:

$$\xi^i(x)\,\frac{\partial}{\partial x^i} + \partial_\mu \xi^k(x)\left(L_k^{\tau\mu}(u)\,\frac{\partial}{\partial u^\tau} + M^{\alpha}_{\beta}{}|^{\mu}_{k}(u)v^{\beta}_{*}\,\frac{\partial}{\partial v^\alpha}\right)$$

The corresponding finite transition laws for $\mathscr{F}$ and $\mathscr{F}_1$ are of the following type:

$$\mathscr{F}_1\left\{\begin{array}{l} \bar{v} = H\left(u, \dfrac{\partial\varphi(x)}{\partial x}\right)\cdot v \\ \left.\begin{array}{l}\bar{u} = G\left(u, \dfrac{\partial\varphi(x)}{\partial x}\right) \\ \bar{x} = \varphi(x)\end{array}\right\}\mathscr{F} \end{array}\right.$$

As the $\mathscr{F}_r$ are constructed inductively, using the sequences: $r \geq 1$

$$S_2T^* \otimes \mathscr{F}_{r-1} \to T^* \otimes \mathscr{F}_r \to \mathscr{F}_{r+1} \to 0$$

we may obtain similar transition laws for them and in particular there is an action of $GL_q(n, \mathbb{R})$ on the fibers of the $\mathscr{F}_r$.

The key idea is now to notice that we have the following commutative diagrams of reciprocal images:

$$\begin{array}{ccc} F_r & \longrightarrow & \mathscr{F}_r \\ \big\downarrow & & \big\downarrow \\ X & \xrightarrow{\ \omega\ } & \mathscr{F} \end{array}$$

This follows easily by induction, using the preceding sequences and the fact that the symbols G_q of R_q or G_1' of R_1' only depend on the section ω of $\mathscr{F}$.

In the sequel we shall use local similar coordinates (x, v) for the F_r, unless we write them more explicitly in a different manner.

We have the finite transition laws:

$$F_0: \qquad \bar{x} = \varphi(x), \quad \bar{v} = \frac{\partial G}{\partial u}\left(\omega(x), \frac{\partial\varphi(x)}{\partial x}\right)\cdot v$$

$$F_1: \qquad \bar{x} = \varphi(x), \quad \bar{v} = H\left(\omega(x), \frac{\partial\varphi(x)}{\partial x}\right)\cdot v$$

REMARK 8.6 The reader may point out that, in a second change of coordinates we have, for F_0 as an example:

$$\bar{\bar{x}} = \psi(\bar{x}), \quad \bar{\bar{v}} = \frac{\partial G}{\partial u}\left(\bar{\omega}(\bar{x}), \frac{\partial \psi(\bar{x})}{\partial \bar{x}}\right) \cdot \bar{v}$$

with

$$\bar{\omega}(\bar{x}) = G\left(\omega(x), \frac{\partial \varphi(x)}{\partial x}\right).$$

The transition laws of the other F_r would be similar to that of F_1.

We need the following technical lemma:

LEMMA 8.7 There is an action of $\mathscr{R}_q(x, x) \subset GL_q(n, \mathbb{R})$ on the fibers $F_{r;x}$ for any $x \in U \subset X$.

Proof It suffices to prove the lemma for F_0 and F_1. We shall adopt the infinitesimal point of view.

Let ξ_q^0 and $\bar{\xi}_q^0$ be any two sections of R_q^0. Then, by definition, we have, with $1 \leq |\mu| \leq q$:

$$\xi_\mu^k(x) L_k^{\tau\mu}(\omega(x)) = 0, \quad \bar{\xi}_\mu^k(x) L_k^{\tau\mu}(\omega(x)) = 0 \qquad \forall x \in U \subset X$$

It follows that we have the commutation relations:

$$\left[\xi_\mu^k(x)\left(\frac{\partial L_k^{\tau\mu}(\omega(x))}{\partial u^\sigma} v^\sigma\right)\frac{\partial}{\partial v^\tau}, \bar{\xi}_\nu^l(x)\left(\frac{\partial L_l^{\tau\nu}(\omega(x))}{\partial u^\sigma} v^\sigma\right)\frac{\partial}{\partial v^\tau}\right]$$

$$\equiv \xi_\mu^k(x)\bar{\xi}_\nu^l(x)\left(\frac{\partial L_k^{\rho\mu}(\omega(x))}{\partial u^\sigma}\cdot\frac{\partial L_l^{\tau\nu}(\omega(x))}{\partial u^\rho} - \frac{\partial L_l^{\rho\nu}(\omega(x))}{\partial u^\sigma}\cdot\frac{\partial L_k^{\tau\mu}(\omega(x))}{\partial u^\rho}\right)v^\sigma \frac{\partial}{\partial v^\tau}$$

$$= \xi_\mu^k(x)\bar{\xi}_\nu^\rho(x)\frac{\partial}{\partial u^\sigma}([L_k^\mu, L_l^\nu])^\tau(\omega(x))v^\sigma \frac{\partial}{\partial v^\tau}$$

because ξ_q^0 and $\bar{\xi}_q^0$ are sections of R_q^0. The lemma is true for F_0 because we know that $[R_q^0, R_q^0] \subset R_q^0$, the bracket being linear and constructed using the fact that the commutation relations of the L_k^μ are the same as those of the $A_k^\mu(q)$.

A similar proof for F_1 can be deduced from the following device. We have:

$$\left[L_k^{\tau\mu}(u)\frac{\partial}{\partial u^\tau} + M_\beta^\alpha|_k^\mu(u)v^\beta \frac{\partial}{\partial v^\alpha}, L_l^{\tau\nu}(u)\frac{\partial}{\partial u^\tau} + M_\beta^\alpha|_l^\nu(u)v^\beta\frac{\partial}{\partial v^\alpha}\right]$$

$$\equiv ([L_k^\mu, L_l^\nu])^\tau(u)\frac{\partial}{\partial u^\tau} + \left(L_k^{\tau\mu}(u)\frac{\partial M_\beta^\alpha|_l^\nu(u)}{\partial u^\tau} - L_l^{\tau\nu}(u)\frac{\partial M_\beta^\alpha|_k^\mu(u)}{\partial u^\tau}\right)v^\beta\frac{\partial}{\partial v^\alpha}$$

$$+ (M_\beta^\gamma|_k^\mu(u)\cdot M_\gamma^\alpha|_l^\nu(u) - M_\beta^\gamma|_l^\nu(u)\cdot M_\gamma^\alpha|_k^\mu(u))v^\beta\frac{\partial}{\partial v^\alpha}$$

As these commutation relations must be identical to that of the L_k^μ, a straightforward computation shows that, for any two sections ξ_q^0 and $\bar{\xi}_q^0$ of R_q^0, we have:

$$\left[\xi_\mu^k(x)M_\beta^\alpha|_k^\mu(\omega(x))v^\beta \frac{\partial}{\partial v^\alpha}, \bar{\xi}_\nu^l(x)M_\beta^\alpha|_l^\nu(\omega(x))v^\beta \frac{\partial}{\partial v^\alpha}\right]$$

$$\equiv \xi_\mu^k(x)\bar{\xi}_\nu^l(x)\left(M_\beta^\gamma|_k^\mu(\omega(x)) \cdot M_\gamma^\alpha|_l^\nu(\omega(x)) - M_\beta^\gamma|_l^\nu(\omega(x)) \cdot M_\gamma^\alpha|_k^\mu(\omega(x))\right)v^\beta \frac{\partial}{\partial v^\alpha}$$

with the same commutation relations as above.

This concludes the proof of the lemma. C.Q.F.D.

Let now $\xi \in \Theta$ be any infinitesimal transformation of Γ and

$$\varphi_t = \exp(t\xi) \subset \Gamma \subset \operatorname{Aut}(X)$$

be the corresponding 1-parameter group of finite transformations of Γ.

As we have $\varphi_t(\omega) = \omega$, according to the last remark, we may use the transition rules of the vector bundle F_r in order to define an action of φ_t on any section η of F_r.

Let us define $\bar{\eta} = \varphi_t(\eta)$ by the following formulas:

For F_0: $$\underset{1}{\bar{\Omega}}{}^\tau(\varphi_t(x)) = \frac{\partial G^\tau}{\partial u^\sigma}\left(\omega(x), \frac{\partial \varphi_t(x)}{\partial x}\right) \cdot \underset{1}{\Omega}{}^\sigma(x)$$

For F_1: $$\underset{1}{\bar{P}}{}^\alpha(\varphi_t(x)) = H_\beta^\alpha\left(\omega(x), \frac{\partial \varphi_t(x)}{\partial x}\right) \cdot \underset{1}{P}{}^\alpha(x)$$

and similarly for the other F_r.

Definition 8.8 $$\mathscr{L}(\xi)\eta = \left.\frac{\partial \varphi_t^{-1}(\eta)}{\partial t}\right|_{t=0} = -\left.\frac{\partial \varphi_t(\eta)}{\partial t}\right|_{t=0} \quad \forall \xi \in \Theta$$

Remark 8.9 The definition is coherent because F_r is a vector bundle and

$$\varphi_t \circ \varphi_t^{-1} = \varphi_t^{-1} \circ \varphi_t = \mathrm{id}.$$

Coming back to the proof of the theorem, we define:

$$\begin{cases} \left(L(\xi_q)\cdot \underset{1}{\Omega}\right)^\tau \equiv -\xi_\mu^k(x)\dfrac{\partial L_k^{\tau\mu}(\omega(x))}{\partial u^\sigma}\underset{1}{\Omega}{}^\sigma + \xi^i(x)\dfrac{\partial \underset{1}{\Omega}{}^\tau}{\partial x^i} \\ \left(L(\xi_q)\cdot \underset{1}{P}\right)^\alpha \equiv -\xi_\mu^k(x)M_\beta^\alpha|_k^\mu(\omega(x))\underset{1}{P}{}^\beta + \xi^i(x)\dfrac{\partial \underset{1}{P}{}^\alpha}{\partial x^i} \end{cases}$$

for any section ξ_q of R_q.

All the properties of the Lie derivative are easy to check except the third. We have the lemma:

LEMMA 8.10 The third property of the Lie derivative holds with the above definition.

Proof We shall give this for the first case only as the second one is very similar.

The checking for the terms of second order is trivial as we have:

$$\xi^i(x)\bar{\xi}^j(x)\frac{\partial^2 \underset{1}{\Omega}^\tau}{\partial x^i \partial x^j} \equiv \xi^j(x)\bar{\xi}^i(x)\frac{\partial^2 \underset{1}{\Omega}^\tau}{\partial x^i \partial x^j}$$

For the terms of first order, we find:

$$\left(\xi^j(x)\frac{\partial \bar{\xi}^i(x)}{\partial x^j} - \bar{\xi}^j(x)\frac{\partial \xi^i(x)}{\partial x^j}\right)\frac{\partial \underset{1}{\Omega}^\tau}{\partial x^i}$$

As for the terms of zero order, we obtain the following coefficient of $\underset{1}{\Omega}^\sigma$:

$$-\xi^i(x)\partial_i\left(\bar{\xi}^l_\nu(x)\frac{\partial L^{\tau\nu}_l(\omega(x))}{\partial u^\sigma}\right) + \bar{\xi}^i\partial_i\left(\xi^k_\mu(x)\frac{\partial L^{\tau\mu}_k(\omega(x))}{\partial u^\sigma}\right)$$

$$+\,\xi^k_\mu(x)\bar{\xi}^l_\nu(x)\left(\frac{\partial L^{\tau\mu}_k(\omega(x))}{\partial u^\rho}\cdot\frac{\partial L^{\rho\nu}_l(\omega(x))}{\partial u^\sigma} - \frac{\partial L^{\tau\nu}_l(\omega(x))}{\partial u^\rho}\cdot\frac{\partial L^{\rho\mu}_k(\omega(x))}{\partial u^\sigma}\right)$$

$$= -\left(\xi^i(x)\partial_i\bar{\xi}^k_\mu(x) - \bar{\xi}^i(x)\partial_i\xi^k_\mu(x)\right)\frac{\partial L^{\tau\mu}_k(\omega(x))}{\partial u^\sigma}$$

$$-\,\xi^k_\mu(x)\bar{\xi}^l_\nu(x)\frac{\partial}{\partial u^\sigma}([L^\mu_k, L^\nu_l])^\tau(\omega(x))$$

keeping in mind that we have now:

$$-\xi^k_\mu(x)L^{\tau\mu}_k(\omega(x)) + \xi^i(x)\frac{\partial \omega^\tau(x)}{\partial x^i} = 0$$

These are the desired terms because $[R_q, R_q] \subset R_q$.

A similar proof applies to the Lie derivative for F_1 and more generally for F_r.

This concludes the proof of our theorem. C.Q.F.D.

REMARK 8.11 The reader will notice that any infinitesimal transformation:

$$\xi^i(x)\frac{\partial}{\partial x^i} + \xi^k_\mu(x)L^{\tau\mu}_k(u)\frac{\partial}{\partial u^\tau}$$

leaves invariant the submanifold $\omega(X) \subset \mathscr{F}$ defined by $u^\tau - \omega^\tau(x) = 0$ whenever ξ_q is a section of R_q.

What we have created is a vector bundle map from $R_q \otimes J_1(F_r)$ to F_r or equivalently a vector bundle map from $J_1(F_r)$ to $R_q^* \otimes F_r$.

Definition 8.12 If F is a vector bundle associated with R_q, we define the vector bundle $\tilde{F} = R_q^* \otimes F$ and denote by $\nabla : F \to \tilde{F}$ the corresponding linear first order operator. We denote by $\Upsilon = \Upsilon(F, R_q)$ the set of solutions of ∇, that is to say the set of sections η of F over $U \subset X$ such that

$$\mathscr{L}(\xi)\eta = 0, \qquad \forall \xi \in \Theta|_U.$$

For the vector bundles F_r of the P-sequence for the Lie operator $\mathscr{D}$, we call $\Upsilon_r = \Upsilon_r(R_q) = \Upsilon(F_r, R_q)$ and introduce $\nabla_r : F_r \to \tilde{F}_r$ as a "transverse" operator, for reasons that will become clear later on.

As R_q is formally transitive, it is easy to see that the symbols of these operators are zero and that we have the commutative and exact diagram, for any vector bundle F associated with R_q:

$$\begin{array}{ccccccc}
 & & 0 & & 0 & & \\
 & & \downarrow & & \downarrow & & \\
0 & \longrightarrow & T^* \otimes F & = & T^* \otimes F & \longrightarrow & 0 \\
 & & \downarrow & & \downarrow & & \\
 & & J_1(F) & \longrightarrow & R_q^* \otimes F & & \\
 & & \downarrow & & \downarrow & & \\
 & & F & \longrightarrow & R_q^{0*} \otimes F & & \\
 & & \downarrow & & \downarrow & & \\
 & & 0 & & 0 & &
\end{array}$$

where the right column is exact because it is the dual of the short exact sequence:

$$0 \longrightarrow R_q^0 \longrightarrow R_q \xrightarrow{\pi_0^q} T \longrightarrow 0$$

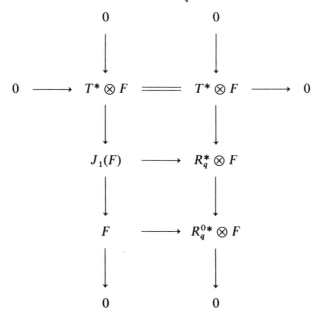

It follows that Υ_r only depends on arbitrary constants, the number of which is $\leq \dim F_r$.

PROPOSITION 8.13 The operator ∇_0 constructed in order to study the normaliser $\tilde{\Gamma}$ of Γ has the same solution as ∇_0 constructed above.

Proof Though this is not evident "*a priori*", the proof however depends on a straightforward computation:

For any solution of ∇_0, we must have:

$$-\xi^k_\mu(x)\frac{\partial L^{\tau\mu}_k(\omega(x))}{\partial u^\sigma}\underset{1}{\Omega^\sigma} + \xi^i(x)\frac{\partial \underset{1}{\Omega^\tau}}{\partial x^i} = 0$$

whenever

$$-\xi^k_\mu(x)L^{\tau\mu}_k(\omega(x)) + \xi^i(x)\frac{\partial \omega^\tau(x)}{\partial x^i} = 0$$

Our problem is now a purely algebraic one.

We shall use the following device, if we set

$$\omega_t(x) = \omega(x) + t\underset{1}{\Omega}(x) + \cdots$$

for small parameter t, we must have:

$$\left.\frac{\partial(\mathscr{D}(t)\cdot\xi)}{\partial t}\right|_{t=0} \equiv \frac{\partial}{\partial t}\left(-\partial_\mu\xi^k(x)L^{\tau\mu}_k(\omega_t(x)) + \xi^i(x)\frac{\partial \omega^\tau_t(x)}{\partial x^i}\right)\bigg|_{t=0} = 0$$

whenever

$$\mathscr{D}\cdot\xi \equiv -\partial_\mu\xi^k(x)L^{\tau\mu}_k(\omega(x)) + \xi^i(x)\frac{\partial\omega^\tau(x)}{\partial x^i} = 0$$

That is to say, we must have:

$$\frac{\partial}{\partial t}\left(-\xi^\tau(x) - \mathscr{E}^\tau_{m+r}(\omega_t(x))\xi^{m+r}(x) + (M^{-1})^\tau_\sigma(\omega_t(x))\frac{\partial\omega^\sigma_t(x)}{\partial x^i}\xi^i(x)\right)\bigg|_{t=0} = 0$$

whenever

$$-\xi^\tau(x) - \mathscr{E}^\tau_{m+r}(\omega(x))\xi^{m+r}(x) + (M^{-1})^\tau_\sigma(\omega(x))\frac{\partial\omega^\sigma(x)}{\partial x^i}\xi^i(x) = 0$$

because

$$\left.\frac{\partial}{\partial t}(M^{-1}(\omega_t(x))\mathscr{D}(t)\cdot\xi)\right|_{t=0} = M^{-1}(\omega(x))\left.\frac{\partial(\mathscr{D}(t)\cdot\xi)}{\partial t}\right|_{t=0} + \left.\frac{\partial M^{-1}(\omega_t(x))}{\partial t}\right|_{t=0}\cdot\mathscr{D}\cdot\xi$$

Thus we must have:

$$R_1''\begin{cases} \dfrac{\partial \mathscr{E}^{\tau}_{m+r}(\omega(x))}{\partial u^\sigma} \cdot \underset{1}{\Omega^\sigma} = 0 \\ \dfrac{\partial \underset{1}{\Omega^\tau}}{\partial x^i} - (M^{-1})^{\rho}_{\alpha}(\omega(x)) \dfrac{\partial \omega^\alpha(x)}{\partial x^i} \dfrac{\partial M^{\tau}_{\rho}(\omega(x))}{\partial u^\sigma} \cdot \underset{1}{\Omega^\sigma} = 0 \end{cases}$$

that is to say $\nabla_0 \cdot \underset{1}{\Omega} = 0$. C.Q.F.D.

Proposition 8.14 The solutions of $\nabla_1 \cdot \underset{1}{P} = 0$ are just

$$\underset{1}{P_*} = 0, \qquad \underset{1}{P_{**}} = \underset{1}{C} = \text{cst}.$$

Proof As in the above proposition, we must have:

$$-\xi^k_\mu(x) M^\alpha_\beta|^\mu_k(\omega(x)) \underset{1}{P^\beta_*} + \xi^i(x) \frac{\partial \underset{1}{P^\alpha}}{\partial x^i} = 0$$

whenever

$$-\xi^k_\mu(x) L^{\tau\mu}_k(\omega(x)) + \xi^i(x) \frac{\partial \omega^\tau(x)}{\partial x^i} = 0$$

First of all, this must be true for any section ξ^0_q of R^0_q. As this is just the rank condition on the matrix that allowed us to study the surjectivity of the map $\pi^{q+1}_q : R^0_{q+1} \to R^0_q$, we must have $\underset{1}{P_*} = 0$.

Finally we must have:

$$\xi^i(x) \frac{\partial \underset{1}{P^\alpha}}{\partial x^i} = 0 \text{ for any section } \xi_q \text{ of } \mathscr{R}_q.$$

As R_q is formally transitive, this implies $\underset{1}{P_{**}} = \underset{1}{C} = \text{cst}$. C.Q.F.D.

We generalise the above situation in the next theorem:

Theorem 8.15 The operators ∇_r are involutive.

Proof As R_q is formally transitive, the symbols of the ∇_r must be zero.

The idea is then to use the preceding diagram in order to study the map $F_r \to R^{0*}_q \otimes F_r$.

We have already proved that ∇_0 is involutive.

As $\nabla_1 \cdot \underset{1}{P} = 0$ is equivalent to

$$\underset{1}{P}_{*} = 0, \quad \frac{\partial \underset{1}{P}_{**}}{\partial x^i} = 0, \quad \frac{\partial \underset{1}{P}_{***}}{\partial x^i} = 0.$$

it is easily seen to be involutive.

Now, any solution of ∇_r is a solution of equations such as:

$$-\xi^k_\mu(x) N^\alpha_\beta|^\mu_k(\omega(x)) \underset{1}{Q}^\beta + \xi^i(x) \frac{\partial \underset{1}{Q}^\alpha}{\partial x^i} = 0$$

whenever ξ_q is a section of R_q.

As above, it follows that we must first have

$$-\xi^k_\mu(x) N^\alpha_\beta|^\mu_k(\omega(x)) \underset{1}{Q}^\beta = 0 \quad \text{for any section } \xi^0_q \text{ of } R^0_q.$$

This gives linear relations between the $\underset{1}{Q}$ and defines the map $F_r \to R^{0*}_q \otimes F_r$.

Now, by construction, we know that the vector fields on F_r:

$$L^{\tau\mu}_k(u) \frac{\partial}{\partial u^\tau} + N^\alpha_\beta|^\mu_k(u) v^\beta \frac{\partial}{\partial v^\alpha}$$

have the same commutation laws as the L^μ_k.

The latter relations, analogous to the compatibility condition of the first kind, are invariant and we can make a linear change of the v, like the one we used in lemma 3.3 in order to construct the P in the case $q = 1$.

The final conditions are similar to that found for ∇_1 and this proves the theorem. C.Q.F.D.

THEOREM 8.16 We have the commutative diagrams:

$$\begin{array}{ccc} F_{r-1} & \xrightarrow{\mathscr{D}_r} & F_r \\ \downarrow{\scriptstyle \mathscr{L}(\xi)} & & \downarrow{\scriptstyle \mathscr{L}(\xi)} \\ F_{r-1} & \xrightarrow{\mathscr{D}_r} & F_r \end{array} \qquad \forall \xi \in \Theta$$

Proof For any infinitesimal transformation ξ of Γ, let

$$\varphi_t = \exp(t\xi) \in \Gamma \subset \mathrm{Aut}(X)$$

be the corresponding 1-parameter family of finite transformations of Γ.

We have $\varphi_t(\omega) = \omega$ because $\varphi_t \in \Gamma$ and we leave the reader to specify as usual the convenient restrictions that are required.

We shall now use a trick.

We first know that T is a fixed bundle. When we are given involutive systems $\mathscr{R}_q$ or R_q, we can construct a bundle $\mathscr{F}$ of geometric objects and we know that F_0 and $\mathscr{D}$ are uniquely determined by the choice of the section ω of $\mathscr{F}$. We refer the reader to their definition.

The same is also true for the vector bundles F_r and the first order operators $\mathscr{D}_r$ as they are constructed inductively from the knowledge of $\mathscr{D}: T \to F_0$.

Working locally for t small, we may consider φ_t either as a particular change of coordinates or better as a vector bundle isomorphism $\varphi_t: F_r \to F_r$ over $\varphi_t: X \to X$.

Using the fact that the operators $\mathscr{D}_r$ keep the same "*form*" because $\varphi_t(\omega) = \omega$, we have the commutative diagrams:

$$\begin{array}{ccc} F_{r-1} & \xrightarrow{\mathscr{D}_r} & F_r \\ \downarrow{\scriptstyle \varphi_t} & & \downarrow{\scriptstyle \varphi_t} \\ F_{r-1} & \xrightarrow{\mathscr{D}_r} & F_r \end{array}$$

where, if η is a section of F_r, $\bar{\eta} = \varphi_t(\eta)$ denotes as usual the transformed section. For example, looking at F_0, we have:

$$\underset{1}{\bar{\Omega}}(\varphi_t(x)) = \frac{\partial G(\omega(x), \partial\varphi_t(x)/\partial x)}{\partial u} \cdot \underset{1}{\Omega}(x)$$

and looking at F_1 we have:

$$\underset{1}{\bar{P}}(\varphi_t(x)) = H\left(\omega(x), \frac{\partial\varphi_t(x)}{\partial x}\right) \cdot \underset{1}{P}(x)$$

Similar formulas hold for the other F_r.

Using the fact that the local expression for $\mathscr{D}_r$ does not contain t explicitly, we can derive with respect to t the formula:

$$\varphi_t \circ \mathscr{D}_r = \mathscr{D}_r \circ \varphi_t$$

and obtain:

$$\left.\frac{\partial\varphi_t}{\partial t}\right|_{t=0} \circ \mathscr{D}_r = \mathscr{D}_r \circ \left.\frac{\partial\varphi_t}{\partial t}\right|_{t=0}$$

But we have:

$$\bar{\eta}(x) = \bar{\eta}(\varphi_t(x)) - (\bar{\eta}(\varphi_t(x)) - \bar{\eta}(x))$$

Developing this series in t, up to order 1, we get:

$$\left.\frac{\partial \varphi_t}{\partial t}\right|_{t=0} = -\mathscr{L}(\xi)$$

and this concludes the proof of the theorem. C.Q.F.D.

REMARK 8.17 Let $\varphi, \psi \in \text{Aut}(X)$ be arbitrary. Using the transition rules of F_r we can construct an action on its sections as in the definition of the Lie derivative, however we shall have in general $(\psi \circ \varphi)(\eta) \neq \psi(\varphi(\eta))$ unless we have $\varphi(\omega) = \psi(\omega) = \omega$. This is the corresponding finite property of the Lie derivative on F_r.

COROLLARY 8.18 For any section ξ of T we have:

$$\mathscr{D}_r \circ \mathscr{L}(\xi) - \mathscr{L}(\xi) \circ \mathscr{D}_r = \left.\frac{\partial \mathscr{D}_r(t)}{\partial t}\right|_{t=0}$$

Proof If ξ is any section of T, the 1-parameter group of transformation

$$\varphi_t = \exp(t\xi) \subset \text{Aut}(X)$$

is such that in general $\omega_t = \varphi_t(\omega) \neq \omega$.

We have proved in theorem 2.18 that $\mathscr{D} : T \to F_0$ with kernel Θ is transformed into an operator $\mathscr{D}(t) : T \to F_0(t)$ with kernel $\Theta(t) = \varphi_t(\Theta)$, which has the same "*form*" as $\mathscr{D}$, except that we have to use ω_t instead of ω when we take $F_0(t) = \omega_t^{-1}(\mathscr{F}_0)$.

Working locally, the formula of the theorem finally has a meaning because $\dim F_0(t) = \dim F_0$ for t small enough. It follows that:

$$\varphi_t \circ \mathscr{D} = \mathscr{D}(t) \circ \varphi_t$$

Differentiating with respect to t for $t = 0$, we get the formula of the corollary. C.Q.F.D.

For the reader who wants to look at concrete examples, we shall detail some useful computations.

1 *Relations between the transition laws of the* F_r *and the symbols of the operators* $\mathscr{D}_r$.

We use local coordinates (x, v^α) for F_{r-1} and (x, w^a) for F_r.

If $A = [A_\alpha^{ai}(\omega(x))]$ is the principal part of $\mathscr{D}_r$, we denote by M and N respectively the set of components of the matrices appearing in the infinitesimal laws of F_{r-1} and F_r.

As in the preceding corollary, we have the commutative diagram:

$$\begin{array}{ccc} T^* \otimes F_{r-1} & \xrightarrow{A} & F_r \\ \Big\downarrow \varphi_t & & \Big\downarrow \varphi_t \\ T^* \otimes F_{r-1}(t) & \xrightarrow{A(t)} & F_r(t) \end{array}$$

where $\varphi_t(\alpha \otimes \eta) = \varphi_t(\alpha) \otimes \varphi_t(\eta)$ for any section α of T^* and η of F_{r-1}.

A straightforward computation which is left to the reader shows that for $2 \leq |\mu| \leq q$:

$$N_b^a|_k^\mu(u) \cdot A_\beta^{bi}(u) - A_\alpha^{ai}(u) \cdot M_\beta^\alpha|_k^\mu(u) \equiv L_k^{\tau\mu}(u) \frac{\partial A_\beta^{ai}(u)}{\partial u^\tau}$$

while, for $\mu = 1$ we have:

$$N_b^a|_k^j(u) A_\beta^{bi}(u) - A_\alpha^{ai}(u) M_\beta^\alpha|_k^j(u) - M_\beta^{aj}(u)\delta_k^i \equiv L_k^{\tau j}(u) \frac{\partial A_\beta^{ai}(u)}{\partial u^\tau}$$

REMARK 8.19 Differentiating the above formula, we get:

$$\mathscr{L}(\xi)(\alpha \otimes \eta) = \mathscr{L}(\xi)\alpha \otimes \eta + \alpha \otimes \mathscr{L}(\xi)\eta$$

and we can extend this formula to $\Lambda^s T^* \otimes F_r$ or $S_s T^* \otimes F_r$.

2 *Direct checking of the commutation formula for $\mathscr{D}_1$:*

Using local coordinates, we have:

$$\mathscr{D}_1 \cdot \underset{1}{\Omega} = \underset{1}{P} \Rightarrow \underset{1}{P^\alpha} = A_\tau^{\alpha i}(\omega(x)) \frac{\partial \underset{1}{\Omega^\tau}}{\partial x^i} + \frac{\partial I^\alpha}{\partial u^\tau} \cdot \underset{1}{\Omega^\tau}$$

Then we have:

$$\mathscr{D}_1 \circ \mathscr{L}(\xi)\underset{1}{\Omega} \equiv A_\tau^{\alpha i}(\omega(x))\partial_i\left(-\partial_\mu \xi^k(x) \frac{\partial L_k^{\tau\mu}(\omega(x))}{\partial u^\sigma} \underset{1}{\Omega^\sigma} + \xi^r(x) \frac{\partial \underset{1}{\Omega^\tau}}{\partial x^r} \right)$$

$$+ \frac{\partial I^\alpha}{\partial u^\tau}\left(-\partial_\mu \xi^k(x) \frac{\partial L_k^{\tau\mu}(\omega(x))}{\partial u^\sigma} \underset{1}{\Omega^\sigma} + \xi^r(x) \frac{\partial \underset{1}{\Omega^\tau}}{\partial x^r} \right)$$

Developing, we get, without writing x explicitly:

$$\begin{cases} \text{order 0:} & -\left[A_\tau^{\alpha i}\partial_i\left(\partial_\mu \xi^k \dfrac{\partial L_k^{\tau\mu}}{\partial u^\sigma}\right) + \dfrac{\partial I^\alpha}{\partial u^\tau}\partial_\mu \xi^k \dfrac{\partial L_k^{\tau\mu}}{\partial u^\sigma}\right]\underset{1}{\Omega^\sigma} \\ \text{order 1} & -A_\tau^{\alpha i}\partial_\mu \xi^k \dfrac{\partial L_k^{\tau\mu}}{\partial u^\sigma}\dfrac{\partial \underset{1}{\Omega^\sigma}}{\partial x^i} + A_\tau^{\alpha i}\partial_i \xi^r \dfrac{\partial \underset{1}{\Omega^\tau}}{\partial x^r} + \xi^r \dfrac{\partial I^\alpha}{\partial u^\tau}\dfrac{\partial \underset{1}{\Omega^\tau}}{\partial x^r} \\ \text{order 2:} & \xi^r A_\tau^{\alpha i}\dfrac{\partial^2 \underset{1}{\Omega^\tau}}{\partial x^i \partial x^r} \end{cases}$$

Now we have the relations, for $2 \leq |\mu| \leq q$:

$$\left(A_\sigma^{\alpha i}\frac{\partial L_k^{\sigma\mu}}{\partial u^\tau} + \frac{\partial A_\tau^{\alpha i}}{\partial u^\sigma}L_k^{\sigma\mu}\right)\frac{\partial u^\tau}{\partial x^i} + \sum_{\lambda + i = \mu} A_\tau^{\alpha i}L_k^{\tau\lambda} + \frac{\partial B^\alpha}{\partial u^\sigma}L_k^{\sigma\mu} \equiv M_\beta^\alpha|_k^\mu(u)\cdot I_*^\beta$$

It follows that:

$$A_\tau^{\alpha i}(u)\frac{d}{dx^i}(\partial_\mu \xi^k(x)L_k^{\tau\mu}(u)) + \frac{\partial I^\alpha}{\partial u^\tau}\partial_\mu \xi^k(x)L_k^{\tau\mu}(u) \equiv \partial_\mu \xi^k(x)M_\beta^\alpha|_k^\mu(u)I_*^\beta$$

Deriving with respect to u and setting $u = \omega(x)$, we get:

$$\frac{\partial A_\tau^{\alpha i}}{\partial u^\sigma}\partial_i(\partial_\mu \xi^k L_k^{\tau\mu}) + A_\tau^{\alpha i}\partial_i\left(\partial_\mu \xi^k \frac{\partial L_k^{\tau\mu}}{\partial u^\sigma}\right) + \frac{\partial^2 I^\alpha}{\partial u^\tau \partial u^\sigma}\partial_\mu \xi^k L_k^{\tau\mu}$$
$$+ \frac{\partial I^\alpha}{\partial u^\tau}\partial_\mu \xi^k \frac{\partial L_k^{\tau\mu}}{\partial u^\sigma} \equiv \partial_\mu \xi^k M_\beta^\alpha|_k^\mu \frac{\partial I_*^\beta}{\partial u^\sigma}$$

Substituting and taking care of the formulas that must be used when $|\mu| = 1$, we obtain:

$$\left[\frac{\partial A_\tau^{\alpha i}}{\partial u^\sigma}\partial_i(\partial_\mu \xi^k L_k^{\tau\mu}) + \frac{\partial^2 I^\alpha}{\partial u^\tau \partial u^\sigma}\cdot \partial_\mu \xi^k L_k^{\tau\mu} - \partial_\mu \xi^k M_\beta^\alpha|_k^\mu \frac{\partial I^\beta}{\partial u^\sigma}\right]\underset{1}{\Omega^\sigma}$$
$$- \partial_\mu \xi^k M_\beta^\alpha|_k^\mu A_\sigma^{\beta i}\frac{\partial \underset{1}{\Omega^\sigma}}{\partial x^i} + \partial_\mu \xi^k L_k^{\tau\mu}\frac{\partial A_\sigma^{\alpha i}}{\partial u^\tau}\frac{\partial \underset{1}{\Omega^\sigma}}{\partial x^i} + \frac{\partial I^\alpha}{\partial u^\sigma}\xi^r \frac{\partial \underset{1}{\Omega^\sigma}}{\partial x^r} + A_\tau^{\alpha i}\xi^r \frac{\partial^2 \underset{1}{\Omega^\sigma}}{\partial x^r \partial x^i}$$
$$\equiv -\partial_\mu \xi^k M_\beta^\alpha|_k^\mu\left(A_\sigma^{\beta i}\frac{\partial \underset{1}{\Omega^\sigma}}{\partial x^i} + \frac{\partial I^\beta}{\partial u^\sigma}\underset{1}{\Omega^\sigma}\right) + A_\tau^{\alpha i}\xi^r \frac{\partial^2 \underset{1}{\Omega^\sigma}}{\partial x^r \partial x^i}$$
$$- \frac{\partial A_\tau^{\alpha i}}{\partial u^\sigma}\partial_i\left(-\partial_\mu \xi^k L_k^{\tau\mu} + \xi^r \frac{\partial \omega^\tau}{\partial x^r}\right)\underset{1}{\Omega^\sigma} + \frac{\partial A_\tau^{\alpha i}}{\partial u^\sigma}\xi^r \frac{\partial^2 \omega^\tau}{\partial x^i \partial x^r}\underset{1}{\Omega^\sigma}$$
$$- \frac{\partial A_\sigma^{\alpha i}}{\partial u^\tau}\left(-\partial_\mu \xi^k L_k^{\tau\mu} + \xi^r \frac{\partial \omega^\tau}{\partial x^r}\right)\frac{\partial \underset{1}{\Omega^\sigma}}{\partial x^i} + \frac{\partial A_\sigma^{\alpha i}}{\partial u^\tau}\xi^r \frac{\partial \omega^\tau}{\partial x^r}\frac{\partial \underset{1}{\Omega^\sigma}}{\partial x^i}$$
$$- \frac{\partial^2 I^\alpha}{\partial u^\sigma \partial u^\tau}\left(-\partial_\mu \xi^k L_k^{\tau\mu} + \xi^r \frac{\partial \omega^\tau}{\partial x^r}\right)\underset{1}{\Omega^\sigma} + \frac{\partial^2 I^\alpha}{\partial u^\sigma \partial u^\tau}\xi^r \frac{\partial \omega^\tau}{\partial x^r}\underset{1}{\Omega^\sigma} + \frac{\partial I^\alpha}{\partial u^\sigma}\xi^r \frac{\partial \underset{1}{\Omega^\sigma}}{\partial x^r}$$

The coefficient of ξ^r is therefore:

$$\frac{\partial}{\partial x^r}\left(A^{\alpha i}_\tau \frac{\partial \underset{1}{\Omega}^\tau}{\partial x^i} + \frac{\partial I^\alpha}{\partial u^\tau}\underset{1}{\Omega}^\tau\right)$$

and finally we have:

$$\begin{aligned}\left([\mathscr{L}(\xi)\circ\mathscr{D}_1 - \mathscr{D}_1\circ\mathscr{L}(\xi)]\cdot\underset{1}{\Omega}\right)^\alpha &\equiv \frac{\partial A^{\alpha i}_\tau}{\partial u^\sigma}\,\partial_i\!\left(-\partial_\mu\xi^k L^{\tau\mu}_k + \xi^r\frac{\partial\omega^\tau}{\partial x^r}\right)\cdot\underset{1}{\Omega}^\sigma\\ &+ \frac{\partial A^{\alpha i}_\sigma}{\partial u^\tau}\left(-\partial_\mu\xi^k L^{\tau\mu}_k + \xi^r\frac{\partial\omega^\tau}{\partial x^r}\right)\frac{\partial\underset{1}{\Omega}^\sigma}{\partial x^i}\\ &+ \frac{\partial^2 I^\alpha}{\partial u^\sigma\partial u^\tau}\left(-\partial_\mu\xi^k L^{\tau\mu}_k + \xi^r\frac{\partial\omega^\tau}{\partial x^r}\right)\cdot\underset{1}{\Omega}^\sigma\end{aligned}$$

This is just the formula of the last corollary, because:

$$\left(\mathscr{D}_1(t)\cdot\underset{1}{\Omega}\right)^\alpha \equiv A^{\alpha i}_\tau(\omega_t(x))\frac{\partial\underset{1}{\Omega}^\tau}{\partial x^i} + \frac{\partial A^{\alpha i}_\tau(\omega_t(x))}{\partial u^\sigma}\frac{\partial\omega^\tau_t(x)}{\partial x^i}\underset{1}{\Omega}^\sigma + \frac{\partial B^\alpha(\omega_t(x))}{\partial u^\sigma}\underset{1}{\Omega}^\sigma$$

Whenever $\mathscr{D}\cdot\xi = 0$ we get $\mathscr{L}(\xi)\circ\mathscr{D}_1 = \mathscr{D}_1\circ\mathscr{L}(\xi)$.

REMARK 8.20 If $\omega_t = \varphi_t(\omega)$ we notice that:

$$\left.\frac{\partial\omega_t(x)}{\partial t}\right|_{t=0} = -\left(-\partial_\mu\xi^k(x)L^{\tau\mu}_k(\omega(x)) + \xi^i(x)\frac{\partial\omega^\tau(x)}{\partial x^i}\right)$$

This is the reason why it is often better to consider $\omega_t = \varphi_t^{-1}(\omega)$ in order to have exactly

$$\left.\frac{\partial\omega_t}{\partial t}\right|_{t=0} = \mathscr{D}\cdot\xi.$$

We now notice that the tangent bundle T is of course associated with $J_1(T)$ because, for any sections ξ and η of T we know that:

$$\mathscr{L}(\xi)\cdot\eta = -\mathscr{L}(\eta)\cdot\xi = [\xi,\eta]$$

T becomes at the same time associated with $J_q(T)$ and, hence, with $R_q \subset J_q(T)$.

There is however a more useful way to look at this association. We have just to notice that, if R_q is a linear system of formally transitive infinitesimal Lie equations, then G_q is a vector bundle.

It is easy to see that:

$$\pi^q_{q-1}([R_q, R_q]) = [\pi^q_{q-1}(R_q), \pi^q_{q-1}(R_q)]$$

It follows that $R^{(1)}_{q-1} = \pi^q_{q-1}(R_q)$ is also a linear system of formally transitive infinitesimal Lie equations and the same thing will be true inductively for

$R_1^{(q-1)}$. However, if R_q is involutive, $R_1^{(q-1)}$ may not be involutive in general, unless $q = 1$. Finally T is associated with $R_1^{(q-1)} \subset J_1(T)$.

We shall denote by $C(\Theta) = \Upsilon(T)$ the set of sections η of T such that $\mathscr{L}(\xi) \cdot \eta = 0$, $\forall \xi \in \Theta$, or equivalently such that $L(\xi_1) \cdot \eta = 0$ for any section ξ_1 of $R_1^{(q-1)}$. We thus have $[\xi, \eta] = 0$, $\forall \xi \in \Theta$, $\forall \eta \in C(\Theta)$. It follows that $C(\Theta)$ is the "*centralizer*" of Θ and we may write $[\Theta, C(\Theta)] = 0$.

As we have already proved for the Υ_r, $C(\Theta)$ only depends on certain constants, the number of which is $\leq \dim T = n$ and can be considered, for this reason, as a vector space.

Moreover, from the Jacobi identity for the usual bracket on T, it follows that, if $\mathscr{L}(\eta) \cdot \xi = 0$ and $\mathscr{L}(\bar{\eta}) \cdot \xi = 0$, $\forall \xi \in \Theta$ then $\mathscr{L}([\eta, \bar{\eta}]) \cdot \xi = 0$, $\forall \xi \in \Theta$. Finally $C(\Theta)$ is in fact a finite dimensional Lie algebra, as this can be deduced using a method similar to that of proposition 6.5.

We may introduce the set $Z(\Theta)$ of sections of $C(\Theta)$ killed by $\mathscr{D}$. We have $Z(\Theta) = \Theta \cap C(\Theta)$ and $Z(\Theta)$ is just the "*center*" of Θ.

Passing to the finite point of view by means of the exponential mapping we may define respectively $C(\Gamma)$ and $Z(\Gamma)$ similarly to $N(\Gamma)$.

THEOREM 8.21 The P-sequence for $\mathscr{D}$ induces a sequence of vector spaces:

$$0 \longrightarrow Z(\Theta) \longrightarrow C(\Theta) \xrightarrow{\mathscr{D}} \Upsilon_0 \xrightarrow{\mathscr{D}_1} \Upsilon_1 \xrightarrow{\mathscr{D}_2} \cdots \xrightarrow{\mathscr{D}_n} \Upsilon_n \longrightarrow 0$$

which is not exact in general, except at $C(\Theta)$.

Proof If we take any section η of F_{r-1} such that $\nabla_{r-1} \cdot \eta = 0$, this means that $\mathscr{L}(\xi) \cdot \eta = 0$, $\forall \xi \in \Theta$ or $L(\xi_q) \cdot \eta = 0$, $\forall \xi_q$ section of R_q. It follows that $\mathscr{D}_r \cdot \mathscr{L}(\xi) \cdot \eta = \mathscr{L}(\xi) \cdot \mathscr{D}_r \cdot \eta = 0$ and thus $\mathscr{D}_r \cdot \eta = \zeta$ is a section of F_r such that $\mathscr{L}(\xi) \cdot \zeta = 0$, $\forall \xi \in \Theta$ or $L(\xi_q)\zeta = 0$, $\forall \xi_q$ section of R_q. This means that $\nabla_r \cdot \zeta = 0$.

We can prove that the ∇_r, as well as ∇_0 and ∇_1, are determined by maps of constant rank from $J_1(F_r)$ to $\tilde{F}_r$.

As the kernels Υ_r of the ∇_r only depend on some constants, we may look at them as vector spaces with maps from Υ_{r-1} to Υ_r induced by $\mathscr{D}_r$ and also denoted simply by $\mathscr{D}_r$. C.Q.F.D.

Moreover, as the operators $\mathscr{D}_r$ involve only $\omega(x)$ and their derivatives $\partial\omega(x)/\partial x$ of some order in their coefficients, the constant maps induced by the $\mathscr{D}_r$ in the latter sequence can only depend on the structure constants c.

In fact, if

$$Q\left(\omega(x), \frac{\partial\omega(x)}{\partial x}\right) = a = \text{cst}$$

is such a coefficient of the latter maps, we must have also

$$Q\left(\bar{\omega}(x), \frac{\partial\bar{\omega}(x)}{\partial x}\right) = a \quad \text{for} \quad \bar{\omega} = \varphi(\omega), \qquad \forall \varphi \in \text{Aut}\,(X).$$

As $\mathscr{R}'_1 \subset J_1(\mathscr{F})$ is involutive, there must be a functional relation between Q and the integrability conditions I, that is to say, a may be expressed by means of the c.

DEFINTION 8.22 We shall define the following finite dimensional vector subspaces of $\Upsilon_r = \Upsilon_r(R_q)$ which are determined by the knowledge of R_q only.

- We call "*r-coboundary*" any element of $B_r = B_r(R_q) = \operatorname{im} \mathscr{D}_r$.
- We call "*r-cocycle*" any element of $Z_r = Z_r(R_q) = \ker \mathscr{D}_{r+1}$. As $\mathscr{D}_{r+1} \circ \mathscr{D}_r = 0$ we have $B_r \subset Z_r \subset \Upsilon_r$.
- We call $H_r = Z_r/B_r = \mathrm{H}_r(R_q)$ the r-group of "*deformation cohomology*" of the involutive system R_q of infinitesimal Lie equations.

It is then easy to rewrite the theorems 7.17 and 7.20 in order to give them the same features as the corresponding ones that can be found in the usual literature on Lie algebras (17) dealing with the Hochschild cohomology.

RIGIDITY THEOREM 8.23 A sufficient condition of rigidity for R_q is that $H_1(R_q) = 0$.

INTEGRABILITY THEOREM 8.24 A sufficient condition of formal integrability for any 1-cocycle of $Z_1(R_q)$ is that $H_2(R_q) = 0$.

We may now identify $Z_0(R_q)$ and $\Delta(\Theta) = \tilde{\Theta}/\Theta$ as the finite dimensional vector space of "*derivations*" of Θ, though this is not the usual point of view.

Of course $\Delta(\Theta)$ is endowed with a natural structure of Lie algebra induced by $\tilde{\Theta} = N(\Theta)$ and we have:

$$\dim \Delta(\Theta) = \dim \tilde{R}_{q+1} - \dim R_{q+1}.$$

the last number being the codimension of Θ into $\tilde{\Theta}$ which is finite.

Using proposition 4.10 we have also the following theorem which was conjectured by D. S. Rim (40):

THEOREM 8.25 There exists a quadratic map:

$$H_1(R_q) \to H_2(R_q) : \underset{1}{C} \to \frac{\partial^2 J(c)}{\partial c \cdot \partial c} \cdot \underset{1}{C} \cdot \underset{1}{C} \qquad \text{(Hessian)}$$

Proof We have already proved that this map takes $Z_1(R_q)$ into $Z_2(R_q)$. It remains to show that it takes $B_1(R_q)$ into $B_2(R_q)$.

Let $\underset{1}{A} \in B_1(R_q)$ be such that $\underset{1}{C} = \mathscr{D}_1 \cdot \underset{1}{A}$. We have

$$-\left(\frac{\partial \mathscr{D}_2(t)}{\partial t}\right)\bigg|_{t=0} \cdot \underset{1}{C} = -\left(\frac{\partial \mathscr{D}_2(t)}{\partial t}\right)\bigg|_{t=0} \circ \mathscr{D}_1 \cdot \underset{1}{A} = \mathscr{D}_2 \cdot \left(\frac{\partial \mathscr{D}_1(t)}{\partial t}\right)\bigg|_{t=0} \cdot \underset{1}{A}$$

and this proves the theorem. C.Q.F.D.

The construction of the P-sequence has thus given us a powerful method in order to study the infinite dimensional Lie algebra Θ such that $[\Theta, \Theta] \subset \Theta$ by means of the deformation cohomology groups $H_r(R_q)$ that are finite dimensional vector spaces, intrinsically defined.

Finally, we shall point out the last striking aspect of this analogy. We have the proposition:

PROPOSITION 8.26 To any transitive pseudogroup Γ of transformations of X, we can associate the Lie group $C(\Gamma)$ acting on X. We have $\dim C(\Gamma) \leq \dim X$ and the action is effective, though it is not always transitive.

EXAMPLE 8.27 $\Gamma: \mathbb{R}^2 \to \mathbb{R}^2 : y^1 = x^1 + a, \quad y^2 = x^2 + f(x^1)$.
The only infinitesimal generator of the action of $C(\Gamma)$ is $\partial/\partial x^2$.

We have the exact sequence of vector spaces:

$$0 \longrightarrow Z(\Theta) \longrightarrow C(\Theta) \xrightarrow{\mathscr{D}} \Delta(\Theta)$$

and we may define

$$B(\Theta) = B_0(R_q) = \operatorname{im} \mathscr{D} \subset \Delta(\Theta).$$

We call $B(\Theta)$ the space of "*inner derivations of* Θ".

PROPOSITION 8.28 The induced map $\mathscr{D}: C(\Theta) \to \Delta(\Theta)$ is a Lie algebra homomorphism with commutative kernel $Z(\Theta)$.

Proof First of all we can consider $\Upsilon_0(R_q)$ as a Lie algebra because it is uniquely determined by the knowledge of the canonical infinitesimal parameters of the label transformations. In fact, if $\underset{1}{A^\alpha} \cdot W^\tau_\alpha(u)(\partial/\partial u^\tau)$ is any infinitesimal label transformation, $\underset{1}{A^\alpha} \cdot W^\tau_\alpha(\omega(x))$ is the corresponding section of F_0 killed by ∇_0.

Now we just need to notice that, if $[\xi, \eta] = 0$, $\forall \xi \in \Theta$, then η is an infinitesimal transformation of $\tilde{\Gamma}$ and thus there exist parameters $a(t)$ for a finite label transformation such that:

$$\exp(t\eta)(\omega) = g(a(t), \omega)$$

Passing to the limit for $t \to 0$, we get our desired homomorphism of Lie algebras, using the fact that the label action is effective because of its construction and that $[L, W] = 0$. C.Q.F.D.

In fact, any element of $C(\Theta)$ determines uniquely a 1-parameter group of transformations of X by the exponential mapping and this is also a finite transformation of $\tilde{\Gamma}$. This result generalizes a theorem of Rim on infinite Lie algebras (40).

EXAMPLE 8.29 Let Γ be the pseudogroup of transformations of X corresponding to the effective action of a Lie group G on X with $\dim G = \dim X$. We shall prove that in this case the notation is coherent.

In this case we know that $\pi_1^1(R_1) = R_1$:

$$R_1 \qquad \omega_k^\rho(x)\frac{\partial \xi^k}{\partial x^i} + \xi^r \frac{\partial \omega_i^\rho(x)}{\partial x^r} = 0$$

We require:

$$-\frac{\partial \xi^k}{\partial x^i}\eta^i + \xi^r \frac{\partial \eta^k}{\partial x^r} = 0$$

whenever ξ is a solution of the involutive system R_1. We must have

$$\xi^r\left(\omega_k^\rho(x)\frac{\partial \eta^k}{\partial x^r} + \frac{\partial \omega_i^\rho(x)}{\partial x^r}\eta^i\right) = 0 \qquad \forall \xi$$

and thus

$$\omega_k^\rho(x)\frac{\partial \eta^k}{\partial x^r} + \frac{\partial \omega_i^\rho(x)}{\partial x^r}\eta^i \equiv \frac{\partial}{\partial x^r}(\omega_k^\rho(x)\eta^k) = 0$$

Finally $\eta^i = \lambda^k \alpha_k^i(x)$ defines $C(\Theta)$.

We know that $C(\Theta)$ is a Lie algebra because, if we set

$$\alpha_k = \alpha_k^i(x)\frac{\partial}{\partial x^i}, \quad \text{then } [\alpha_j, \alpha_k] = c_{jk}^i \alpha_i.$$

However the isomorphism, in this case, between Θ and $C(\Theta)$ "*is just a coincidence*".

If now we substitute in order to compute $\mathscr{D}\cdot\eta$, we have:

$$\lambda^k\left(\omega_j^\rho(x)\frac{\partial \alpha_k^j(x)}{\partial x^i} + \alpha_k^j(x)\frac{\partial \omega_i^\rho(x)}{\partial x^j}\right) = \lambda^k \alpha_k^j(x)\left(\frac{\partial \omega_i^\rho(x)}{\partial x^j} - \frac{\partial \omega_j^\rho(x)}{\partial x^i}\right)$$
$$= \lambda^k c_{rk}^\rho \cdot \omega_i^r(x)$$

and finally $\underset{1}{A}{}_r^\rho = \lambda^k c_{rk}^\rho$ because the sections of Υ_0 killed by $\mathscr{L}_0$ are given by $\underset{1}{\Omega}{}_i^\rho(x) = A_r^\rho \omega_i^r(x)$.

In the theory of finite dimensional Lie algebras, the corresponding elements of $B_0(R_1)$ are called "*inner derivations*".

We finally recall that the bundle $\mathscr{F}$ of geometric objects has been determined by the knowledge of the involutive system R_q only. If we take R_{q+1} instead of R_q, we shall obtain new vector spaces $B_r(R_{q+1})$, $Z_r(R_{q+1}$, $H_r(R_{q+1})$. However we keep the same $C(\Theta) = \Upsilon(T)$ because R_q is involutive.

It remains to study the relations between these new vector spaces and the former ones.

The first thing to do is to construct the new P-sequence with

$$\hat{F}_0 = \frac{J_{q+1}(T)}{R_{q+1}} = R'_1 \quad \text{and} \quad \hat{F}_r = \frac{\Lambda^r T^* \otimes \hat{F}_0}{\delta(\Lambda^{r-1} T^* \otimes G'_2)}.$$

Let the new P-sequence be:

$$0 \longrightarrow \Theta \longrightarrow T \xrightarrow{\hat{\mathcal{D}}} \hat{F}_0 \xrightarrow{\hat{\mathcal{D}}_1} \hat{F}_1 \xrightarrow{\hat{\mathcal{D}}_2} \cdots \xrightarrow{\hat{\mathcal{D}}_n} \hat{F}_n \longrightarrow 0$$

where $\hat{\mathcal{D}}$ is now an involutive operator of order $q + 1$.

Using the following commutative and exact diagrams:

$$\begin{array}{ccccccccccc}
0 & \longrightarrow & R_{q+r+1} & \longrightarrow & J_{q+r+1}(T) & \longrightarrow & J_r(\hat{F}_0) & \longrightarrow \cdots \longrightarrow & \hat{F}_r & \longrightarrow & 0 \\
 & & \downarrow & & \downarrow{\scriptstyle \pi^{q+r+1}_{q+r}} & & \downarrow{\scriptstyle J_r(\psi_0)} & & \downarrow{\scriptstyle \psi_r} & & \\
0 & \longrightarrow & R_{q+r} & \longrightarrow & J_{q+r}(T) & \longrightarrow & J_r(F_0) & \longrightarrow \cdots \longrightarrow & F_r & \longrightarrow & 0 \\
 & & \downarrow & & \downarrow & & \downarrow & & \downarrow & & \\
 & & 0 & & 0 & & 0 & & 0 & &
\end{array}$$

we can construct epimorphisms $\psi_r : \hat{F}_r \to F_r$ with kernels K_r.

THEOREM 8.30 The following diagram:

$$\begin{array}{ccccccccccccc}
 & & & & & & 0 & & 0 & & & 0 & \\
 & & & & & & \downarrow & & \downarrow & & & \downarrow & \\
 & & & & 0 & \longrightarrow & K_0 & \longrightarrow & K_1 & \longrightarrow & \cdots \longrightarrow & K_n & \longrightarrow 0 \\
 & & & & \downarrow & & \downarrow & & \downarrow & & & \downarrow & \\
0 & \longrightarrow & \Theta & \longrightarrow & T & \xrightarrow{\hat{\mathcal{D}}} & \hat{F}_0 & \xrightarrow{\hat{\mathcal{D}}_1} & \hat{F}_1 & \xrightarrow{\hat{\mathcal{D}}_2} & \cdots \xrightarrow{\hat{\mathcal{D}}_n} & \hat{F}_n & \longrightarrow 0 \\
 & & \| & & \| & & \downarrow{\scriptstyle \psi_0} & & \downarrow{\scriptstyle \psi_1} & & & \downarrow{\scriptstyle \psi_n} & \\
0 & \longrightarrow & \Theta & \longrightarrow & T & \xrightarrow{\mathcal{D}} & F_0 & \xrightarrow{\mathcal{D}_1} & F_1 & \xrightarrow{\mathcal{D}_2} & \cdots \xrightarrow{\mathcal{D}_n} & F_n & \longrightarrow 0 \\
 & & & & \downarrow & & \downarrow & & \downarrow & & & \downarrow & \\
 & & & & 0 & & 0 & & 0 & & & 0 &
\end{array}$$

is commutative. The columns are exact and the top row is exact.

Proof The commutativity and the exactness of the columns follows from the preceding diagram that allowed us to construct the epimorphism ψ_r. Moreover we know that the P-sequence for $\mathscr{D}$ and the P-sequence for $\hat{\mathscr{D}}$ are both formally exact.

It follows that we have the exact sequences:

$$0 \longrightarrow G'_{r+1} \longrightarrow J_r(K_0) \longrightarrow J_{r-1}(K_1) \longrightarrow \cdots \longrightarrow K_r \longrightarrow 0$$

and that $K_0 = G'_1$. Thus we obtain the commutative diagram with exact columns:

$$\begin{array}{ccccccccccc}
 & & & & 0 & & 0 & & & 0 & \\
 & & & & \downarrow & & \downarrow & & & \downarrow & \\
 & & 0 & \longrightarrow & K_0 & \longrightarrow & K_1 & \longrightarrow \cdots \longrightarrow & & K_n & \longrightarrow 0 \\
 & & \downarrow & & \downarrow{\scriptstyle j_{r+n}} & & \downarrow{\scriptstyle j_{r+n-1}} & & & \downarrow{\scriptstyle j_r} & \\
0 & \longrightarrow & G'_{r+n+1} & \longrightarrow & J_{r+n}(K_0) & \longrightarrow & J_{r+n-1}(K_1) & \longrightarrow \cdots \longrightarrow & & J_r(K_n) & \longrightarrow 0 \\
 & & \downarrow{\scriptstyle \delta} & & \downarrow{\scriptstyle D} & & \downarrow{\scriptstyle D} & & & \downarrow{\scriptstyle D} & \\
0 & \longrightarrow & T^* \otimes G'_{r+n} & \longrightarrow & T^* \otimes J_{r+n-1}(K_0) & \longrightarrow & T^* \otimes J_{r+n-2}(K_1) & \longrightarrow \cdots \longrightarrow & & T^* \otimes J_{r-1}(K_n) & \longrightarrow 0 \\
 & & \downarrow{\scriptstyle \delta} & & \downarrow{\scriptstyle D} & & \downarrow{\scriptstyle D} & & & \downarrow{\scriptstyle D} & \\
 & & \vdots & & \vdots & & \vdots & & & \vdots & \\
 & & \downarrow{\scriptstyle \delta} & & \downarrow{\scriptstyle D} & & \downarrow{\scriptstyle D} & & & & \\
0 & \longrightarrow & \Lambda^n T^* \otimes G'_{r+1} & \longrightarrow & \Lambda^n T^* \otimes J_r(K_0) & \longrightarrow & \Lambda^n T^* \otimes J_{r-1}(K_1) & \longrightarrow \cdots & & & \\
 & & \downarrow & & \downarrow & & \downarrow & & & & \\
 & & 0 & & 0 & & 0 & & & &
\end{array}$$

An easy chase ends the proof of the theorem. C.Q.F.D.

Corollary 8.31 The two P-sequences have isomorphic cohomologies.

The vector bundles $\hat{F}_r$ are associated now with R_{q+1}. However, as the vector bundles F_r are associated with R_q, they are also associated with

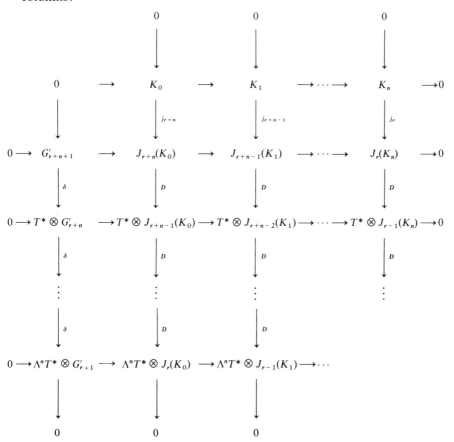

R_{q+1} because $\pi_q^{q+1}: R_{q+1} \to R_q$ is surjective and we may consider all the vector bundles appearing in the diagram of theorem 8.30 as associated with R_{q+1}.

$\forall \xi \in \Theta$ we have $\mathscr{L}(\xi) \circ \mathscr{D} = \mathscr{D} \circ \mathscr{L}(\xi)$ and $\mathscr{L}(\xi) \circ \hat{\mathscr{D}} = \hat{\mathscr{D}} \circ \mathscr{L}(\xi)$ and it is easily seen that the epimorphism $\psi_r: \hat{F}_r \to F_r$ induce maps $\psi_r: H_r(R_{q+1}) \to H_r(R_q)$ which are in general neither monomorphisms nor epimorphisms.

If we consider the tridimensional diagram obtained by letting $\mathscr{L}(\xi)$ act on the diagram of theorem 8.30, a chase which is left to the reader shows that we have the exact sequences:

$$0 \longrightarrow H_0(R_{q+1}) \xrightarrow{\psi_0} H_0(R_q) \longrightarrow 0$$

$$0 \longrightarrow H_1(R_{q+1}) \xrightarrow{\psi_1} H_1(R_q)$$

In particular, if $H_1(R_q) = 0$ then $H_1(R_{q+1}) = 0, \ldots$ and we may say that Θ or Γ is rigid.

It follows, in particular, that $\forall r \geq 0$:

$$\dim H_0(R_{q+r+1}) = \dim H_0(R_{q+r}), \quad 0 \leq \dim H_1(R_{q+r+1}) \leq \dim H_1(R_{q+r})$$

Thus $\dim H_1(R_{q+r})$ does not depend on r when r is big enough, as it is of course an integer.

It is an open problem to know whether such relations can be found for the other cohomology groups or not!

The later approach to the deformation cohomology was used in order to link the non-linear and the linear aspects. We shall now provide a direct and intrinsic approach that must be used when dealing with the linear machinery.

As a matter of fact, we have $[J_q(T), J_q(T)] \subset J_q(T)$ where the bracket on $J_q(T)$ is defined in the usual way by the formula

$$[\flat(\xi_q, \flat(\bar{\xi}_q)] = \flat([\xi_q, \bar{\xi}_q]) \quad \text{on } X$$

or equivalently by the formula

$$[\#(\eta_q), \#(\bar{\eta}_q)] = \#([\eta_q, \bar{\eta}_q]) \quad \text{on } Y.$$

As we shall deal with $J_q(T)$ only, changing slightly our notations and keeping the first point of view, it is easy to check that:

$$[\xi_q, \eta_q] = \{\xi_{q+1}, \eta_{q+1}\} + i(\xi)D\eta_{q+1} - i(\eta)D\xi_{q+1}$$

for any two sections ξ_{q+1} and η_{q+1} of $J_{q+1}(T)$ with $\xi_q = \pi_q^{q+1} \circ \xi_{q+1}$ and $\eta_q = \pi_q^{q+1} \circ \eta_{q+1}$.

Denoting by $\mathscr{C}(\xi, \eta, \zeta)$ the sum over a given cycle, we have the Jacobi's identities:

$$\begin{aligned} \mathscr{C}(\xi, \eta, \zeta)[[\xi_q, \eta_q], \zeta_q] &= 0 \\ \mathscr{C}(\xi, \eta, \zeta)\{\{\xi_{q+1}, \eta_{q+1}\}, \zeta_q\} &= 0 \end{aligned} \qquad \text{(Jacobi)}$$

We now recall, because of their constant use, the following formulas, already proved:

$$\begin{aligned} i(\zeta)D\{\xi_{q+1}, \eta_{q+1}\} &= \{i(\zeta)D\xi_{q+1}, \eta_q\} + \{\xi_q, i(\zeta)D\eta_{q+1}\} \\ i(\zeta)D[\xi_{q+1}, \eta_{q+1}] &= [i(\zeta)D\xi_{q+1}, \eta_q] + [\xi_q, i(\zeta)D\eta_{q+1}] \\ &\quad + i(L(\eta_1)\zeta)D\xi_{q+1} - i(L(\xi_1)\zeta)D\eta_{q+1} \end{aligned}$$

We understand at this time the definition of $L(\xi_1)$ and an easy computation shows that T is associated with $J_1(T)$.

LEMMA 8.32 $J_q(T)$ is associated with $J_{q+1}(T)$.

Proof We define

$$\begin{aligned} L(\xi_{q+1})\eta_q &= [\xi_q, \eta_q] + i(\eta)D\xi_{q+1} \\ &= \{\xi_{q+1}, \eta_{q+1}\} + i(\xi)D\eta_{q+1} \end{aligned}$$

All the properties of the Lie derivative are easy to check but the third. We prove it using a direct computation:

$$\begin{aligned} [L(\xi_{q+1}), L(\bar{\xi}_{q+1})]\eta_q &= L(\xi_{q+1})(\{\bar{\xi}_{q+1}, \eta_{q+1}\} + i(\bar{\xi})D\eta_{q+1}) \\ &\quad - L(\bar{\xi}_{q+1})(\{\xi_{q+1}, \eta_{q+1}\} + i(\xi)D\eta_{q+1}) \\ &= \{\xi_{q+1}, \{\bar{\xi}_{q+2}, \eta_{q+2}\}\} + \{\xi_{q+1}, i(\bar{\xi})D\eta_{q+2}\} \\ &\quad + i(\xi)D\{\bar{\xi}_{q+2}, \eta_{q+2}\} - \{\bar{\xi}_{q+1}, \{\xi_{q+2}, \eta_{q+2}\} \\ &\quad - \{\bar{\xi}_{q+1}, i(\xi)D\eta_{q+2}\} - i(\bar{\xi})D\{\xi_{q+2}, \eta_{q+2}\} \\ &\quad + i(\xi)D(i(\bar{\xi})D\eta_{q+2}) - i(\bar{\xi})D(i(\xi)D\eta_{q+2}) \\ &= \{\{\xi_{q+2}, \bar{\xi}_{q+2}\}, \eta_{q+1}\} + \{i(\xi)D\bar{\xi}_{q+2}, \eta_{q+2}\} \\ &\quad - \{i(\bar{\xi})D\xi_{q+2}, \eta_{q+1}\} + i(\xi)D(i(\bar{\xi})D\eta_{q+2}) \\ &\quad - i(\bar{\xi})D(i(\xi)D\eta_{q+2}) \\ &= \{[\xi_{q+1}, \bar{\xi}_{q+1}], \eta_{q+1}\} + i(\xi)D(i(\bar{\xi})D\eta_{q+2}) \\ &\quad - i(\bar{\xi})D(i(\xi)D\eta_{q+2}) \end{aligned}$$

The two last terms are detailed as follows, using local coordinates:

$$\begin{aligned} &\xi^i\left(\frac{\partial}{\partial x^i}\left(\bar{\xi}^j\left(\frac{\partial \eta_\mu^k}{\partial x^j} - \eta_{\mu+1_j}^k\right)\right) - \bar{\xi}^j\left(\frac{\partial \eta_{\mu+1_i}^k}{\partial x^j} - \eta_{\mu+1_i+1_j}^k\right)\right) \\ &\quad - \bar{\xi}^j\left(\frac{\partial}{\partial x^j}\left(\xi^i\left(\frac{\partial \eta_\mu^k}{\partial x^i} - \eta_{\mu+1_i}^k\right)\right) - \xi^i\left(\frac{\partial \eta_{\mu+1_j}^k}{\partial x^i} - \eta_{\mu+1_i+1_j}^k\right)\right) \\ &= \left(\xi^j \frac{\partial \bar{\xi}^i}{\partial x^j} - \bar{\xi}^j \frac{\partial \xi^i}{\partial x^j}\right)\left(\frac{\partial \eta_\mu^k}{\partial x^i} - \eta_{\mu+1_i}^k\right) = ([\xi, \bar{\xi}])^i\left(\frac{\partial \eta_\mu^k}{\partial x^i} - \eta_{\mu+1_i}^k\right) \end{aligned}$$

We obtain finally:

$$[L(\xi_{q+1}), L(\bar{\xi}_{q+1})]\eta_q = L([\xi_{q+1}, \bar{\xi}_{q+1}])\eta_q \qquad \text{C.Q.F.D.}$$

We shall now list important formulas:

LEMMA 8.33 We have the formula:

$$L(\xi_q)\{\eta_q, \zeta_q\} = \{L(\xi_{q+1})\eta_q, \zeta_q\} + \{\eta_q, L(\xi_{q+1})\zeta_q\}$$

Proof

$$\begin{aligned}\{\{\xi_{q+1}, \eta_{q+1}\} &+ i(\xi)D\eta_{q+1}, \xi_q\} + \{\eta_q, \{\xi_{q+1}, \zeta_{q+1}\} + i(\xi)D\zeta_{q+1}\} \\ &= \{\{\xi_{q+1}, \eta_{q+1}\}, \zeta_q\} + \{\eta_q, \{\xi_{q+1}, \zeta_{q+1}\}\} \\ &\quad + \{i(\xi)D\eta_{q+1}, \zeta_q\} + \{\eta_q, i(\xi)D\zeta_{q+1}\} \\ &= \{\xi_q, \{\eta_{q+1}, \zeta_{q+1}\}\} + i(\xi)D\{\eta_{q+1}, \zeta_{q+1}\} \\ &= L(\xi_q)\{\eta_q, \zeta_q\}. \qquad \text{C.Q.F.D.}\end{aligned}$$

LEMMA 8.34 We have the formula:

$$\begin{aligned}L(\xi_{q+1})[\eta_q, \zeta_q] = [L(\xi_{q+1})\eta_q, \zeta_q] + [\eta_q, L(\xi_{q+1})\zeta_q] + i([\eta, \zeta])D\xi_{q+1} \\ - [i(\eta)D\xi_{q+1}, \zeta_q] - [\eta_q, i(\zeta)D\xi_{q+1}]\end{aligned}$$

Proof Using the Jacobi's identities we get:

$$\begin{aligned}[\xi_q, [\eta_q, \zeta_q]] &= [[\xi_q, \eta_q], \zeta_q] + [\eta_q, [\xi_q, \zeta_q]] \\ &= [L(\xi_{q+1})\eta_q - i(\eta)D\xi_{q+1}, \zeta_q] + [\eta_q, L(\xi_{q+1})\zeta_q - i(\xi)D\xi_{q+1}]\end{aligned}$$

C.Q.F.D.

LEMMA 8.35 We have the formula:

$$i(\zeta)(DL(\xi_{q+1})\eta_q - L(\xi_q)D\eta_q) = L(i(\zeta)D\xi_{q+1})\eta_{q-1}$$

Proof According to remark 8.19 we have:

$$i(\zeta)(L(\xi_q)D\eta_q) = L(\xi_q)(i(\zeta)D\eta_q) - i(L(\xi_1)\zeta)D\eta_q$$

and

$$\begin{aligned}i(\zeta)D[\xi_q, \eta_q] &+ i(\zeta)D(i(\eta)D\xi_{q+1}) \\ &= [i(\zeta)D\xi_q, \eta_{q-1}] + [\xi_{q-1}, i(\zeta)D\eta_q] - i(L(\xi_1)\zeta)D\eta_q \\ &\quad + i(L(\eta_1)\zeta)D\xi_q + i(\zeta)D(i(\eta)D\xi_{q+1}) \\ &= L(i(\zeta)D\xi_{q+1})\eta_{q-1} - i(\eta)D(i(\zeta)D\xi_{q+1}) + L(\xi_q)(i(\zeta)D\eta_q) \\ &\quad - i(i(\zeta)D\eta_1)D\xi_q + i(L(\eta_1)\zeta)D\xi_q - i(L(\xi_1)\zeta)D\eta_q \\ &\quad + i(\zeta)D(i(\eta)D\xi_{q+1}) \\ &= i(\zeta)L(\xi_q)D\eta_q + L(i(\zeta)D\xi_{q+1})\eta_{q-1} + i(L(\eta_1)\zeta)D\xi_q \\ &\quad - i(\eta)D(i(\zeta)D\xi_{q+1}) + i(\zeta)D(i(\eta)D\xi_{q+1}) - i(i(\zeta)D\eta_1)D\xi_q\end{aligned}$$

We shall show that the last four terms vanish, using local coordinates:

$$\left(-\eta_i^j\zeta^i+\eta^i\frac{\partial\zeta^j}{\partial x^i}\right)\left(\frac{\partial\xi_\mu^k}{\partial x^j}-\xi_{\mu+1_j}^k\right)-\zeta^j\left(\frac{\partial\eta^i}{\partial x^j}-\eta_j^i\right)\left(\frac{\partial\xi_\mu^k}{\partial x^i}-\xi_{\mu+1_i}^k\right)$$
$$+\zeta^j\left(\frac{\partial}{\partial x^j}\left(\eta^i\left(\frac{\partial\xi_\mu^k}{\partial x^i}-\xi_{\mu+1_i}^k\right)\right)-\eta^i\left(\frac{\partial\xi_{\mu+1_j}^k}{\partial x^i}-\xi_{\mu+1_i+1_j}^k\right)\right.$$
$$\left.-\eta^i\left(\frac{\partial}{\partial x^i}\left(\zeta^j\left(\frac{\partial\xi_\mu^k}{\partial x^j}-\xi_{\mu+1_j}^k\right)\right)-\zeta^j\left(\frac{\partial\xi_{\mu+1_i}^k}{\partial x^j}-\xi_{\mu+1_i+1_j}^k\right)\right)\equiv 0$$

C.Q.F.D.

From the above lemmas we deduce the following useful formulas for the Lie derivative:

$$\mathscr{L}(\xi)\eta_q=[j_q(\xi),\eta_q]$$
$$\mathscr{L}(\xi)\{\eta_q,\zeta_q\}=\{\mathscr{L}(\xi)\eta_q,\zeta_q\}+\{\eta_q,\mathscr{L}(\xi)\zeta_q\}$$
$$\mathscr{L}(\xi)[\eta_q,\zeta_q]=[\mathscr{L}(\xi)\eta_q,\zeta_q]+[\eta_q,\mathscr{L}(\xi)\zeta_q]$$
$$\mathscr{L}(\xi)\circ D=D\circ\mathscr{L}(\xi)$$

We shall use the preceding lemmas in order to construct the deformation cohomology.

Let us start with a given involutive and formally transitive system $R_q\subset J_q(T)$ of infinitesimal Lie equations. We have

$$[R_{q+r},R_{q+r}]\subset R_{q+r},\qquad \forall r\geq 0$$

and $J_{q+r}(T)$ becomes associated with R_{q+r+1}.

Proposition 8.36 We have the short exact sequence:

$$0\longrightarrow \Upsilon(J_q^0(T))\longrightarrow \Upsilon(J_q(T))\xrightarrow{\pi_0^q}\Upsilon(T)\longrightarrow 0$$

Proof We have only to prove the surjectivity at $\Upsilon(T)$. Let $\eta\in\Upsilon(T)$ be a section of T such that $L(\xi_1)\eta=0$ for any section ξ_1 of $R_1=\pi_1^q(R_q)$.

Using lemma 8.35 and the fact that $D\circ j_q=0$ we obtain:

$$i(\zeta)D(L(\xi_2)j_1(\eta))=L(i(\zeta)D\xi_2)\eta=0$$

because

$$R_2\subset(R_1)_{+1}\quad\text{and}\quad DR_2\subset T^*\otimes R_1.$$

The later notation has a meaning because we use the same letter for a bundle and its set of local sections.

In fact

$$\pi_0^1(L(\xi_2)j_1(\eta)) = L(\xi_1)\eta = 0 \quad \text{and} \quad L(\xi_2)j_1(\eta) \in T^* \otimes T.$$

An elementary property of the δ-cohomology shows that $L(\xi_2)j_1(\eta) = 0$. We shall use an induction on q in order to prove that $L(\xi_{q+1})j_q(\eta) = 0$ and thus $j_q(\eta) \in \Upsilon(J_q(T))$ projects onto $\eta \in \Upsilon(T)$.

According to lemma 8.35 we have:

$$i(\zeta)D(L(\xi_{q+2})j_{q+1}(\eta)) = L(i(\zeta)D\xi_{q+2})j_q(\eta) = 0$$

It follows that $L(\xi_{q+2})j_{q+1}(\eta) \in S_{q+1}T^* \otimes T$ and we deduce the proposition from the exact sequence:

$$0 \longrightarrow S_{q+1}T^* \otimes T \xrightarrow{\ \delta\ } T^* \otimes S_qT^* \otimes T$$

C.Q.F.D.

Let now F, F', F'' be vector bundles associated with R_q and $\Phi : F \to F'$, $\psi : F' \to F''$ be morphisms invariant by R_q such that we have the short exact sequence:

$$0 \longrightarrow F \xrightarrow{\ \Phi\ } F' \xrightarrow{\ \Psi\ } F'' \longrightarrow 0$$

Then, in general, we only have the exact sequence:

$$0 \longrightarrow \Upsilon(F) \xrightarrow{\ \Phi\ } \Upsilon(F') \xrightarrow{\ \Psi\ } \Upsilon(F'')$$

unless R_q has particular properties.

More precisely, if F is a vector bundle associated with R_q, let E be the sub-bundle of F invariant by R_q^0. We have the exact sequence of vector bundles:

$$0 \longrightarrow E \longrightarrow F \longrightarrow R_q^{0*} \otimes F$$

where the last morphism is induced by ∇. Then, of course, E is associated with R_q and $\Upsilon(E) = \Upsilon(F) = \Upsilon$.

Proposition 8.37 We have the short exact sequence:

$$0 \longrightarrow \Upsilon \xrightarrow{\ \Phi\ } \Upsilon' \xrightarrow{\ \Psi\ } \Upsilon'' \longrightarrow 0$$

if and only if we have the short exact sequence:

$$0 \longrightarrow E \xrightarrow{\ \Phi\ } E' \xrightarrow{\ \Psi\ } E'' \longrightarrow 0$$

Proof $\eta \in \Upsilon \Leftrightarrow L(\xi_q)\eta = 0,\ \forall \xi_q \in R_q \Leftrightarrow \nabla \cdot \eta = 0$

where $\nabla : F \to R_q^* \otimes F$ is a first order involutive operator already defined.

Let now $\alpha : T \to R_q$ be a map such that $\pi_0^q \circ \alpha = \text{id}$. Then we have

$$\eta \in \Upsilon \Leftrightarrow \eta \in E, \quad L(\alpha(\xi))\eta = 0 \qquad \forall \xi \in T$$

because $R_q = R_q^0 \oplus \alpha(T)$. Moreover we may integrate ∇ as we did for $\#(R_q)$ when dealing with the third fundamental theorem. It follows that we have first to find out $E \subset F$ and then to solve the restriction of ∇ to E, that is to say $\nabla : E \to T^* \otimes E$.

Moreover, as $[R_q, R_q] \subset R_q$ and $[R_q^0, R_q^0] \subset R_q^0$, this operator is equivalent to the standard Frobenius case and thus involutive. It also does not depend on the map α because, if $\bar{\alpha} : T \to R_q$ is another map such that $\pi_0^q \circ \bar{\alpha} = \text{id}$ then $\bar{\alpha} - \alpha : T \to R_q^0$ and $L(\bar{\alpha}(\xi))E - L(\alpha(\xi))E = L((\bar{\alpha} - \alpha)(\xi))E = 0$ because E is invariant by R_q^0. Finally we have the relation $\dim \Upsilon = \dim E$ which proves the proposition. C.Q.F.D.

DEFINITION 8.38 Such an operator $\nabla : E \to T^* \otimes E$ is called a "*covariant derivative*".

Looking back to the killing equations for a section of $S_2 T^*$, the reader will discover, using local coordinates, why we used this name.

It is easy to check that:

$$\nabla(f \cdot \eta) = df \otimes \eta + f \cdot \nabla \eta \qquad \forall f \in C^\infty(X), \eta \in E$$

We may extend this operator to an operator:

$$\nabla : \Lambda^r T^* \otimes E \to \Lambda^{r+1} T^* \otimes E$$

by the formula:

$$\nabla(\omega \otimes \eta) = d\omega \otimes \eta + (-1)^r \omega \wedge \nabla \eta \qquad \forall \omega \in \Lambda^r T^*, \eta \in E$$

From the commutation relations $[R_q, R_q] \subset R_q$ we obtain $\nabla^2 = \nabla \circ \nabla = 0$ and this gives rise to the important ∇-sequence:

$$0 \longrightarrow \Upsilon \longrightarrow E \xrightarrow{\nabla} T^* \otimes E \xrightarrow{\nabla} \cdots \xrightarrow{\nabla} \Lambda^n T^* \otimes E \longrightarrow 0$$

According to proposition 6.3.15 the first ∇ is involutive and equivalent to a Frobenius type operator. The later ∇-sequence is locally and formally exact as it is locally isomorphic to the d-sequence.

REMARK 8.39 Using the well known Whitehead algebraic lemmas for R_q^0-modules, one sees that the conditions of proposition 8.37 are satisfied when R_q^0 is a semi-simple Lie algebra, as we know that

$$R_{q;x}^0 \approx R_{q;y}^0 \qquad \forall x, y \in X.$$

EXAMPLE 8.40 When $q = 1$, $R_1^0 = G_1 = 0$ then $E = F$ and the conditions of proposition 8.37 are always satisfied. We let the reader prove that the only such possible case is that of Lie groups already considered.

PROPOSITION 8.41 R_q is associated with R_{q+1}.

Proof Let $\xi_{q+1} \in R_{q+1}$ and $\eta_q \in R_q$. We have

$$L(\xi_{q+1})\eta_q = [\xi_q, \eta_q] + i(\eta)D\xi_{q+1} \in R_q \qquad \text{C.Q.F.D.}$$

As R_q and $J_q(T)$ are associated with R_{q+1}, then $F_0 = J_q(T)/R_q$ is associated with R_{q+1}. In fact it is easy to check that R_q^0 and J_q^0 are associated with R_q only and we deduce that F_0 is associated with R_q from the following exact commutative diagram.

$$\begin{array}{ccccccccc}
 & & 0 & & 0 & & 0 & & \\
 & & \downarrow & & \downarrow & & \downarrow & & \\
0 & \longrightarrow & R_q^0 & \longrightarrow & J_q^0(T) & \longrightarrow & F_0 & \longrightarrow & 0 \\
 & & \downarrow & & \downarrow & & \| & & \\
0 & \longrightarrow & R_q & \longrightarrow & J_q(T) & \longrightarrow & F_0 & \longrightarrow & 0 \\
 & & \downarrow & & \downarrow & & \downarrow & & \\
0 & \longrightarrow & T & = & T & \longrightarrow & 0 & & \\
 & & \downarrow & & \downarrow & & & & \\
 & & 0 & & 0 & & & &
\end{array}$$

Moreover, as G_{q+r} and $S_{q+r}T^* \otimes T$ are associated with R_1 and thus with R_q, it follows that G'_r is associated with R_1 and thus with R_q. Finally

$$F_r = \frac{\Lambda^r T^* \otimes F_0}{\delta(\Lambda^{r-1} T^* \otimes G'_1)}$$

becomes associated with R_q.

We shall now prove directly the following key propositions of the deformation cohomology.

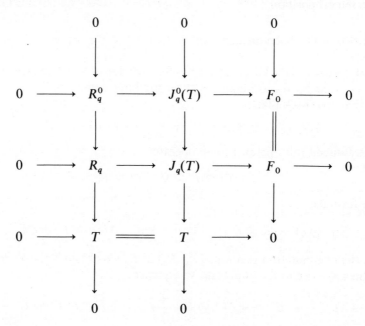

PROPOSITION 8.42 There is an homomorphism of Lie algebras

$$\mathscr{D} : \Upsilon(T) \to \Upsilon(F_0)$$

induced by the operator $\mathscr{D} : T \to F_0$.

Proof Changing slightly the formula of lemma 8.34 we get:

$$\begin{aligned} L(\xi_{q+1})[\eta_q, \zeta_q] &= [L(\xi_{q+1})\eta_q, \zeta_q] + [\eta_q, L(\xi_{q+1})\zeta_q] \\ &\quad + L(i(\zeta)D\xi_{q+2})\eta_q - L(i(\eta)D\xi_{q+2})\zeta_q \end{aligned}$$

where we have used the formal integrability of R_q.

If $L(\xi_1)\eta = 0$ and $L(\xi_1)\zeta = 0$ then $L(\xi_1)[\eta, \zeta] = 0$, that is to say, if η and $\zeta \in \Upsilon(T)$ then $[\eta, \zeta] \in \Upsilon(T)$ and $\Upsilon(T)$ may be considered as a finite dimensional Lie algebra with dim $\Upsilon(T) \leq n$.

The same property is held by $\Upsilon(F_0)$ with a bracket induced by that of $J_q(T)$. In fact, whenever $L(\xi_{q+1})\eta_q \in R_q$ and $L(\xi_{q+1})\zeta_q \in R_q$, from the above formula we deduce $L(\xi_{q+1})[\eta_q, \zeta_q] \in R_q$.

Now, whenever $\eta \in \Upsilon(T)$, we know, according to proposition 8.36, that $L(\xi_{q+1}) \cdot j_q(\eta) = 0$ and this proves the proposition because we may take $j_q(\eta)$ as a representative of $\mathscr{D} \cdot \eta$.

More precisely, setting $\underset{1}{\Omega} = \mathscr{D} \cdot \eta$ we obtain:

$$[\xi_q, j_q(\eta)] + i(\eta)D\xi_{q+1} = 0$$

and it follows that $L([\xi_q, j_q(\eta)])\omega = 0$ because $i(\eta)D\xi_{q+1} \in R_q$. Finally we get:

$$L(\xi_q)L(j_q(\eta))\omega - L(j_q(\eta))L(\xi_q)\omega = L(\xi_q)\underset{1}{\Omega} = 0.$$

and we know that $j_q([\eta, \zeta]) = [j_q(\eta), j_q(\zeta)]$. C.Q.F.D.

PROPOSITION 8.43 There is a map $\mathscr{D}_r : \Upsilon_{r-1} \to \Upsilon_r$ induced by the first order operator $\mathscr{D}_r : F_{r-1} \to F_r$.

Proof A straightforward computation, which is left to the reader as an exercise, shows that, $\forall A_{q+1}^{r-1} \in \Lambda^{r-1}T^* \otimes J_{q+1}(T)$:

$$\begin{aligned} & i(\zeta_{(1)}) \cdots i(\zeta_{(r)})(DL(\xi_{q+2})A_{q+1}^{r-1} - L(\xi_{q+1})DA_{q+1}^{r-1}) \\ &\qquad = \sum_{s=1}^{r} (-1)^{s+1} i(\zeta_{(1)}) \cdots i(\hat{\zeta}_{(s)}) \cdots i(\zeta_{(r)}) \cdot L(i(\zeta_{(s)})D\xi_{q+2})A_q^{r-1} \end{aligned}$$

where A_q^{r-1} is the projection of A_{q+1}^{r-1} on $\Lambda^{r-1}T^* \otimes J_q(T)$.

According to the second diagram that can be found in the proof of theorem 5.5.3, if A_{q+1}^{r-1} is such that A_q^{r-1} is a representative of a section of Υ_{r-1}, then DA_{q+1}^{r-1} is a representative of the section of F_r we want to study.

By hypothesis we have; $\forall s = 1, \ldots, r$:

$$L(i(\zeta_{(s)})D\xi_{q+2})A_q^{r-1} \in \Lambda^{r-1}T^* \otimes R_q \cup \delta(\Lambda^{r-2}T^* \otimes S_{q+1}T^* \otimes T)$$

From the commutative and exact diagram:

$$\begin{array}{ccccc} T^* \otimes \Lambda^{r-2}T^* \otimes S_{q+1}T^* & \xrightarrow{(-1)^{r-2}\delta} & \Lambda^{r-1}T^* \otimes S_{q+1}T^* & \longrightarrow & 0 \\ \downarrow \delta & & \downarrow \delta & & \\ T^* \otimes \Lambda^{r-1}T^* \otimes S_qT^* & \xrightarrow{(-1)^{r-1}\delta} & \Lambda^rT^* \otimes S_qT^* & \longrightarrow & 0 \end{array}$$

resulting from the relations in local coordinates:

$$\omega_{\mu+1_i,I,j}(dx^i \wedge dx^I) \wedge dx^j = \omega_{\mu+1_i,I,j}\, dx^i \wedge (dx^I \wedge dx^j)$$

we deduce that:

$$DL(\xi_{q+2})A^{r-1}_{q+1} - L(\xi_{q+1})DA^{r-1}_{q+1} \in \Lambda^rT^* \otimes R_q \cup \delta(\Lambda^{r-1}T^* \otimes S_{q+1}T^* \otimes T)$$

and this concludes the proof of the proposition, because an easy chase in the following commutative diagram on page 391 shows that:

$$L(\xi_{q+2})A^{r-1}_{q+1} \in D(\Lambda^{r-2}T^* \otimes J_{q+2}(T)) \cup \Lambda^{r-1}T^* \otimes R_{q+1} \cup \Lambda^{r-1}T^* \otimes S_{q+1}T^* \otimes T$$

C.Q.F.D.

This approach, though it is intrinsic, does not go very far indeed because we cannot introduce the structure constants c and the Jacobi relations $J(c) = 0$ of the non-linear point of view.

REMARK 8.44 The two relations we have found between $H_r(R_q)$ and $H_r(R_{q+1})$ seem to be the only ones that can be found in the general case. Of course, according to proposition 8.37, in the case of example 8.40 we have $H_r(R_1) = H_r(R_q)$, $\forall_q \geq 1$ and this is just the Hochschild cohomology for a Lie algebra of dimension n. However, this relation does not exist in general when Γ corresponds to the action of Lie group G on X. The reader will check that in this case $\tilde{\Gamma}$ corresponds to the action of a Lie group $\tilde{G}$ on X. Looking at the infinitesimal point of view and using proposition 8.42, if $\mathscr{G}$ is the Lie algebra of G and $\tilde{\mathscr{G}}$ is the Lie algebra of $\tilde{\mathscr{G}}$ we have, in general, $\tilde{\mathscr{G}} \neq \mathscr{G} \oplus \Delta(\mathscr{G})$ unless $\dim G = \dim X$.

To end this paragraph, we shall use the later methods in order to give a direct proof of the following non-trivial theorem, already proved by means of other methods.

THEOREM 8.45 If R_q is formally integrable and G_q is 2-acyclic, then $\tilde{R}_{q+1}$ is formally integrable.

Proof According to lemma 8.32, $\tilde{\xi}_{q+1}$ is a section of $\tilde{R}_{q+1}$ whenever $L(\tilde{\xi}_{q+1})\xi_q$ is a section of R_q for any section ξ_q of R_q. As R_q is formally

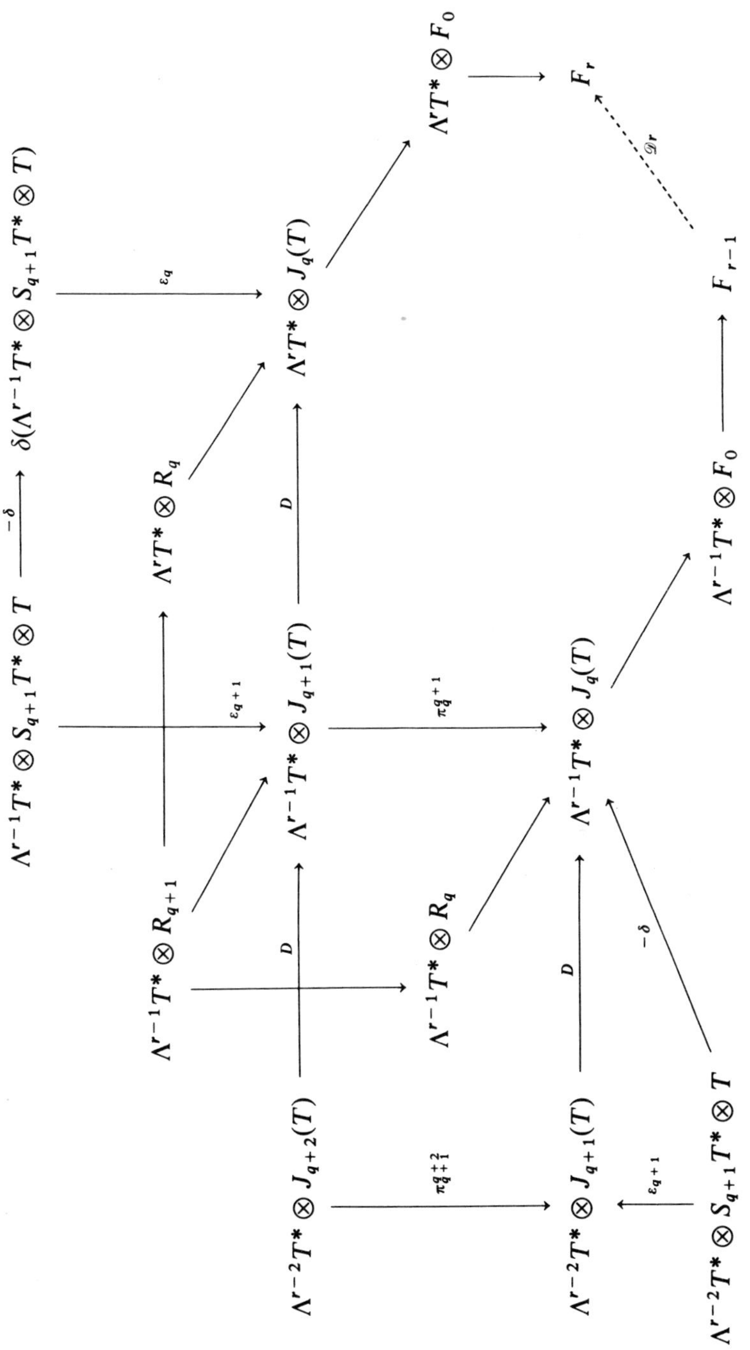

$\Lambda^{r-1}T^* \otimes S_{q+1}T^* \otimes T$
$-\delta$
$\delta(\Lambda^{r-1}T^* \otimes S_{q+1}T^* \otimes T)$
$\Lambda^{r-1}T^* \otimes R_{q+1}$
$\Lambda^{r}T^* \otimes R_q$
ε_{q+1}
ε_q
$\Lambda^{r-2}T^* \otimes J_{q+2}(T)$
D
$\Lambda^{r-1}T^* \otimes J_{q+1}(T)$
D
$\Lambda^{r}T^* \otimes J_q(T)$
π^{q+2}_{q+1}
π^{q+1}_q
$\Lambda^{r-1}T^* \otimes R_q$
$\Lambda^{r}T^* \otimes F_0$
$\Lambda^{r-2}T^* \otimes J_{q+1}(T)$
D
$\Lambda^{r-1}T^* \otimes J_q(T)$
F_r
ε_{q+1}
$-\delta$
$\mathscr{D}_r$
$\Lambda^{r-2}T^* \otimes S_{q+1}T^* \otimes T$
$\Lambda^{r-1}T^* \otimes F_0$
F_{r-1}

integrable this condition is equivalent to $\{\tilde{R}_{q+1}, R_{q+1}\} \subset R_q$. In particular we must have $\{\tilde{G}_{q+1}, R_{q+1}\} \subset G_q$ and thus $\tilde{G}_{q+1} = G_{q+1}$.

Let us define $\tilde{R}_{q+2}$ by the similar condition $\{\tilde{R}_{q+2}, R_{q+2}\} \subset R_{q+1}$. Then, according to the criterion of formal integrability, the proof of the theorem depends on the two following propositions:

PROPOSITION 8.46 $\pi^{q+2}_{q+1} : \tilde{R}_{q+2} \to \tilde{R}_{q+1}$ is an epimorphism.

Proof This is the hard step. Let us introduce splitting maps

$$\alpha : T \to R_q, \quad \beta : T \to R_{q+1}, \quad \gamma : T \to R_{q+1},$$
$$\varphi : R_q \to R_{q+1}, \quad \psi : R_{q+1} \to R_{q+2}$$

such that

$$\beta = \varphi \circ \alpha, \quad \gamma = \psi \circ \beta, \quad \pi^q_0 \circ \alpha = \mathrm{id}_T, \quad \pi^{q+1}_0 \circ \beta = \mathrm{id}_T, \quad \pi^{q+2}_0 \circ \gamma = \mathrm{id}_T.$$

We define $Z \in \Lambda^2 T^* \otimes G_q$ by the formula:

$$Z(\xi, \eta) = \{\tilde{\xi}_{q+1}, \{\gamma(\xi), \gamma(\eta)\}\} - \{\varphi(\{\tilde{\xi}_{q+1}, \beta(\xi)\}), \beta(\eta)\} - \{\beta(\xi), \varphi(\{\tilde{\xi}_{q+1}, \beta(\eta)\})\}$$

We have

$$\delta Z(\xi, \eta, \zeta) = \mathscr{C}(\xi, \eta, \zeta)\{Z(\xi, \eta), \alpha(\zeta)\}$$

Using the Jacobi's identities we obtain:

$$\begin{aligned}
&\mathscr{C}(\xi, \eta, \zeta)\{\{\tilde{\xi}_{q+1}, \{\gamma(\xi), \gamma(\eta)\}\}, \alpha(\zeta)\} \\
&\quad = \mathscr{C}(\xi, \eta, \zeta)\{\{\tilde{\xi}_{q+1}, \beta(\xi)\}, \{\beta(\eta), \beta(\zeta)\}\} \\
&\quad = \mathscr{C}(\xi, \eta, \zeta)\{\{\varphi(\{\tilde{\xi}_{q+1}, \beta(\xi)\}), \beta(\eta)\}, \alpha(\zeta)\} \\
&\qquad + \mathscr{C}(\xi, \eta, \zeta)\{\beta(\xi), \varphi(\{\tilde{\xi}_{q+1}, \beta(\eta)\})\}, \alpha(\zeta)\}
\end{aligned}$$

It follows that $\delta Z = 0$ and $Z = \delta B$ with $B \in T^* \otimes G_{q+1}$ because G_q is 2-acyclic. Thus:

$$Z(\xi, \eta) = \{B(\xi), \beta(\eta)\} + \{\beta(\xi), B(\eta)\}$$

Using once more the Jacobi's identities one gets:

$$\begin{aligned}
&\{\{\tilde{\xi}_{q+2}, \gamma(\xi)\} - \varphi(\{\tilde{\xi}_{q+1}, \beta(\xi)\}) - B(\xi), \beta(\eta)\} \\
&\qquad + \{\beta(\xi), \{\tilde{\xi}_{q+2}, \gamma(\eta)\} - \varphi(\{\tilde{\xi}_{q+1}, \beta(\eta)\}) - B(\eta)\} = 0
\end{aligned}$$

where $\tilde{\xi}_{q+2} \in J_{q+2}(T)$ projects onto $\tilde{\xi}_{q+1} \in J_{q+1}(T)$.

We define $C \in T^* \otimes S_{q+1}T^* \otimes T$ by the formula:

$$C(\xi) = \{\tilde{\xi}_{q+2}, \gamma(\xi)\} - \varphi(\{\tilde{\xi}_{q+1}, \beta(\xi)\}) - B(\xi)$$

We have $\delta C = 0$ and $C = \delta A$ with $A \in S_{q+2}T^* \otimes T$ because $S_{q+1}T^* \otimes T$ is 1-acyclic. Thus $C(\xi) = \{A, \gamma(\xi)\}$. It follows that we may choose $\tilde{\xi}_{q+2}$ in such a way that:

$$\{\tilde{\xi}_{q+2}, \gamma(\xi)\} = \varphi(\{\tilde{\xi}_{q+1}, \beta(\xi)\}) + B(\xi) \in R_{q+1}$$

With an arbitrary section ξ_{q+2} of R_{q+2}, let us define $M \in T^* \otimes S_q T^* \otimes T$ by the formula:

$$M(\eta) = \{\{\bar{\xi}_{q+2}, \xi_{q+2}\} - \varphi(\{\bar{\xi}_{q+1}, \xi_{q+1}\}), \beta(\eta)\}$$

We have:

$$\{\{\bar{\xi}_{q+2}, \xi_{q+2}\}, \beta(\eta)\} = \{\bar{\xi}_{q+1}, \{\xi_{q+2}, \gamma(\eta)\}\} + \{\{\bar{\xi}_{q+2}, \gamma(\eta)\}, \xi_{q+1}\} \in R_q$$

and thus $M \in T^* \otimes G_q$. Using the Jacobi's identity it is easy to check that $\delta M = 0$. We have $M = \delta N$ with $N \in G_{q+1}$ because G_q is 1-acyclic, that is to say $M(\eta) = \{N, \beta(\eta)\}$. As G_{q+1} is the 1-prolongation of G_q, we obtain, because R_q is formally transitive:

$$\{\bar{\xi}_{q+2}, \xi_{q+2}\} = \varphi(\{\bar{\xi}_{q+1}, \xi_{q+1}\}) + N \in R_{q+1}$$

This ends the proof of the proposition. C.Q.F.D.

PROPOSITION 8.47 $\tilde{R}_{q+2}$ is the 1-prolongation of $\tilde{R}_{q+1}$.

Proof According to lemma 2.29 we have:

$$\{i(\eta)D\bar{\xi}_{q+2}, \xi_{q+1}\} = i(\eta)D\{\bar{\xi}_{q+2}, \xi_{q+2}\} - \{\bar{\xi}_{q+1}, i(\eta)D\xi_{q+2}\} \in R_q$$

It follows that $\tilde{R}_{q+2} \subset (\tilde{R}_{q+1})_{+1}$.

The proposition can be deduced from a chase in the commutative and exact diagram:

$$\begin{array}{ccccccc}
 & & 0 & & 0 & & \\
 & & \downarrow & & \downarrow & & \\
0 & \longrightarrow & G_{q+2} & = & G_{q+2} & \longrightarrow & 0 \\
 & & \downarrow & & \downarrow & & \\
0 & \longrightarrow & \tilde{R}_{q+2} & \longrightarrow & (\tilde{R}_{q+1})_{+1} & & \\
 & & \downarrow & & \downarrow & & \\
0 & \longrightarrow & \tilde{R}_{q+1} & = & \tilde{R}_{q+1} & \longrightarrow & 0 \\
 & & \downarrow & & \downarrow & & \\
 & & 0 & & 0 & &
\end{array}$$

C.Q.F.D.

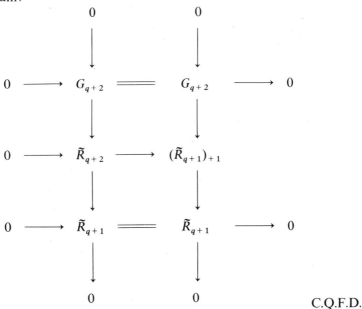

9 Theorem of analytic realization

Let x_0 be a given point of a manifold X.

DEFINITION 9.1 The "*symbol*" of a vector space $V_q \subset J_q(T)_{x_0}$ is the vector space $G_q = V_q \cap (S_q T^*_{x_0} \otimes T_{x_0})$.

DEFINITION 9.2 A vector space $V_{q+1} \subset J_{q+1}(T)_{x_0}$ is called a "*prolongation*" of a vector space $V_q \subset J_q(T)_{x_0}$ if $G_{q+1} = (G_q)_{+1}$ and $\pi_q^{q+1}(V_{q+1}) \subset V_q$.

One has the exact sequence:

$$0 \longrightarrow G_{q+1} \longrightarrow V_{q+1} \xrightarrow{\pi_q^{q+1}} V_q$$

and

$$\dim V_{q+1} \leq \dim V_q + \dim G_{q+1}$$

DEFINITION 9.3 A prolongation V_{q+1} of V_q is said to be "*maximum*" if $V_q = \pi_q^{q+1}(V_{q+1})$.

DEFINITION 9.4 A vector space $V_q \subset J_q(T)_{x_0}$ is called "*transitive*" if $V_0 = \pi_0^q(V_q) = T_{x_0}$.

We denote V_q^0 the kernel of the projection of V_q on V_0.

THEOREM 9.5 If a transitive vector space V_q admits a maximum prolongation V_{q+1} such that $\{V_{q+1}, V_{q+1}\} \subset V_q$ and if its symbol G_q is 3-acyclic, then one can find an infinite sequence of transitive vector spaces V_{q+r} such that V_{q+r+1} is a maximum prolongation of V_{q+r} and $\{V_{q+r+1}, V_{q+r+1}\} \subset V_{q+r}$, $\forall r \geq 0$.

REMARK 9.6 We have $\{V_\infty, V_\infty\} \subset V_\infty$ and, as the bracket $\{\ \}$ verifies the Jacobi identity, we may speak of V_∞ as an "*infinite dimensional Lie algebra*".

Proof For simplicity we do not write explicitly the point x_0.
Let

$$\alpha : T \to V_q, \quad \beta : T \to V_{q+1} \quad \text{and} \quad \varphi : V_q \to V_{q+1}$$

be splitting maps such that

$$\pi_0^q \circ \alpha = \mathrm{id}_T, \quad \pi_0^{q+1} \circ \beta = \mathrm{id}_T \quad \text{and} \quad \beta = \varphi \circ \alpha.$$

Let us choose a map

$$\bar{\gamma} : T \to J_{q+2}(T) \quad \text{such that} \quad \pi_{q+1}^{q+2} \circ \bar{\gamma} = \beta.$$

We may define

$$A \in \Lambda^2 T^* \otimes S_{q+1} T^* \otimes T$$

by the formula:

$$A(\xi, \eta) \equiv \{\bar{\gamma}(\xi), \bar{\gamma}(\eta)\} - \varphi(\{\beta(\xi), \beta(\eta)\})$$

Using the Jacobi identity we obtain:

$$\delta A(\xi, \eta, \zeta) \equiv -\mathscr{C}(\xi, \eta, \zeta)\{\varphi(\{\beta(\xi), \beta(\eta)\}), \beta(\zeta)\} = -B$$

Thus

$$\delta B = 0 \quad \text{and} \quad B \in \Lambda^3 T^* \otimes G_q.$$

Therefore there exists

$$C \in \Lambda^2 T^* \otimes G_{q+1} \quad \text{such that} \quad B = \delta C.$$

Now we have $\delta(A + C) = 0$ and there exists

$$D \in T^* \otimes S_{q+2} T^* \otimes T \qquad \text{such that} \quad A + C = \delta D.$$

This means that, setting $\gamma = \bar{\gamma} - D$, we have

$$\{\gamma(\xi), \gamma(\eta)\} = \varphi(\{\beta(\xi), \beta(\eta)\}) - C \in V_{q+1}$$

for any $\xi, \eta \in T$.

For any $\xi_{q+1} \in V_{q+1}$ let $\bar{\xi}_{q+2} \in J_{q+2}(T)$ be such that $\pi^{q+2}_{q+1}(\bar{\xi}_{q+2}) = \xi_{q+1}$. With a map γ as above, let us now define $K \in T^* \otimes S_{q+1} T^* \otimes T$ by the formula:

$$K(\eta) \equiv \{\bar{\xi}_{q+2}, \gamma(\eta)\} - \varphi(\{\xi_{q+1}, \beta(\eta)\})$$

Using the Jacobi identity we obtain:

$$\begin{aligned}\delta K(\eta, \zeta) \equiv {} & \{\xi_{q+1}, \{\gamma(\eta), \gamma(\zeta)\}\} - \{\varphi(\{\xi_{q+1}, \beta(\eta)\}), \beta(\zeta)\} \\ & + \{\varphi(\{\xi_{q+1}, \beta(\zeta)\}), \beta(\eta)\}\end{aligned}$$

and thus $\delta K \in \Lambda^2 T^* \otimes G_q$.

From the following commutative and exact diagram:

$$\begin{array}{ccccccccc}
0 \longrightarrow & G_{q+2} & \xrightarrow{\delta} & T^* \otimes G_{q+1} & \xrightarrow{\delta} & \Lambda^2 T^* \otimes G_q & \xrightarrow{\delta} & \Lambda^3 T^* \otimes S_{q-1} T^* \otimes T \\
 & \downarrow & & \downarrow & & \downarrow & & \| \\
0 \longrightarrow & S_{q+2} T^* \otimes T & \xrightarrow{\delta} & T^* \otimes S_{q+1} T^* \otimes T & \xrightarrow{\delta} & \Lambda^2 T^* \otimes S_q T^* \otimes T & \xrightarrow{\delta} & \Lambda^3 T^* \otimes S_{q-1} T^* \otimes T
\end{array}$$

we deduce that $K = L + \delta M$ with $L \in T^* \otimes G_{q+1}$ and $M \in S_{q+2} T^* \otimes T$. Taking $\xi_{q+2} = \bar{\xi}_{q+2} - M$ we obtain:

$$\{\xi_{q+2}, \gamma(\eta)\} = L(\eta) + \varphi(\{\xi_{q+1}, \beta(\eta)\}) \in V_{q+1}.$$

Finally, if $\xi_{q+1}, \eta_{q+1} \in V_{q+1}$ and if we choose the corresponding ξ_{q+2}, η_{q+2} as above, we may define $P \in S_{q+1}T^* \otimes T$ by means of the formula:

$$P \equiv \{\xi_{q+2}, \eta_{q+2}\} - \varphi(\{\xi_{q+1}, \eta_{q+1}\})$$

Using the Jacobi identity we see easily that $\delta P \in T^* \otimes G_q$. Thus $P \in G_{q+1}$ and $\{\xi_{q+2}, \eta_{q+2}\} \subset V_{q+1}$.

Using the above process with a basis of V_{q+1}, we may construct a splitting map $\psi: V_{q+1} \to J_{q+2}(T)$ such that $\{\psi(V_{q+1}), \psi(V_{q+1})\} \subset V_{q+1}$ and define $V_{q+2} = \psi(V_{q+1}) \oplus G_{q+2}$.

The proof the theorem then follows from an easy induction. C.Q.F.D.

PROPOSITION 9.7 Under the hypothesis of the preceding theorem, the infinite Lie algebra constructed is unique, up to an isomorphism.

Proof One needs just to construct an homomorphism of Lie algebra between two such infinite Lie algebras V_∞ and $\bar{V}_\infty$. In fact

$$G_{q+r} = (G_q)_{+r} = \bar{G}_{q+r}, \qquad \forall r \geq 0$$

and we see that $\dim V_{q+r} = \dim \bar{V}_{q+r}$ from an easy inductive argument. The proof is left to the reader, as in exercise, because it is similar to the proof of the last theorem and uses only the 2-acyclicity of $G_q = \bar{G}_q$. C.Q.F.D.

The key idea is now to look for a Lie equation, the prolongations of which could allow one to study the later algebra of formal power series.

THEOREM OF ANALYTIC REALIZATION 9.8 If a transitive vector space V_q admits a maximum prolongation V_{q+1} such that $\{V_{q+1}, V_{q+1}) \subset V_q$ and if its symbol G_q is 3-acyclic, then one can find on a neighbourhood of x_0 an analytic Lie equation $R_q \subset J_q(T)$ formally integrable and such that $R_{q+1,x_0} = V_{q+1}$.

REMARK 9.9 From the above theorem and proposition it follows that any infinite Lie algebra prolonging V_q is isomorphic to R_{∞, x_0}.

Proof For every section ξ_q and η_q of $J_q(T)$ we have:

$$[\xi_q, \eta_q] = \{\xi_{q+1}, \eta_{q+1}\} + i(\xi)D\eta_{q+1} - i(\eta)D\xi_{q+1}$$

and we deduce that

$$[V_q^0, V_q^0] = \{V_{q+1}^0, V_{q+1}^0\} \subset V_q^0.$$

As our problem is a purely local one, we may take $X = \mathbb{R}^n$ with $n = \dim V_0$ and choose a fixed point $x_0 = y_0$, for example the origin of a cartesian frame. Working as in paragraph 4, we construct the distribution $\#(R_q^0)$ with

$R^0_{q,x_0} = V^0_q$. We denote by $\Phi^\tau(y_0, p)$ a fundamental set of analytic invariants of that distribution.

Our problem will be now to determine a section ω of $\mathscr{F}$ in order to get the desired Lie equation:

$$R_q \qquad -L_k^{\tau\mu}(\omega(x))\xi_\mu^k + \xi^i \frac{\partial\omega^\tau(x)}{\partial x^i} = 0$$

First of all we shall define $\omega(x_0)$ by means of the values of the later invariants for the identity at x_0 in order to have $R^0_{q,x_0} = V^0_q$. The Cramer's theorem then shows that $j_1(\omega)(x_0)$ is uniquely determined by the condition $R_{q,x_0} = V_q$.

Now the equations defining the prolongation R_{q+1} of R_q are constituted by the ones defining R_q and by the new following ones:

$$-L_k^{\tau\mu}(\omega(x))\xi^k_{\mu+1_j} - \frac{\partial L_k^{\tau\mu}(\omega(x))}{\partial u^\sigma}\frac{\partial\omega^\sigma(x)}{\partial x^j}\xi^k_\mu + \frac{\partial\omega^\tau(x)}{\partial x^i}\xi^i_j + \xi^i\frac{\partial^2\omega^\tau(x)}{\partial x^i \partial x^j} = 0$$

The knowledge of $j_1(\omega)(x_0)$ determines R^0_{q+1,x_0}.

Lemma 9.10 $R^0_{q+1,x_0} = V^0_{q+1}$.

Proof Let ξ^0_q be a section of R^0_q and ξ^0_{q+1} be a section of $J^0_{q+1}(T)$ projecting onto ξ^0_q and such that $\xi^0_{q+1}(x_0) \in V^0_{q+1}$. From the preceding formulas, we obtain $(i(\eta)D\xi^0_{q+1})(x_0) \in V_q$ for any section η_q of R_q because R_q is a Lie equation.

Therefore

$$L_k^{\tau\mu}(\omega(x))\xi^k_\mu(x) = 0$$

and

$$L_k^{\tau\mu}(\omega(x))\frac{\partial\xi^k_\mu(x)}{\partial x^j} + \frac{\partial L_k^{\tau\mu}(\omega(x))}{\partial u^\sigma}\frac{\partial\omega^\sigma(x)}{\partial x^j}\xi^k_\mu(x) = 0$$

But

$$L_k^{\tau\mu}(\omega(x_0))\left(\frac{\partial\xi^k_\mu(x_0)}{\partial x^j} - \xi^k_{\mu+1_j}(x_0)\right) + \frac{\partial\omega^\tau(x_0)}{\partial x^i}\xi^i_j(x_0) = 0$$

and we deduce that $\xi^0_{q+1}(x_0) \in R^0_{q+1,x_0}$. Thus $V^0_{q+1} \subset R^0_{q+1,x_0}$ and these vector spaces are identical because they have the same symbol and the same projection V^0_q from the fact that V_{q+1} is a maximum prolongation of V_q.

C.Q.F.D.

Constructing the integrability conditions, we know from the preceding lemma that the map $\pi_q^{q+1} : R^0_{q+1, x_0} \to R^0_{q, x_0}$ is surjective and thus

$$I_*\left(\omega(x_0), \frac{\partial\omega(x_0)}{\partial x}\right) = 0.$$

Let us define

$$c = I_{**}\left(\omega(x_0), \frac{\partial\omega(x_0)}{\partial x}\right).$$

LEMMA 9.11 $J(c) = 0$.

Proof According to the lemma 9.10, V_{q+1} is defined by certain equations among which are the following ones:

$$-L_k^{\tau\mu}(\omega(x_0))\xi^k_{\mu+1_j} - \frac{\partial L_k^{\tau\mu}(\omega(x_0))}{\partial u^\sigma}\frac{\partial\omega^\sigma(x_0)}{\partial x^j}\xi^k_\mu + \frac{\partial\omega^\tau(x_0)}{\partial x^i}\xi^i_j + \xi^i A^\tau_{ij} = 0$$

If now ξ_{q+1}, η_{q+1} are sections of $J_{q+1}(T)$ projecting respectively onto the sections ξ_q, η_q of R_q and such that $\xi_{q+1}(x_0) \in V_{q+1}, \eta_{q+1}(x_0) \in V_{q+1}$ then one has:

$$(i(\xi)D\eta_{q+1} - i(\eta)D\xi_{q+1})(x_0) \in V_q$$

A tedious but straightforward computation then shows that one must have $\xi^i(x_0)\eta^j(x_0)(A^\tau_{ij} - A^\tau_{ji}) = 0$ and thus $A^\tau_{ij} = A^\tau_{ji}$. Finally we may choose ω such that $\partial^2\omega^\tau(x_0)/\partial x^i\partial x^j = A^\tau_{ij}$ and we shall have $j_2(\omega)(x_0) \in \mathscr{R}'_2$ because $\pi_q^{q+1} : V_{q+1} \to V_q$ is surjective.

Now, according to the proof of the Lie third theorem, we know that $\mathscr{R}'_2 \to \mathscr{R}'_1$ is surjective if and only if $\mathscr{R}'_2 \to \mathscr{F}$ is surjective. But this is the case because the matrix $[L_k^{\tau\mu}(\omega(x_0))]$ has maximum rank and therefore $J(c) = 0$.

C.Q.F.D.

Finally $\mathscr{R}'_1$ is formally integrable because its symbol is 2-acyclic when that of R_q is 3-acyclic. We may use the Cartan–Kähler theorem in order to find an analytic solution ω with $j_2(\omega)(x_0)$ given as above. To this section corresponds an analytic Lie equation R_q which is formally integrable because its symbol is 2-acyclic and $\pi_q^{q+1} : R_{q+1} \to R_q$ is surjective.

C.Q.F.D.

Problems

1) Problem

We define the Lie pseudogroup of contact transformations of $X = \mathbb{R}^3$, by the condition:

$$f(w) = \rho(x)w \quad \text{with} \quad w = dx^1 - x^3\, dx^2$$

1°) Show that there exists a "*generating function*" $W(x)$ such that

$$\xi^1(x) = x^3\frac{\partial W(x)}{\partial x^3} - W(x), \quad \xi^2(x) = \frac{\partial W(x)}{\partial x^3}, \quad \xi^3(x) = -\frac{\partial W(x)}{\partial x^2} - x^3\frac{\partial W(x)}{\partial x^1}$$

and

$$\mathscr{L}(\xi)w = -\frac{\partial W(x)}{\partial x^1}\cdot w$$

2°) Find the special finite equations of this Lie pseudogroup.
3°) Find the corresponding special infinitesimal equations.
4°) Take prolongations in order to get an involutive system of order 1.
5°) Construct the P-sequence.
6°) Determine the general finite equations and a bundle $\mathscr{F}$ of geometric objects.

(Hint: There is a very nice symmetric expression

$$\frac{\omega^1(y)(\partial y^1/\partial x^i) + \omega^2(y)(\partial y^2/\partial x^i) + \omega^3(y)(\partial y^3/\partial x^i)}{\sqrt{\partial(y^1, y^2, y^3)/\partial(x^1, x^2, x^3)}} = \omega^i(x) \qquad i = 1, 2, 3\Big)$$

7°) Determine the corresponding general infinitesimal equations.
8°) Find out the integrability conditions:

(Hint: 1 condition of the second kind:

$$\omega^1(x)\left(\frac{\partial\omega^3(x)}{\partial x^2} - \frac{\partial\omega^2(x)}{\partial x^3}\right) + \omega^2(x)\left(\frac{\partial\omega^1(x)}{\partial x^3} - \frac{\partial\omega^3(x)}{\partial x^1}\right) + \omega^3(x)\left(\frac{\partial\omega^2(x)}{\partial x^1} - \frac{\partial\omega^1(x)}{\partial x^2}\right) = c\Big)$$

9°) What is the specialisation giving rise to the Lie pseudogroup of contact transformations.
(Hint: $\omega^1(x) = 1, \quad \omega^2(x) = -x^3, \quad \omega^3(x) = 0 \Rightarrow c = 1$)
10°) Find the infinitesimal generators of the label transformations.

$$\left(\text{Hint:}\quad u^1\frac{\partial}{\partial u^1} + u^2\frac{\partial}{\partial u^2} + u^3\frac{\partial}{\partial u^3}\right)$$

11°) What is the corresponding action on the structure constant.
12°) What is the codimension of Γ in $\tilde{\Gamma}$ for the specialisations

- $\omega^1(x) = 1, \quad \omega^2(x) = -x^3, \quad \omega^3(x) = 0$ (Hint: 0)
- $\omega^1(x) = 1, \quad \omega^2(x) = 0, \quad \omega^3(x) = 0$ (Hint: 1)

13°) Study the similar general problem for $n > 3$, n odd. (Use proposition 2.5.1)

2) Problem

We shall study the bundle $S_2 T^*$ of geometric objects.

1°) For $n = 4$, what are the specialisations giving rise to the Lie pseudogroup of Lorentz transformations plus translations or to the Lie pseudogroup of Euclidean motions.

2°) Show that the finite equations corresponding to $S_2 T^*$ are not involutive in general.

3°) Extend the bundle $S_2 T^*$ in order to find the general finite equations:

$$\mathcal{R}_2 \begin{cases} \omega_{kl}(y) \dfrac{\partial y^k}{\partial x^i} \dfrac{\partial y^l}{\partial x^j} = \omega_{ij}(x) \\ \dfrac{\partial x^i}{\partial y^l} \cdot \dfrac{\partial^2 y^l}{\partial x^j \partial x^k} + \gamma^l_{mp}(y) \dfrac{\partial y^m}{\partial x^j} \dfrac{\partial y^p}{\partial x^k} \dfrac{\partial x^i}{\partial y^l} = \gamma^i_{jk}(x) \end{cases}$$

4°) Find out the corresponding general infinitesimal equations.

5°) What is the condition on ω_{ij} for constancy of rank:
(Hint: $\det [\omega_{ij}(x)] \neq 0$)

6°) Find out the integrability conditions ($\omega^{ik}(x)\omega_{jk}(x) = \delta^i_j$):

$$\gamma^i_{jk} - \frac{1}{2}\omega^{il}(x)\left(\frac{\partial \omega_{kl}(x)}{\partial x^j} + \frac{\partial \omega_{jl}(x)}{\partial x^k} - \frac{\partial \omega_{jk}(x)}{\partial x^l}\right) = 0$$

$$\rho^i_{jkl}(x) - \frac{1}{n-1}(\delta^i_l \sigma_{jk}(x) - \delta^i_k \rho_{jl}(x)) = 0$$

$$\frac{1}{n(n-1)} \rho(x) = c$$

with

$$\rho^i_{jkl}(x) = \frac{\partial \gamma^i_{jk}(x)}{\partial x^l} - \frac{\partial \gamma^i_{jl}(x)}{\partial x^k} + \gamma^r_{jk}(x)\gamma^i_{rl}(x) - \gamma^r_{jl}(x)\gamma^i_{rk}(x)$$

$$\rho_{jk}(x) = \rho^i_{jki}(x)$$

$$\rho(x) = \omega^{jk}(x)\rho_{jk}(x)$$

7°) Use the P-sequence in order to compute the number of compatibility conditions

$$\left(\text{Hint: } \dim F_1 = \frac{n^2(n^2-1)}{12}, \dim F_2 = \frac{n^2(n+1)(n-1)(n-2)}{24}\right)$$

3) Problem

Take $\dim X = n \equiv 2m$ and $\bar{\mu} = m + \mu = m+1, \ldots, 2m$.

1°) Show that the following system defines a Lie pseudogroup:

$$\begin{cases} \dfrac{\partial y^{\bar{\mu}}}{\partial x^{\bar{\nu}}} - \dfrac{\partial y^{\mu}}{\partial x^{\nu}} = 0 \\ \\ \dfrac{\partial y^{\mu}}{\partial x^{\bar{\nu}}} + \dfrac{\partial y^{\bar{\mu}}}{\partial x^{\nu}} = 0 \end{cases} \qquad \text{(Cauchy–Riemann)}$$

2°) Show that the corresponding special infinitesimal equations are those of an involutive system of order 1.

3°) Construct the P-sequence and prove that it is just the real Dolbeant sequence. (Hint: 1) Show that the P-sequence stops at F_{m-1}

2) Show that $\dim F_r = 2m \dfrac{m!}{(m-r-1)!(r+1)!}$

3) Use the formal exactness.)

4°) In the case $n = 4$ compute the dimensions of C_r, $C_r(T)$, F_r.
(Hint: $\dim F_0 = 8$, $\dim F_1 = 4$, $\dim C_0(T) = 20$, $\dim C_1(T) = 40$, $\dim C_2(T) = 40$, $\dim C_3(T) = 20$, $\dim C_4(T) = 4$, $\dim C_0 = 12$, $\dim C_1 = 36$, $\dim C_2 = 40$, $\dim C_3 = 20$, $\dim C_4 = 4$).

5°) Prove that the later system is elliptic.
(Hint: Use the Euler–Poincare formula.)

6°) Find out the general finite equations:

$$\left(\text{Hint:} \quad \omega_l^k(y)\frac{\partial y^l}{\partial x^j}\frac{\partial x^i}{\partial y^k} = \omega_j^i(x) \quad \text{with} \quad \omega_k^i(x)\omega_j^k(x) = -\delta_j^i\right)$$

7°) Show that there are $2m^2$ unconnected corresponding general infinitesimal equations:

$$\left(\text{Hint:} \quad \underset{1}{\Omega_j^i} \equiv \omega_k^i(x)\frac{\partial \xi^k}{\partial x^j} - \omega_j^k(x)\frac{\partial \xi^i}{\partial x^k} + \xi^r \frac{\partial \omega_j^i(x)}{\partial x^r} = 0\right)$$

8°) Find out the integrability conditions:

$$\left(\text{Hint:} \quad \omega_j^r(x)\left(\frac{\partial \omega_r^i(x)}{\partial x^k} - \frac{\partial \omega_k^i(x)}{\partial x^r}\right) - \omega_k^r(x)\left(\frac{\partial \omega_r^i(x)}{\partial x^j} - \frac{\partial \omega_j^i(x)}{\partial x^r}\right) = 0\right)$$

9°) Find a fundamental set of differential invariants.
(Hint: Look at a half part of the matrix $[\omega_j^i(x)]$.)

4) Problem

Fill in the details of the following example:

$$\Gamma : \mathbb{R}^3 \to \mathbb{R}^3 : y^1 = f^1(x^1), \quad y^2 = f^2(x^1, x^2, x^3), \quad y^3 = f^3(x^1, x^2, x^3)$$

Special finite equations $\qquad p_3^1 = 0, \quad p_2^1 = 0$

Special infinitesimal equations $\partial_3 \xi^1 = 0, \quad \partial_2 \xi^1 = 0$

P-sequence: $0 \longrightarrow \Theta \longrightarrow \underset{③}{T} \xrightarrow{\mathscr{D}} \underset{②}{F_0} \xrightarrow{\mathscr{D}_1} \underset{①}{F_1} \longrightarrow 0$

General finite equations:

$$\mathscr{R}_1 \begin{cases} \dfrac{p_3^1 + \omega^1(y)p_3^2 + \omega^2(y)p_3^3}{p_1^1 + \omega^1(y)p_1^2 + \omega^2(y)p_1^3} = \omega^1(x) \\[2ex] \dfrac{p_2^1 + \omega^1(y)p_2^2 + \omega^2(y)p_2^3}{p_1^1 + \omega^1(y)p_1^2 + \omega^2(y)p_1^3} = \omega^2(x) \end{cases}$$

General infinitesimal equations:

$$R_1 \begin{cases} \underset{1}{\Omega^2} \equiv \dfrac{\partial \xi^1}{\partial x^3} + \omega^1(x)\dfrac{\partial \xi^2}{\partial x^3} + \omega^2(x)\dfrac{\partial \xi^3}{\partial x^3} - \omega^2(x)\dfrac{\partial \xi^1}{\partial x^1} - \omega^1(x)\omega^2(x)\dfrac{\partial \xi^2}{\partial x^1} \\[2ex] \qquad - (\omega^2(x))^2 \dfrac{\partial \xi^3}{\partial x^1} + \xi^i \dfrac{\partial \omega^2(x)}{\partial x^i} = 0 \\[2ex] \underset{1}{\Omega^1} \equiv \dfrac{\partial \xi^1}{\partial x^2} + \omega^1(x)\dfrac{\partial \xi^2}{\partial x^2} + \omega^2(x)\dfrac{\partial \xi^3}{\partial x^2} - \omega^1(x)\dfrac{\partial \xi^1}{\partial x^1} - (\omega^1(x))^2\dfrac{\partial \xi^2}{\partial x^1} \\[2ex] \qquad - \omega^1(x)\omega^2(x)\dfrac{\partial \xi^3}{\partial x^1} + \xi^i \dfrac{\partial \omega^1(x)}{\partial x^i} = 0 \end{cases}$$

Integrability conditions: (first kind only)

$$\frac{\partial \omega^1(x)}{\partial x^3} - \frac{\partial \omega^2(x)}{\partial x^2} + \omega^1(x)\frac{\partial \omega^2(x)}{\partial x^1} - \omega^2(x)\frac{\partial \omega^1(x)}{\partial x^1} = 0$$

Normaliser: $\Gamma = \tilde{\Gamma}$.

Compare these results with the ones of problem 1.

5) Problem

Automorphic systems

Let X, Y be C^∞ connected manifolds with $\dim X = n \neq m = \dim Y$ and Γ any transitive Lie pseudogroup of transformations of Y, of order q.

Definition An automorphic system $\mathscr{R}_q \subset J_q(X \times Y)$ is a non-linear system of order q on $X \times Y$ such that, if f is a solution of $\mathscr{R}_q$, the other solutions are $\psi \circ f$ where ψ is any finite transformation of Γ.

Remark As in the case $n = m$, there is a better formal definition using the composition of jets $J_q(Y \times Y) \times J_q(X \times Y) \to J_q(X \times Y)$ but the above definition will be sufficient for our purpose in the analytic case.

1°) Define, as in the case $n = m$, an involutive distribution on $J_q(X \times Y)$ and compute, for any $r \geq 0$, the maximum number of unconnected differential invariants of order $q + r$. Do they always exist?
(Hint: Use a combinatorial computation involving the characters of an involutive defining system of Γ of order q in $J_q(Y \times Y)$.)

2°) When such invariants exist, find out a Lie form for $\mathcal{R}_{q+r}$ and exhibit a bundle $\mathcal{F}$ of geometric objects as in the case $n = m$, with section ω.
(Hint: Use proposition 6.3.26.).

3°) Show that

$$\mathcal{R}_{q+r+1} \subseteq (\mathcal{R}_{q+r})_{+1}$$

and that the maps

$$\pi^{q+r+1}_{q+r} : \mathcal{R}_{q+r+1} \to \mathcal{R}_{q+r}, \quad (\mathcal{R}_{q+r})_{+1} \to \mathcal{R}_{q+r}$$

are surjective.
(Hint: Use proposition 6.5.20.)

4°) Prove that

$$G_{q+r+1} \subseteq (\pi^{q+r+1}_{q+r})^{-1}(G_{q+r})_{+1}$$

(Hint: Use question 3.)

5°) Prove that there exists, under certain regularity conditions, an r big enough such that:

$$\mathcal{R}_{q+r+1} = (\mathcal{R}_{q+r})_{+1}$$

(Hint: Use proposition 2.5.9 in order to find r big enough such that

$$G_{q+r+1} = (\pi^{q+r+1}_{q+r})^{-1}(G_{q+r})_{+1}.$$

Then use question 3.)

6°) Find out the integrability conditions for ω.

$$\left(\text{Hint: Study the matrix } \frac{\partial \Phi^\tau}{\partial p^k_\mu}(x, y, p) \quad \text{with} \quad |\mu| = q\right)$$

7°) *Application* Study the case $n = 1$, $\Gamma : \mathbb{R}^m \to \mathbb{R}^m : \bar{y}^l = a^l_k y^k$ considered by Vessiot and Drach.
For more details we refer the reader to (24).

6) Problem

$X = \mathbb{R}^3$, let $\alpha = dx^1$, $\beta = dx^2 \wedge dx^3$ and consider $\Gamma : f(\alpha) = \alpha, f(\beta) = \beta$.

1°) Find the finite transformations of Γ.
2°) Determine the finite and infinitesimal equations of Γ.
3°) Find the bundle $\mathcal{F}$ and the inequality conditions on its fiber.
4°) Find out the corresponding general finite and infinitesimal equations.

5°) What are the integrability conditions?
6°) What are the Jacobi conditions.
(Hint: If $\alpha = \omega_i(x)\, dx^i, \beta = \omega_{ij}(x)\, dx^i \wedge dx^j, \alpha \wedge \beta \neq 0$ we must have $d\alpha = c^1 \cdot \beta$, $d\beta = c^2\alpha \wedge \beta$ and $c^1 \cdot c^2 = 0$.)
7°) Find out a specialisation for which $c^1 = 0, c^2 = 0$.
8°) Study the deformation theory of the structures involved in question 1°) and 7°).

7) Problem

Prove that the deformation cohomology is just the Hochschild cohomology in the case of a Lie group G acting on X with $\dim G = \dim X$. (Hint: Use example 5.9)

Why is there a change of grading?

8) Problem

1°) In Problem 5, when r is big enough, prove that the system of the integrability conditions for ω, called "*resolvent system*", is formally integrable.
2°) Deduce from this fact a new proof of proposition 4.4 and show that the Jacobi conditions may be exhibited without any term of degree zero in the polynomials.

BIBLIOGRAPHY

The bibliography contains the only references that are used throughout the book.

1. U. Amaldi, Introduzione alla teoria dei gruppi infiniti di trasformazioni (libreria dell'Universita, Roma, 1942).
2. ———, Sulla teoria dei gruppi infiniti continui (Annali di Matematica, 1897).
3. J. E. Campbell, Introductory treatise on Lie's theory of finite continuous transformation groups (Chelsea, 1903, 1966).
4. C. Caratheodory, Calculus of variations and partial differential equations of the first order I, II (Teubner, 1935; Holden Day, 1965).
5. E. Cartan, Leçons sur les invariants integraux (Hermann, 1922).
6. ———, Les systèmes differentiels exterieurs et leurs applications scientifiques (Hermann, 1946).
7. P. M. Cohn, Lie groups. Cambridge tracts No. 46 (Cambridge University Press, 1965).
8. J. Drach, Sur le problème logique de l'integration des equations differentielles (Annales Toulouse, 1908–1912).
9. ———, Sur l'integration logique des equations differentielles ordinaires (International congress of mathematicians, Cambridge, 1912).
10. M. Engel, Ueber die definitionsgleichungen der continuirlichen transformationsgruppen (Math. Annalen, t. XXVII).
11. M. Gerstenhaber, On the deformation of rings and algebras (Ann. of Math. 79, 1961, p. 59–104).
12. H. Goldschmidt, Existence theorems for analytic linear partial differential equations (Ann. of Math., Vol. 86, 1962, p. 246–270).
13. ———, Prolongations of linear partial differential equations:
    I: A conjecture of E. Cartan (Ann. Scient. Ec. Norm. Sup., 4° série, t. 1, 1968, p. 417–444).
    II: Inhomogeneous equations (Ann. Scient. Ec. Norm. Sup., 4° série, t. 1, 1968, p. 617–625).
14. ———, Integrability criteria for systems of non-linear partial differential equations (J. Differential geometry, 1, 1969, p. 269–307).
15. ———, Sur la structure des equations de Lie:
    I: Le troisiéme théoreme fondamental (J. Differential geometry, 6, 1972, p. 357–373).
    II: Equations formellement transitives (J. Differential geometry, 7, 1972, p. 67–95).
16. V. Guillemin, S. Sternberg, Deformation theory of pseudogroup structures (Mem. Amer. Math. Soc., 64, 1966).

16a. ———, The Lewy counterexample and the local equivalence problem for G structures (J. differential geometry, 1, 1967, p. 127–131).

17. G. Hochschild, On the cohomology groups of an associative algebra (Ann. of Math, 46, 1945, p. 58–67).

18. E. Inonü, E. P. Wigner, On the contraction of Lie groups and Lie algebras (Proc. Nat. Acad. Sci. U.S.A., 39, 1953, p. 510).
19. M. Janet, Sur les systèmes d'equations aux derivées partielles (These de Doctorat-Gauthier-Villars, 1920).
20. ———, Les systèmes d'equations aux derivées partielles (J. Math. pures et appl., t. 3, 1920, p. 65).
21. ———, Leçons sur les systèmes d'equations aux derivées partielles (Cahiers scientifiques, fax IV, Gauthier-Villars, Paris, 1929).
22. ———, Les systèmes d'equations aux derivées partielles (Memorial Sc. Math. fasc. XXI, Gauthier-Villars).
23. ———, Equations aux derivées partielles (Formulaire de Math, C.N.R.S., fax. VII, 1956).
24. A. K. Kumpera, Invariants differentiels d'un pseudogroupe de Lie I, II (J. Differential geometry, 10, 1975, p. 347–416).
25. ———, D. C. Spencer, Lie equations: I, general theory (Annals of Math. Studies No. 73, Princeton University Press, 1972).
26. M. Kuranishi, Lectures on involutive systems of partial differential equations (Publ. Soc. Mat. São Paulo, 1967).
27. S. Lie, Die grundlager für die theorie der unendlichen gruppen (Leipziger Berichte, 1891, p. 391, 1895, p. 282) (Gesam. Abh. Bd. VI).
28. B. Malgrange, Pseudogroupes de Lie elliptiques (Seminaire J. Leray, Collège de France, Paris, 1969–1970).
29. ———, Equations de Lie: I, II (J. Differential geometry: 6, 1972, p. 503–522; 7, 1972, p. 117–141).
30. P. Medolaghi, Sulla teoria dei gruppi infiniti continui (Ann. Mat. Pura Appl. 25, 1897, p. 179–218).
31. W. S. Piper, Algebraic deformation theory (J. Differential geometry, 1, 1967, p. 133–168).
32. J. F. Pommaret, Etude interne des systèmes lineaires d'equations aux derivées partielles. (Ann. Inst. Henri Poincaré, Vol. XVII, No. 2, 1972, p. 131–158).
33. ———, Theorie des deformations de structures (Ann. Inst. Henri Poincaré, Vol. XVIII, No. 4, 1973, p. 285–352).
34. ———, Sur certaines derivations des algébres de Lie infinies (C. R. Acad. Sc. Paris, t. 280, 1975, p. 1495).
35. ———, Sur l'exactitude des complexes associes à des operateurs de Lie de type fini (C. R. Acad. Sc. Paris, t. 282, 1976, p. 587).
36. ———, Etude de certains complexes associes à des operateurs de Lie de type fini (C. R. Acad. Sc. Paris, t. 282, 1976, p. 635).
37. ———, Pseudogroupes de Lie algebriques (C. R. Acad. Sc. Paris, t. 280, 1975, p. 1693).
38. ———, Theory of deformation of structures (In lecture notes in Mathematics No. 484, Springer, 1975, p. 57–78).
39. D. G. Quillen, Formal properties of over-determined systems of linear partial differential equations (Thesis, Harvard University, 1964).
40. D. S. Rim, Deformation of transitive Lie algebras (Ann. of Math. 83, 1966, p. 339–357).

41. C. H. Riquier, Les systèmes d'equations aux derivées partielles (Gauthier–Villars, Paris, 1910).
42. ———, La methode des fonctions majorantes et les systèmes d'equations aux derivées partielles (Memorial Sc. Math. fax. XXXII, Gauthier–Villars).
42. J. F. Ritt, Differential algebra (Dover, 1950, 1966).
44. D. C. Spencer, Overdetermined systems of linear partial differential equations (Bull. Amer. Math. Soc., 75, 1965, p. 1–114).
45. ———, Deformation of structures on manifolds defined by transitive continuous pseudogroups.
I: (Annals of Math., 76, 1962, p. 306–445).
II, III: (Annals of Math., 81, 1965, p. 389–450).
46. J. W. Sweeny, The D. Neumann poblem (Acta Math. Vol. 120, 1968, p. 223–251).
47. J. M. Thomas, Riquier's theory (Annals of Math., t. 30, 1929, t. 35, 1934).
48. ———, Systems and roots (W. Byrd Press, 1962).
49. E. Vessiot, Theorie des groupes continus (Ann. Sci. Ecole Norm. Sup., 20, 1903, p. 411–451).
50. ———, Sur la theorie de Galois et ses diverses generalisations (Ann. Sci. Ec. Norm. Sup., 21, 1904, p. 9).
51. ———, Sur l'integration des systèmes differentiels qui admettent des groupes continus de transformations (Acta. Math., 28, 1904, p. 307–350).
52. G. Vivanti, Leçons elementaires sur la theorie des groupes de transformations (Gauthier–Villars, 1904).

Recommended textbooks

Differential Geometry

53. R. Abraham, Foundations of mechanics (W. A. Benjamin, 1967).
54. F. W. Warner, Foundations of differentiable manifolds and Lie groups (Scott, Foresman and Co., 1971).
55. G. Sorani, An introduction to real and complex manifolds (Gordon and Breach, 1969).

Algebra

56. M. F. Atiyah, I. G. MacDonald, Introduction to commutative algebra (Addison-Wesley, 1969).
57. S. T. Hu, Introduction to homological algebra (Holden-Day, 1968).
58. S. Lang, Algebra (Addison-Wesley, 1965).
59. D. G. Northcott,* Lessons on rings, modules and multiplicities (Cambridge University Press, 1968).
60. * An introduction to homological algebra (Cambridge University Press, 1966).
61. B. L. van der Waerden, Algebra I, II (Springer-Verlag, F. Ungar, 1970).

Analysis

62. E. Goursat, Cours d'analyse mathematique (Gauthier–Villars, Paris, 1911).

INDEX OF DEFINITIONS